Computerized Modeling of Sedimentary Systems

Springer
Berlin
Heidelberg
New York
Barcelona
Budapest
Hong Kong
London
Milan
Paris
Singapore
Tokyo

Jan Harff, Wolfram Lemke,
Karl Stattegger (Eds.)

Computerized Modeling of Sedimentary Systems

With 212 Figures and 37 Tables

Springer

Editors:
Professor Dr. Jan Harff
Dr. Wolfram Lemke
Baltic Sea Research Institute
Seestrasse 15
D-18119 Rostock-Warnemuende
Germany
e-mail: jan.harff@io-warnemuende.de
e-mail: wolfram.lemke@io-warnemuende.de

Professor Dr. Karl Stattegger
University of Kiel
Institute of Geology and Paleontology
Olshausenstraße 40
D-24118 Kiel
Germany
e-mail: ks@gpi.uni-kiel.de

ISBN 3-540-64109-2 Springer-Verlag Berlin Heidelberg New York

Library of congress Cataloging-in-Publication Data
Computerized modeling of sedimentary systems/Jan Harff, Wolfram Lemke, Karl Stattegger, eds.
p. cm.
Includes bibliographical references.
ISBN 3-540-64109-2 (hardcover)
1. Sedimentary basins – environmental aspects – Computer simulation.
2. Sedimentation and deposition – Environmental aspects – Computer simulation.
I. Harff, Jan. II. Lemke, Wolfram, 1955-. III. Stattegger, Karl.
QE571.C555 1998
551.3´03--dc21

Cover design: design & production, Heidelberg
Typesetting: Medio, Berlin

SPIN: 10568759 32/3020 - 5 4 3 2 1 0 - Printed on acid-free paper

Preface

Sedimentation processes depend on a complex interaction of environmental factors and react to changes of these factors sensitively. Variations in climatic, oceanographic, but also in anthropogenically induced environmental variables are reflected in the sedimentary record at varying scales. The understanding of these processes gives way to the reconstruction of the past and is a key to the prediction of future trends .

During the past decades computerized modeling has become a powerful tool to describe the complicated interrelations between measured data and the dynamics of sedimentary systems. Sedimentary modeling and quantitative dynamic stratigraphy have become widely known keywords of high acceptance. Especially in cases where observations are limited and/or expensive, computer simulations may substitute for the lack of data. State-of-the-art research work requires a thorough knowledge of processes at the interfaces between atmosphere, hydrosphere, biosphere and lithosphere, and therefore an interdisciplinary approach to sedimentary systems by oceanographers, biologists, geoscientists, mathematicians and computer scientists.

In October 1996, about 100 scientists of these special fields from 20 different countries met in Güstrow, Germany, for an international symposium to discuss state-of-the-art approaches in the computerized modeling of sedimentary systems at the invitation of the Baltic Sea Research Institute Warnemünde. Posters and oral and software presentations were assigned to specific thematic foci reflecting the interdisciplinary character of the conference.

This book, a product of the conference, contains 23 chapters that are allocated to three main topics. The first topic climatic, oceanographic and biological forcing of sedimentary systems, covers papers on theories and methods describing particular processes of sediment accumulation. Wind- and current-driven transport and deposition-forming sediment bodies of various scales are discussed here. Special emphasis is given to nearshore processes where the system reacts very sensitively to the variation of the driving atmospheric and oceanographic forces. The biological component is discussed here, focusing on the influence of biological productivity of the ocean on sedimentation and the benthos activity on the uppermost sediment layers. The modeling of global changes in the environment forced by climate changes which are recorded from sedimentary sequences is used to reconstruct the past and to extrapolate trends for the future. Under the topic trends and periodicity in the sedimentary record as a response to environmental change, authors discuss this question with special regard to cyclic processes. Process studies of sediment formation and knowledge of the process pattern in longer-termed time dimensions as principles of lithosphere dynamics also have to be taken into account for the development of space-time models of the basin fill. Besides chapters dealing with basin modeling, an outlook is given on the reliability of model results that can be generalized to the methodology of modeling.

This book is addressed to graduate students and professionals engaged in the field of sedimentary basin analysis, and sedimentary modeling in general. In the future, communication between groups of different expertise will become even more important than today. This includes the discussion and evaluation of an increasing number of models. Different modeling approaches must be assembled to coupled or nested models, allowing more comprehensive and realistic simulations of natural systems. This book should contribute to force that communication on quantification of environmental research methods and on the development of computer techniques with the general aim of managing and protecting natural resources.

The editors would like to express their thanks to all who have contributed to this volume, particularly to the authors for preparing the manuscripts but also to the following scientists who provided their expertise and time in reviewing the manuscripts: P.L. de Boer, H. Dietrich, R. Endler, B.W. Flemming, F. Gradstein, G. Gust, B.J. Haupt, M. Kersten, R.-O. Niedermeyer, C. Schäfer-Neth, M. Schulz, W. Schwarzacher, D. Seidov, R. Slingerland, F. Tauber, J. Tipper, J. Wendebourg

Special thanks are due to the sponsoring organizations, especially to the German Science Foundation (DFG), the International Union of Geological Sciences (IUGS) and the Scientific Committee of Ocean Research (SCOR), who helped to lead the symposium to success. Without their support it would not have been possible to bring scientists from different parts of the world together to share their ideas during fruitful scientific discussions.

Oktober 1998 Jan Harff, Wolfram Lemke, Karl Stategger

List of Contributors

Agterberg, Frederik P.
Geological Survey of Canada, 601 Booth Street, Ottawa,Ontario K1A 0E8, Canada
E-mail agterberg@gsc.emr.ca

Ainsaar, Leho
Institute of Geology, University of Tartu, Venmuise 46, Tartu EE2400, Estonia
E-mail lainsaar@math.ut.ee

Barceló, Carlos
Department of Informatics and Applied Mathematics, University of Girona,
Av. L1. Santaló s/n, 17071 Girona, Spain
E-mail barcelo@ima.udg.es

Bechstaedt, T. (cf. Leyrer)

Bell, Robert
Centre of Excellence in Coastal Oceanography and Marine Geology, Department of Earth Sciences, Waikato University Hamilton, and National Institute of Water and Atmospheric Research, Silverdale Road, P.O. Box 11-115, Hamilton, New Zealand

Berger, W.H.
Scripps Institution of Oceanography, UCSD, La Jolla, California, USA

Black, Kerry
Centre of Excellence in Coastal Oceanography and Marine Geology, Department of Earth Sciences, Waikato University Hamilton, and National Institute of Water and Atmospheric Research, Silverdale Road, P.O. Box 11-115, Hamilton. New Zealand
E-mail black@hamilton.niwa.cri.nz

Bohling, Geoff C.
Mathematical Geology Section, Kansas Geological Survey, University of Kansas,
1930 Constant Avenue, Lawrence, Kansas 66047-3726, USA

Cao, S.
Twange International Inc., Regency Square Tower, 6200 Dr., Ste. 410, Houston, Texas 77036, USA

Chizova, I.
Central Research Institute of Geological Prospecting for Base & Precious Metals, Warshawsky Road 129b, 113545 Moscow,Russia

Davis, John C.
Kansas Geological Survey, University of Kansas, 1930 Constant Avenue, Lawrence Kansas 66047-3726, USA

Fennel, Katja
Baltic Sea Reserach Institute, University of Rostock, Seestrasse 15, 18119 Rostock-Warnemünde, Germany
E-mail katja.fennel@io-warnemuende.de

Fohrmann, Herrmann
SFB 313, University of Kiel, Heinrich-Hecht-Platz 10, 24118 Kiel, Germany

Gorelikova, Nina V.
IGEM, Russian Academy of Sciences, Staromonetny Per. 35, 19017 Moscow, Russia
E-mail ngor@igem.msk.su

Gradstein, Felix M.
Saga Petroleum a.s., Postboks 490, 1301 Sandvika, Norway

Green, Malcolm
Centre of Excellence in Coastal Oceanography and Marine Geology, Department of Earth Sciences, Waikato University Hamilton, and National Institute of Water and Atmospheric Research, Silverdale Road, P.O. Box 11–115, Hamilton, New Zealand

Gustavs, Inka
Zum Fohlenhof 4, 18147 Rostock, Germany

Healy, Terry
Centre of Excellence in Coastal Oceanography and Marine Geology, Department of Earth Sciences, Waikato University Hamilton, and National Institute of Water and Atmospheric Research, Silverdale Road, P.O. Box 11–115, Hamilton, New Zealand

Hoselmann, Christian
Niedersächsisches Landesamt für Bodenforschung, Stilleweg 2, 30655 Hannover, Germany
E-mail n231hoselmann@rzvax.hannover.bgr.de

Hume, Terry
Centre of Excellence in Coastal Oceanography and Marine Geology, Department of Earth Sciences, Waikato University Hamilton, and National Institute of Water and Atmospheric Research, Silverdale Road, P.O. Box 11–115, Hamilton, New Zealand

Kendall, Christopher G.St.C.
Department of Geology, University of South Carolina, Columbia, South Carolina 29208, USA
E-mail kendall@sc.edu

Kruger, Joe
Department of Geology, Idaho State University, Pocatello, Idaho 83209, USA

Lerche, Ian
Dept. of Geological Sciences, University of South Carolina, Columbia, South Carolina 29208-0001, USA
E-mail black@geol.sc.edu

Leyrer, Karl
Institute of Geology and Paleontology, University of Heidelberg, Im Neuenheimer Feld 234, 69120 Heidelberg, Germany
E-mail gt7@ix.urz.uni-heidelberg.de

Lou, Jing
Institute of Hydrophysics, GKSS Research Center, Max-Planck-Strasse, 21502 Geesthacht, Germany
E-mail lou@glerl.noaa.gov

Lowrie, A.
230 F.Z. Goss Road, Picayune, Mississippi 39466, USA

Malloy, S.
South Carolina Governor's School of Science and Mathematics, 305 East Home Ave., Hartsville
South Carolina 29550, USA

Martma, Tõnu
Institute of Geology, University of Tallinn, Estonia pst. 7, Tallinn EE0001, Estonia

Meidla, Tõnu
Institute of Geology, University of Tartu, Vanemuise 46, Tartu EE2400, Estonia

Morrill, Carrie
Carleton College, Department of Geology, Northfield, Minnesota 55057, USA

Neumann, Thomas
Baltic Sea Research Institute, University of Rostock, Seestrasse 15, 18199 Rostock-Warnemünde, Germany
E-mail thomas.neumann@io-warnemuende.de

Oldman, John
Centre of Excellence in Coastal Oceanography and Marine Geology, Department of Earth Sciences, Waikato University Hamilton, and National Institute of Water and Atmospheric Research, Silverdale Road, P.O. Box 11–115, Hamilton, New Zealand

Olea, Ricardo A.
Kansas Geological Survey, University of Kansas, 1930 Constant Avenue, Lawrence, Kansas 66047-3726, USA

Patyk-Kara, N.
IGEM, Russian Academy of Sciences, Staromonetny Per. 35, 109017 Moscow, Russia

Paul, André
Geosciences, University of Bremen, P.O. Box 330 440, 28334 Bremen, Germany
E-mail apau@palmod.uni-bremen.de

Pawlowsky, V.
Department of Applied Mathematics, Polytechnic University of Catalonia, Barcelona, Spain

Petersen, K
Unocal Corporation, P.O. Box 4551, Sugarland, Texas 77210, USA

Plakht, J.
Ramon Science Center, P.O. Box 194, Mizpe Ramon, 80600 Israel

Pohlmann, Thomas
Institute of Oceanography, University of Hamburg, Troplowitzstrasse 7, 22529 Hamburg, Germany

Puls, Walter
GKSS Research Center, 21502 Geesthacht, Germany
E-mail Walter.Puls@gkss.de

Ritzrau, Will
SFB 313, University of Kiel, Heinrich-Hecht-Platz 10, 24118 Kiel, Germany
E-mail will@sfb313.uni-kiel.de

Rockenbauch, K. (cf. Leyrer)

Rosenthal, Wolfgang
Institute of Hydrophysics, GKSS Research Center, Max-Planck-Strasse, 21502 Geesthacht, Germany

Rubel, Madis
Institute of Geology, University of Tartu, Vanemuise 46, Tartu EE2400, Estonia

Schäfer-Neth, Christian
SFB 313, University of Kiel, Heinrich-Hecht-Platz 10, 24118 Kiel, Germany
E-mail risto@sfb313.uni-kiel.de

Schwartz, Gregory J.
Department of Geography, University of Colorado, Boulder, Colorado 80309, USA

Schwarzacher, Walter
School of Geosciences, Queen's University, Belfast BT7 1NN, Northern Ireland, United Kingdom
E-mail w.schwarzacher@qub.ac.uk

Seidov, Dan
Earth System Science Center, Pennsylvania State University, University Park, Pennsylvania, USA

Sen, Animikh
Department of Geology, University of South Carolina, Columbia, South Carolina 29208, USA

Sidaravičienė, Nijolė
Lithuanian Institute of Geology, Devėenkos 13, Vilnius 2600
Lithunia

Sirotinskaya, Susanna
VIEMS Mathematcial Methods Laboratory, All-Russian Research Institute of Mineral Ressources, Economics and Mineral Land Use, 3-ya Magistralnaja Str. 38, 123853 Moscow, Russia

Sivtsov, A.
IGEM, Russian Academy of Sciences, Staromonetny Per. 35, 109017 Moscow, Russia

Streif, Hansjörg
Niedersächsisches Landesamt für Bodenforschung, Stilleweg 2, 30655 Hannover, Germany

Strohmenger, C. (cf. Leyrer)

Sündermann, Jürgen
Institute of Oceanography, University of Hamburg, Troplowitzstrasse 7, 22529 Hamburg, Germany

Watney, W. Lynn
Kansas Geological Survey, University of Kansas, 1930 Constant Avenue, Lawrence, Kansas 66047-3726, USA
E-mail lwatne@pcmail.kgs.ukans.edu

Wold, Christopher N.
Platte River Associates, Inc, 2790 Valmont Road, Boulder, Colorado 80304, USA
E-mail chris@platte.com

Wolf, Thomas
Institute of Hydrophysics, GKSS Research Center, Max-Planck-Strasse, 21402 Geesthacht, Germany

Zhou, Di
South China Sea Institute of Oceanology, Chinese Academy of Sciences, Laboratory of Marine Tectonics and Geophysics, 164 West Xingang Road, Guangzhou 510301, P.R. China
E-mail itnhlib@scut.edu.cn

Contents

Topic A:
Climatic, Oceanographic and Biological Forcing of Sedimentary Systems

Lagrangian Modelling Techniques Simulating Wave and Sediment Dynamics Determining Sand-Body Equilibria

K. Black, M. Green, T. Healy, R. Bell, J. Oldman and T. Hume

1 Introduction

Estuarine and continental shelf sand bodies develop states which, in natural unsteady conditions, equilibrate, transpose, oscillate, decay and regenerate. Dynamic adjustments may lead to long-term stability of a feature (e.g. the establishment of an estuarine flood-tidal delta) or the feature may simply metamorphose into another state (e.g. a cross-shore migrating sand bar which welds onto the shore to become part of an open-ocean or estuarine beach). Traditionally, the equilibrium and geomorphological features in estuaries have been mostly explained by tidal circulation alone. However, there are estuarine features which are governed equally by local wave processes, especially in the intertidal zones.

To be effective, numerical simulations of estuarine sedimentary systems must therefore treat waves, currents and sometimes their interactions. For the wave component, the models must consider wave growth in limited fetches and the transformations in shallow water, while deformation by strong tidal streams may also be a first-order process.

Wave resuspension of sediment inside estuaries has been mostly neglected in favour of tidal mechanisms, particularly in comparison to the many studies of wave-driven sediment transport on open-coast beaches and ebb-tidal deltas. In this paper, we numerically hindcast measurements made with a self-contained instrumented tripod in an estuary. As measurements were made on a sand bank within a much larger physical environment, the data challenges present predictive capacity. The challenge is to predict suspended sediment concentrations at a location which is within this much larger environment using models only. We numerically predict all of the forcing parameters; namely height, period and near-bed orbital motion of wind-generated waves, tidal and wind-driven currents, sediment reference concentrations and concentration profiles. Locally measured current and wave data are employed for validation only.

Sand bank emergence/submergence, currents, wave conditions and suspended sediment concentration must all be simulated at the full estuarine spatial scales of tens of kilometres in order to predict the sediment concentrations a few centimetres above the sea bed at a single point. Waves have a profound influence on sediment suspension on the sand banks and then sediment is transferred into the channels by tidal currents. As such, the wave-induced sediment loads confound the simple notions of flood and ebb dominance in a tidal channel, when sediment concentrations may be more a function of dynamics on the adjacent sand flat than direct responses to tidal currents.

Modelling of such data is not trivial. At the least, it requires a hydrodynamic tidal circulation model, a wave generation and dissipation model and a sediment transport model. The "surf zone" in shallow water on the sand bank may need to be treated as well with a radiation-stress-driven simulation. We adopt three models in this chapter which are coupled to jointly predict hydrodynamics (wind and tidal circulation), wave generation and transformation, and suspended sediment load.

Because of the high spatial variability in factors controlling transport, only numerical models provide the potential for prediction of sediment dynamics in such an environment. The aim in this chapter is to test the capacity of numerical models to determine which physical processes can be predicted effectively and what remains to be achieved.

2 Numerical models

We chose three computer models: hydrodynamic and advection/dispersion model 3DD (Black 1995); wave generation model WGEN3DD (Black 1997); and sediment transport model POL3DD (Black 1996). A surf-zone model 2DBEACH (2-dimensional BEACH wave transformation and circulation model) (Black and Rosenberg, 1992a) was also used, but limitations on bathymetry resolution precluded detailed analysis of the results.

Model 3DD (3-dimensional dynamics) contains four process models in a single computer code. These are: (i) 2- and 3-dimensional hydrodynamics, (ii) advection/diffusion of salinity and/or temperature, (iii) surface gravity waves in shallow water using a Boussinesq approximation and (iv) ocean/atmosphere heat transfers. Because of its general capacity, 3DD has been applied to a wide range of vertically stratified and homogeneous ocean, continental shelf and shallow water environments (e.g. Young et al. 1993; Middleton and Black 1994; Black et al. 1996).

The wave generation model WGEN3DD (wave generation coupled to 3DD) was developed for fetch-limited water bodies and treats plan shapes which change during the tidal cycle with the submergence and emergence of intertidal sand banks. WGEN3DD applies the JONSWAP (joint North Sea wave project) equations assuming pseudo-steadiness and is therefore most useful in small estuaries of up to about 40 km maximum fetch. Since the original version presented by Black and Rosenberg (1992b), the model has been extended to include depth-limited breaking, shoaling and bed friction in the JONSWAP formulae. WGEN3DD has been linked to the hydrodynamic model 3DD so that nonlinear wave-current interactions in the bed friction term can be treated, while coupling with a sediment dynamics model provides for calculation of sediment transport in wave and current environments.

WGEN3DD acts by subdividing the fetch along the wind direction into short lengths of order 10–100 m. In each segment, empirical formulae for wave generation are applied to find the theoretical heights and periods. The model incorporates linear wave shoaling and then truncates the height if it exceeds the breaking depth limitation. Steepness limited truncation of the spectrum can be simulated (Black 1997) but was not applied in this study. The effect of shoaling, breaking and friction is accumulated in each segment to produce the hindcast wave heights along the fetch. The frictional dissipation is calculated by adopting the formulae of Bretschneider and Reid (1954). To obtain bottom orbital currents from sea surface wave height, the theoretical JONSWAP spectrum is decomposed into frequency intervals Δf and the linear theory wave transform is applied to each frequency band. A regular time series of bed orbital currents is reconstituted at 2 Hz by linear summation of the transformed spectral components (assuming a random phase for each component). For sediment transport calculations, the third velocity moment defined by Black and Rosenberg (1991) is calculated from the reconstituted time series. This is given by:

$$U_3 = \sqrt{2}\left(\frac{1}{N}\sum_{i=1}^{N}\left|u_i^3\right|\right)^{1/3}.$$

The dispersal model POL3DD (pollution dispersal coupled to 3DD) (Appendix 1) tracks suspended "particles" to simulate sediment transport or larval dispersal. The model includes a range of larval behaviours, an oil spill simulation, outfall modelling capacity and an estuarine or beach sediment transport simulation. The sediment model (e.g. Black 1987; Black 1994) uses Lagrangian techniques which are particularly useful near sharp concentration gradients common to sediment suspension cases, as they exhibit minimal numerical diffusion/dispersion because the particle positions are exactly known and particle advection is calculated directly from the currents (Black and Gay 1990).

All of the models use the same rectangular bathymetry grids to determine depth, fetches or shape of the estuary. A 200-m square horizontal grid is adopted. Model 3DD solves the conservation and momentum equations explicitly on a Eulerian grid to specify in two dimensions the vertically averaged currents and the sea levels. With WGEN3DD, the same grid is adopted and the model is coupled to 3DD via sea levels transferred to account for the large sea gradients which occur in the estuary. POL3DD is Lagrangian and is coupled to the other models via the transferred sea levels, currents and near-bed orbital motion, which are interpolated within the Eulerian grids of the other models at the positions of the particles in POL3DD each time step. When using output from a 2-dimensional hydrodynamic model, a logarithmic shape of the velocity profile is assumed, having a vertically averaged speed given by the currents from the hydrodynamic model. Being Lagrangian, the particles in POL3DD are not bound to a grid. However, average concentration is obtained by summing the volumes and masses carried by all particles within layers, and dividing by the layer volume V ($=200 \times 200 \times 0.05$ m^3); a layer thickness of 0.05 m provided sufficient resolution for the present study. Examples of the vertical particle distribution from POL3DD over bedforms are presented by Black et al. (1997).

3 Field Measurements

The field site lies in the eastern sector of the Manukau Harbour, which is a large (368 km^2) estuary on the North Island of New Zealand (Fig. 1). The hydrodynamic behaviour is dominated by the barotropic tide (tidal range during the measurement period was 2 m), although local wind-driven circulation is significant on the intertidal flats. Over fetches of up to 25 km, waves in excess of 1.5 m height may be generated at high tide. However, around 40% of the estuary is exposed at low tide, which strongly limits generation of wind waves by the prevailing southwesterly winds (Bell et al. 1998). The site has notable similarities with the southern German Bight where small wind waves have a strong influence on sediment transport and sediment redistribution (Haupt, pers. comm.).

The instrumented tripod RALPH (operated by the Geological Survey of Canada) was deployed on the intertidal flat in the eastern sector of the harbour (Fig. 1), some 500 m from the adjacent tidal channel. RALPH is a 2.5 m-high tripod equipped with sensors for measuring waves, currents, suspended sediment concentration (SSC) and bedforms. In this paper, we use data from (i) two pressure sensors at z=0 and 150 cm height above the sea bed (ii) an Optical

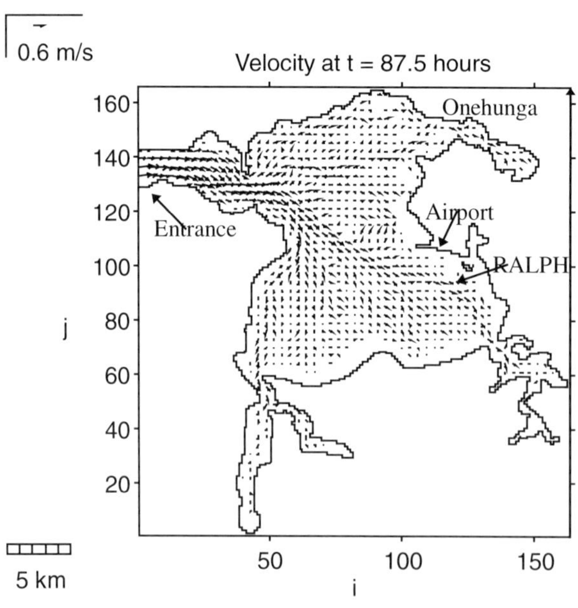

Fig. 1 Manukau Harbour (North Island, New Zealand) and simulated currents near the time of peak flood.

Backscatterance Sensor (OBS) at 7 cm and (iii) electromagnetic current meters at 48 and 97 cm height above the bed. All instruments sampled at 5 Hz and 4400 samples were stored each burst. Data treatment is described by Green et al. (1998).

Sediment dynamics on the intertidal flat is controlled by waves, which occur intermittently (Green et al. 1998; Bell et al. 1998). However, when waves are present, there is a characteristic continuous variation over the tidal cycle in the controlling processes. Both the surface wave heights and the bed orbital currents vary considerably over the tidal cycle in response to changing sand bank emergence and water depth respectively. Wave heights are largest at high tide. However, as depth profoundly influences the penetration of orbital motion to the bed, SSC is much lower around high tide than at lower tide levels, due to this factor alone.

Green et al. (1998) also noted a change in "kind", which occurred with the arrival at the measurement site of the "turbid fringe", the narrow, highly turbid edge of the estuarine water body. The turbid fringe initially travels with the rising tide across the sand flat, spreads more uniformly with time and ultimately leads to an overall increase in turbidity throughout the estuary as storm wave-energy develops. The turbid fringe exhibited concentrations which were larger than expected for the measured orbital currents. The precise mechanisms for this were not isolated, although factors such as turbulence penetration due to wave breaking in the following winds were considered.

RALPH was situated on the crest of a subtle linear bar, in a field of bars which emerge from the water at spring low tide. The crests are 0.5–1.0 km long with average spacing of 50 m and amplitude of approximately 0.05 m. Surface sediments on the intertidal flat were found by Dolphin (1992) to comprise 96% sand and 4% mud, while anaerobic sediments are typically encountered 1–2 cm below the surface. The low mud content and constant stirring of the bed allows the bed to remain non-cohesive.

Sediments at the RALPH site have mean grain size of 0.1 mm which compares to median grain sizes of 0.2 mm in shallower water further up the intertidal flat (Dolphin 1992). Muds were observed to intermittently accumulate as a fine veneer, but do not persist (Green et al. 1998). Symmetrical wave-generated ripples (height 1–2 cm; length 10–50 cm) were frequently observed. The sediments can therefore be described as predominantly noncohesive fine sands mixed with small amounts of surficial and interstitial muds. They experience a depth of disturbance of around 1–2 cm (Dolphin 1992); similar to the ripple height, although greater disturbance depths (up to 8 cm) have been noted in some parts of the intertidal flat (Dolphin 1992).

To examine all relevant processes, a storm was chosen for model simulation. This was the period consisting of hourly data bursts 2152-2188 from 8:00 on May 9 to 03:00 on May 10, 1995. (Note that a burst number increment of 1 corresponds to a half-hour time interval). To test the model, measurements are compared with model predictions from two adjacent model cells at the RALPH site. These are cells I,J=(119,95) and (118,96), where I is positive to the east and J is positive to the north (Fig. 1).

4
Model Comparisons

4.1
Water Levels and Currents

No water level measurements were made at the model boundary in the entrance to the Manukau Harbour during the RALPH deployment. To overcome this deficiency, model boundary levels were generated using water levels recorded within the estuary at Onehunga (Fig. 1) as follows. First, we modelled the estuary using Onehunga tides at the open model boundary. The phase lag and amplitude adjustment was then obtained using a lagged regression analysis of model-predicted tides at Onehunga and the original measurements. The lag was found to be 0.58 hours and amplitude factor was 0.784, plus a mean sea level (MSL) shift of 0.48 m was identified due to superelevation of MSL and a datum shift. These factors were then used to generate sea level boundary conditions from the Onehunga records. Winds were taken from the nearby Auckland airport anemometer and were assumed to be spatially-uniform throughout the model grid. To allow for land boundary layer effects, speeds were multiplied by 1.1 (a 10% augmentation), although the effect of this factor could not be separated from the effect of bed frictional resistance, later set by calibration.

Water levels measured at RALPH were found to be in agreement with the model predictions after a small additional correction was applied; the amplitude of the sea level oscillation at the boundary was increased by 6% and the mean vertical shift was lowered by 8 cm from 0.48 m to 0.40 m. The 0.4 m vertical shift was subsequently found to be a result of a datum shift between the eastern and western segments of the harbour, as applied by the Navy hydrographer. The spatial variation in the datum shift remains unknown and is later shown to be highly important in the very shallow conditions being simulated. Accentuation of the overtides (evident in Fig. 2a) is caused by the simple linear regression adopted to find the relationship between Onehunga and open boundary water levels. In essence, the overtides are represented twice; once in reality as the tides travel to Onehunga and then again in the model during the simulation.

To identify any boundary layer velocity gradients, current measurements at elevations of 48 and 97 cm are presented. The currents from the model are mostly in good agreement (Fig. 2b). The speeds are least well predicted during the windy ebb phase when the model over-estimates the magnitudes. This may be caused by unrepresented wave-driven currents on the intertidal flat or wind-induced shear through the water column which is not simulated by a 2-dimensional model. Notably, the measurements and the model are in close agreement during the previous two half cycles when waves were not present and the measurements at the two instrument elevations are also mostly similar except during the windy ebb phase.

The measured current directions rotate more smoothly than the model but are generally in good agreement (Fig. 2c). Small deviations in direction are undoubtedly caused by poor bathymetry resolution at the 200 m grid scale adopted

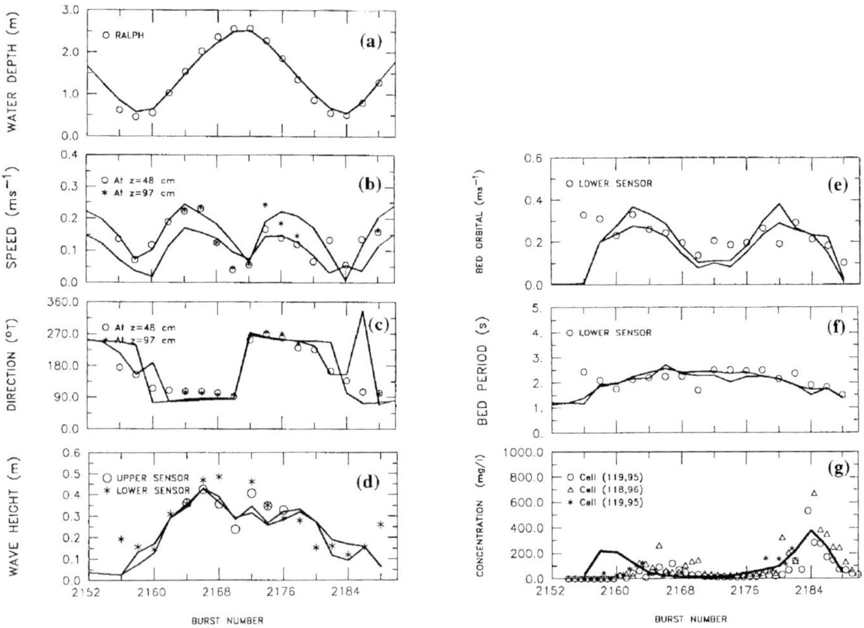

Fig. 2a–e Comparison of the measured tidal levels, current speeds, current directions, wave height and bed orbital currents with the model from cells (119,95) and (118,96). The *solid lines* are the model prediction at the two adjacent model cells and the *symbols* show field measurements. **2f, g** Comparison of the measured bed orbital peak spectral period and SSC with the model from cells (119,95) and (118,96). Note that two large values at cell (118,96) are not included in order to optimise readability. The *solid lines* are the model prediction at the two adjacent model cells and the *symbols* show field measurements

in the model. This is exacerbated by the relative coarseness of the published hydrographic survey of the region used to form the model bathymetry and difficulties with the datum. For example, flow directions at the two adjacent model cells show somewhat different behaviour as would be expected in model runs with realistic bathymetry (Fig. 2c). This highlights the critical importance of very accurate bathymetry.

4.2 Waves

We used the same wind measurements and the tides at Onehunga (with a 0.4 m datum shift in accordance with the 3DD modelling) to hindcast the wave conditions recorded at RALPH. Pressures recorded at both the upper and lower sensors were converted to sea-surface elevations using the linear wave theory transformation for each frequency band in the spectrum. It was found that the two sensors gave similar (significant) wave heights around low-mid tide, but the lower sensor indicated surface wave heights some 10–20% larger than the upper sensor at high tide. This error is caused by accentuation of the high frequency sensor noise in the spectrum by the linear theory transform for the deeper sensor (Hutt and Black 1997). The error occurs when depth is about 2 m for 0.4 m waves with peak spectral periods of approximately 3 s and would increase for smaller waves with shorter wavelengths or in deeper water.

Comparison of the measured sea surface wave heights and the model-predicted values indicates a close agreement (Fig. 2d). The deviation at the peak relates to the three largest "measured" values which are over-estimated heights from the lower pressure sensor. The model shows good agreement at this time with the heights calculated from the upper sensor records. The only major deviation occurs around bursts 2156 and 2158. The wind speed was very low (2 ms^{-1}) and so heights are under-predicted. Possibly, the total variance in the pressure records used to calculate the "inferred" significant wave height may include variance from other (non-wind-wave) processes, sensor noise or lower frequency seiching. One of the authors (Green) observed nonlinear solitary waves at the site of unknown origin. Irrespective of the cause of the deviation, the wave model cannot generate the inferred wave height, given the very low wind speeds. Because of the low wind speeds, near-bed orbital currents predicted by the model are correspondingly low at the start of the period (Fig. 2e). Near-bed peak spectral periods show good agreement with the measured values, except around burst 2156 when height predictions were low (Fig. 2f).

All processes (winds, sand bank emergence, fetch, shoaling, breaking, frictional dissipation, surface spectral shape and bed orbital transformation) are calculated directly by the model, but the friction coefficient needs to be calibrated. Best fit to the measurements was obtained with a friction coefficient $C_f =$ 0.025. This value is physically acceptable for the low relief sandy beds experienced in the shallow parts of the estuary, compared with higher values of around 0.06–0.09 obtained on coastal rocky terrain (McComb et al. 1997; Hutt 1997). The friction is highly important because of the long travel distances in shallow water, particularly around low tide, and the model indicates that wave heights are strongly influenced by friction.

4.3 Suspended Sediment Concentrations

In the next phase, the model predictions of currents and waves are used in the sediment transport model POL3DD. While some deviations between models and data have been noted, the model predictions are used here rather than actual measurements in order to use numerical models only to predict SSC at the RALPH site. Pick-up function techniques [used by Black and Rosenberg (1991), Black (1994), Black et al. (1995)] for the entrainment of sediment under waves have been applied here without modification and the third velocity moment is adopted as the "representative" orbital current, in accordance with the recommendations in these publications. Similarly, the vertical eddy diffusivity for sediment is taken as constant and equal to 0.0015 $m^2.s^{-1}$. A grain size popu-

Table 1 Fall velocity, equivalent grain size for a sphere with quartz density of 2650 kg.m^{-3} and layer thicknesses.

Fall vel. ($m.s^{-1}$)	Grain size (mm)	Grain size (phi)	Layer thickness (m)
0.0005	0.0255	5.2944	0.0001
0.0025	0.0580	4.1080	0.0004
0.0078	0.1070	3.2245	0.02
0.0220	0.2014	2.3120	0.001
0.0590	0.4181	1.2582	0.001
0.0005	0.0255	5.2944	0.0001
0.0025	0.0580	4.1080	0.0004

lation split into 7 fractions was simulated with sizes ranging from muds to medium/coarse sands (Table 1). Notably, the grain size and fall velocity distribution were not measured by Green et al. (1998) and so the sizes were selected to reflect the dominance of 0.1 mm sands and to limit the mud content to approximate the observed values (Dolphin 1992).

The model keeps track of the grain sizes at the sea bed in each model cell and adjusts the size if fine fractions are preferentially suspended. The mean grain size (D) calculated by the model during operation is used in the formula by Swart (1974) to calculate the wave friction factor,

$$f_w = \exp\left[5.213\left(2.5D / a_b\right)^{0.194} - 5.977\right],$$

where a_b is the orbital semi-excursion defined as $U_3 T_p / 2\pi$ and T_p is the peak spectral period. (Notably, $C_f = 0.5 f_w$). The friction factor therefore varies during the tidal cycle in response to both sea bed grain size changes and depth-induced changes in orbital currents at the bed.

The availability of sediment is set by allocating initial thicknesses of each sediment fraction as a notional layer on the bed. The model assumes that each fraction is independent but the total thickness of the layers acts as a depth of disturbance limitation. By treating the fractions independently, one fraction may be totally suspended while the others remain "available" for suspension from the bed. Any settled mass becomes available for resuspension (Black 1996). With the Lagrangian model, settlement is treated independently of suspension, thereby allowing for any hysteresis in the entrainment and settlement processes.

For the simulations presented here, the waves and currents are also treated as independent; the waves determine entrainment concentrations while the currents are used to specify advection. While wave-current interaction at the bed would modify the combined friction velocity, it has been shown in the studies noted above that entrainment on beaches is best predicted by the orbital motion alone. This is because the suspension process in the wave boundary layer is mostly driven by turbulence generated by the waves, even though currents and current-generated friction velocity remains relatively important beyond the wave boundary layer. While this current-induced turbulence is important in the water column, we are dealing with a small elevation above the bed of 7 cm only. Diffusion is modelled as a random walk and a horizontal eddy diffusivity of 0.015 $m^2.s^{-1}$ has been chosen (both along and across the flow streamlines). This value is ten times larger than the vertical eddy diffusivity in order to incorporate advective and turbulent processes associated with short-duration current oscillations and wave breaking. The selected value remains uncalibrated but its influence is relatively small on the model grid of 200 m cells. In the absence of bedload transport data for model confirmation, we do not treat bedload in this chapter.

Model output is compared to the measurements in Fig. 2g which shows predictions from 2 adjacent cells and a prior run with a slightly altered depth. The magnitude of the predictions are comparable with the measurements, i.e. 2 peaks in SSC with magnitudes of about 300-400 $mg.l^{-1}$ and a period of low concentration of less than 50 $mg.l^{-1}$ around the high tide. The second peak is quite well modelled, even though adjacent model cells give different results. The first peak is less well predicted, but this is primarily because of the low wind speeds around burst 2156, which resulted in a low orbital motion prediction. The results are highly sensitive to depth, primarily because of the strong dependence of bottom orbital motion on water column attenuation and the interaction between fetch, bed friction and bank emergence. Green et al. (1998) also found mostly good agreement between measured SSC and predictions by adopting the same methods as those applied here. However, they were unable to explain the unexpectedly high concentrations in the turbid fringe, which were larger than expected for the measured orbital motion. With the full spatial model and its capacity to treat the process of grain size winnowing, changes in roughness length and horizontal advection, we are able to predict the magnitude and the phase of peak SSC, even though it does not coincide with peak orbital motion. Secondly, Green et al. (1998) were unable to explain concentrations recorded around peak flood tide which were lower than expected for the measured orbital currents (bursts 2162-2166). The model, on the contrary, predicts low concentrations at this time which match the field measurements (Fig. 2g). The modelling shows advection by the flood currents of lower SSC from the deep channel as the cause of the anomaly noted by Green et al. (1998).

5 Discussion

One must conclude that if the inputs can be specified accurately, effective numerical simulations of currents, water levels, surface waves, wave orbital currents and SSC are possible in an estuary, even with complex morphology, strong tidal currents and unsteady wave conditions.

While Green et al. (1998) were able to correctly predict SSC for some stages of the tidal cycle using point measurements, the numerical models were needed to address spatial variability in SSC, advection and changing sea bed conditions. Several additional critical factors determining SSC have been identified.

First, the advection of less turbid channel water onto the intertidal zone during the flood evidently explains anomalously low values of SSC during peak flood flow. Second, it appears that changes in grain size at the bed, when fines are suspended, lead to a small increase in the wave friction factor f_w. This causes the Shields stress and entrainment to increase and a feedback to f_w ensues. Coarsening of the grain size and a corresponding lowering of the mean bed level is shown in Fig. 3 in a period when fines have been lifted into the water column.

Third, the concept of layer thickness defining sediment availability was of paramount importance to the success of the simulations. Although the waves are small, Shields stress is more than adequate to lift fine material into the water column such that entrainment could have continued if sediment were available. If suspension was allowed to continue up to the maximally possible sediment transport capacity of the fluid, the predicted concentrations were much larger than those measured. The settlement time scales of the fines are long, much longer than the time scales for wave generation and development of critical bed shear stress. By imposing a layer thickness of 0.1 mm for the finest sediment (Table 1), their entrainment was limited by availability a short time after waves first developed. The model showed that SSC would have been much higher if more mud were available at the bed and suspension had been allowed to continue without imposing availability limitations.

This concept of availability-limited entrainment over beds of mixed sediment grain sizes is fundamental to the methodology employed in numerical models and to understanding estuarine sediment dynamics. In the field, it means that bed sediments must be carefully sampled over the depth of disturbance, while retaining both the surficial and interstitial muds, if model accuracy is to be improved.

The pick-up function which allows the processes of entrainment and settlement to be separated (Black 1994) makes it possible in a Lagrangian model to

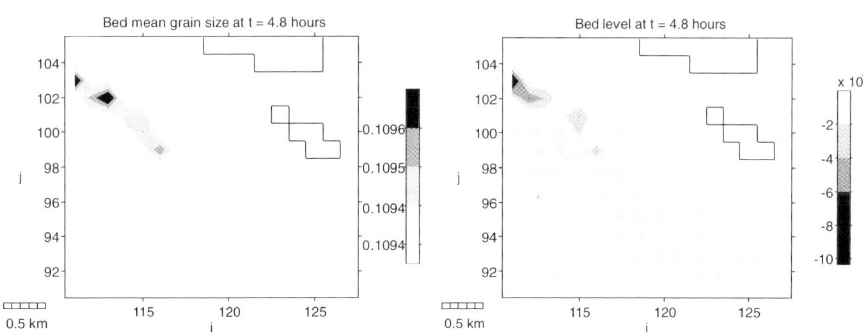

Fig. 3 Mean sediment grain size at the bed and bed erosion/accretion at 4.8 h after 8 am on May 9. The legend on the *left* shows mean grain size, while the legend to the *right* shows bed level change (x10^{-4})

realistically simulate hysteresis. The Lagrangian model assigns mass to particles of exactly known position and so mass conservation is always guaranteed. This is not always true of Eulerian solution schemes which are mostly not suited to the modelling of a field of sources and sinks (Black 1987).

With the long settlement times of the fines, the model shows the turbid fringe initiating in a narrow band parallel with the channel margin of the sand flat near RALPH where wave orbital currents are at a maximum. This maximum migrates across the sand flat with the incoming or out-going tide, causing the turbid fringe to move in synchrony (Fig. 4). Because of their relatively long settlement times, the suspended sediments (particularly the fines) spread more widely by diffusion and secondary entrainment. This leads to a broad, low-SSC region covering the intertidal zone at high tide (Fig. 4 and 17.6 hours) and a transfer of fines from the banks into the channels (Fig. 4) every tidal cycle when waves are present for initial entrainment.

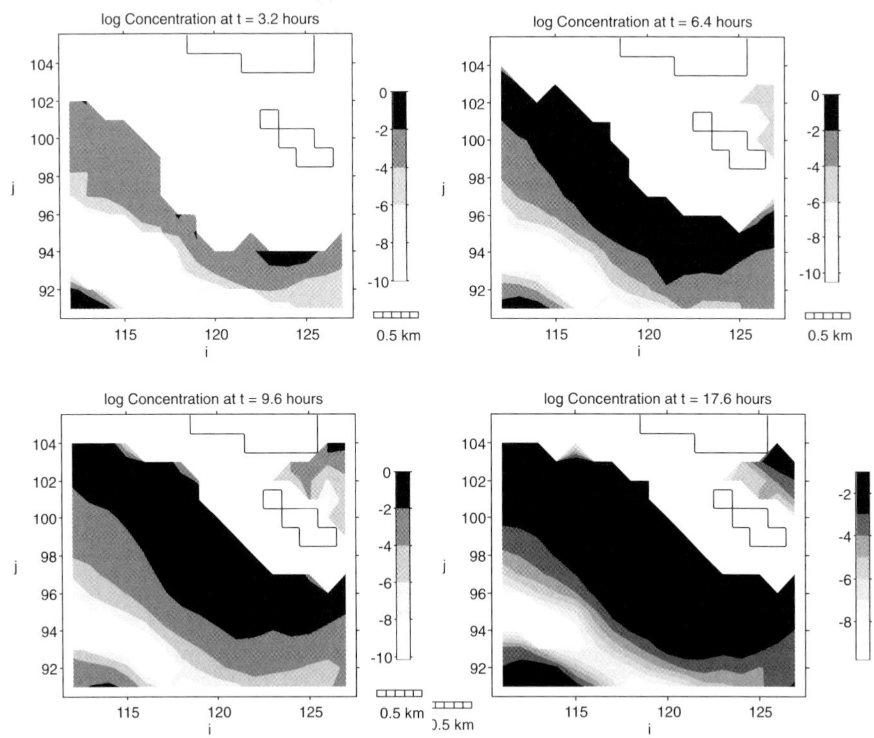

Fig. 4 Log_{10} of SSC around the RALPH site at 3.2, 6.4, 9.6 and 17.6 h after 8 am on May 9. Note the drop in average concentration at 17.6 h and the tendency for muds to remain in suspension at this time after the storm has passed. The legends show the log of the concentrations C to the base 10. They vary from 10^0 (=1 kg.m^{-3}) to 10^{-10} kg.m^{-3}. Note that the maximum concentration of 10^0 kg.m^{-3} at 9.6 h drops to $10^{-1.5}$ kg.m^{-3} at 17.6 hours.

Thus, the cause of low mud content on the exposed banks in the Manukau Harbour is associated with intermittent entrainment of the fines and losses from shallow banks where wave orbital currents are largest. In general, the distribution of muds may therefore be in balance with (i) the input volumes and source locations (river sources are often located up harbour), (ii) settlement time scales as a function of grain size, (iii) the wave energy levels at each location in the estuary, (iv) tidal current strength and direction on the sand banks (including the residual current) and (v) the flushing time of the estuary overall. Local wave energy determines the entrainment rate. Tidal currents and horizontal diffusion move the fines into the channels while the flushing time determines the amount of mud expelled through the estuary entrance when waves are stirring the bed internally. The settlement time scales determine the excursion distance per event for each grain size fraction. However, the key factor determining mud content on intertidal banks is wave energy.

To simulate the SSC, it was necessary to firstly generate accurate predictions of water levels, currents and wave energy. Water levels and currents were effectively predicted at the RALPH site, notwithstanding input data limitations including (i) no boundary sea level measurements, (ii) coarse hydrographic survey data confounded by a spatially-varying datum shift and (iii) no representation in the 200 m model grid of the meso-scale sand waves observed at the site. Improved predictions must therefore be possible with finer model grids and more accurate measurements of sea level and bathymetry. While tidal modelling is one of the most proven numerical simulations, the use of these models on intertidal sand flats in a storm is less well documented. The need for high-resolution modelling is highlighted by the findings of Bell et al. (1998) that bed slope on the intertidal flat, which partly determines surf zone width, is also very important.

Wave conditions at the RALPH site were also effectively predicted, even though the waves were considerably influenced by bathymetry during generation. The smaller heights as the tide drops were shown by the model to relate to increased effect of bed friction, wave breaking around low tide and shortened fetches as the banks emerge. The correctly modelled height oscillation at high tide, for example, is due to an oscillation in the wind strength. Thus, pseudo-steadiness, as assumed in the model, is evidently adequate at the 1-h time-scale of the measured winds over the fetches of up to 20 km. The assumption is also confirmed by the prevalence of single-peaked narrow spectra recorded at the site. Other oscillations in the wave heights (e.g. 2184 and 2186) relate to changes in bank submergence and these are being accurately predicted. The only significant deviations occurred at the start of the model period when wind speeds were lower than expected for the measured orbital currents. The wave heights, bed orbital motion and SSC prediction were all then affected by this anomalous wind strength. The effective height predictions conversely indicate that deformation by tidal currents (which is not being modelled) is a secondary process.

The challenge to model a comprehensive dataset from a site within a complex estuary was initially established as a test of present numerical modelling capacity. The results lead us to suggest that the models appear to be operating effectively. Most of the deviations between models and measurements have related to difficulties with the collection of comprehensive datasets at a broad spatial scale as well as at the local site. Of the errors encountered, the water depth throughout the estuary appears to be the most significant. Because of the short wave periods, depth strongly influences wave generation and bed orbital motion. The biggest deviation however stemmed from the low measured wind speeds at a time when measured orbital motion was significant.

Additional processes where more information would have been useful include wave breaking frequency at the site which could cause average turbulence levels to vary and fall velocity and grain size spatial variability. Unsteady sea surface gradients in tidal estuaries need to be accounted for but this can be treated by coupling the wave generation and tidal circulation models (the latter providing accurate water levels), as undertaken here.

In summary, wave breaking under following winds in shallow water, subtle changes to skin friction at intra-tidal time scales and preferential entrainment of sediment fractions within a population will need to be studied further before numerical simulations are able to model all aspects of SSC time series in estuaries. However, the comparison here, using only numerical models to simulate all of the processes, has shown that good progress is being made.

Acknowledgements

The study was funded by the Foundation for Research, Science and Technology in the project "Wave and Sediment Dynamics in Coastal and Estuarine Systems". The first author thanks Dr Jan Harff for his kind invitation to attend the conference.

References

Bell, R.G., Hume, T.M., Dolphin, T.J., Green, M.O. and Walters, R.A. (1998). Characterisation of physical environmental factors on an intertidal sandflat, Manukau Harbour, New Zealand. Journal of Experimental Marine Biology and Ecology. 216: 11–31.

Black, K.P. (1987). A numerical sediment transport model for application to natural estuaries, harbours and rivers. In: 'Numerical modelling applications to marine systems'. ed: J. Noye. North Holland/Elsevier. Mathematics Studies No. 145, p. 77–105.

Black, K.P. (1994). Suspended sediment load during an asymmetric wave cycle over a plane bed. Coastal Engineering. 23: 95–114.

Black, K.P. (1995). The numerical hydrodynamic model 3DD and support software. Occasional Report No. 19. Department of Earth Sciences, University of Waikato, New Zealand. 53 pp.

Black, K.P. (1996). Lagrangian dispersal and sediment transport model POL3DD. Occasional Report No. 21. Department of Earth Sciences, University of Waikato, New Zealand. 69 pp.

Black, K.P. (1997). The numerical wave generation model WGEN3DD. Occasional Report Department of Earth Sciences, University of Waikato, in press.

Black, K. P. and Gay, S. L. (1990). A numerical scheme for determining trajectories in particle models. In 'Acanthaster and the Coral Reef: A Theoretical Perspective'. (Ed. R. Bradbury.) Lecture notes in Biomathematics, Springer-Verlag. p. 151–156.

Black, K.P. and Rosenberg, M.A. (1991). Suspended load at three time scales. Coastal Sediments '91. Waterway, Port, Coastal and Ocean Div. (ASCE). 1: 313–327.
Black, K.P. and Rosenberg, M.A. (1992a). Natural stability of beaches around a large bay. Journal of Coastal Research. 8(2): 385-397.
Black, K.P. and Rosenberg, M.A. (1992b). Semi-empirical treatment of wave transformation outside and inside the breaker line. Coastal Engineering. 16: 313–345.
Black, K.P., Gorman, R.M. and Symonds, G. (1995). Sediment transport near the break point associated with cross-shore gradients in vertical eddy diffusivity. Coastal Engineering. 26: 153–175.
Black, K.P., Bell, R.G., Oldman, J.W. and Gorman, R. (1996). Challenges encountered modelling discharge options for Mangere WTP. New Zealand Waste Water Association Annual Conference, Nelson, September, 1996, pp. 264–268.
Black, K.P., Osborne, P., Villard, P. and Grant, D. (1997). Numerical simulation of high-resolution turbulence, orbital current and suspended sediment concentrations under swell at intra-wave time scales. Pacific Coasts and Ports'97, Christchurch, 7–11 September 1997.
Bretschneider, C.L. and Reid, O.R. (1954). Modification of the wave height due to bottom friction, percolation and refraction. Technical Memorandum 45, Beach Erosion Board, US Army Corps of Engineers. 36 p.
Dolphin, T.J. (1992). Low amplitude multiple bar systems in a fetch limited intertidal environment. MSc thesis, Univ. of Auckland, New Zealand. 143 pp. (Unpubl.)
Green, M.O., Black, K.P. and Amos, C.L. (1998). Control of estuarine sediment dynamics by interactions between currents and waves at several scales. Marine Geology. 144: 97–116.
Hutt, J. and Black, K.P. (1997). Vertical attenuation of wave-induced pressure. Pacific Coasts and Ports'97, Christchurch, 7-11 September 1997.
Hutt, J. (1997). Bathymetry and wave parameters defining the surfing quality of five adjacent reefs. MSc. Thesis, University of Waikato, New Zealand. 170 pp.
Jenkins, G.P. and Black, K.P. (1994). Temporal variability in settlement of a coastal fish (Sillaginodes punctata) determined by low-frequency hydrodynamics. Limnology and Oceanography. 39(7): 1744–1754.
McComb, P.J., Black, K.P., Atkinson, P.N., Healy, T.R. and Bell, R.G. (1997). High-resolution wave transformation on a coast with complex bathymetry. Pacific Coasts and Ports'97, Christchurch, 7-11 September 1997.
Middleton, J.F. and Black, K.P. (1994). The low frequency circulation in and around Bass Strait: a numerical study. Continental Shelf Research 14 (13/14): 1495–1521.
Nielsen, P. (1986) Suspended sediment concentrations under waves. Coastal Engineering. 10:23–31.
Swart, D.H. (1974). A schematization of onshore-offshore transport. Proc. 14th Int. Conf. Coastal Eng. (ASCE), 884-900.
Young, I.R., Black, K.P. and Heron, M.L. (1993). Circulation in the ribbon reef region of the Great Barrier Reef. Continental Shelf Research. 14 (2/3): 117–142.

Appendix 1: Lagrangian model POL3DD

For the horizontal advection, a 2nd order accurate scheme is used as described by Black and Gay (1989). In the Monte Carlo (random walk) diffusion stage, all particles in the water column move a small random distance horizontally or vertically (ΔX_n) where $\Delta X_n = R_n (6E_s \Delta t)^{1/2}$ and R_n is a random number in the uniform range (-1,1), Es is the lateral, longitudinal or vertical eddy diffusivity, and Δt the time step. Notably, a different vertical diffusion procedure is adopted for buoyant plumes. As the position of each particle is specified by its (x,y,z) coordinate in a sigma transformed coordinate system, it is possible to apply chosen vertical velocity profiles around the mean (vertically-averaged) current for those cases when a 2-dimensional hydrodynamic model is used to generate the flow patterns. Several alternatives are offered but the most commonly-used profile is logarithmic. The speed at any elevation is,

$$u(z) = U\left[\log_{10}(z/z_0)/\log_{10}(0.37d/z_0)\right],$$

where U is the vertically-averaged speed, z the elevation above the bed, d the depth and z_0 the roughness length.

The vertical particle motion is,

$$\Delta Z_n = R_n\left(6E_z\Delta t\right)^{1/2} + W_f\,\Delta t + W_s\,\Delta t,$$

where W_f is the fall velocity (positive upward). W_s is the vertical current strength (positive upward) taken from the hydrodynamic model in 3-dimensional simulations and E_z is the vertical eddy diffusivity.

Bedload transport is calculated at positions along the cell walls after interpolating gridded arrays for local velocity, bed conditions and grain size characteristics. By specifying sediment transport on cell walls, mass changes in the two bounding cells can be calculated directly without any need for interpolation or averaging. The positions along the cell walls are generated randomly. Thus, with unbiased locations and many repetitions, the model accounts for horizontal current shear along the wall, which strongly influences the sediment loads in the non-linear sediment transport equations. This is contrary to a Eulerian scheme which always calculates the sediment load at the cell mid-point only. A user-selected number of positions are generated each time step so that

$$M_T = \left(\frac{\Delta x}{N}\sum_{n=1}^{N} Q_n\right)\Delta t,$$

where M_T is the total mass transported across the wall, N is the number of positions generated randomly, and Q_n is the mass flux normal to the wall ($kg.m^{-1}.s^{-1}$) at each position. The bed level change associated with the bedload transport in the cells adjacent to the wall is, then

$$\Delta T_b = \pm M_T / \left(\rho_s P \Delta x \Delta y\right),$$

where ρ_s is the sediment density and P the pore space correction. The losses/gains of mass from the adjacent cells across the cell walls are calculated and used to update a cell-by-cell erosion/accumulation matrix. For a distribution of grain sizes, the total load is found by calculating the mass fluxes for each size class in succession and weighting by the fraction of mass in each class (see below).

When sediment is suspended naturally by currents, the sediment concentration near the seabed is composed of two independent components: (i) a "source" of sediment being entrained in response to instantaneous currents, turbulence and pressure gradients at the bed, and (ii) a "sink" associated with the settling of previously entrained sediments In steady flows where bed availability is constant, these two components are equal, but they are often different in non-steady environments (Black, 1994). Thus, it is physically important and notionally simpler to treat the two as independent.

In POL3DD, suspended sediment entrainment is calculated at multiple positions within each cell. The model selects random positions throughout the cells because: (i) cell wall mid-point releases would put the sediment onto particular streamlines and (ii) random releases on the cell walls only (as for bedload) are insufficient because SSC can travel larger distances than the bedload, thereby connecting the walls to the internal domain of the cells.

In the model, entrained particles are released at bed level, each carrying a concentration equal to the local suspended sediment concentration calculated after interpolating for local current strength, bed conditions, bed roughness, grain fall velocity and grain size. The mass carried by each entrained particle is an equal fraction of the total mass suspended during the time step. Knowing concentration and mass, the particle "volume" is calculated.

Specifically, a sediment flux boundary condition is applied (Black, 1994). The mass suspended is,

$$M_T = C_Z A w_s \Delta t,$$

where A is bed area and C_z is a near-bed concentration. Thus, the total mass entrained in a time step over a cell with dimensions $\Delta x, \Delta y$ is

$$M_T = C_0 w_s \left(m\Delta t\right)\Delta x \Delta y,$$

where C_0 is the near-bed concentration calculated using one of the empirical sediment suspension formulae (e.g. Nielsen, 1986) and m provides a time splitting option and is the number of time steps between simulated entrainment events. If N_R is the number of particles entrained within each model cell, the mass and volume per particle is, therefore

$$M_p = M_T / N_R \qquad V_p = M_p / C_o,$$

while the bed level change in a model cell is

$$\Delta Z_S = -M_T / \left(\rho_s P \Delta x \Delta y\right).$$

As for other aspects of POL3DD, particles are advected by currents derived from a hydrodynamic model simulation, while the horizontal and vertical eddy diffusivity are modelled as random walks.

Particles near the bed which fall out of suspension are removed each time step and the mass accumulation matrix is updated. The array storing sediment fall velocity distribution in each cell is also updated. All particles within a distance Δd, given by $\Delta d = w_s\, m\Delta t$, are said to have settled.

To fulfil the sediment flux boundary condition, diffusive random-walk excursions downwards which would carry a particle to the bed are disallowed by setting jump length to zero. This has implications for the entrainment simulation. During the entrainment phase, particles are placed at bed level with a calculated mass M_p, as described above. In disallowing downward jumps, a fraction of these particles remain at the bed level after the diffusion stage and will therefore fall out of suspension when their vertical advection is calculated next. Such deposition is a numerical artefact as the particles should be remaining in the water column with the newly-entrained component. Numerically, the problem can be treated by adding more particles or making a mass adjustment. The following approach was adopted in POL3DD.

If the eddy diffusivity at bed level is E_S, the maximum "jump" a particle can make is

$$E_j = \left(6E_s m\Delta t\right)^{1/2}.$$

Thus, the particle will jump somehere in the range $\pm Ej$ for a uniform random diffusivity. The actual jump size is

$$E_{sa} = R_n E_j,$$

where R_n is a random number uniformly distributed in the range $-1<R_n<1$.

For a particle starting at bed level, all downward jumps will be disallowed. In addition, all upward jumps which are less than the settling advection distance ($\Delta d = w_s . m\Delta t$) will still result in particle settlement in the next vertical advection phase. That is, of the total number of particles released at the bed, all those which experience negative jumps or positive jumps less than Δd will fall out of suspension during the following advection stage. Thus, the fraction of mass lost by this process is

$$S_f = \left(E_j + \Delta d\right) / 2E_j = \frac{1}{2}\left(1 + \Delta d / E_j\right),$$

where $2E_j$ is the total range of all possible jumps. The mass compensation factor to be applied to particles remaining after the first time step is therefore $1/S_f$.

As particles cannot pass through the water surface, a reflective boundary condition is applied at the surface in POL3DD. Numerically, particle jumps through the surface are simply disallowed by setting the jump length to zero. Jumps onto land are similarly prevented by shortening the jump length.

The total mass and concentration are determined in each model cell by accumulating the masses and volumes carried by the particles resident within the cells. The concentration at any elevation is the sum of the mass of all particles lying in a vertical cell, divided by the cell volume,

$$C_z = \sum_{i=1}^{n} M_{pi} / \left(dx\, dy\, dz\right),$$

where n is the number of particles while dx, dy and dz are the dimensions of the averaging area.

Grain size classes are selected by the user and defined by median fall velocity (and the erosion thresholds for bedload and suspended load). An "accumulation" file then specifies the relative abundance of mass in each grain size class in each model cell at the commencement of the model run. The model updates the mass in each class as the simulation proceeds so that changes to the grain size distribution with time are modelled.

Specifically, the model considers the bed to consist of an erodible layer of a known depth. The fraction of mass for each size class in each model cell is then specified in the "accumulation" file as a fraction of this total depth with units of metres. Assuming that the sediment is uniformly mixed, the depths have the same distribution as the masses in the grain size classes. By treating each size individually, winnowing, which may cause the fines to be lost leaving an armoured or lagged surface, is modelled.

Modeling Sediment Resuspension in Coastal Areas

J. Lou, T. Wolf and W. Rosenthal

A quasi-3D suspended sediment transport model under waves and current has been developed. The nonlinear wave-current interaction has been taken into account by the changes in turbulent viscosity and enhancement in bottom stresses.The model has been applied in Cleveland Bay, Australia, to study suspended sediment distribution in January and March 1993 when two high turbidity events (> 100 mg/l) were observed. The model was able to predict these high sedimentconcentration events. Swell was found to be the dominant suspension mechanism in the bay. In addition, the model was improved and applied to the Oder Estuary, which forms the border between Germany and Poland. The concentration levels are clearly related to the higher waves and water depth. The model results agree with known values and data images in the Oder Estuary.

1 Introduction

Understanding sediment transport processes is of fundamental importance in coastal dynamics and environmental evolution. Numerical modeling is one method that may be used to study sediment transport. The research in this field is still in its infancy and the model results are therefore usually of rather limited value. Although the vertical concentration structure can be explicitly taken into account by two-dimensional vertical (2DV) models (Van Rijn 1986; Celik and Rodi 1988), 2D horizontal (2DH) sediment transport models based on the depth averaged formula are still widely used (De Vriend 1987; Nairn and Southgate 1993; Ziegler and Nisbet 1994), with the disadvantage of neglecting the concentration profiles. Since suspended sediment transport in coastal areas is a three dimensional phenomenon, it should be described by three-dimensional models. However, most 3D models are expensive to run and they are thus commonly used to predict short-term sediment transport only (Koutitas and O'Connor 1980; O'Connor and Nicholson 1988; Miller 1984). An alternative to 3D modeling is the application of a depth-integrated approach (Galappatti and Vreugdenhil 1985).

This chapter focuses on the development of a quasi-3D numerical model for suspended sediment transport in coastal waters under wave-current movement. The wave-current interactions and the quasi-3D sediment concentration model will be briefly described. The new numerical model was then applied to Cleveland Bay, Australia, and the Oder Estuary, which straddles the border between Germany and Poland. In this chapter the model predictions are discussed and tested against the field data.

2 Wave-Current Bottom Stresses

Bottom stresses required for sediment transport modeling are calculated by a bottom boundary layer model. The calculated stresses are then used in the bottom boundary condition for the sediment transport model. The influence of wave-current interaction on the bottom shear stress is calculated using an iterative procedure based on the concept of Grant and Madsen (1979). A similar method was used by Signell et al. (1990), but with a different definition of wave friction.

The maximum bottom stress $\tau_{b,max}$ for wave-current combinations is defined as:

$$\tau_{b,max} = \frac{1}{2} f_{cw} \rho (U_\delta^2 + u_c^2 + 2 U_\delta u_c cos\phi_c), \tag{1}$$

in which ρ is the water density, f_{cw} is an unknown effective friction factor under combined wave-current movement, U_δ is the maximum near bottom wave orbital velocity which may be determined from linear wave theory, u_c is the current velocity near the sea bed, and ϕ_c is the angle between wave propagation and the current direction.

The calculation of the effective friction coefficient f_{cw} begins by determining the oscillatory component of the stress τ_w, whereby:

$$\tau_w = \rho u_{*w}^2 = \frac{1}{2} \rho f_w U_\delta^2, \tag{2}$$

with f_w as the wave friction factor whose value can be obtained by using the relationship of Jonsson (1966).

With u_{*w} determined, an iterative procedure is used to calculate f_{cw} at the upper edge of the wave-current boundary layer as briefly described below:

(i) Starting with an initial guess of f_{cw}, the steady shear velocity component u_{*c} is obtained by:

$$u_{*c} = \sqrt{f_{cw}} u_c. \tag{3}$$

(ii) The combined wave-current friction velocity u_{*cw} is defined as:

$$u_{*cw} = \sqrt{\tau_{b,max}/\rho} \tag{4}$$

and can be obtained by the wave-related and current-related friction velocities:

$$u_{*cw} = (u_{*c}^2 + u_{*w}^2 + 2u_{*c}u_{*w}cos\phi_c)^{1/2}. \tag{5}$$

(iii) To determine the effective bottom roughness k_b:

$$k_b = k_s\left[24\frac{u_{*cw}}{U_\delta}\frac{A_\delta}{k_s}\right]^{\left(1-\frac{u_{*c}}{u_{*cw}}\right)}, \tag{6}$$

where k_s is the physical bottom roughness, and A_δ is the near bottom excursion amplitude.

(iv) The effective roughness is then used to calculate the velocity profile in the boundary layer (Signell et al. 1990):

$$u = \frac{u_{*c}}{\kappa}ln\left(\frac{z}{k_b/30}\right). \tag{7}$$

(v) Solving for the velocity at a reference level, a new estimate of the friction coefficient is obtained:

$$f_{cw} = \left[\frac{\kappa}{ln(30\delta_{cw}/k_b)}\right]^2. \tag{8}$$

The above procedure (i-v) is repeated until the successive estimates of f_{cw} differ by less than a preset error value (10^{-6} in this case).

3 Suspended Sediment Concentration Model

A quasi-3D suspended sediment concentration model based on the advection diffusion equation has been developed. It starts from the work of Galappatti and Vreugdenhil (1985), who introduced an asymptotic solution to a 2DV model under steady current. In this study, a quasi-3D model, developed for more complicated flow fields, was generalized for application to combined wave-current movement.

The wave effect is taken into account by assuming an analogy of mixing profile on a wave-averaged and turbulence-averaged scale and the modified eddy viscosity coefficient, as well as by introducing an enhanced bed stress. In cases where the suspended load is the main mode of sediment transport, an asymptotic solution of the advection diffusion equation is presented. The vertical concentration profile has been shown to depend only on the vertical velocity profile and the mixing coefficient. It can therefore be calculated in advance. The three-dimensional concentration is represented in terms of depth-averaged concentration and its horizontal derivatives. As a result, the three-dimensional suspended sediment concentration problem is separated into two parts: a two-dimensional depth-averaged model and vertical sediment concentration profiles solved in advance. As an example, the first order asymptotic solution is expressed as:

$$c = \left\{ a_{11}(\zeta) + a_{23}(\zeta) \frac{h}{w_s} \left(\frac{\partial \bar{u}}{\partial x} + \frac{\partial \bar{v}}{\partial y} \right) \right\} \bar{c} + a_{21}(\zeta) \frac{h}{w_s} \frac{\partial \bar{c}}{\partial t} + a_{22}(\zeta) \frac{\bar{u} h}{w_s} \frac{\partial \bar{c}}{\partial x}$$
$$+ a_{22}(\zeta) \frac{\bar{v} h}{w_s} \frac{\partial \bar{c}}{\partial y} - a_{21}(\zeta) \frac{h}{w_s} \frac{\partial}{\partial x} \left(\varepsilon_x \frac{\partial \bar{c}}{\partial x} \right) - a_{21}(\zeta) \frac{h}{w_s} \frac{\partial}{\partial y} \left(\varepsilon_y \frac{\partial \bar{c}}{\partial y} \right), \tag{9}$$

in which $\bar{c}(x, y)$ is the depth-averaged concentration, $\bar{u}$ and $\bar{v}$ are the depth-averaged velocities in the x and y directions, h is the water depth, w_s is the sediment settling velocity in still water at a specified kinematic viscosity, $\zeta = (z - a)/h$ is a new vertical coordinate, a is a reference level where the bottom boundary condition is applied, $a_{11}(\zeta), a_{21}(\zeta), a_{22}(\zeta), a_{23}(\zeta)$ are profile functions determined only by the velocity profiles and the vertical mixing coefficient. They can therefore be solved in advance.

The model solves the 3D advection diffusion concentration equation with almost the same efficiency as the 2DH models. To reduce numerical dispersions, the second order upwind difference scheme has been applied to the horizontal advection terms. A hybrid Crank-Nicolson and ADI solution scheme was developed to calculate the SSC results. More details of the model have been given in Lou (1995) and Lou and Ridd (1997).

Two significant effects of wave-current interactions on SSC distributions have been taken into account: (i) changes in the intensity of turbulence, and (ii) enhancement of bottom stresses. The latter has been solved by the boundary layer model. The first effect will be discussed in the following section.

3.1 Sediment Mixing Coefficient Under Wave-Current

Derived from wave-induced sediment concentration distribution data (Bosman 1982), the following three-layer wave diffusion coefficient was proposed (Van Rijn 1986):

$$\varepsilon_{s,w} = \begin{cases} \varepsilon_{w,bed} = 6.5 \times 10^{-4} \alpha_b \delta D_*^2 \hat{u}_{orb} & z \le \delta \\ \varepsilon_{w,max} = 3.5 \times 10^{-2} \alpha_b \frac{h H_s}{T} & z \ge 0.5h \\ \varepsilon_{w,bed} + (\varepsilon_{w,max} - \varepsilon_{w,bed}) \frac{z-\delta}{0.5h-\delta} & \delta < z < 0.5h, \end{cases} \tag{10}$$

in which δ is the thickness of the near bed mixing layer (or wave bottom boundary layer thickness), h is the water depth, H_s is the significant wave height, T is the wave period, D_* is a dimensionless particle size parameter, and α_b is the wave breaking coefficient.

The sediment mixing coefficient due to the combination of waves and current is assumed to be given by the sum of the squares of the current-related and wave-related values as suggested by Van Rijn (1989):

$$\varepsilon_s^2 = \varepsilon_{s,c}^2 + \varepsilon_{s,w}^2. \tag{11}$$

The current-related turbulent eddy coefficient $\varepsilon_{s,c}$ is calculated numerically from the 3D circulation model. This approach corresponds to the summation of kinetic energy of both wave and current motions as $\varepsilon \sim l \cdot E^{1/2}$, in which E is the kinetic energy and l is the mixing length scale.

3.2 Bottom Boundary Conditions in the SSC Model

It is assumed that the location of the fluid-sediment interface has been averaged over the wavelength of bedforms such as ripples or dunes. Let z denote a coordinate upward normal to this boundary. If the reference location, $z = a$, is chosen sufficiently close to the boundary, the normal fluid velocity w can be neglected. As a result, the upward flux of suspended sediment, evaluated at a reference distance $z = a$ above the bed, can be given by:

$$F_a = w_s(E_s - c_a cos\theta), \tag{12}$$

where c_a is the near bed concentration of suspended sediment, $w_s c_a cos\theta$ is the deposition rate per unit bed area due to the fall velocity w_s, and E_s is a dimensionless coefficient describing the entrainment of bottom sediment into suspension due to turbulence. If the bed slope is small, the net upward flux at the bed can be approximated by:

$$F_a = w_s(E_s - c_a). \tag{13}$$

To solve the distribution of suspended sediment, it is necessary to specify the values of E_s and c_a at the reference bottom boundary.

3.2.1 *Entrainment Coefficient*

Natural coastal sediments are often not uniformly sorted. In the model, the bed sediment mixture is divided into N size ranges such that p_k denotes the volume fraction of the bed material of the kth range.Values of p_k have been obtained in practice by measuring the particle size distributions of surficial sediments. Furthermore, D_k denotes the characteristic grain size of the sediment in the kth range.

Garcia and Parker (1991) developed an empirical relation of the entrainment coefficient. This new relation has been generalized to sediment mixtures with the aid of field data. It was demonstrated that this empirical fit was superior to other approaches and that it provided reasonable estimates of the entrainment coefficient. As a result, a similar approach is adopted in our study:

$$E_{s,k} = \frac{AZ_k^5}{1 + \frac{A}{0.3} Z_k^5}, \tag{14}$$

where

$$Z_k = \frac{u'_*}{w_{s,k}} R_k^{0.6} \left(\frac{D_k}{D_{50}} \right)^{0.2}, \tag{15}$$

and

$$R_k = \frac{\sqrt{(s-1)\, gD_k}}{\nu} D_k, \tag{16}$$

with A being a constant of 1.3×10^{-7}, D_{50} being the median size of bottom sediment, u'_* being the bottom shear velocity, D_k being the characteristic size of sediment in the kth range, $w_{s,k}$ being the settling velocity of the sediment in kth range at a specified kinematic viscosity. The suspension rate of class k should be normalized by the per unit volume fraction content of material in the kth range on the surficial bed as $p_k E_{s,k}$.

3.2.2
Bottom Concentration

The suspended sediment concentration at reference level $z = a$ above the bed is expressed as (Van Rijn 1989):

$$c_a = 0.015 \frac{D_{50}}{a} \frac{T_B^{1.5}}{D_*^{0.3}}, \tag{17}$$

in which T is a bed shear stress parameter, $T_B = (\tau_b - \bar{\tau}_{b,cr})/\bar{\tau}_{b,cr}$, τ_b is the effective bed shear stress under combined waves and currents, $\bar{\tau}_{b,cr}$ is the time-averaged critical bed shear stress of Shields (1936).

This is the bottom boundary condition for unlimited bed sediment source. If the sediment erosion on the bed is limited, the probability function representing this sediment suspension rate should be multiplied in the above equation.

4
Model Application to Cleveland Bay

Cleveland Bay is near latitude 19°S on the north coast of Australia (see Fig.1). On its northwest side is Magnetic Island. The bay is shallow, reaching a maximum depth of 15 m at its seaward edge. The main approach for ships entering the bay is an artificial channel known as Platypus Channel extending northwards. Platypus Channel is often dredged to a depth of 12–13 m and the dredged material is dumped in the northern part of the bay near the 10-m isobath. Tides in Cleveland Bay have a large spring-neap asymmetry, with a maximum range of about 3.8 m at springs and 0.5–0.8 m at neaps. The waves in the bay consist of two components: local wind waves and swell. The sediments commonly show a gradational size distribution from fine silt to coarse sand.

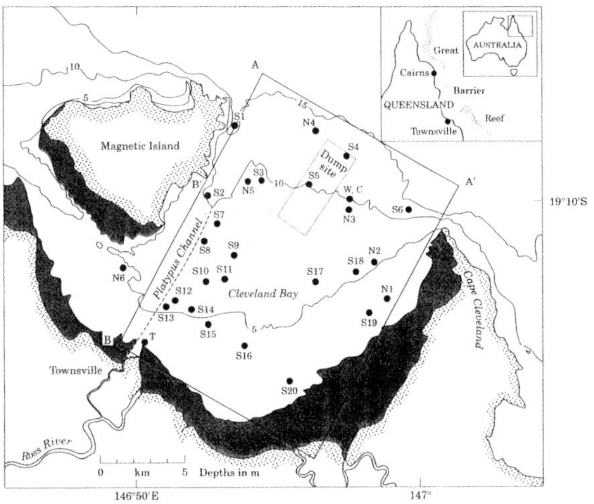

Fig. 1 Cleveland Bay and field measurement locations. SSC sites are denoted as *N*, current sites as *C*, sediment sample traps as *S*, the wave recording site as *W*, and the tide level site as *T*

From 1 January to 30 April 1993, a 4-month field study in Cleveland Bay was carried out (Comarine 1993). The sediment concentrations at six locations (N1,2...6 in Fig. 1) were recorded using self-cleaning optical backscatter nephelometers (Ridd and Larcombe 1994) positioned at 30 cm above the seabed. Waves and currents near the dump site were recorded over the same period. The wave spectrum was arbitrarily separated at a period of 7 seconds into "swell" (with periods greater than 7 sec) and "wind waves" (short waves with periods less than 7 sec) (Lou et al. 1996). The tidal elevation data were obtained from Townsville Port.

During the 4 months, two high sediment concentration events with peak SSC data larger than 100 mg/l were observed. The first was between 9 and 11 January, the second was between 12 and 14 March (days 71–73 in 1993).

From 9–11 January, a high SSC event corresponding to a high swell and spring tide was observed. During this period, a spring tide with a maximum tidal current velocity of 0.53 m/s occurred between 8–10 January. With increasing swell, SSC levels rapidly increased to greater than ambient values (>100 mg/l). The SSC peaked at about 1800 h on 9 January with the maximum SSC level above 200 mg/l, and then decreased to the background value below 10 mg/l. This high SSC event lasted for about 4 days from 8 to 12 January.

A similarly high SSC event was observed between 12 and 14 March (days 71–73) 1993. This turbid water layer also occurred under an increasing swell, but this time during a neap tide rather than a spring tide. Compared with the first event, the swell action was weaker and the SSC level in this period was lower, reading a peak value of about 100 mg/l. This high SSC event lasted for about 2 days.

Using the SSC model with measured current and wave data, sediment concentrations in Cleveland Bay were simulated for two periods: i) 5–15 January 1993, and ii) 6–16 March 1993.

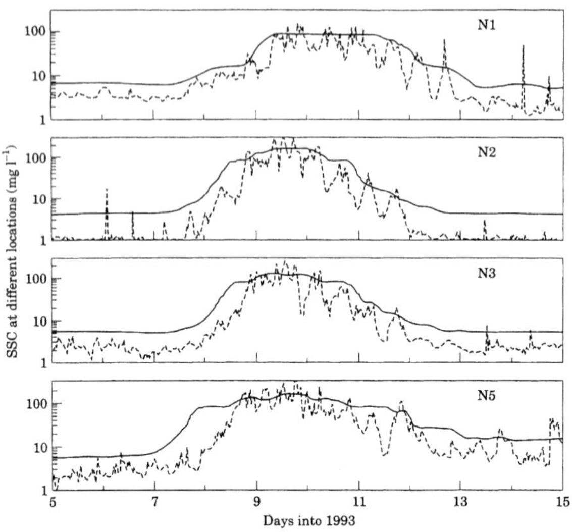

Fig. 2 SSC model results for 5–15 Jan. *Solid line* Model result; *dashed line* field data

It has been shown by Lou and Ridd (1996) that the most important factor causing sediment suspension in Cleveland Bay is swell rather than local wind waves. Therefore, only the swell data measured near the dump site were used in the model simulation, wind waves being neglected.

SSC data at station N4 (near offshore boundary) and N6 (around southwest boundary) have been used as the concentration open boundary conditions.

The modeled SSC results from 5 to 15 January 1993 are given in Fig. 2. These results apply to a height of 30 cm above the sea bottom. Because SSC data from N4 and N6 were used as boundary conditions in the model simulation, no results for these two sites are shown. It can be seen that the high SSC event from 9 to 12 January was reproduced by the model. Generally, the predicted high SSC patterns agree well with field measurements at all four stations. The differences between simulated SSC results and field measurements are within an order of magnitude. For the most part of high SSC about 100 mg/l, the difference between model predictions and field data is less than a factor of 5. For the background SSC values before 8 January and after 12 January, the model results are higher than the data. Compared with field data, the model predictions are essentially smooth, no peaks or troughs in the field data having been simulated. These peaks or troughs in SSC measurements result partly from tidal advections near the concentration front where either clear or turbid waters flow across the nephelometers by flood or ebb tides respectively.

Between 12 to 14 March (days 71–73 into 1993), a period of moderate swell (with the significant wave height of 0.5 m) occurred over neap tide. During this period another high SSC event was observed. In this case, the wind and swell intensity were smaller, and the observed SSC magnitudes were lower and lasted for a shorter time interval than the first event in January. To simulate this high SSC event, a 10-day model run from 6 to 16 March 1993 was carried out. A

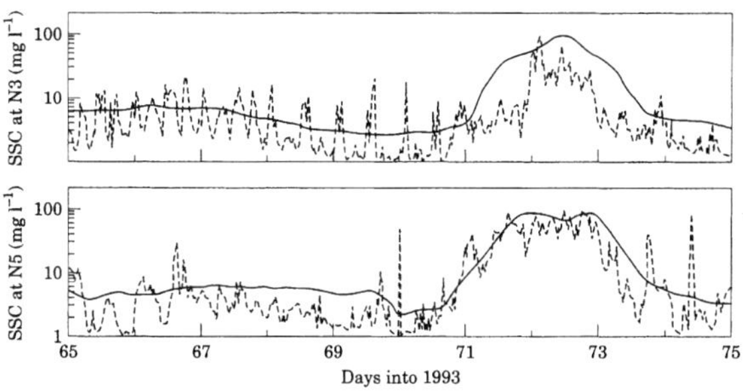

Fig. 3 SSC model results for 6–16 March. *Solid line* Model result; *dashed line* field data

comparison of measurements with model results is presented in Fig. 3. Because of instrument problems, no SSC data were recorded at N1 and N2 during this period. Only the results of N3 and N5 have thus been considered. The general trends of the model predictions at station N5 agree well with the measured data. For the elevated SSC period in days 71–73, the difference between model prediction and field data at N5 is less than a factor 4. Before and after this high SSC event, the model calculations mostly overestimate the data with a difference within a factor of 5. At station N3, from days 71–74, the model results are higher than the data. The observed high SSC greater than 10 mg/l at N3 lasted for 1 day (days 72–73), whereas the predicted high SSC lasted for 2.5 days.

5 Suspended Sediment Distribution in the Oder Estuary

The Oder Estuary forms the border between Germany and Poland (see Fig. 4) and drains into the Baltic Sea, where tidal effects are negligible. The Oder Estuary, which connects with the Baltic Sea through three straits: Peene, Swina and Dziwna, is a very shallow bay with an average water depth of about 4 m. Only the Swina strait reaches depth of 10–17 m. The local wind field is an important factor for generating an unsteady circulation and waves. Some results of wave and current modeling in the Oder Estuary can be found in Gunther and Rosenthal (1983), Winkel (1994) and Wolf and Gayer (1996).

The sediment distributions in the Oder Estuary are non-homogeneous. The composition of the sediments (granulometric and chemical) varies over a wide range. Silts and sandy silts cover above 60 % of the bottom and there are high concentrations of organic matter (14–35 %) in the silts. The sediments in the western half ofthe estuary (German side) are somewhat coarser than those in the eastern half (Polish side). The fine sediments (silts and sandy silts) are mainly located in the inner part of the estuary, whereas the sands are mostly found close to the coastline. Detailed descriptions of sediment characteristics in the Oder Estuary can be found in Osadczuk et al. (1996) and Lampe and Meyer (1995).

As a preliminary model application, suspended sediment concentrations in the Oder Estuary were simulated for different wave and current conditions. An

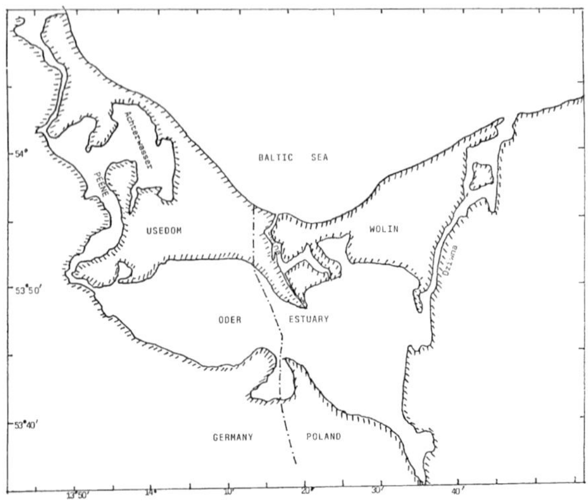

Fig. 4 Location of the Oder Estuary

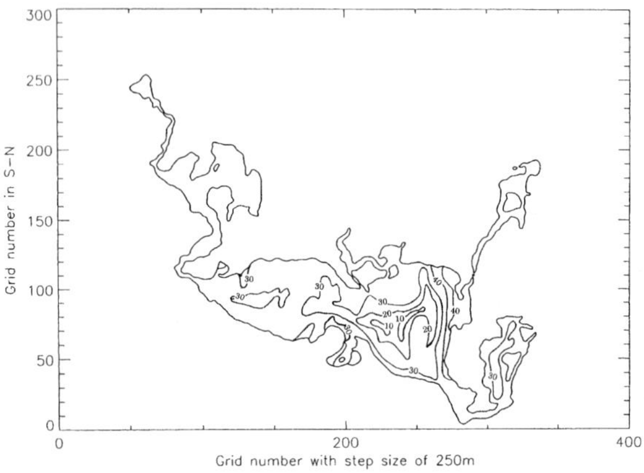

Fig. 5 Suspended sediment concentrations in mg/l in the Oder Estuary under calm wave climate

initial uniform depth-averaged concentration of 40 mg/l was assumed for the whole study area.

Figure 5 shows the SSC model result under a calm wave climate without currents, a uniform significant wave height of 0.2 m, and a peak wave period of 3.0 sec. The result indicates that during this calm weather period the wave-induced bottom stresses did not excess the critical bottom stress and that the sediment particles would settle down to the sea bed, resulting in clearer water. It also shows that most of the settling is predicted to occur in the eastern part of the Estuary (Polish side).

By contrast, with a wave height of 0.6 m and a wave period of 4.0 sec, the model predicts resuspension into the water column to produce turbid water (see Fig. 6). In the eastern part, the concentration is slightly higher than the initial background value; substantially higher suspension was mainly found in the western part (German side).

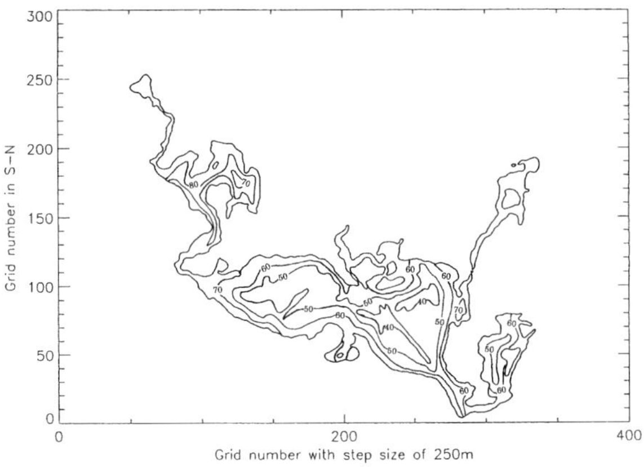

Fig. 6 Suspended sediment concentrations in mg/l in the Oder Estuary under moderate wave climate

The concentration levels are clearly related to the higher waves and greater water depth. Though the model needs further careful calibrations with more field data, the agreement with known values and data images are very promising.

6 Conclusions

The new sediment transport model presented in this chapter integrates the instantaneous vertical structure of the flow, wave-current interaction, and turbulent eddy coefficient changes due to wave/current coexistence. It is superior to semi-empirical sediment transport models and to most other models which are based on Bagnold's energetics concept using vertically integrated equations. It is more efficient than the fully 3D models, while all kinds of boundary conditions for a 3D model can be used.

The new model was applied in Cleveland Bay in Australia where the model predictions were compared with the field data during two high SSC events in January and April 1993. The following results were achieved: (1) High concentration events of over 50 mg/l were predicted with an error less than a factor of 3. (2) Fluctuations in SSC appearing in the data, probably caused by advection near concentration fronts, were not predicted by the model. (3) During periods of low concentration (less than 10 mg/l), the model results were often higher than the field data.

The preliminary model results of suspended sediment concentrations in the Oder Estuary agree with known values and data images. However, further model calibrations with higher quality field data is still considered necessary.

Considering the state of the art of sediment transport modeling (with model errors of larger than one order of magnitude being quite common), the results achieved with the new model using real data are very encouraging.

References

Bosman, J., 1982, Concentration measurements under oscillatory motion, *Delft Hydraulics Report M1695-II.*

Celik, I., and Rodi, W., 1988, Modeling suspended sediment transport in nonequilibrium situations, *Journal of Hydraulic Engineering*, **114** (10), 1157–1189.

Comarine, 1993, Environmental Monitoring Program: Oceanographic and sediment data, Townsville Port Authority, Australia.

De Vriend, H. J., 1987, 2DH mathematical modelling of morphological evolutions in shallow water, *Coastal Engineering*, **11** (1), 1–27.

Galappatti, G., and Vreugdenhil, C. B., 1985, A depth-integrated model for suspended sediment transport, *Journal of Hydraulic Research*, **23** (4), 359–376.

Garcia, M. H., and Parker, G., 1991, Entrainment of bed sediment into suspension, *Journal of Hydraulic Engineering*, **117** (4), 414–435.

Grant, W. D., and Madsen, O. S., 1979, Combined wave and current interaction with a rough bottom, *Journal of Geophysical Research*, **84** (C4), 1797–1808.

Gunther, H., and Rosenthal, W., 1983, Shallow water surface wave model based on the Texel-Marsen-Arsloe (TMA) wave spectrum, *proceedings of the 20th Congress of IAHR, Moscow.*

Koutitas, C., and O'connor, B. A., 1980, Numerical modelling of suspended sediments, *Advances in Water Resources*, **3** (6), 51–57.

Lampe, R., and Meyer, H., 1995, Sedimentqualitat und hydrographische Bedingungen ein Vergleich verschiedener Boddengewässer (in German), *Bodden*, **2**, 7–26.

Lou, J., 1995, Modelling of hydrodynamics and suspended sediment transport in coastal areas (Ph. D thesis), *James Cook University, Australia*, 243 pp.

Lou, J., Ridd, P., Mayocchi, C., and Heron, M., 1996, Wave induced benthic velocity variations in shallow waters, *Estuarine, Coastal and Shelf Science*, **42**, 787–802.

Lou, J., Ridd, P., 1996, Wave-current bottom shear stresses and sediment resuspension mechanisms in Cleveland Bay, Australia. *Coastal Engineering*, **29**, 169–186.

Lou, J., Ridd, P, 1997, Modeling of suspended sediment transport in coastal areas under waves and currents. *Estuarine, Coastal and Shelf Science*, **45**, 1–16.

Miller, H. P., 1984, Three-dimensional free surface suspended particles transport in the South Biscayne Bay, Florida, *International Journal of Numerical Methods in Fluids*, **4**, pp. 901–914.

Nairn, R. B., and Southgate, H. N., 1993, Deterministic profile modelling of nearshore processes. Part 2: Sediment transport and beach profile development, *Coastal Engineering*, **19**, pp. 57–96.

O'connor, B. A., and Nicholson, J., 1988, A three-dimensional model of suspended particulate sediment transport, *Coastal Engineering*, **12**, pp. 157–174.

Osadczuk, A., Lampe, R., Meyer, H., and Sobieraj, D., 1996, Characteristics of recent sediments from Szczecin Bay, Hydrotechnical transactions, Polish academy of sciences, Institute of hrdro-engineering, Gdansk, pp. 119–128,

Ridd, P. V., and Larcombe, P., 1994, Biofouling control of optical backscatter suspended sediment sensors. *Marine Geology*, **116**, pp. 255–258.

Shields, A., 1936, Anwendung der Ähnlichkeitsmechanik und der Turbulenzforschung auf die Geschiebebewegung, *Mitteilungen der Preussischen Versuchsanstalt fur Wasserbau und Schiffbau*, Berlin, Heft 26, 26 pp.

Signell, R. P., Beardsley, R. C., Graber, H. C., and Capotondi, A., 1990, Effect of wave-current interaction on wind-driven circulation in narrow, shallow embayments, *Journal of Geophysical Research*, **95** (C6), pp. 9671–9678.

Van Rijn, L. C., 1986, Mathematical modelling of suspended sediment in nonuniform flows, *Journal of Hydraulic Engineering*, **112** (6), pp. 433–455.

Van Rijn, L. C., 1989, Sediment transport by currents and waves, *Report H 61, Delft Hydraulics Laboratory.*

Wolf, T., and Gayer, G., 1996, Coupled modeling of hydrodynamical parameters in the Greifswalder Bodden, Hydrotechnical transactions, Polish academy of sciences, Institute of hrdro-engineering, Gdansk, pp. 65–76,

Winkel, N., 1994, Modellierung von Seegang in extremem Flachwasser (in German), *Report GKSS 94/E/66.*

Ziegler, C. K., and Nisbet, B., 1994, Fine-grained sediment transport in Pawtuxet River, Rhode Island, *Journal of Hydraulic Engineering*, **120** (5), pp. 561–576.

Model Simulation of Temporal Variability of Suspended Matter in the North Sea

W. Puls, T. Pohlmann and J. Sündermann

1 Introduction

The North Sea is a shelf sea with maximum water depths of 50 m in its southern half and between 50 and 120 m in its northern half. Special geographic features are the Norwegian Trench and the Skagerrak with maximum water depths between 300 m and 600 m. Figure 1 shows the bathymetry of the North Sea together with some principal locations.

In the North Sea, the average water circulation resulting from non-linear tides, from mean wind and from horizontal density differences is anticlockwise, with North Atlantic water entering from the Orkney-Shetland area and with English Channel water entering through the Straits of Dover. The water masses leave the North Sea towards the north along the Norwegian coast. In the south-

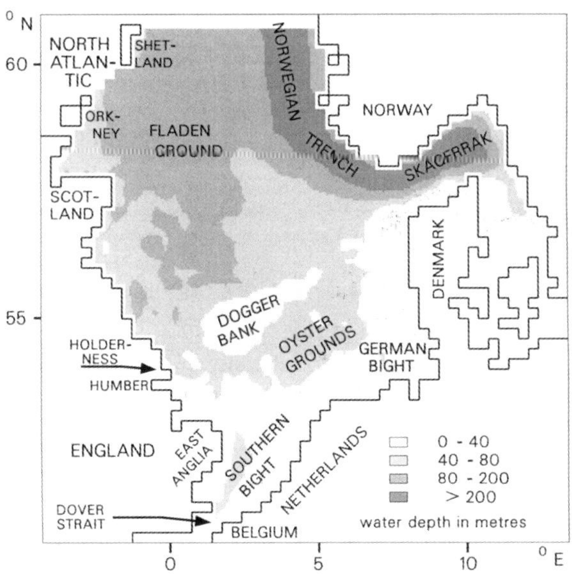

Fig. 1 North Sea model domain with bathymetry and some principal locations

ern and the northern North Sea residual currents are typically in the order of 5 and 1 cm s^{-1}, respectively. Maximum tidal currents are 1.2 m s^{-1}. A comprehensive description of the hydrography of the North Sea is given by Otto et al. (1990).

The suspended particulate matter (SPM) regime in the North Sea is described by several authors, e.g. by Eisma and Irion (1988), Howarth et al. (1993) or Dyer and Moffat (1998). SPM is supplied to the North Sea (a) from the adjacent seas, above all the North Atlantic (mainly the Orkney-Shetland area) and the English Channel via the Straits of Dover, (b) from cliff erosion, particularly along the English coasts of East Anglia and Holderness, (c) from rivers and dumping, and (d) from seafloor erosion. The major SPM flux out of the North Sea goes northward towards the North Atlantic via the Norwegian coastal current. Major SPM sedimentation occurs in the Skagerrak and the Norwegian Trench.

In winter SPM consists of terrigeneous material plus detritus (10–20% by weight). Typical (depth averaged) SPM concentrations during winter are 2–5 mg l^{-1} and 0.2–1.0 mg l^{-1} in the southern and the northern North Sea, respectively. The approximate line separating the SPM regime of the North Sea into a northern and a southern part is between the Humber in the west and North Denmark in the east. During summer, SPM concentrations in the northern North Sea may increase significantly due to phytoplankton production with maximum phytoplankton concentrations in the order of 0.5 mg l^{-1}. In the southern North Sea phytoplankton concentrations also increase to about 0.5 mg l^{-1}, but phytoplankton does not become dominant here because terrigeneous SPM (plus detritus) concentrations are still in the order of 1–2 mg l^{-1}. The grain size of terrigeneous SPM in the North Sea is typically in the clay fraction or in the fine silt fraction (Eisma and Kalf 1979).

The temporal variability of SPM concentrations in the North Sea at a fixed site and a fixed water depth is due to

- (tidal) advection of an SPM concentration gradient past the site, e.g. König et al. (1994), Jones et al. (1994)
- cyclic resuspension and deposition by tidal currents, e.g. Jago et al. (1993)
- resuspension/erosion (and subsequent deposition) by storm induced waves and currents (Green et al. 1995)
- production and decay of phytoplankton, e.g. Reid et al. (1990)
- vertical diffusion and settling, e.g. Jones et al. (1994), Puls et al. (1995).

The mostly pronounced variability of the SPM concentration in the shallow southern North Sea is due to storm events. König et al. (1994), for instance, measured an increase from 4 mg l^{-1} to 20 mg l^{-1} during a storm in the northern part of the German Bight (sample depth: 10 m, total water depth: 20 m, bottom sediment: sand).

The erosion of sediment during storms is an important transport mechanism for particulate bound substances (contaminants) in the sediment. To give an example: the content of *anthropogenic* lead in the (fine) sediments of the German Bight including the German Wadden Sea is estimated to be 35 000 tons (Puls et al. 1997a). This amount has accumulated during the last 100 years. Within the last decade, the input of lead into the North Sea was substantially reduced; but there is still a lead impact on the German Bight environment because of the lead in the sediment, which leads to the question of how long the lead in the sediment will remain in the German Bight. One reduction mechanism is the erosion of fine sediment by storms, the following transport with the current over a certain distance and finally the redeposition of the fine sediment. Stepwise, fine sediment will be transported further "until it eventually reaches a point where the base of storm-induced waves will not reach the seabed any more. In the case of the North Sea this point can be found in the Skagerrak" (cited from Liebezeit 1991).

This chapter deals with the temporal variability of the SPM regime in the North Sea, especially the southern North Sea. The temporal variability is calculated (1) for the data collected by the North Sea Project (NSP) of the British Natural Environment Council (NERC), and (2) for the data computed by a 3-dimensional SPM transport model. The comparison of the model results with measured data gives an impression of the model's quality.

2 SPM Transport Model

The model domain of the SPM transport model for the North Sea is shown in Fig. 1. The horizontal grid size, indicated by the zigzagging coastlines, is 20 km. The thickness of the vertical layers is 5 m from the surface down to 50 m (i.e. 10 layers); below 50 m the layer thickness increases. The maximum number of layers is 19. The horizontal grid lines subdivide the model domain into 1350 horizontal grid cells. The additional subdivision by the vertical layers results in 13851 single grid boxes. The time step for the model simulations presented here is 80 min.

The sea bed below each horizontal grid cell has a thickness of 20 cm, it is vertically divided into 43 layers. The thickness of the layers increases from 0.1 mm just below the sediment surface up to 24 mm in the deeper parts.

Most 3-dimensional models for SPM transport are based on solving a 3-dimensional transport-diffusion equation using the finite difference method where SPM is treated as a continuum (e.g. O'Connor and Nicholson 1988). In contrast to such an approach, the SPM transport model presented here uses a stochastic, or Monte-Carlo, procedure. A description of a tracer particle transport method is for instance given by Al-Rabeh and Gunay (1992).

The movement of SPM in the model domain is simulated by the migration and the random walk of tracer particles (Lagrangian tracer technique). Up to 100 000 particles are used, each being loaded with a certain mass of SPM. The SPM masses on the single particles must not be identical, each particle can have a different settling velocity. During a model run these particles are injected at grid points where SPM input takes place. A particle carries only the mass of SPM from one single source, e.g. "Rhine river". When SPM is deposited on the sea bed, SPM mass is subtracted from the particles. When there is erosion of fine sediment, the SPM mass on the particles is increased.

A typical number of particles in the southern North Sea is 100 per horizontal grid cell. This means a relative standard deviation of the estimated SPM concentration in that horizontal grid cell of ± 10 %, resulting from the randomness of the particle number.

What follows is a brief description of the general concept of the model simulations, of the prerequisites for SPM modelling and of the simulated processes. For a more comprehensive model description see Pohlmann and Puls (1994).

2.1 General Concept

Sediment transport in waters is induced by near-bed water movements, generated by currents and by surface water waves. The forces on the sediment bed are expressed by the bed shear stress τ or by the bed shear velocity $u_* := (\tau \rho^{-1})^{0.5}$, with ρ being the water density. The relationship between u_* and the mean current velocity u_{100} at 1 m above bed is $u_{100} = C\, u_*$, with C being a Chézy coefficient in the order of 15.

In hydraulic engineering, sediment transport is usually quantified by transport formulae, relating u_* to the mass of transported sediment. Transport formulae exist for sediment transported near the bed (bed load), sediment in suspension (suspended load) or the sum of both (total load), they are mostly valid for non-cohesive sediment.

The model presented here deals only with suspended sediment. A transport formula is not applied because such a formula implies that the SPM mass or the SPM concentration in the water column is in equilibrium with the local, instantaneous bed shear velocity u_*. SPM from the North Atlantic, for instance, settles only 10 m within 50 days. Regarding a time scale of several days, the content of Atlantic-borne SPM in a water column is independent of u_*.

The grain size of SPM in the North Sea is typically in the clay fraction or in the fine silt fraction (Eisma and Kalf 1979; Eisma and Kalf 1987). The North Sea model simulates the grain size fraction < 20 µm only.

In the presented model version, the model simulates terrigeneous SPM (plus detritus) only, i.e. phytoplankton dynamics are not simulated.

The model is designed to simulate the large-scale, long-term SPM transport in the North Sea during several years. At present, the years 1976–1995 can be simulated because both wave data and current data are available for this time period.

2.2 Prerequisites for SPM Modelling

For a realistic simulation of the transport of fine sediment in the North Sea, the SPM model needs information about (1) bottom sediment, (2) SPM sources, (3) currents and diffusion coefficients, and (4) surface water waves.

1. Information about the content of fine sediment (fraction < 20 μm) in the sea bed sediment is given in Fig. 2.
2. The input of SPM *(fraction* < 20 μm) into the model domain is dominated by the residual fluxes from the North Atlantic and the Straits of Dover. At the seaward model boundaries, the SPM flux into the model domain is the product of a prescribed SPM concentration (constant in time) and the instantaneous water flux. The prescribed SPM concentrations are 0.6 mg l^{-1} between Scotland and Orkney, 0.4 mg l^{-1} between Orkney and Shetland and 0.3 mg l^{-1} between Shetland and Norway. This data is mainly based on measurements of Eisma and Kalf (1987). In the Straits of Dover, prescribed SPM concentrations are 7.2 mg l^{-1} in the northern half and 4.6 mg l^{-1} in the southern half, based on data of Lafite et al. (1997). The residual SPM flux into the North Sea between Scotland and Shetland is calculated to be 9 Mt a^{-1} as an average of 1986–1993. The SPM flux through the Straits of Dover is calculated to be 17 Mt a^{-1} for the same time period.
   SPM supply from the cliffs of eastern England (Holderness, East Anglia) is 4.7 Mt a^{-1}(Odd and Murphy 1992). In order to achieve a good agreement of their simulated SPM concentrations with measured data, Odd and Murphy (1992) assumed that 90 % of the annual cliff erosion occurs in the four month period November–February.
   The SPM masses entering the North Sea from rivers and by dumping (sewage sludge, fly ash) can be taken from Pohlmann and Puls (1994). The input from rivers and by dumping is comparatively small, the sum is about 2 Mt a^{-1}; the highest mass flux comes from the Rhine/Meuse/Scheldt with 0.75 Mt a^{-1}.

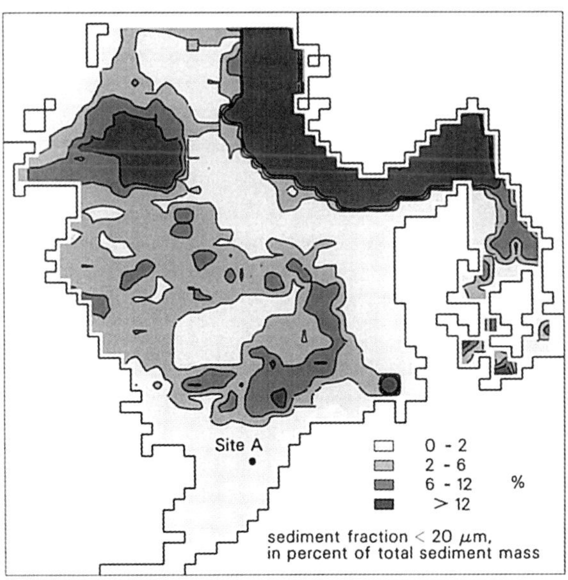

Fig. 2 North Sea model domain with the distribution of fine sediment content (< 20 μm) in the bottom sediment

3. The currents are provided by a separate 3-dimensional North Sea water circulation model (Pohlmann 1996). The water circulation model has the same grid structure as the SPM transport model. The provided simulated currents are:
   - M2/S2 tidal currents
   - daily residual currents, generated by water density gradients and wind.

   The water circulation model also provides daily averaged coefficients of vertical diffusion which includes information about stratification (thermocline, halocline).
4. Simulated 2-h surface wave data is given at 60 points in the North Sea. The data is provided by the German Amt für Wehrgeophysik, Traben-Trarbach (Helbig 1985).

The near-bed currents and the surface water waves are combined to calculate the bed shear velocity u_*, (Puls et al. 1997b). For each of the 1350 horizontal grid cells of the model domain, u_* is calculated every 80 min.

For a model run starting at the time t_1, there is a "hot start" concerning the distribution of SPM in the model domain and concerning the fine sediment in the bed. This hot start is achieved by premodelling the time period of 2 years before t_1, using the actual currents and waves of that time period.

2.3 Simulated Processes

Maximum SPM concentrations in the North Sea are about 100 mg l^{-1}. This is a comparatively small concentration which does not affect vertical mixing or SPM settling.

Erosion of the bed sediment happens if the bottom shear velocity u_* increases above the threshold shear velocity of the sediment. The threshold shear velocity $u_*{}^E{}_{CR}$ is 2.8 cm s^{-1} both for sandy *and* muddy sediment, which is the result of recent model tunings. The concept for the erosion of a sediment bed is: a surface sediment layer of the thickness D_E is stirred up; D_E is proportional to $u_*{}^2 - (u_*{}^E{}_{CR})^2$. The present mass of the sediment fraction < 20 μm in the disturbed layer is entrained into the water column, the coarser sediment is not regarded. Erosion is thus not simulated as a rate (i.e. a time-dependent) process, but as an instantaneous process. Maximum D_E-values are in the order of 2 cm. The increase of the SPM concentration in the water column during such an extreme event is in the order of 100 mg l^{-1}.

Newly deposited SPM clothes the bed in a thin veneer. This mud blanket is *resuspended* instantaneously (en bloc) if the bottom shear velocity, u_*, increases above the threshold shear velocity of the veneer material. According to the measured data of Jago et al. (1993) the threshold shear velocity for resuspension should be in the order of 2 cm s^{-1}, but in the present model version the same value is used as for sediment *erosion*, i.e. 2.8 cm s^{-1}.

SPM is advected by the currents provided by the water circulation model of Pohlmann (1996). As a by-product, the spatial gradients of horizontal velocities provide for horizontal diffusion. Diffusion of tracer particles is simulated by a random walk process, such as described in Stronach et al. (1993). The coefficients of vertical diffusion are provided by the water circulation model, a typical value near the water surface during stormy weather is 0.03 $m^2 s^{-1}$.

The settling velocity of SPM depends on the SPM's origin. SPM from the North Atlantic Ocean settles with 2×10^{-3} mm s^{-1}, SPM eroded from the sea bed settles with 3×10^{-1} mm s^{-1}. SPM from the English Channel, from rivers, cliff erosion and dumping is divided into three fractions with settling velocities of 1×10^{-1}, 2×10^{-2} and 4×10^{-3} mm s^{-1}. The information about settling velocities comes from in situ measurements (Puls et al. 1995), it is deduced from mud accumulation rates (e.g. Johnson and Elkins 1979) or is obtained by tuning the model results via comparison with observed time series of SPM concentrations.

Deposition of SPM takes place (1) by settling and (2) by water filtration by bottom dwelling suspension feeders. Deposition by settling is possible only if the bottom shear velocity, u_*, is below a threshold shear velocity for deposition (Krone 1962), which is between 0.8 and 2.8 cm s^{-1}, depending on the SPM's settling velocity. A typical water filtration rate, being valid for sandy bottom in the southern North Sea (seasonal average) is 100 litres m^{-2} day^{-1} (Rachor, pers. comm.).

Most of the SPM deposited by settling forms a mud blanket on the bed surface. If this mud blanket is not resuspended, it is gradually consumed by bottom-dwelling deposit feeders and stored in the upper 5 cm of the sediment bed, now being an integral part of the bed. SPM deposited by suspension feeders is immediately stored in the sub-bottom layer 0–5 cm.

Vertical mixing in the sediment bed by bioturbation (0–20 cm sub-bottom depth) is simulated by simple diffusion (Boudreau 1986). The effective diffusion coefficient is 10 $cm^2 a^{-1}$ at the bed surface and decreases linearly down to 1 $cm^2 a^{-1}$ with increasing sub-bottom depth.

2.4 SPM Time Series

To give an impression of the short-term variability of the SPM regime in the southern North Sea, Fig. 3 shows time series of the bed shear velocity, u_*, and of depth-averaged SPM concentrations (total SPM and eroded SPM). The time series is simulated for December 1988 at "site A" (see Fig. 2), a location in the Southern Bight, water depth 27 m. The bed shear velocity shows four maxima per day which reflects the maxima of the M2 tidal currents. During storms, surface water waves generate a substantial increase of u_*, e.g. on December 6-7, 1988. The semidiurnal variability of the total SPM concentration (on December 1-2 and December 20-22) results from tidal advection of a concentration gradient past the site. Obviously, this spatial gradient exists only for non-eroded SPM, because eroded SPM does not show the semidiurnal variability.

When u_* exceeds the threshold shear velocity for erosion (2.8 cm s^{-1}), fine sediment is picked up from the bed by removing the fine sediment (as long as it

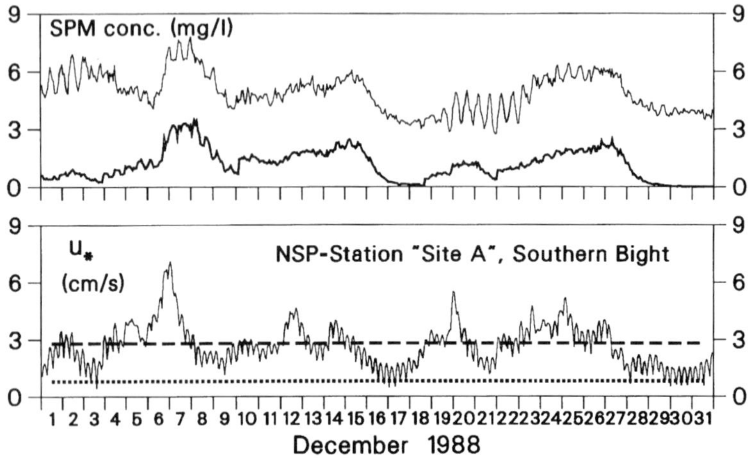

Fig. 3 Simulated time series of bed shear velocities u_* and depth-averaged SPM concentrations at site A (see Fig. 2) during December 1988. In the *lower diagram*, the *dashed horizontal line* is the threshold bed shear velocity for deposition of fast settling SPM and for erosion of fine sediment (2.8 cm s^{-1}), the *dotted horizontal line* is the threshold bed shear velocity for deposition of slowly settling SPM (0.8 cm s^{-1}). The *thick curve in the upper diagram* is the SPM concentration of *eroded* SPM, the thin curve is *total* (including eroded) SPM

is not exhausted) from the sediment layer which is stirred up by the high bed shear velocity. The result is an increase in the concentration of eroded SPM and a corresponding increase in the total SPM concentration.

Deposition of eroded SPM takes place as soon as u_* remains below the threshold for deposition, which is 2.8 cm s^{-1} for eroded, fast-settling SPM, and 0.8 cm s^{-1} for slowly settling SPM. According to the u_* time series in Fig. 3, deposition of slowly settling SPM seldomly occurs at site A.

The general level of SPM at site A is about 5 mg l^{-1}. It is mainly supplied to the North Sea through the Straits of Dover. The maximum concentration of eroded fine sediment (thick curve in Fig. 3) is 3 mg l^{-1} on December 7–8 which is small regarding the high u_*-value of 7 cm s^{-1}. The mass of eroded fine sediment is small because the content of fine sediment is small in the sandy sea bed of the Southern Bight.

3 SPM Regimes During Extreme Weather Conditions

As examples for the temporal variability of the SPM regime in the North Sea, two extreme situations are presented: (a) calm weather conditions and (b) stormy weather conditions. Two dates from the North Sea project (NSP) survey period (Sect. 4.1) are selected: August 9, 1988 (calm) and February 14, 1989 (storm). The "extremity" of the two situations can be assessed from Fig. 10 which shows a long-term time series of u_* at site A (see Fig. 2) in the Southern Bight.

3.1 Calm Conditions

On August 9, 1988, Beaufort 3 (weak breeze) was measured in the central North Sea, Beaufort 2 in the Southern Bight and Beaufort 1 in the German Bight. The near bed (1 m above bed) current velocities, calculated by the water circulation model (Pohlmann 1996), are shown in Fig. 4. The currents are generated by den-

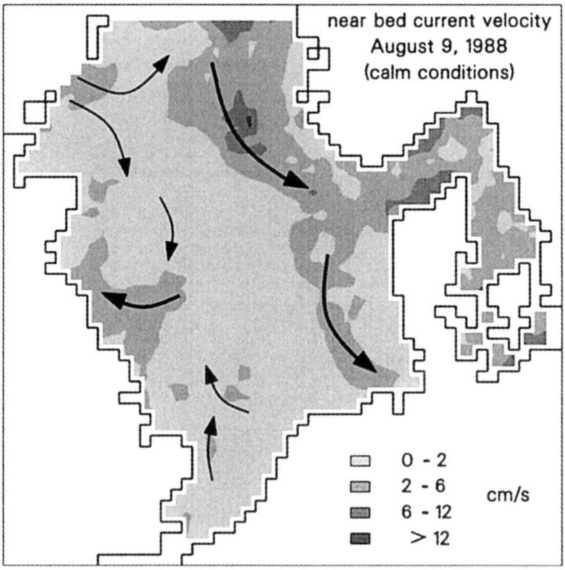

Fig. 4 Near-bed current velocities (1 m above bed), computed by a 3-dimensional North Sea water circulation model. Daily average on 09-08-1988 (calm weather conditions)

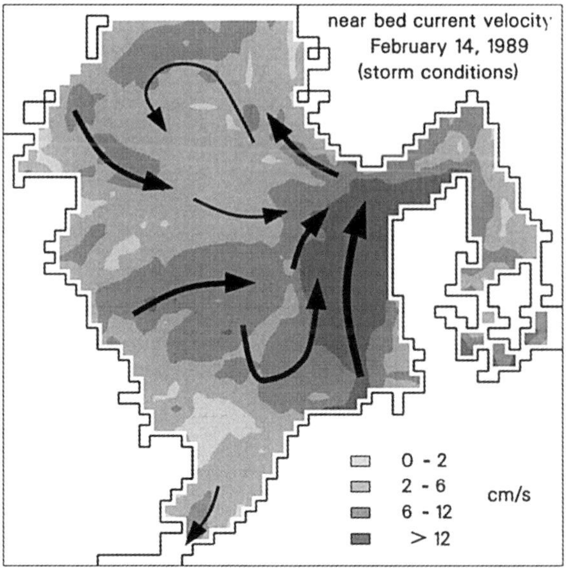

Fig. 5 Near-bed current velocities (1 m above bed), computed by a 3-dimensional North Sea water circulation model. Daily average on 14-02-1989 (stormy conditions)

sity gradients and wind only, tidal currents are not shown. Maximum near-bed current velocities are between 2 and 6 cm s^{-1}; in the southern North Sea they are below 2 cm s^{-1}.

The bed shear velocity u_* during calm weather (Fig. 6) is generated mainly by tidal currents. In the Southern Bight, u_* is between 2 and 4 cm s^{-1} which corresponds to near bed (tidal) current velocities between 30 and 60 cm s^{-1}.

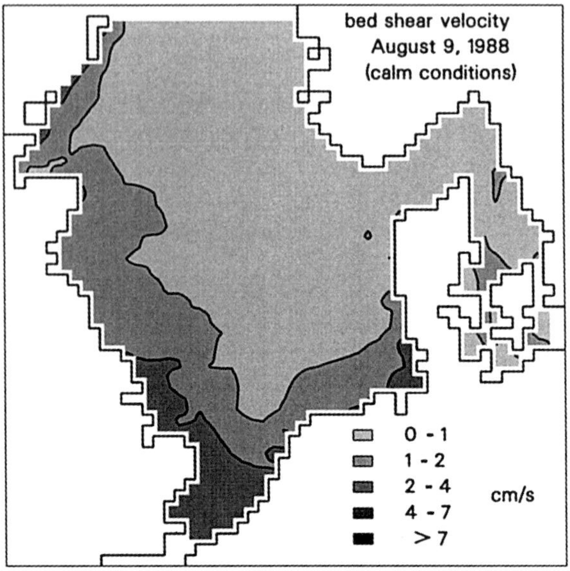

Fig. 6 Computed bed shear velocities in the North Sea, generated by the combined action of currents and waves. Snapshot picture on 09-08-1988, 12:00 UTC (calm weather conditions)

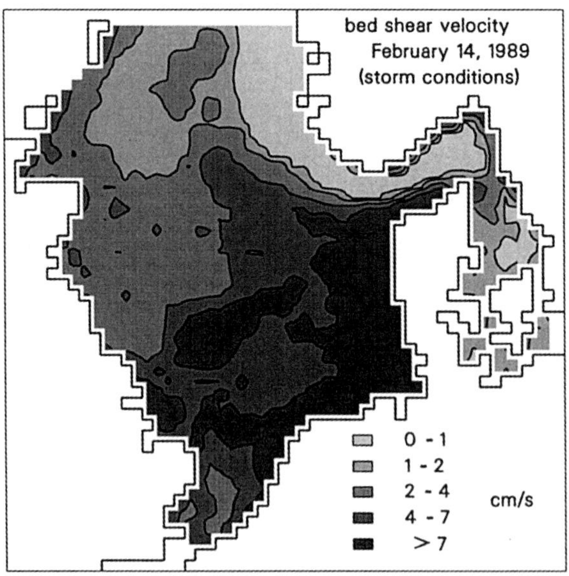

Fig. 7 Computed bed shear velocities in the North Sea, generated by the combined action of currents and waves. Snapshot picture on 14-02-1989, 12:00 UTC (stormy conditions)

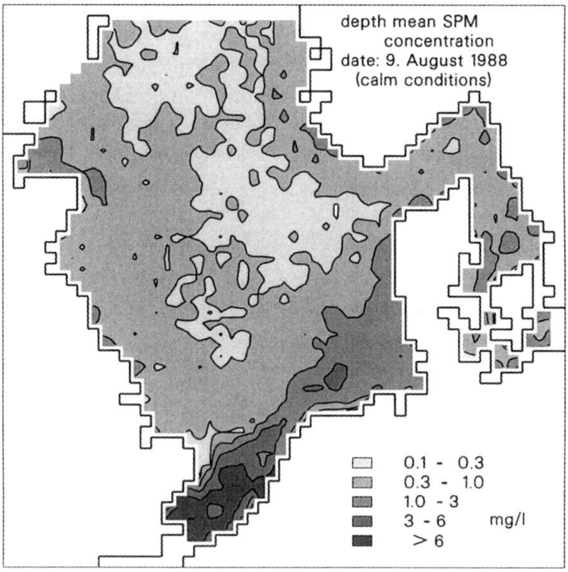

Fig. 8 Computed SPM concentrations (depth-averaged) in the North Sea, snapshot picture on 09-08-1988, 12:00 UTC, calm weather conditions

The SPM concentration in the North Sea is generally low during calm weather as it is shown in Fig. 8. In most parts the depth averaged SPM concentration is around 0.3 mg l^{-1}, highest values appear in the Southern Bight with concentrations > 6 mg l^{-1}. Figure 8 presents the spatial distribution of minimum or background (depth-averaged) SPM concentrations in the North Sea. The comparison with measured data, however, shows that the computed concentrations plotted in Fig. 8 are too high. König et al. (1994) consider the minimum SPM content in the German Bight to be 0.5 ± 0.2 mg l^{-1}. Puls et al. (1995) measured a background SPM concentration of 1.5–2.5 mg l^{-1} in January 1993 in the German Bight. The relatively high modelled SPM concentrations in the southern North Sea are due to high input concentrations in the Straits of Dover. We will be faced again with this problem in Section 4.2.

3.2
Stormy Conditions

Weather conditions on February 14, 1989 were severe: there was a prominent storm field above the North Sea. There was only Beaufort 5 (fresh wind) in the Southern Bight, but Beaufort 12 (hurricane force) during gusts in the German Bight and still heavier gusts in the night before in the Orkney-Shetland area. The wind was from the northwest, there was a storm flood warning for the German coast. Near-bed current velocities (Fig. 5) in the eastern part of the North Sea were above 12 cm s^{-1}. The bed shear velocity u_* (Fig. 7) exceeded 7 cm s^{-1} in the shallow part of the North Sea where the seabed was severely affected by waves. The contour of the Dogger Bank is clearly visible.

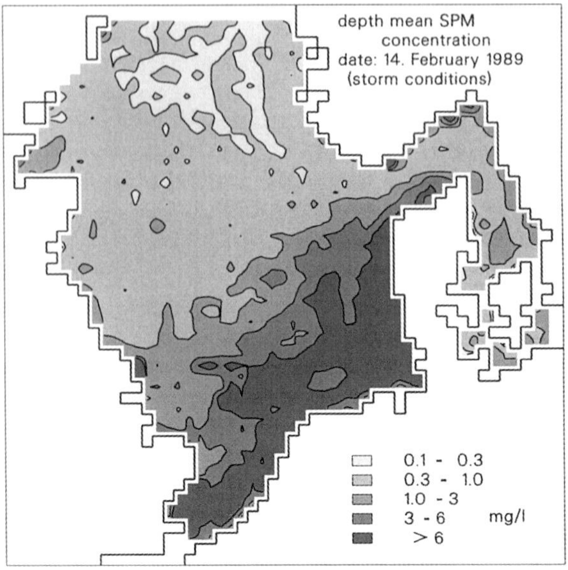

Fig. 9 Computed SPM concentrations (depth-averaged) in the North Sea, snapshot picture on 14-02-1989, 12:00 UTC, stormy weather conditions

The SPM concentration on February 14, 1989 (Fig. 9) is above 6 mg l^{-1} in most parts of the shallow southern North Sea. In the deeper northern part, the SPM concentration does not increase substantially; it is still below 1 mg l^{-1}.

The comparison of simulated and measured SPM concentrations during storm conditions is not straightforward: during storms, also coarse grains (sand) are suspended in the water column, contributing to the SPM concentration, while the model still simulates the grain fraction < 20 μm only.

4 Comparison of Measured and Simulated SPM Data

The quality of the SPM transport model is assessed by comparing the SPM concentrations measured in the southern North Sea during the NSP survey period with the SPM concentrations simulated by the model for the same time period.

4.1 NERC NSP Measurements

The time period August 1988–October 1989 is selected for presenting the SPM variability calculated by the model. This time period was the core survey period of the North Sea Project (NSP) of the British Natural Environment Research Council (NERC), see Simpson (1994) or Howarth et al. (1993).

Between August 1988 and October 1989, 15 monthly cruises were undertaken around a network of fixed stations in the southern North Sea, each cruise lasting about 12 days (Fig. 10). The locations of the NSP stations are indicated in Fig. 11

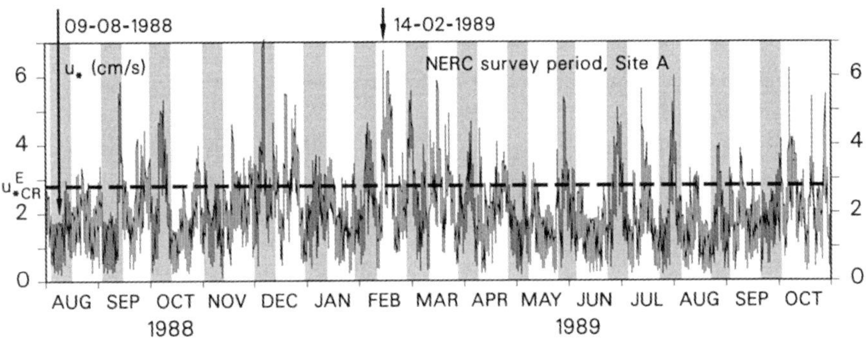

Fig. 10 Computed time series of the bed shear velocity u_* at site A (see Fig. 2) during the complete NERC NSP period August 1988 – October 1989. The *shaded time sections* indicate the 15 NSP survey periods. The *dashed horizontal line* indicates the threshold bed shear velocity for erosion (= 2.8 cm s^{-1})

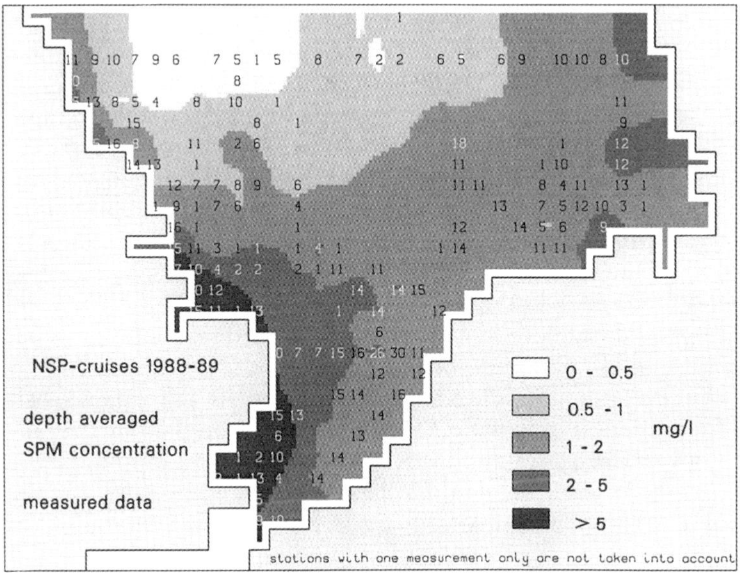

Fig. 11 Horizontal distribution of depth- and time-averaged SPM concentrations in the southern North Sea, *measured* between August 1988 and October 1989 within the framework of the British NERC North Sea Project. The *positions of the numbers* indicate the locations of the NSP stations within the grid of the SPM transport model. The *number itself* gives the quantity of measured SPM concentrations which enter into temporal averaging at each single station. The *shading between the stations* is obtained from inter- or extrapolation. The *numbers* are plotted as *white characters* if the shading is too dark for black characters.

as numbers. SPM samples were taken 2 m below the sea surface, 2 m above the seabed (in rough weather : about 5 m above seabed) and at about mid-depth. A total of 4672 water samples were taken (Dyer and Moffat 1998). These samples were analysed for SPM concentrations by filtering through GF/C glass microfibre filters.

For having an impression of the intensity of the NSP observations, Fig. 10 shows the survey periods covered by the NSP cruises. This is done in a plot with a time series of the (simulated) bed shear velocity at site A (Fig. 2).

The SPM concentrations measured during the NSP cruises were used for comparison with the computed results of the numerical SPM transport model.

4.2
Mean SPM Concentrations

Figure 11 shows the spatial distribution of the temporal means of the depth averaged SPM concentrations collected during the NSP core period August 1988–October 1989. The plotted data were derived as follows: in a first step, the temporal means of the depth-averaged SPM concentrations were calculated separately for each station. The number of vertical means entering into this temporal averaging procedure is given in Fig. 11 for each station. The temporal averaging is done in logarithmic terms. The reason for using a logarithmic scale is shown by an example: In linear terms, an increase from 1 to 2 mg l^{-1} has the same "importance" as an increase from 10 to 11 mg l^{-1}. For an SPM regime with concentrations in the order of 1 mg l^{-1}, however, an increase by 1 mg l^{-1} is substantial, while for a regime with concentrations in the order of 10 mg l^{-1}, the same increase is of subordinate importance. The importance of single SPM concentrations with respect to the general concentration level (i.e. the SPM regime) at a specific site is more adequately described by using a logarithmic scale than a linear scale.

After temporal averaging over the NSP study period, there is one depth averaged SPM concentration for each station. In a second step these data are spatially inter- or extrapolated (in logarithmic terms) in order to obtain a contour plot as shown in Fig. 11. In larger areas with no data (that means: no stations), the shown shading is doubtful, of course. This is especially true for the stationless area in the centre of the southern North Sea and in coastal areas.

The computed data corresponding to the measured data is obtained as follows. For each measured (depth averaged) SPM concentration, there is the information about the time of sampling t and the station number S. When the simulation run passes the time t, the depth-averaged SPM concentration computed at station S is recorded and stored in a file. After having simulated the whole NSP period, the file contains for each measured depth-averaged SPM concentration a corresponding computed value. For the contour plot of computed SPM concentrations, shown in Fig. 12, the computed SPM concentrations are handled in the same way as the measured data.

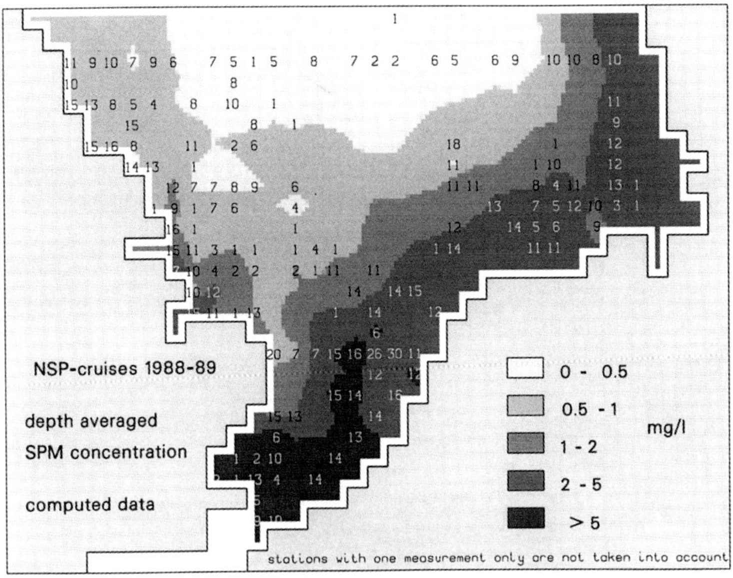

Fig. 12 Horizontal distribution of depth- and time-averaged SPM concentrations in the southern North Sea, *computed* for comparison with the measured NSP data shown in Fig. 11. The *numbers* are the same as given in Fig. 11; they give the quantities of computed SPM concentrations at each single station. For each station, the computed SPM concentrations are averaged in time. The *shading between the stations* is obtained from inter- or extrapolation.

The comparison of measured (Fig. 11) and computed (Fig. 12) SPM concentrations shows: the dominant SPM source for the *computed* concentrations is the Straits of Dover. With a depth-averaged SPM concentration of 2–5 mg l^{-1}, the Dover "plume" extends along the Dutch coast towards the German Bight. For the *measured* SPM concentrations, highest values were observed around East Anglia, where the measured values are above 5 mg l^{-1}. Just here, the computed values remain under 1 mg l^{-1}.

In view of the discrepancies between measured and computed SPM concentrations, the quality of the SPM transport model is assessed to be *unsatisfactory.* The reason for this situation is not assumed to be the wrong simulation of SPM supply or the incorrect simulation of SPM transport processes, but the insufficient simulation of currents in the coastal zones of the southern North Sea: the model's grid is much too coarse (both horizontally and vertically) to be able to simulate the SPM-typical transport processes ("trapping mechanisms") in the coastal zones correctly (Postma 1984). This is especially true for the "coast-dominated" Southern Bight (Eisma and Kalf 1979). A more detailed discussion of the model results is given in Puls et al. (1997b).

4.3
Temporal Variation of SPM Concentrations

The temporal variation of measured and computed depth averaged SPM concentrations c at each separate NSP station is determined from the same measured or computed data which are the basis for the mean SPM concentrations shown in Figs. 11 and 12. The evaluation is done in logarithmic terms log c, resulting in a mean value mean(log c) and a standard deviation σ(log c). After having taken the anti-logarithm, the final result, i.e. the temporal variation, is expressed as a standard deviation factor $10^{\sigma(\log c)}$. The mean concentration is then (i) divided by this factor and (ii) multiplied with this factor, resulting in a concentration range which is somehow equivalent to a "normal" standard deviation. To give an example: The original SPM concentration time series is {0.2, 2, 20} mg l^{-1}. In logarithmic terms, the data is {–0.70, 0.30, 1.30}. The mean and the standard deviation of the log-data is:

$$\text{mean}(\log c) \pm \sigma(\log c) = 0.30 \pm 1.$$

Taking the anti-logarithm, the mean SPM concentration is $10^{0.30}$ = 2 mg l^{-1}, the std. dev. factor is 10^1 = 10. The variability range of SPM concentrations is thus from 2/10 = 0.2 mg l^{-1} to 2 × 10 = 20 mg l^{-1}.

There is a principal problem with the temporal variability of SPM concentrations. SPM concentrations are regarded on a logarithmic scale (see Sect. 4.2). SPM erosion, however, rather acts on a linear scale, because the increase in the SPM concentration in the water column by erosion does (of course) not depend on the SPM background concentrations at the specific site. To give an example: there are two SPM regimes with background concentrations of 1 mg l^{-1} and 4 mg l^{-1}. The increase of the SPM concentration during an erosion event is 5 mg l^{-1} in both regimes. There are three measurements during non-erosion conditions and one measurement during the erosion event. The time series of SPM concentrations are thus {1,1,1,6} and {4,4,4,9} for the two regimes. Now, in logarithmic terms the std. dev. factors are 2.45 and 1.50, respectively, i.e. there is a substantial difference. In linear terms the standard deviation is ± 2.5 mg l^{-1} for *each* regime, which is reasonable because the increase of the SPM is the same for each regime. The example shows that processing SPM concentrations in *linear* terms is more adequate for pure erosion events. For the sake of conformity, however, the temporal variability of SPM concentrations is calculated in logarithmic terms, as it is done for the temporal-averaged SPM concentrations in Section 4.2. It must be kept in mind that the use of a logarithmic scale means decreased std. dev. factors in SPM regimes having high concentrations and increased standard deviations in SPM regimes having low concentrations.

The distributions of std. dev. factors of measured and computed SPM concentrations are shown in Figs. 13 and 14, respectively. Just as in Figs. 11 and 12, the space outside the stations is inter- or extrapolated. The spatial means of the factors (simple average of the 477 horizontal grid cells representing the model do-

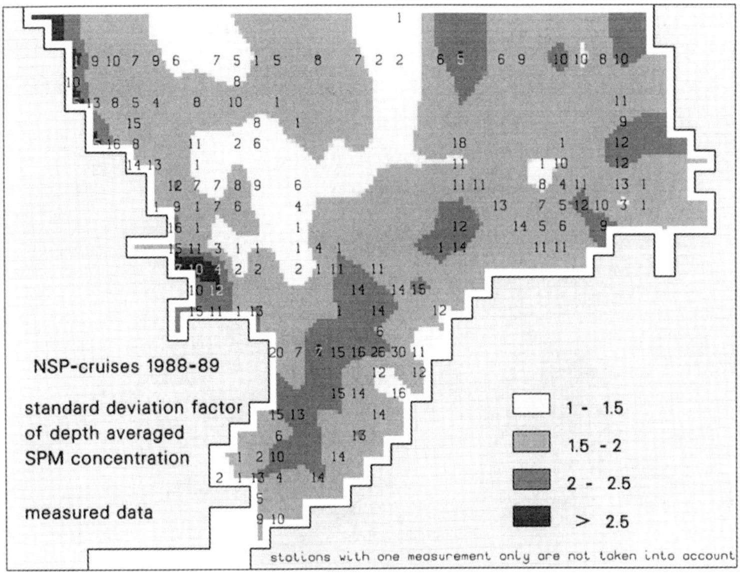

Fig. 13 Horizontal distribution of the temporal variation of depth-averaged SPM concentrations in the southern North Sea, determined from SPM concentrations *measured* between August 1988 and October 1989 within the framework of the British NERC North Sea Project. The temporal variation is expressed by a so-called standard deviation factor, defined in the text. For the *meaning of the numbers* see Fig. 11. The *shading between the stations* is obtained from inter- or extrapolation

main as shown in the two figures) are 1.76 for the measured data and 1.74 for the computed data. For a typical mean concentration of 2 mg l^{-1}, the typical variability range is thus from 1.15 to 3.5 mg l^{-1}.

The std. dev. factors exhibit a patchy distribution both in Fig. 13 and in Fig. 14. On the whole there is an approximate agreement of measured and computed std. dev. factors, except for two parts of the southern North Sea: in the southern part of the Southern Bight the computed std. dev. factors are too low, around East Anglia the computed std. dev. factors are too high. As explained above, this is a consequence of calculating the std. dev. factors *in logarithmic terms* in SPM regimes having comparatively high and low SPM concentrations, respectively.

The temporal variability of SPM concentrations in the southern North Sea is dominated by successions of calm periods and storms, see also Section 5.3. The approximate agreement of measured (Fig. 13) and computed (Fig. 14) std. dev. factors means that (a) the occurrence of surface waves is well represented in the model, and that (b) the processes of erosion and deposition are modelled fairly well.

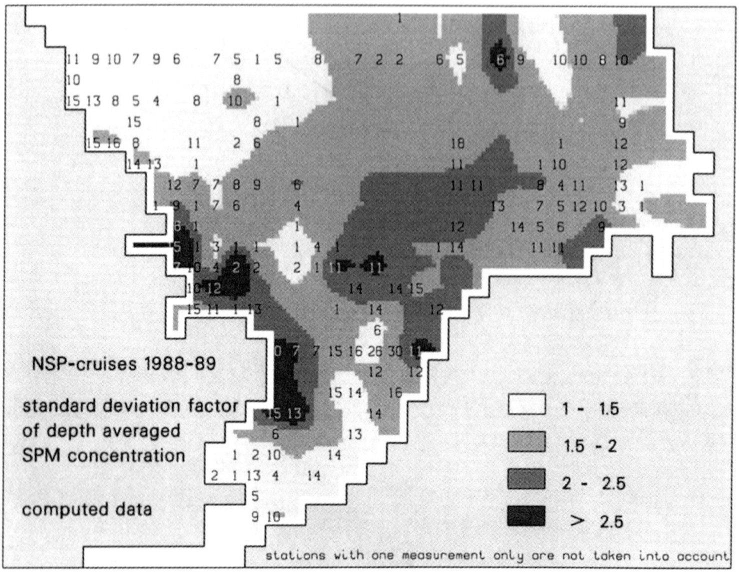

Fig. 14 Horizontal distribution of the temporal variation of depth-averaged SPM concentrations in the southern North Sea, *computed* for comparison with the measured SPM data shown in Fig. 13. The temporal variation of SPM concentrations is expressed by a so-called standard deviation factor, defined in the text. The *position of the numbers* in the plot are the positions of the NSP stations. Each *number* indicates the number of SPM concentrations entering into the determination of the station's standard deviation factor. The *shading between the stations* is obtained from inter- or extrapolation

5
Representativity of NSP SPM Data

The measured distribution of the temporal variation of SPM concentrations shown in Fig. 13 is based on a limited number of stations and on a limited number of measurements per station. There are two questions concerning the distribution given in Fig. 13:

1. Was the number of measurements per station (i.e. the frequency of NSP survey cruises) sufficient for being representative for the whole period August 1988–October 1989 (time representativity)?
2. Was the number of stations and their spatial arrangement adequately chosen in order to yield a sufficient spatial coverage of the whole area of the southern North Sea (space representativity)?

In Puls et al. (1997b) these questions were answered positively for the *mean* SPM concentration. Here the questions will be answered for the temporal variation (expressed as standard deviation factors) of the SPM concentrations. In the following, the *computed* SPM concentrations will be regarded only. Finally, the result for the computed data will be assumed to be also valid for the measured data.

5.1 Time Representativity

Question no. 1 is answered as follows: for each NSP station (with more than one measurement), the std. dev. factor of depth averaged computed SPM concentrations is determined, based on the SPM data of the *complete* NSP period of 15 months. The model time step is 80 min, so more than 8000 computed SPM concentrations enter into this determination for each separate station. These std. dev. factors are shown in Fig. 15 – the values outside the stations are inter- or extrapolated. The positions of the stations are indicated by crosses. The making of Fig. 15 is thus the same as the making of Fig. 14. The only difference is the data basis, i.e. the number of SPM concentrations at each station: between 2 and 30 in Fig. 14, indicated by the numbers within the contour plot of Fig. 14, and more than 8000 in Fig. 15.

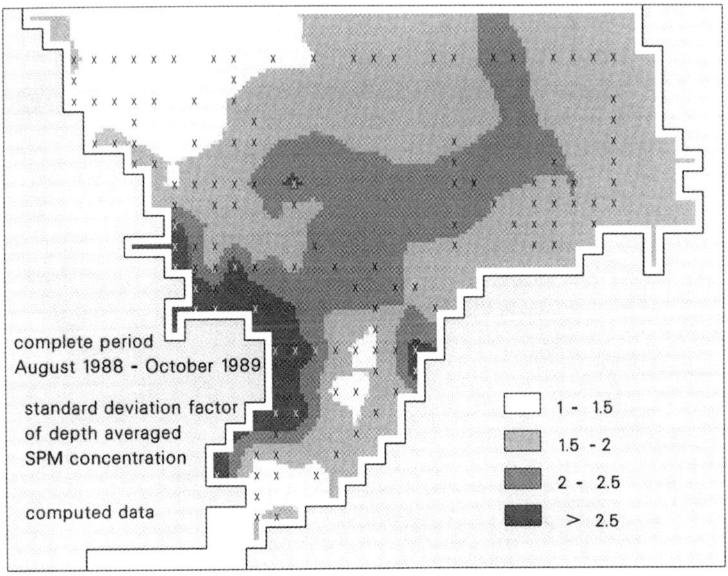

Fig. 15 Horizontal distribution of the temporal variation of depth averaged SPM concentrations in the southern North Sea, *computed* for comparison with the *computed* data shown in Fig. 14. At each location indicated by an x (NSP stations), the standard deviation factor is obtained from the *complete* time series of computed depth averaged SPM concentrations between August 1988 and October 1989. The *shading between the x-locations* is obtained from inter- or extrapolation

The comparison of Figs. 14 and 15 shows: the two patterns of std. dev. factors are nearly identical. Only in the Dogger Bank area, the temporal variability was insufficiently registered by the NSP cruises. The spatial means are 1.74 and 1.89 in Figs. 14 and 15, respectively. Answering question no. 1, it is concluded that, with regard to the temporal variation of depth averaged SPM concentrations, the frequency of NSP survey cruises was sufficient to yield a representative picture of the complete NSP period August 1988–October 1989.

5.2 Space Representativity

Next, question no. 2 is answered. For this purpose, Fig. 16 shows the complete temporal *and* spatial distribution of std. dev. factors (one value for each of the 477 horizontal grid cells of the model domain of the southern North Sea) computed by the SPM transport model. For the assessment of the spatial representativity of the NSP network of stations, Fig. 16 is compared with Fig. 15.

The std. dev. factors of both Fig. 15 and Fig. 16 are based on complete information in time: on more than 8000 SPM concentrations (i) at each separate NSP

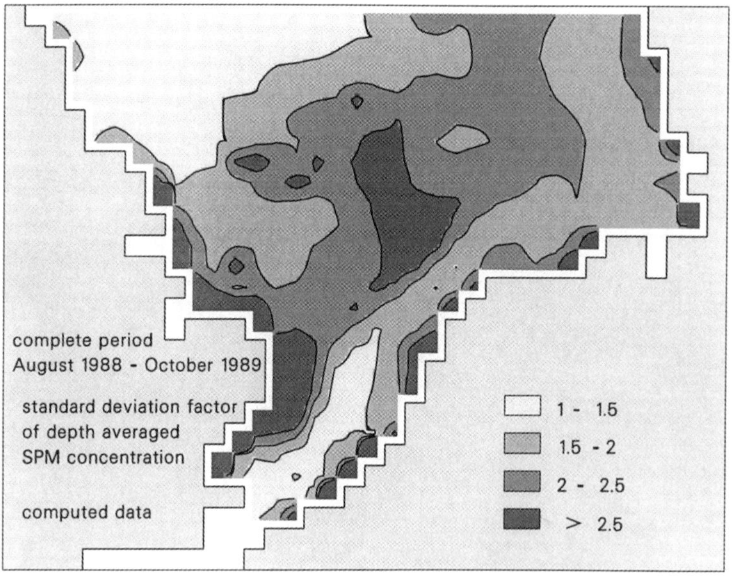

Fig. 16 Horizontal distribution of the temporal variation of depth-averaged SPM concentrations in the southern North Sea, *computed* for comparison with the *computed* data shown in Fig. 15. The shown standard deviation factors are based on the complete data set (complete with respect to time *and* space) of SPM concentrations computed between August 1988 and October 1989. There is one standard deviation factor at each horizontal grid cell of the model domain. Therefore, no "stations" are indicated in this plot, there is no inter- or extrapolation of data

station in Fig. 15 (stations marked as crosses) and (ii) at each of the 477 horizontal grid cells covering the model domain of the southern North Sea in Fig. 16. The difference between the two figures is: in Fig. 15 the std. dev. factors outside the stations must be inter- or extrapolated in order to obtain a contour plot, while in Fig. 16 there is already one computed std. dev. factor for *each* of the 477 horizontal grid cells, inter- or extrapolation is not necessary.

The comparison of Figs. 15 and 16 shows that two areas with high std. dev. factors are not realized by the NSP network of stations: (i) along the Dutch and the German coast and (ii) in the central southern North Sea above the Oyster Grounds. The recording of the near-coast "details" was beyond the scope of the NERC North Sea Project. The non-consideration of the Oyster Grounds patch with std. dev. factors > 2.5 in the central southern North Sea, however, is a problem. It leads to the conclusion that, with regard to the temporal variation of SPM concentrations, the spatial arrangement of NSP stations does *not* yield a representative picture for the whole southern North Sea. It is "bad luck" that the patch is located in just that area where there is a "hole" in the network of NSP stations.

It must be kept in mind that the std. dev. factor is a measure for the *relative* variation of the SPM concentration. The high standard deviations of the Oyster Grounds patch are due to (i) a low basic level of SPM concentrations during calm weather and (ii) a comparatively intense increase of the SPM concentration during storms because of the high mud content in the Oyster Grounds sediment (see Fig. 2).

5.3 Conclusion for the Measured Data

The above considerations and conclusions are valid for the *computed* std. dev. factors. Regarding the approximate agreement of the measured and the computed distributions of std. dev. factors shown in Figs. 13 and 14, the above conclusions are assumed to hold also for the *measured* (i.e. the *true*) std. dev. factors.

An additional remark concerning the std. dev. factors shown in Fig. 16: the SPM concentration used for the determination of this plotted distribution has a time frequency of 18 values per day (time step: 80 min). The spatial mean of the std. dev. factors in Fig. 16 is 2.37. If the 18 SPM concentrations per day are averaged, and the std. dev. factors are determined from these daily averages, the horizontal mean is 2.29. The small difference of the two horizontal means shows that the SPM variation within a day (i.e. due to tidal currents and due to the random number of particles within a horizontal grid cell) plays an insignificant role. The dominant reason for the temporal variability of SPM concentrations is the succession of storm events and calm periods.

6 Meteorological Conditions During the NSP Period

This final section answers the question after the representativity of weather conditions during the single months of the NSP period August 1988–October 1989:

were weather conditions "normal" or were they extraordinary? This consideration also gives an impression of the seasonal variability of storm events. As we know, the frequency and the intensity of storms directly affect the SPM regime.

The data used for the assessment of the NSP period's weather conditions are computed time series of the bed shear velocity, u_*, at site A in the Southern Bight (Fig. 2) for the 10 years 1985–1994. The frequency of the u_*-data is 72 values per day (time step: 20 min). In order to suppress the daily tidal variation, daily averages of u_* were calculated. These daily averages are used for producing box and whisker plots (Figs. 17 and 18).

Figure 17 shows the box and whiskers for each single month (e.g. August 1988) of the NSP period (light shading). For comparison, the box and whiskers of the nine *remaining* months between 1985 and 1994 (e.g. all the Augusts between 1985 and 1994, excluding August 1988) are shown beside it (dark shading). For each month, the minimum of u_* is in the order of 1 cm s^{-1}; this is the daily average of u_* at site A during *calm* weather conditions, i.e. when there are tidal currents only. The increased frequency and severity of storm events in winter is obvious from Fig. 17.

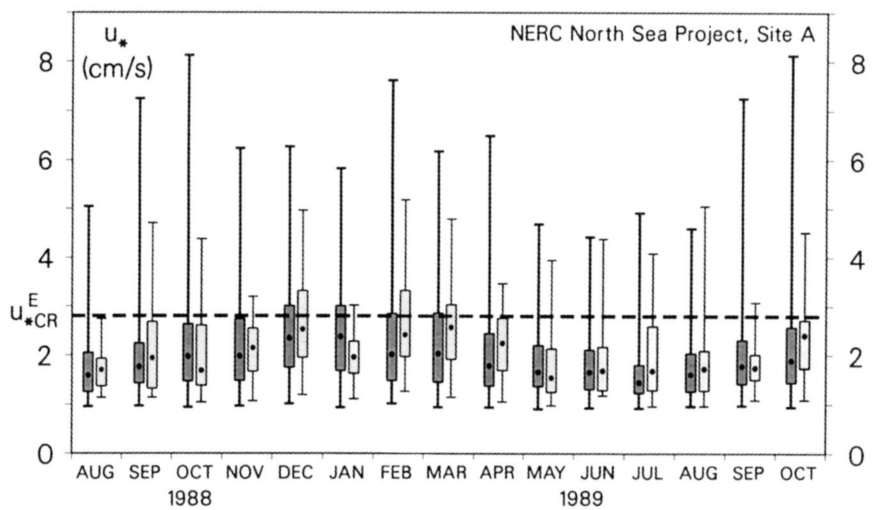

Fig. 17 Box and whiskers presenting the characteristics of computed u_* time series at site A (see Fig. 2). The data basis are daily averages of the u_* time series. The top and the bottom of each box are the 75% and the 25% quartiles, respectively, the dot within each box is the median of the u_*-sample. The ends of the whiskers are the minimum and the maximum values of the u_*-samples. There are two box and whiskers for each month of the NSP period August 1988–October 1989. The *light-shaded boxes* represent the *single* months of the NSP period. The *dark-shaded boxes* aside represent the 9 remaining months of the 10-year time period 1985–1994. For example: the first pair of box and whiskers stands for August 1988, representing (a) August 1988 itself (*light shading*) and (b) all the 9 remaining Augusts from 1985 to 1887 and from 1989 to 1994 (*dark shading*). The *dashed horizontal line* indicates the threshold bed shear velocity for erosion (= 2.8 cm s^{-1})

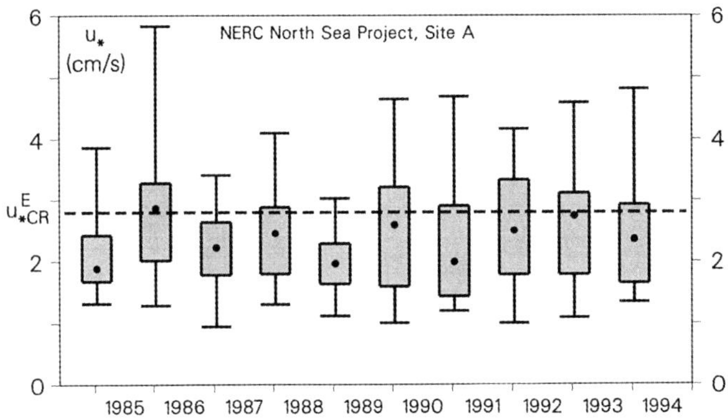

Fig. 18 Box and whiskers representing the characteristics of the u_* time series (at site A) of each January between 1985 and 1994. The data basis are daily averages of the u_* time series. The meaning of the box and whiskers is explained in the caption of Fig. 17. The comparison of the 10 Januaries shows that weather conditions were exceptionally calm during January 1989. The *dashed horizontal line* indicates the threshold bed shear velocity for erosion (=2.8 cm s^{-1})

By comparing the detailed u_* data of Fig. 10 with the box and whisker parameters presented in Fig. 17, it must always be considered that the box and whiskers are computed from daily averages of u_*. The maximum u_* in December 1988, for instance, was computed to be 7 cm s^{-1} in Figs. 10 and 3, but only 5 cm s^{-1} in Fig. 17. The reason is obvious from Fig. 3: the maximum u_* of 7 cm s^{-1} just appears around midnight of December 6. The daily u_* averages of December 6 and December 7 are therefore appreciably less than 7 cm s^{-1}.

The question is: which months of the NSP period deviate significantly (95% level) from average conditions, i.e. from the average of the 9 remaining months? This question is answered by a Chi^2 median test (Lozan 1992). The test tells that only 3 months differ significantly from their corresponding 9-month average: January, February and March 1989. While February and March 1989 were extraordinarily stormy, January 1989 was extremely calm. The exceptional calmness of January 1989 is also pointed out in Fig. 18 which shows the box and whiskers for u_* at site A separately for each January between 1985 and 1994.

The high temporal variation of u_* as shown in Figs. 17 and 18 induces a correspondingly high temporal variation of SPM concentrations in the southern North Sea by erosion and deposition. For a better assessment of the frequency of storm events, the threshold bed shear velocity for erosion $u_*{}^E{}_{CR}$ (Sect. 2.3) is plotted as a dashed line.

It is obvious that one single cruise cannot give a reliable image of the SPM regime in the southern North Sea – for this purpose repeated survey cruises are necessary, as carried out within the British NERC North Sea Project in the southern North Sea or within the German so-called TUVAS project in the German Bight (Heinrich 1994).

7
Summary and Conclusion

The SPM regime in the southern North Sea is dominated by successions of storm events and calm periods. For the adequate simulation of this SPM regime, a numerical SPM transport model needs input information about both the current field and the surface waves. The SPM's temporal variability in the North Sea is demonstrated by the comparison of two extreme weather situations.

The computed results of the model are compared with the measured SPM data of the British North Sea Project (NSP). The distribution of *mean* SPM concentrations in the southern North Sea is not satisfactorily simulated, but the *temporal variation* of SPM concentrations is modelled quite well. It is concluded that erosion and deposition of SPM are well simulated by the SPM transport model.

The model results are used to assess the representativity of the NSP data with regard to its temporal variation. It is concluded that the time frequency of NSP survey cruises was, in fact, sufficient to give a representative picture of the complete NSP period August 1988–October 1989. The spatial representativity of the measured NSP data, however, was shown to be imperfect.

Finally a recommendation is given: the planning of future projects with repeated survey cruises around a network of stations may include representativity tests with a numerical model. Such tests will decrease the risk of getting incomplete results.

Acknowledgements

This study was funded by the participating institutions and the Bundesminister für Bildung, Wissenschaft, Forschung und Technologie under grant 03F0111C. The authors wish to thank the Bundesamt für Wehrtechnik und Beschaffung (Koblenz) and the Amt für Wehrgeophysik (Traben-Trarbach) for making available the wave data of the North Sea.

References

Al-Rabeh AH, Gunay N (1992) On the application of a particle dispersion model. Coastal Eng., vol. 17, 195–210

Boudreau BP (1986) Mathematics of tracer mixing in sediments: I.Spatially-dependent, diffusive mixing. Amer. J. Sci., vol. 286, 161–198

Dyer KR, Moffat TJ (1998) Fluxes of suspended matter in the East Anglian plume, southern North Sea. Continental Shelf Research (accepted)

Eisma D, Kalf J (1979) Distribution and particle size of suspended matter in the Southern Bight of the North Sea and the eastern Channel. Netherlands Journal of Sea Research 13 (2), 298–324

Eisma D, Kalf J (1987) Distribution, organic content and particle size of suspended matter in the North Sea. Netherlands Journal of Sea Research 21 (4), 265–285

Eisma D, Irion G (1988) Suspended matter and sediment transport. In: Salomons W, Bayne BL, Duursma EK, Förstner U (eds) Pollution of the North Sea – An Assessment. Springer-Verlag Berlin Heidelberg New York, 20–35

Green MO, Vincent CE, McCave IN, Dickson RR, Rees JM, Pearson ND (1995) Storm sediment transport: observations from the British North Sea shelf. Continental Shelf Research 15, No. 8, 889–912

Heinrich H (1994) Optimierung des Nordsee-Monitorings: TUVAS-Ergebnisse. Deutsche Hydrographische Zeitschrift, Supplement 1, 71–83

Helbig M (1985) Numerische Seegangsvorhersage im Nordseegebiet und die mögliche Nutzung für den Umweltschutz. Symposium "Bundeswehr und Umweltschutz" der Bundesakademie für Wehrverwaltung und Wehrtechnik, 25.11.–27.11. 1985, Mannheim, p. 11.01–11.27

Howarth MJ, Dyer KR, Joint IR, Hydes DJ, Purdie DA, Edmunds H, Jones JE, Lowry RK, Moffat TJ, Pomroy AJ, Proctor R (1993) Seasonal cycles and their spatial variability. Phil. Trans. R. Soc. Lond. A 343, 383–403

Jago CF, Bale AJ, Green MO, Howarth MJ, Jones SE, McCave IN, Millward GE, Morris AW, Rowden AA, Williams JJ (1993) Resuspension processes and seston dynamics, southern North Sea. Phil. Trans. Roy. Soc. Lond. A 343, 475–491

Johnson TC, Elkins SR (1979) Holocene deposits from the northern North Sea: evidence for dynamic control of their mineral and chemical composition. Geol. Mijnb. 58, 353–366

Jones S, Jago CF, Prandle D, Flatt D (1994) Suspended sediment dynamics: Measurement and modelling in the Dover Strait. In: KJ Beven, PC Chatwin, JH Millbank (eds) Mixing and Transport in the Environment. John Wiley & Sons Ltd., 183–201

König P, Frohse A, Klein H (1994) Measurements of suspended matter dynamics in the German Bight. In: Sündermann J (ed) Circulation and contaminant fluxes in the North Sea. Springer-Verlag Berlin Heidelberg New York, 250–270

Krone RB (1962) Flume studies of the transport of sediment in estuarial shoaling processes. Final Report, Hydraulic Eng. Lab. and Sanitary Eng. Res. Lab. Univ. California, Berkeley, 111 pp

Lafite R, Shimwell SJ, Grochowski N, Dupont J-P, Nash L, Salomon J-C, Cabioch L, Collins MB (1997) Suspended particulate matter fluxes through the Strait of Dover. Oceanologica Acta (submitted)

Liebezeit G (1991) Kohlenhydrate in marinen Sinkstoffen und Sedimenten - Umsetzungen und Biomarkerkriterien. Habilitationsschrift, Fachbereich Geowissenschaften, Universität Hamburg, 153 pp.

Lozan JL (1992) Angewandte Statistik für Naturwissenschaftler. Verlag Paul Parey, Berlin Hamburg, 237 pp

O' Connor BA, Nicholson J (1988) A three-dimensional model for suspended particulate sediment transport. Coastal Engineering 12, 157–174

Odd NVM, Murphy DG (1992) Particulate pollutants in the North Sea. Calibration of a 20 km gridded 3D model simulating a representative annual cycle of mud transport. HR Wallingford, Report SR 292, 15 pp

Otto L, Zimmerman JTF, Furnes GK, Mork M, Saetre R, Becker G (1990) Netherlands Journal of Sea Research 26 (2-4), 161–238

Pohlmann T, Puls W (1994) Currents and transport in water. In: Sündermann J (ed) Circulation and contaminant fluxes in the North Sea. Springer-Verlag Berlin Heidelberg New York, 345–402

Pohlmann T (1996) Predicting the thermocline in a circulation model of the North Sea - Part I: model description, calibration and verification. Continental Shelf Research 16, No. 2, 131–146

Postma H (1984) Introduction to the symposium on organic matter in the Wadden Sea. Netherlands Institute for Sea Research - Publication Series 10, 15–22

Puls W, Kühl H, Frohse A, König P (1995) Measurements of the suspended matter settling velocity in the German Bight (North Sea). Deutsche Hydrographische Zeitschrift 47, No. 4, 259–276

Puls W, Gerwinski W, Haarich M, Schirmacher M, Schmidt D (1997a) Lead budget for the German Bight. Marine Pollution Bulletin 34, No.6, 410–418

Puls W, Pohlmann T, Sündermann J (1997b) Suspended particulate matter in the southern North Sea: application of a numerical model to extend NERC North Sea Project data interpretation. Deutsche Hydrographische Zeitschrift, 49, No. 2/3

Reid PC, Lancelot C, Gieskes WWC, Hagmeier E, Weichart G (1990) Phytoplankton of the North Sea and its dynamics: a review. Netherlands Journal of Sea Research 26 (2-4), 295–331

Simpson JH (1994) Introduction to the North Sea Project. In: Charnock H, Dyer KR, Huthnance JM, Liss PS, Simpson JH, Tett PB (eds) Understanding the North Sea system. Chapman & Hall, London Glasgow New York, 1–4

Stronach JA, Webb AJ, Murty TS, Cretney WJ (1993) A three-dimensional numerical model of suspended sediment transport in Howe Sound, British Columbia. Atmosphere-Ocean 31 (1), 73–97

Icebergs in the North Atlantic: Modelling Circulation Changes and Glacio-Marine Deposition

C. Schäfer-Neth and K. Stattegger

1 Introduction

Meltwater events (MWE) during and at the end of the last glaciation led to dramatic changes in the oceanic circulation of the North Atlantic combined with a strong reduction in deepwater formation or even a breakdown of the conveyor belt (Bond 1995; Bond and Lotti 1995; Broecker 1991; Seidov et al. 1996). These changes depend on (i) the amount of freshwater brought in and (ii) the locations of freshwater input and its transport path; that means the conditions to form one or several large freshwater lenses on top of the water column in the central Greenland, Iceland, and Norwegian (GIN) Seas.

Several episodes of rapid discharge of icebergs triggering MWEs are documented by the meltout and sedimentation of debris transported within these icebergs, recognized as coarse-grained ice-rafted detritus (IRD) and building up the well-known Heinrich layers on the seafloor (Heinrich 1988; Bond et al. 1992; Dowdeswell et al. 1995).

Recent investigations of Heinrich layers claim major iceberg discharge at the end of the last glaciation (Termination I) mainly from the Hudson Strait drainage basin of the North American Laurentide ice sheet (Dowdeswell et al. 1995; Andrews et al. 1994; Grousset et al. 1993) and, to a minor extent, from the breakdown of the Barents ice sheet (Bischof 1994; Rosell-Melé and Koç 1997; Svendsen et al. 1996; Vorren et al. 1990). On the other hand, numerical experiments demonstrate clearly that a significant weakening of the conveyor belt predominantly would take place if a major freshwater injection entered the northern North Atlantic (NNA) from the northeast caused by the breakdown of the Barents-shelf ice sheet (Schäfer-Neth and Stattegger 1997).

In this chapter, we present OGCM-based numerical experiments of the release, transport and melting of icebergs from various sources along the NNA margins. Starting scenario is the Last Glacial Maximum (LGM) stage which has been modelled successfully by Schäfer-Neth (1994, 1997; Haupt et al. 1994).

2
The Ocean Circulation Model

For the experiments, we use SCINNA ('Sensitivity and Circulation of the Northern North Atlantic', Schäfer-Neth 1994), a three-dimensional prognostic Ocean General Circulation Model employing the primitive equations based on the Modular Ocean Model by Pacanowski et al. 1993). On a rotated spherical grid where the model's north pole is located at 180 W/30 N, the model domain covers the GIN Seas and parts of the neighboring basins (Fig. 3). The horizontal grid spacing is 0.5° ($\approx$ 55 km) in latitude and longitude, and vertically there are 17 levels with thicknesses increasing from 50 m at the top to 1000 m at the bottom of the deepest basins, enabling a realistic representation of topography. Test runs forced with modern sea surface temperatures, salinities, and wind stress reproduced the modern oceanography reasonably well (Haupt et al. 1994, 1995). After aiding the paleoceanographic reconstruction of the LGM, that is, the reconstruction of physically consistent temperature and salinity distributions, with the associated circulation patterns from the proxy data (Schäfer-Neth 1994,1997), SCINNA is now being applied to study the deglaciation phases since the LGM.

3
Idealized Icebergs

Real icebergs exist in a great variety of sizes and shapes that are too complex to be implemented in a general circulation model. Instead, to keep the model code simple and to save computation time, icebergs are implemented as idealized pie-shaped blocks of ice (Fig. 1). The iceberg portion above water is neglected (i) because it is generally small compared to the immersed part, and (ii) because for typical water and wind velocities water drag is generally the most important force acting upon an iceberg (cf. White et al. 1980, Eq. (31) for typical wind and water velocities; Matsumoto 1996). Due to the high ice albedo, the iceberg's heat gain from atmosphere and solar irradiation (below 100 W m^{-2}; Lemke et al. 1989) is much smaller than the heat fluxes (Fig. 2) from the

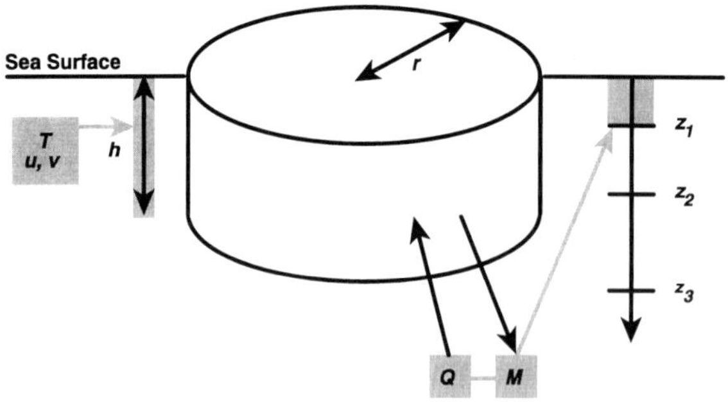

Fig. 1 In the model, icebergs are represented by idealized pie-shaped blocks of ice. Explanation of symbols: T water temperature; u, v horizontal velocities; h, r iceberg height and radius; Q heat flowing from water into ice; M meltwater runoff from iceberg to water; $z_{1...3}$ depths of model layer interfaces

surrounding water (Russel-Head 1980). Therefore the heat exchange at the iceberg top is neglected, too.

In the current implementation, icebergs of predefined height h and radius r (Fig. 1) are released at prescribed locations at fixed time intervals. Once an iceberg has been generated, its drift and decay are computed as follows. First, the horizontal velocities are averaged from the model grid over the total iceberg height (Fig. 1, shaded boxes to the left) to the current location of the iceberg. The new position of the iceberg is then calculated by

$$\begin{pmatrix}\lambda\\ \phi\end{pmatrix}^{new} = \begin{pmatrix}\lambda\\ \phi\end{pmatrix}^{old} + \text{time step} \times \frac{180}{\pi \text{ Earth radius}} \begin{pmatrix}\bar{u}\cos\phi^{old}\\ \bar{v}\end{pmatrix}, \tag{1}$$

where λ and ϕ denote longitude and latitude, respectively, and the averaged zonal and meridional velocities are denoted by $\bar{u}$ and $\bar{v}$. We favour this Lagrangian approach because it more clearly reveals how the ocean might respond to swarms of individual icebergs compared with an Eulerian technique. The latter would have to be implemented in a fashion similar to one of the commonly used sea-ice models, using some measure of iceberg coverage as a tracer. This would require a high diffusion to maintain numerical stability, which can be expected to broaden the iceberg drift paths and to yield results closer to those obtained with pure meltwater inputs.

Like $\bar{u}$ and $\bar{v}$, the average temperature $\bar{T}$ at the iceberg position is computed from the surrounding model grid points, thus yielding the iceberg's melt rate μ by the empirical relation from tank experiments:

$$\mu = 0.018\,(\bar{T} + 1.8)^{1.5} \tag{2}$$

(Fig. 2; Russel-Head 1980). Then the iceberg's new dimensions are given by

$$\begin{pmatrix}h\\ r\end{pmatrix}^{new} = \begin{pmatrix}h\\ r\end{pmatrix}^{old} - \text{time step} \times \mu. \tag{3}$$

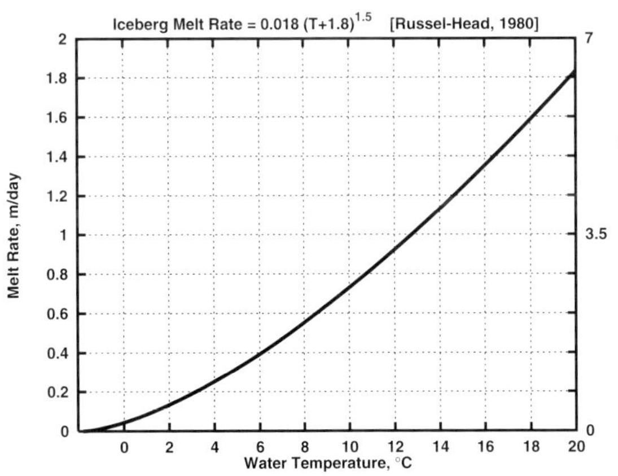

Fig. 2 Iceberg decay as function of water temperature, after laboratory experiments by Russel-Head (1980)

If either height or radius shrink below a certain limit (25 m for the current study), the assumption of an iceberg predominantly influenced by the surrounding waters is no longer met and the iceberg is completely removed from the system.

The change of the iceberg volume is proportional (i) to the mass of freshwater added to the ocean (Fig. 1):

$$M = \rho_{ice}\,\pi\big[(hr^2)^{old} - (hr^2)^{new}\big], \tag{4}$$

and (ii) the amount of heat taken from the ocean required for melting the ice:

$$Q = \kappa_{ice}\,M, \tag{5}$$

with the ice density $\rho_{ice} = 0.91\,\mathrm{g/cm^3}$ and the heat of fusion $\kappa_{ice} = 334\,\mathrm{J/g}$. These heat and freshwater fluxes lead to a temperature and salinity stratification in the ultimate vicinity of the iceberg (Foldvik et al. 1980; Ohshima et al. 1994) that cannot be resolved in the circulation model. Assuming that the meltwater ascends along the iceberg's sides and spreads at the sea surface, the fluxes are applied over the entire top level of the model grid box containing the iceberg (Fig. 1, shaded boxes to the right).

Apart from the water to ice heat transfer, there are many processes that may deteriorate an iceberg, such as wave erosion, calving of overhanging ice, wind-induced convection, and heat transfers induced by water flows relative to the ice (White et al. 1980). These processes that increase the ice melt rate cannot be resolved by the circulation model used here and must therefore be parameterized. For the experiment discussed here in further detail, the melt rate was computed according to Eq. (2) and (3). To examine the consequences of the additonal deterioration mechanisms, we repeated this experiment with a ten fold melt rate increase: $\mu = 0.18\,(\bar{T} + 1.8)^{1.5}$. Although this is a substantial change, the resulting circulation patterns and temperature-salintity distributions did not change very much, except for strengthened density gradients and intensified currents.

4 Experiments and Results

4.1 The Experiments

Our experiments are based on the reconstructed glacial summer scenario (Schäfer-Neth 1994, 1997), the time slice being best documented by sediment core measurements. To arrive at this scenario, SCINNA was driven by restoring to sea-surface temperatures reconstructed from faunal assemblages (Weinelt et al. 1996; Pflaumann et al. 1996) and estimates of sea-surface salinities derived from these temperatures and oxygen isotope measurements taken from various publications. For wind forcing, a glacial wind field was modelled with the EC-

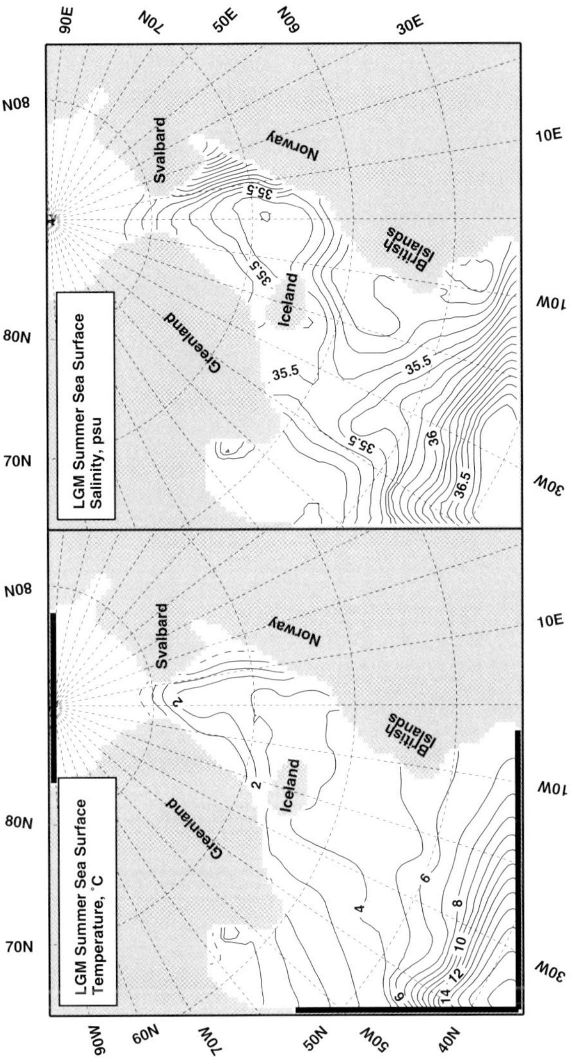

Fig. 3 Sea-surface temperature (*left*) and salinity (*right*) for the glacial summer, reconstructed from sediment-core based temperature estimates and oxygen isotope measurements (Schäfer-Neth 1994, 1997). The coastline differs from the modern one due to the lower glacial sea level. Contour intervals: 1 °C and 0.1 psu. *Heavy black lines* along the southwestern and northern boundaries indicate restoring zones

HAM T42 atmospheric model (Hoffmann, pers. comm.) using the same temperature data as bottom boundary conditions. With these forcing data sets, SCINNA was set in motion for some 400 years. The resulting temperature, salinity, density, and velocity distributions exhibit a glacial summer scenario that is quite similar to modern winter conditions, comprising a relatively warm and salty surface inflow (Figs. 3, 4, left) from the Atlantic into the Norwegian Sea that is balanced by outflows via the East Greenland Current and a deep overflow over the Iceland-Scotland Ridge (Fig. 4, right), and a formation of deep water in the GIN Seas (Fig. 5). These three-dimensional fields were extracted

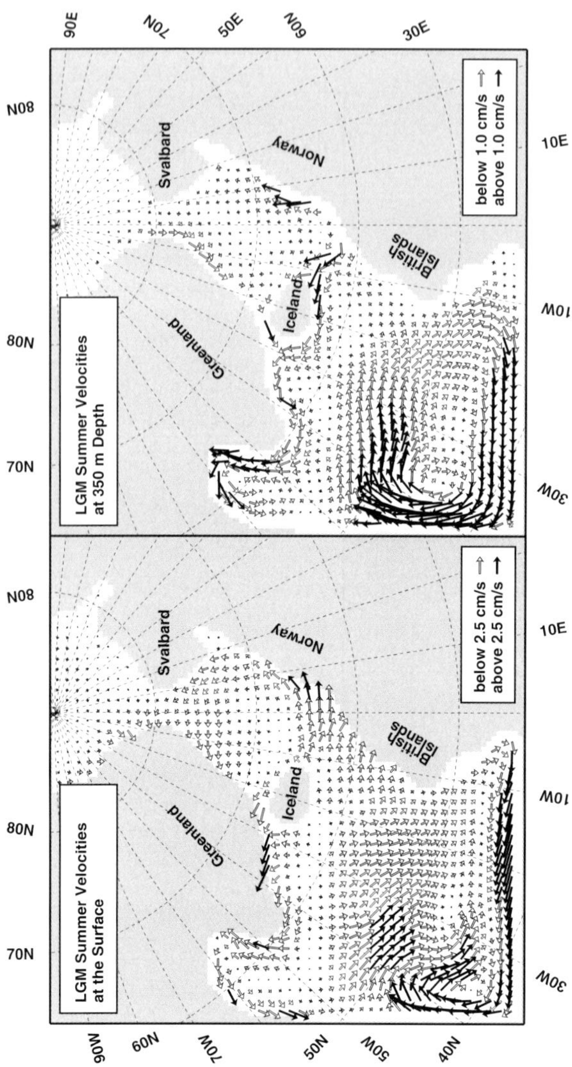

Fig. 4 Modelled glacial summer circulation at the surface (*left*) and 350 m depth (*right*). Only one vector in two is displayed for clarity. The currents, especially the flows into and out of the GIN Seas, are quite similar to modern conditions

for initialization and *T/S* restoring (Fig. 3) at the southwestern and northernmost boundaries of the meltwater and iceberg experiments.

For forcing the meltwater studies, we used the surface heat and freshwater fluxes that were diagnosed from the LGM scenario and modified by additional freshwater sources along the margins of the glacial ice sheets of Labrador, Greenland, and/or Europe. Depending on the source area of the meltwater input, the ocean reacts in two distinctly different ways (Schäfer-Neth and Stattegger 1997): If the meltwater originates from Greenland only, even intense pulses of 1 Sv (= 10^6 t/s), which are equivalent to a complete melting of the present-day Greenland ice within about 200 years, cause no major change in

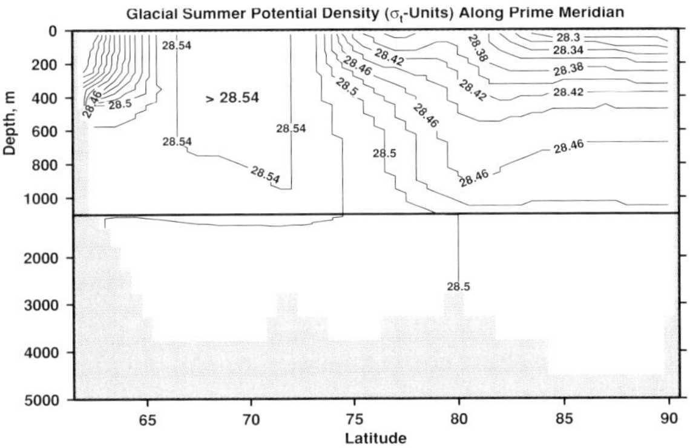

Fig. 5 Meridional section of potential density along Prime Meridian for the glacial reconstruction. At about 69 N, there is a chimney of dense water with $\sigma_t \geq 28.54\ \mathrm{g\ cm^{-3}}$ extending from the surface down to 1200 m. In resemblance to modern conditions, this gives clear evidence of deep convection. Contour interval: 0.02 g/cm^3

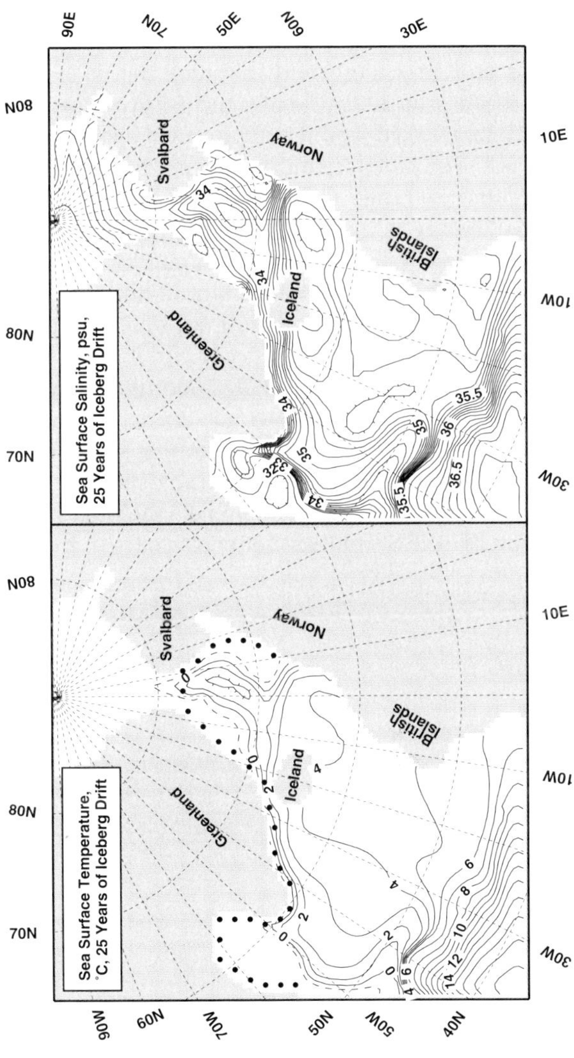

Fig. 6 After 25 years of integration with icebergs, both sea-surface temperature (*left*) and salinity (*right*) are greatly reduced all over the Greenland, Norwegian and Labrador Seas, and a cold and fresh tongue extends towards Europe along 50 N. Contour intervals: 1 °C, 1 psu below 33, 0.2 psu between 33 and 35, and 0.1 psu above 35 psu. *Dots* indicate location of iceberg generation

the current system except for a strengthening of the East Greenland Current. On the other hand, meltwaters from the European coast has severe effects. By shifting the warm and salty inflow away from Britain over to Iceland, they can spread over the whole GIN Seas, thereby stopping the deep water formation in this region and pushing the circulation from its normal cyclonal-antiestuarine into ananticyclonal-estuarine mode.

Our iceberg experiments extend these meltwater studies and address the following questions: (i) Do icebergs cause circulation changes comparable to those induced by continental meltwater runoff? (ii) Can Heinrich events be simulated with SCINNA, and what are their consequences for the ocean circulation? (iii) Can the IRD deposits in Heinrich layers be correlated with distinct iceberg source regions?

To summarize the results of numerous experiments with different regions and intensities of iceberg input, we discuss here one study in which icebergs were released at the coasts of Europe, Greenland, and Labrador. In this experiment, every 25th day, a huge iceberg of 300 m height and 5 km radius was launched at each of the 32 locations marked by dots in Fig. 6. On average, this amounts to a 0.29 Sv input of ice. For comparison, height estimates for the glacial ice dome over the Barents Sea range from 1000 m (Peltier 1994) to 3400 m (Lambeck 1997). Taking 2000 m as a mean value [Saltzman and Verbitzky 1992; Eq. (8)] yields an ice mass of about 3×10^{16} t, and a melting of this mass over 2000 years would result in a meltwater input of almost 0.5 Sv. The lifetime of the icebergs is highly variable, depending on water temperature. Icebergs entering warmer regions at about 50 N decay within 2–4 years, whereas those transported to polar regions under freezing conditions may last for some decades. Of course, the distribution of warmer and colder areas changes with time, because the icebergs compose a heat sink that is not fixed in space. After about 20 years, the model reaches a new steady state with almost all icebergs drifting to regions warm enough for complete melting, yielding a constant freshwater input of 0.25 Sv that is compensated for by the boundary restoring zones.

4.2 Iceberg-Induced Circulation Changes

This new state is marked by distinctly decreased temperatures (Fig. 6, left) reaching the freezing point in the Labrador Sea and at the coasts of Greenland and Norway. The salinitiy is lowered to values of 30 psu in these regions (Fig. 6, right), and in the central GIN Seas it drops by 0.8 to 34.2 psu. Most prominent are the front from Norway to Iceland along 67 N and the cold low-salinity tongue pointing from Canada to Europe at 50 N.

The front corresponds to a strong westward current (Fig. 7, left) north of Iceland. As a short cut of the GIN Seas cyclone (Fig. 4, left), this current isolates the GIN Seas very effectively from the warmer North Atlantic and feeds an intensified East Greenland Current. In the Greenland Sea, the fomerly exist-

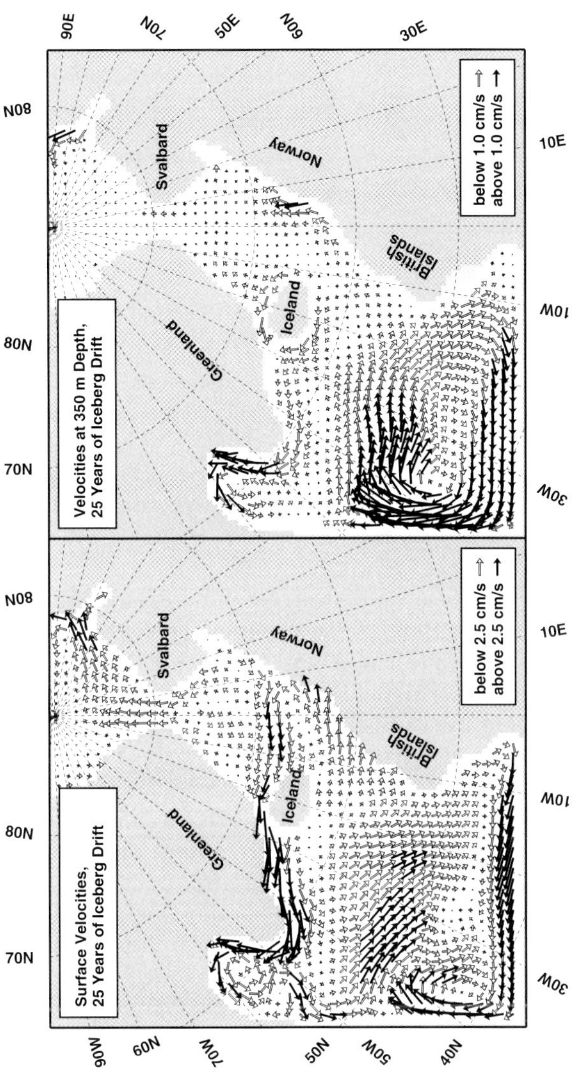

Fig. 7 Under the influence of drifting and melting icebergs, the current system changes drastically in the GIN Seas. Instead of feeding a basin-wide cyclone, the inflow from the Atlantic turns westward towards Iceland (*left*). The deep outflow over the Iceland-Scotland Ridge ceases, and a weak inflow can be found instead (*right*). The circulation of the North Atlantic, however, is hardly affected by the meltwater, except for intensified currents around the southern tip of Greenland and in the Labrador Sea

ing cyclone is replaced by an anticyclone. At 350 m depth (Fig. 7, right), the outflow from the GIN Seas (Fig. 4, right) has stopped and instead a weak inflow has been established. This reversal is linked to the salinity reduction at the surface that causes the deep water formation to stop (Fig. 8). These changes were found as well in additional experiments with icebergs originating off the European coasts only, whereas studies with icebergs released only around Greenland showed essentially unmodified circulation patterns, except for a stronger East Greenland Current. Thus the eastern part of the GIN Seas can be regarded as the region most sensitive to massive iceberg generation. In addition, icebergs from this area seem to be more effective than pure meltwater

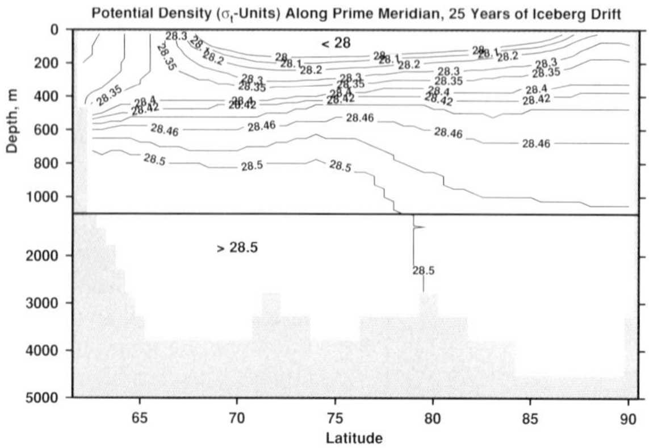

Fig. 8 Meridional potential density section along Prime Meridian for the iceberg experiment. The melting icebergs impose a lid of relatively fresh water over the GIN Seas and stop the convection there. Contour interval: 0.02 g/cm^3

inputs of equivalent amount. According to our numerical experiments, a meltwater inflow of 0.1 Sv hardly affects the circulation, but an equivalent production of icebergs clearly does – in the results shown here, only 0.06 Sv come from the icebergs released near Europe.

The tongue of cold and fresh surface waters corresponds remarkably well with the distribution of IRD found in Heinrich layers (Bond et al. 1992; Dowdeswell et al. 1995). In the experiment presented here, the icebergs do not drift past 30 W, but other experiments with higher iceberg input (above 0.5 Sv) in the Labrador Sea produced iceberg tracks ending at the eastern Atlantic coast. In these midlatitudes, there is some tendency for the icebergs to focus their tracks onto a distinct path. This is due to the meltwater release along the track that reinforces the density gradient which defines the current axis. That is, given the subtropical and subpolar gyres, the icebergs cannot do anything but drift eastward along about 50 N, regardless of their origin. To illustrate this, Fig. 9 shows two snapshots of iceberg locations at integration day 3600 (top) and 8500 (bottom). The positions of icebergs released at four different locations (colored arrows) are displayed. At day 3600, it is easy to tell which iceberg came from which starting point, but at day 8500, icebergs from all starting points have reached 50 N. This is consistent with the results of Robinson et al. (1995), who reconstructed iceberg paths from magnetic susceptibility. The iceberg movements are very irregular. Especially near the Barents Sea and northeast of the Denmark Strait, they sometimes form swarms about the size of Iceland. Due to the local salinity minimum associated with these swarms, they rotate anticyclonally, gathering nearby icebergs and releasing them randomly. This randomness becomes very clear when comparing the distributions of green and yellow points in the lower panel of Fig. 9. As indicated by the respective arrows, source locations 2 (green) and 3 (yellow) are very close, but still icebergs from 3 drift in two opposite directions, whereas those from 2 go only to the west. The tracks starting at location 4 (blue) are split into a westward and a northward branch, too. Gwiazda et al. (1996a, b), who examined the possible origin of IRD found in Heinrich layers 2 and 3, tracked parts

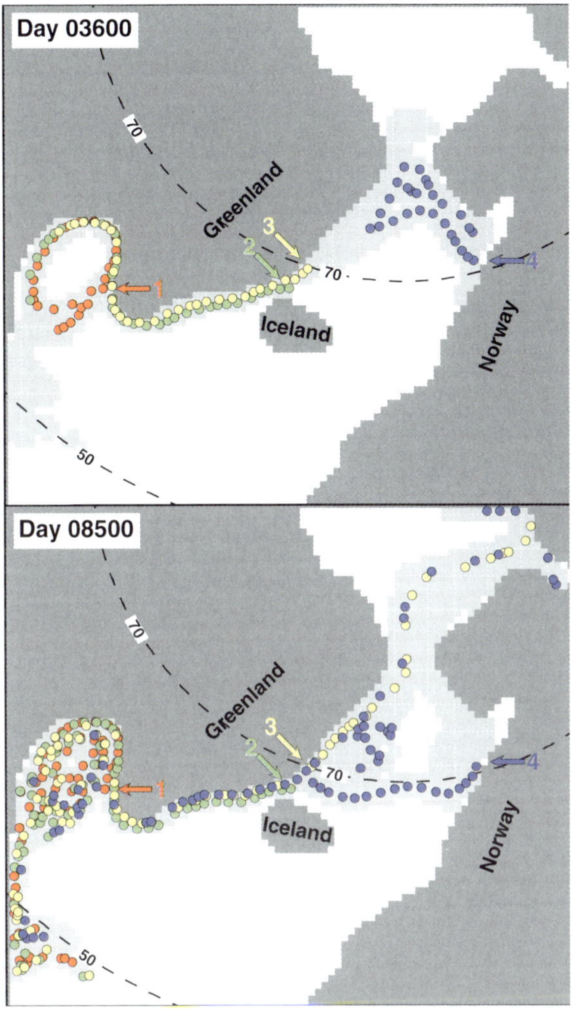

Fig. 9 Two snapshots of iceberg positions after 3600 (*top panel*) and 8500 (*bottom*) days. Model grid boxes containing icebergs are marked by *light shading*. *Dots* denote the positions of icebergs released at four different locations that are identified by the *colored arrows*. At day 3600, the drift paths are clearly correlated with the respective source regions. This is no longer valid a couple of years later. Although the icebergs launched at location 2 (*green*) still drift towards Labrador, those starting nearby at location 3 (*yellow*) have developed an additional branch entering the Arctic. Similarly, icebergs from the Barents shelf (*blue*) can be found everywhere. It should be noted, too, that icebergs from all sources reach the tongue at 50 N

of the IRD back to even Scandinavia. Thus the chaotic iceberg movements must be regarded not as a specific feature of the model but as a realistic behaviour.

4.3 Sedimentation of Heinrich Layers

From the melting experiments we obtain an average melting rate and freshwater influx from icebergs (Fig. 10, left). By adding sediment of coarse silt or coarser size to the melting ice which sinks down rapidly without much lateral transport, we can produce IRD layers on the sea floor. The thickness of these IRD layers is governed by (i) the transport paths of icebergs, (ii) the melting

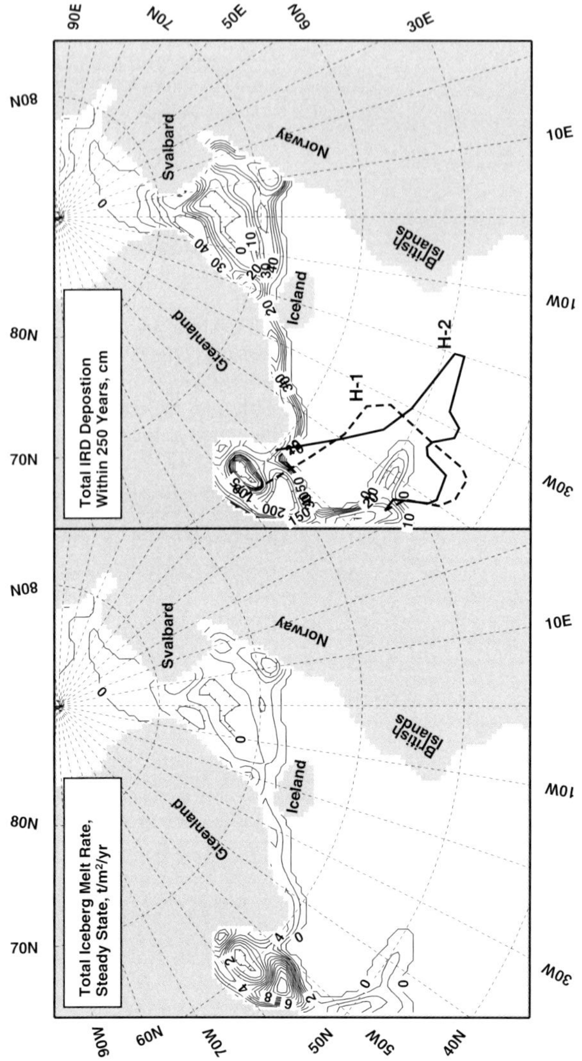

Fig. 10 Distribution of steady-state iceberg melt rates (*left*) and deduced thickness (*right*) of the IRD layer deposited within 250 years using a sediment concetration of 1 $^0/_{00}$. Contour intervals: *left* 1 t m^2 year^{-1}, *right* 10 cm below 50, 25 cm between 50 and 100, and 50 cm above 100 cm. *Heavy dashed and solid lines*: 10 cm isopachs of Heinrich layer 1 and 2 after Dowdeswell et al. (1995)

rate dependent on seawater temperature, and (iii) the concentration of initial iceberg sediment load.

The sediment load of an iceberg is concentrated mainly in the basal part (Dowdeswell and Murray 1990), most probably within the lowermost 10 m. Here, the maximum sediment concentration may reach up to 10 % of the total volume, yielding average sediment concentrations of 1 % in icebergs of 100 m height, 0.5 % for 200 m, and so on. However, in the experiment presented here, the initial height was fixed to 300 m for all icebergs, giving a maximum concentration of about 0.3 %. To account for the natural variability, we adopted a mean overall sediment content of 0.1 % for our estimation of IRD deposition. A uniform distribution of IRD within the ice is consistent with the model results of Matsumoto (1996), who could not appropriately simulate the IRD sedimentation during isotope stage 5e by confining the sediment load to the base of the icebergs, but had to use an almost uniform distribution. However, the IRD layer thickness not only depends on the amount and distribution of sediment within the ice, but also on the duration of a Heinrich event.

Running the model for 250 years of steady state with constant iceberg meltwater inflow (see notes above on the steady state), that is within the estimates of Heinrich event durations (Andrews et al. 1994; Manighetti et al. 1995), we yield the characteristic distribution pattern of IRD shown in Fig. 10. This figure contains also the 10-cm isopachs of Heinrich layer 1 and 2 from Dowdeswell et al. (1995) for comparison. The modelled IRD distribution exhibits a tongue extending eastward along 50 N, resembling the natural distributions. The modelled tongue, however, is quite short, especially compared to H-2. According to our other model experiments with icebergs of different initial heights, the maximum extent of the modelled IRD belt critically depends on iceberg height. Higher icebergs can travel far more eastward and eventually reach the European continent. Thus, good estimates of typical postglacial iceberg dimensions are essential for a realistic simulation of Heinrich events. Intercomparison of numerical model results and sediment core data will help to reconstruct typical dimensions of the icebergs that were released during the Heinrich event.

5 Conclusions

It has been shown that melting icebergs have effects on the circulation system similar to pure meltwater inputs from the coasts. Especially the different circulation changes caused by icebergs from Europe or Labrador are comparable. However, icebergs can be more effective than continental meltwater runoff because they are moving sources of freshwater. They can transport freshwater over a long distance and subsequently release it within a small area. Thus the region of meltwater input by icebergs can be more confined than is the case with continental runoff, yielding a more severe influence on density field and circulation.

These first experiments using freely drifting and melting icebergs gave promising results for further modelling of Heinrich events. Both extent and thickness of the Heinrich layers could be approximately reproduced. Sensitivity studies with different iceberg sizes and production rates will yield estimates of typical iceberg dimensions and freshwater inputs during the Heinrich events.

According to our results, icebergs can drift from a given source region to almost any place in the North Atlantic. Thus, to fully understand Heinrich events, it should not only be mapped where IRD was deposited, it is equally important to determine where the material came from.

Due to the temperature decrease by the icebergs, the heat flux from ocean to atmosphere must be reduced, and it might be argued that this should be included in the surface forcing fields of the experiments. In the present study, the heat loss to ice was generally of the same order of magnitude as the surface heat flux or even larger, and a surface heat-flux reduction would not have had significant effects. This would be even more the case if additional iceberg deterioration mechanisms (White et al. 1980) were included in the model, thereby causing higher heat loss of the ocean to the icebergs.

Acknowlegements

We wish to thank A. Paul for his thorough and fruitful review. This work was supported by the Deutsche Forschungsgemeinschaft within the framework of Sonderforschungsbereich 313, University of Kiel.

References

Andrews, J. T., Erlenkeuser, H., Tedesco, K., Aksu, A., and Jull, A. (1994) Late Quaternary (Stage 2 and 3) Meltwater and Heinrich Events, Northwest Labrador Sea. *Quaternary Research*, 41: 26–34.

Bischof, J. (1994) The Decay of the Barents Ice Sheet as Documented in Nordic Seas Ice-Rafted Debris. *Marine Geology*, 117, 35–55.

Bond, G. C. and 13 others. (1992) Evidence for Massive Discharges of Icebergs into the North Atlantic Ocean During the Last Glacial Period. *Nature*, 360: 245–249.

Bond, G. C. (1995) Climate and Conveyor. *Nature*, 377: 383–384.

Bond, G. C. and Lotti, R. (1995) Iceberg Discharges into the North Atlantic on Millenial Time Scales During the Last Glaciation. *Science*, 267: 1005–1009.

Broecker, W. S. (1991) The Great Ocean Conveyor. *Oceanography*, 1: 79–89.

Dowdeswell, J. A. and Murray, T. (1990) Modelling Rates of Sedimentation from Icebergs. *In* Dowdeswell, J. A. and Scourse, J. C., eds., Glacimarine Environments: Processes and Sediments. *Geological Society of London Special Publication*, 53: 121–137.

Dowdeswell, J. A., Maslin, M., Andrews, J., and McCave, I. N. (1995) Iceberg Production, Debris Rafting, and the Extent and Thickness of Heinrich Layers (H-1, H-2) in North Atlantic Sediments. *Geology*, 23: 301–304.

Foldvik, A., Gammelsrød, T., and Gjessing, Y. (1980) Flow Around Icebergs. *Annals of Glaciology*, 1: 67–70.

Grousset, F. E., Labeyrie, L., Sinko, J., Cremer, M., Bond, G., Duprat, J., Cortijo, E., and Huon, S. (1993) Patterns of Ice-Rafted Detritus in the Glacial North Atlantic (40–55 °N). *Paleoceanography*, 8: 175–192.

Gwiazda, R. H., Hemming, S., and Broecker, W. (1996 a) Tracking the Sources of Icebergs with Lead Isotopes: The Provenance of Ice-Rafted Debris in Heinrich Layer 2. *Paleoceanography*, 11: 77–93.

Gwiazda, R. H., Hemming, S., and Broecker, W. (1996 b) Provenance of Icebergs During Heinrich Event 3 and the Contrast to their Sources During Other Heinrich Episodes. *Paleoceanography*, 11: 371–378.

Haupt, B. J., Schäfer-Neth, C., and Stattegger, K. (1994) Modelling Sediment Drifts; A Coupled Oceanic Circulation-Sedimentation Model of the Northern North Atlantic. *Paleoceanography*, 9: 897–916.

Haupt, B. J., Schäfer-Neth, C., and Stattegger, K. (1995) Three-Dimensional Numerical Modelling of Late Quaternary Paleoceanography and Sedimentation in the Northern North Atlantic. *Geologische Rundschau*, 84: 137–150.

Heinrich, H. (1988) Origin and Consequences of Cyclic Ice-Rafting in the Northeast Atlantic Ocean During the Past 130,000 Years. *Quaternary Research*, 29: 142–152.

Lambeck, K. (1996) Limits on the Areal Extent of the Barents Sea Ice Sheet in Late Weichselian Time. *Global and Planetary Change*, 12: 41–51.

Lemke, P., Owens, W., and Hibler III, W. (1989) A Coupled Sea Ice–Mixed Layer–Pycnocline Model for the Weddell-Sea. *Report Max-Planck-Institut für Meteorologie Hamburg*, 28: 26 pp.

Manighetti, B., Maslin, M., McCave, I. N., and Shackleton N. (1995) Chronology for Climate Change: Developing Age Models for the BOFS Cores. *Paleoceanography*, 10: 513–526.

Matsumoto, K. (1996) An Iceberg Drift and Decay Model to Compute the Ice-Rafted Debris and Iceberg Meltwater Flux: Application to the Interglacial North Atlantic. *Paleoceanography*, 11: 729–742.

Ohshima, K. I., Kawamura, T., Takizawa, T., and Ushio, S. (1994) Step-Like Structure in Temperature and Salinity Profiles, Observed near Icebergs Trapped by Fast Ice, Antarctica. *Journal of Oceanography*, 50: 365–372.

Pacanowski, R., Dixon, K. D., and Rosati, A. (1993) The G.F.D.L Modular Ocean Model Users Guide. *GFDL Ocean Group Technical Report No. 2*, Geophysical Fluid Dynamics Laboratory / NOAA, Princeton University.

Peltier, W. R. (1994) Ice Age Paleotopography. *Science*, 265: 195–201.

Pflaumann, U., Duprat, J., Pujol, C, and Labeyrie, L. (1996) SIMMAX, a Transfer Technique to Deduce Atlantic Sea Surface Temperatures from Planctonic Foraminifera – the EPOCH Approach. *Paleoceanography*, 11: 15–35.

Robinson, S. G., Maslin, M., and McCave, I. N. (1995) Magnetic Susceptibility Variations in Upper Pleistocene Deep-Sea Sediments of the NE Atlantic: Implications for Ice Rafting and Paleocirculation at the Last Glacial Maximum. *Paleoceanography*, 10: 221–250.

Rosell-Melé, A. and Koç, N. (1997) Paleoclimatic Significance of the Stratigraphic Occurrence of Photosynthetic Biomarker Pigments in the Nordic Seas. *Geology*, 25: 49–52.

Russel-Head, D. S. (1980) The Melting of Free-Drifting Icebergs. *Annals of Glaciology*, 1: 119–122.

Saltzman, B. and Verbitzky, M. Y. (1992) Astenospheric Ice-Load Effects in a Global Dynamical-System Model of the Pleistocene Climate. *Climate Dynamics*, 8: 1–11.

Schäfer-Neth, C. (1994) Modellierung der Paläoozeanographie des nördlichen Nordatlantiks zur Zeit der letzten Maximalvereisung. PhD Thesis, University of Kiel, Germany, 105pp.

Schäfer-Neth, C. (1997) Changes in the Seawater Salinity-Oxygen Isotope Relation Between Last Glacial and Present: Sediment Core Data and OGCM Modelling. *Paleoclimates*, in press.

Schäfer-Neth, C. and Stattegger, K. (1997) Meltwater Pulses in the Northern North Atlantic: Retrodiction and Forecast by Numerical Modelling. *Geologische Rundschau*, 86: 492–498

Seidov, D., Sarnthein, M., Stattegger, K., Prien, R., and Weinelt, M. (1996) North Atlantic Ocean Circulation During the Lat Glacial Maximum and Subsequent Meltwater Event: A Numerical Model. *Journal of Geophysical Research*, 101: 16 305–16 332.

Svendsen, I. J., Elverhøi, A., and Mangerud, J. (1996) The Retreat of the Western Barents Sea Ice Sheet on the Western Svalbard Margin. *Boreas*, 25: 224–256.

Vorren, T. O., Lebesbye, E., and Larsen, K. (1990) Geometry and Geneseis of the Glacigenetic Sediments in the Southern Barents Sea. *In:* Bleil, U.and Thiede, J. , eds., Geological History of the Polar Oceans: Arctic Versus Antarctic. *Kluwer Academic Publishers*, 269–288.

Weinelt, M., Sarnthein, M., Pflaumann, U., Schulz, H., Jung, S., and Erlenkeuser, H., (1996) Ice-Free Nordic Seas During the Last Glacial Maximum? Potential Sites of Deepwater Formation. *Paleoclimates*, 3: 23–57.

Weinelt, M., Sarnthein, M., Pflaumann, U., Schulz, H., Jung, S., and Erlenkeuser, H., (1996) Ice-Free Nordic Seas During the Last Glacial Maximum? Potential Sites of Deepwater Formation. *Paleoclimates*, 3: 23–57.

White, F. M., Spaulding, M. L., and Gominho, L., (1980) Theoretical Estimates of the Various Mechanisms Involved in Iceberg Deterioration in the Open Ocean Environment, *U. S. Coast Guard Research and Development Rep.*, CG-D-62-80: 126 pp.

Numerical Study of Glacial and Meltwater Global Ocean Thermohaline Conveyor

D. Seidov, B. J. Haupt

1 Introduction

The ocean thermohaline circulation is often referred to as a global conveyor (Gordon 1986; Broecker and Denton 1989; Broecker 1991; review in Gordon et al. 1992). It is common knowledge that the global ocean thermohaline circulation is strongly controlled by the production of the North Atlantic Deep Water (NADW). Warm and salty subtropical water is carried to the high latitudes in the North Atlantic (NA) by the North Atlantic Current. It is cooled there and descends to set forth the deep ocean current system which is believed to be a global feature, a conveyor. Since the conveyor is mainly driven by latitudinal density gradients, which in high latitudes are controlled primarily by salinity, the density-driven conveyor is also referred to as the global salinity conveyor belt (Broecker 1991). The intriguing part of the problem is that the driving mechanism of change is thought to be very localized, with the key area of convection in the northern NA being surprisingly small with respect to the global ocean volume.

Both high-latitudinal salinity and temperature varied dramatically during the past several hundred years during the major glacial-interglacial cycles, and many believe that the global thermohaline ocean circulation was radically different at various stages during the Late Quaternary. A strong positive feedback between a curtailed conveyor and ongoing glaciation or deglaciation may be expected. As sediment is transported by ocean currents, the conveyor history is thought to be imprinted in the seabed sediment. Hence, the ocean past circulation can be reconstructed, in principle, on the basis of sediment accumulation records (Einsele 1992; Hsü 1989). Alternatively, one can try to reconstruct sediment transport using simulated past ocean circulation. Finally, the two approaches may be tied together by a Lagrangian approach to compare sediment transport and water-volume motion. The results of such a multithread approach to past ocean climatology can help both sedimentologists and oceanographers to predict future change of the sediment transport and ocean climate in basis of top analogues given by past circulation patterns.

This chapter reviews some of our recent paleoceanographic modeling and presents some of our new results on the modeling of the global and North Atlantic paleocirculation and sedimentation regimes. The emphasis is on sedimentation and water motion inferred using two specially designed models; a three-dimensional (3-D) sediment transport model and a semi-Lagrangian water transport model which are used as add-ons to traditional ocean general circulation models (OGCMs).

In the following sections we give an overview of the setup of past ocean sea-surface boundary conditions, a brief description of the numerical models employed, a short review of the results of our recent regional modeling of the NA circulation and sedimentation, and a display of some global conveyor simulations at two time slices, these being at and after the last glacial maximum.

2
Time Slices and Data

The global, and even regional basin-scale numerical modeling of ocean circulation necessitates a regular-grid coverage of the sea-surface with hydrological data comprising sea-surface boundary conditions – a requirement not easily met in paleoceanographic modeling. In fact only the glacial-to-interglacial cycle of the last 20 000 years has the sea-surface data coverage that might be seriously considered as suitable for ocean circulation simulations based on proxy data. Moreover, only the last glacial maximum (LGM) has a global sea-surface temperature (SST) array compiled by CLIMAP (1981). This data set has recently received much criticism because it is thought that the tropics are to warm (Guilderson et al. 1994; Beck et al. 1997; Webb et al. 1997). However, this data set remains the only global SST compilation currently available.

Thermohaline circulation cannot be properly modeled without knowing salinity distribution. Hence, sea-surface salinity (or equivalent freshwater fluxes across the sea-surface needed to maintain the observed salinity), should be known along with the SST to provide complete thermohaline sea-surface boundary conditions for computing ocean circulation. Moreover, freshwater discharges in the high latitudes of the North Atlantic may have enough power to profound by affect the circulation in this basin and perhaps worldwide. The freshwater fluxes are therefore thought to have beena major cause of past ocean climate changes linked to the global salinity conveyor belt operation.

There was a substantial increase of salinity of the World Ocean at the LGM due to large amount of freshwater deposited in the continental and shelf ice sheets. Some authors assume values as high as 1 psu (Duplessy et al. 1988; Fichefet et al. 1994). The removal of the freshwater stored in the ice sheets may be considered as more or less a global feature over large areas because freshwater was mainly evaporated from the ocean. Therefore the glacial increase of sea-surface salinity (SSS) may be assumed, although to the first approximation only, as a uniform increase. On the contrary, the deglacial freshwater discharges were tied to high latitudes, mostly in the North Atlantic and perhaps also in the Southern Ocean (as there are many indications of an interhemispheric synchrony of deglaciations (e.g., Bard et al. 1997), and therefore cannot be inserted into a model as a homogeneous salinity decrease everywhere. Furthermore, the distribution and intensity of such freshwater fluxes may become the most important paleoceanographic information in computerized studies of ancient ocean conveyors. In essence, the high-latitude SSS data become the major unknown and their absence would jeopardize any systematic use of an OGCM in ocean paleoclimate studies.

The situation is not so hopeless, however. The conveyor operation is thought to be hampered by suppression of the deep convection in the northern NA and in the Nordic Seas because of major meltwater discharges. Fortunately, these areas are relatively well covered by proxy data and corresponding paleoreconstructions which provide both SST and SSS for LGM and a subsequent meltwater event near 13 000 ^{14}C years here abbreviated as MWE. Based on these proxies for the LGM and MWE time slices, SST and SSS to the north of 40°N were reconstructed (Duplessy et al. 1988; Sarnthein et al. 1995). These reconstructions provided a basis for setting up thermohaline sea-surface conditions on a 1°× 1° regular grid in the entire NA north of 10°N (Seidov et al. 1996). Such boundary conditions were compiled for both LGM and MWE. Seidov et al. (1996) used these data in their simulations of the NA circulation during these two time slices. To compare the emerging circulation patterns to the present-day circulation, a control (modern) run was also carried out. In the control run modern sea-surface climatology from Levitus (1982) was used. The control run is henceforth referred to as the Holocene/Modern (HM) experiment.

To simulate the ocean circulation one needs not only the thermohaline sea-surface forcing, but the wind stress data as well. The wind stresses for LGM and for the control run were extracted from the output of the Hamburg atmosphere general circulation model (Lautenschlager and Herterich 1991; Lorenz et al. 1996). At LGM this model is driven by the CLIMAP (1981) SST, whereas the modern run was driven by present-day sea-surface climatology. The MWE wind was assumed to be the same as at LGM.

Thermohaline sea-surface conditions and the wind-stress distributions comprise the necessary boundary conditions allowing simulation of the past ocean annual mean circulation. Table 1 gives the sources of the data used to compile the regional NA regular arrays of boundary conditions.

Based on these regional NA data, global distributions of SST from CLIMAP (1981) were updated in the northern NA and the Nordic Seas to form LGM and MWE SST distributions. Present-day sea-surface salinity was increased everywhere by 0.8 psu (to compromise between the upper limit estimate of 1 psu and smooth connection of the low and high latitudes in the NA; see Seidov et al. (1996) and the regional NA LGM SSS replaced the modified SSS to the north of 40°N to form the global LGM and MWE SSS distribution. Modern mean annual SST and SSS are specified from new ocean climatological data sets (Levitus and Boyer 1994; Levitus et al. 1994), except for the area south of Greenland where some additional cooling of up to 2 °C mimicking cold spells from Greenland was needed to obtain present-day annual NADW production.

Table 1 Surface data sources for different time slices in the numerical experiments

Data \ time	Modern/Holocene	LGM (18000-15000 ^{14}C year B.P.)	MWE (14200–13200 ^{14}C year B.P.)
Wind stress	T42 wind stress	T42 wind stress calculated using CLIMAP (1981) surface conditions (Lorenz et al. (1996)	as for LGM
sea-surface temperature (SST)	Levitus (1982) and Dietrich (1969) in the NGS (see text and Fig.1)	CLIMAP (1981) and LGM data of Schulz (1994) and data of Sarnthein et al. (1992,1995) in the northern NA and NGS	CLIMAP (1981) and data of Schulz (1994) in the northern NA and NGS
sea-surface salinity (SSS)	Levitus (1982) and Dietrich (1969) in the NGS (see text)	Levitus (1982) south of 40°N and recalculated SSS using δ^{18}O from Duplessy et al. (1991) north of 40°N, and Sarnthein et al. (1995) north of 50°N in the northern NA and NGS for LGM (see Fig. 1)	LGM SSS and recalculated SSS using δ^{18}O data of Sarnthein et al. (1995)in the northern NA and NGS for MWE north of 50°N

Abbreviations: LGM last glacial maximum; MWE meltwater event; NA North Atlantic; NGS Norwegian-Greenland Seas. The data in the NA are combined with the modified present-day sea-surface climatology to provide the basis for the global ocean modeling (see text).

3 Numerical Models

The core of our approach is a combination of the traditional, albeit rather coarse, resolution simulations of the global ocean circulation and the NA circulation with somewhat better resolution using an OGCM together with a new technique of tracing the water-volume transport based on the output from the OGCM. This approach was first suggested in Seidov and Haupt (1997) in regional NA modeling and then was applied to past global conveyor simulations. An analogous approach has been advanced recently by Drijfhout et al. (1996), who traced the present-day conveyor. Another new feature is sediment transport modeling which takes into account the major conveyor control, that is the deep ocean convection.

3.1 OGCM

Two different numerical models of the general ocean circulation were used in the regional NA and global conveyor studies. For the regional NA studies we employed a planetary geostrophic ocean circulation model especially designed for coarse-resolution, large-scale ocean circulation studies. The basics of this model are described in Seidov (1996) and Seidov and Prien (1996). The model uses the fact that the momentum balance away from the equator is predominantly geostrophic, and therefore it belongs to the "intermediate" or "planetary geostrophic" class of models (Hasselmann 1982; Seidov 1986 1996; Colin de Verdière 1988; Maier-Reimer et al. 1991 1993; Zhang et al. 1992). Except for the linear dynamics inherent to planetary geostrophic models, such models retain all features of the most advanced primitive equation models (PEM) that are the present-day standards in physical oceanography (e.g., Bryan 1969, Cox 1984, Semtner 1986). Planetary geostrophic formalism allows some simplifications of the momentum and vorticity balance equations, and a longer time step, and consequently becomes computationally more effective. Parallel runs have been made on a coarse-resolution grid using this model and the Geophysical Fluid Dynamics Laboratory (GFDL) (Bryan 1969; Cox 1984; Pacanowski et al. 1993), a PEM. Seidov and Prien (1996) have shown that results of this planetary geostrophic model match well with the performance of the GFDL model. However, since the planetary-geostrophic approach may lead to significant discrepancies in the equatorial region, the global ocean circulation was simulated using the GFDL-Modular Ocean Model (MOM, version 1.1) (Cox 1984; Pacanowski et al. 1993).

Any OGCM contains a procedure to allow convective mixing that develops due to hydrostatic instability if denser water is formed (or advected) over lighter water. This is the key process for deep water production and transport. As done commonly in an OGCM, this instability is accommodated by mixing water vertically until it regains complete hydrostatic stability (Cox 1984). This vertical adjustment takes place as a step process with several successive mixings of adjacent layers until complete hydrostatic stability is restored. As the water convects, the particles associated with the mixing volumes exchange their positions vertically, and hence the convection facilitates deep ventilation.

3.2 Ocean Sedimentation Model

The sedimentation model employed here was designed by Haupt (1995) and tested in Haupt et al. (1994 1995). It is a large-scale sedimentation model consisting of two components: (1) a 3-D sediment transport model in the ocean interior, and (2) a 2-D sediment transport model in a thin near-bottom layer following smoothed bottom topography. The 3-D component models the advection-diffusion of sediment similar to the equations of advection-diffusion of heat and salt in the OGCM, and includes an added term to compute the settling of sediment in the water column.

The 3-D component of the sedimentation model simulates the lateral transport and the entry of sediment particles at the sea-surface, including the entry of ice-rafted sedimentary material. The supply of large-grain sediment at the sea-surface can be increased or decreased depending on the amount of icebergs injected into the open ocean due to decay of major ice sheets. The icebergs distribution over the NA in the past can be inferred from analyses of proxy data from foraminifera (e.g. Maslin et al. 1995). The mass of sediment covering the sea floor depends only on the balance of sources and sinks, whereas the spatial variation of sedimentation rates depends on the circulation pattern and particle grain size. The 2-D model is initialized at every time step by the exchange of sediment between the ocean body and the ocean floor. The sediment in the bottom layer is transported by a corrected benthic flow which is largely a projection of the OGCM velocity field onto the smoothed bottom layer (1 cm thick). Additionally, the near-bottom velocities are reduced to take bottom friction into account (Miller et al. 1977; Zanke 1978; Sündermann 1983).

The impact of convection was not included in the previous version of the sedimentation model (Haupt et al. 1994). In the updated version the convection is incorporated in the form of convection depths from the OGCM (Seidov and Haupt 1997). Vertical mixing, similar to that employed in the OGCM, is applied to concentration. The sediment transport model 'knows' when to mix water because we have encoded the convection depths in the velocity field as an additional parameter. When used in the sedimentation model, this information is decoded to enable vertical mixing in the grid points where convection occurs (see below in the discussion of convection patterns).

3.3 Particle-Tracing Model

The water-volume trajectory-tracing model was developed by Haupt (1995) and was employed to trace particle drifts in the northern North Atlantic (Haupt et al. 1994 1995). This model exercises a hybrid Eulerian-Lagrangian (or semi-Lagrangian) approach; the velocity components are interpolated to the current positions of the Lagrangian particles from the nearby grid points of a Eulerian numerical grid. As in the sediment transport model described above, the Eulerian velocity field is provided by the OGCM, whereas the coordinates of Lagrangian particles are calculated straightforwardly, using the Lagrangian velocity along the trajectory.

As we have emphasized above, ventilating convection induced by hydrostatic instability is included in all three components of our simulations. The semi-Lagrangian trajectory-tracing model of Haupt et al. (1994) was upgraded in Seidov and Haupt (1997) to facilitate vertical ventilation in convective "chimneys" (Send and Marshall (1995) show that ventilating convection occurs as water mixing in water columns or "chimneys"). Here the velocity field from the OGCM is supplemented by the convection depths showing where and to what depth the vertically mixing volume should be propelled in the turbulent chimney. A detailed discussion of the parameterization of chimney mixing in the particle-tracing model is given in Seidov and Haupt (1997).

More details of the sediment transport model and the particle-tracing model can be also found in Haupt et al. (1997; see this Vol.).

4 The Setup of Numerical Experiments

The numerical experiments in both the regional NA and the global circulation cases were carried out in two steps. First, the OGCM was run using the appropriate boundary conditions. Integration in time was continued for several thousand years in the regional and extended over 10000 years in the global runs to gain absolutely stationary solutions. Then for both modern and LGM conditions, the steady-state current velocities were used to calculate the trajectories of the particles that move with the water-volumes. The MWE velocity field is the result of integration over only 500 years from the LGM steady-state with the LGM sea-surface conditions replaced by the MWE ones (the duration of this meltwater event is estimated to be of several hundreds to a thousand years; Sarnthein et al. 1995).

The regional NA study includes modeling of sediment transport, that is, these simulations comprise a threefold approach. The global ocean circulation modeling at this stage employs only the OGCM and trajectory tracing models, and therefore forms a twofold approach to water transport and ventilation problem.

For the regional NA studies we used a $2° \times 2°$ grid with 12 vertical levels. In view of uncertainties of the past sea-surface conditions, and in order to facilitate multiple runs extended over 10 years with different parameter settings, we employed a coarse resolution of $6° \times 4°$ in longitude and latitude respectively with 12 vertical levels. Although all major currents, except for the largely wind-driven Antarctic Circumpolar Current (ACC), are two to three times weaker than the observed ones, most of these currents are still clearly seen on the vector maps. At the same time, the meridional thermohaline overturning, which largely depends on the deep convection and isopycnal outcrop, is modeled far better than the horizontal flows. For the present-day overturning we have obtained a value that agrees well with experiments of much finer resolution. The overturning in the NA, which comprises the NADW production, amounts to 23 Sv (1 Sv=10^6 m^3s^{-1}), for comparison Antarctic Bottom Water (AABW) inflow into the Atlantic Ocean is about 9 Sv. Since we are mainly interested in the major changes in conveyor operation and rely only on the relative changes of the conveyor intensity, the coarse resolution is legitimate for global paleocirculation studies such as that presented here. We therefore consider it to be sufficient for the semi-Lagrangian simulations which are the core of our investigation.

In contrast to sensitivity studies which operate with so-called mixed boundary conditions (e.g., Manabe and Stouffer 1994; Rahmstorf 1995), we restore the upper-layer thermohaline fields to the specified sea-surface temperature (SST) and sea-surface salinity (SSS). The sea-surface boundary conditions are described the in previous section. In our study, the important reason for the chosen approach is the local nature of the freshwater driving. Indeed, the LGM sea-surface conditions were disturbed by the meltwater invasion only in the small area in the northern NA and Norwegian-Greenland Seas where reliable proxy data are available (see above). This restoring technique implies that the freshwater fluxes were those that maintained the reconstructed SSS in these areas. In fact, we merely diagnose the circulation regimes which would satisfy the observed sea-surface conditions.

5 North Atlantic Sediment Transport and Water Motion

5.1 OGCM Results

Here, we briefly review our regional NA modeling. More details can be found in Seidov et al. (1996) and Seidov and Haupt (1997). The meridional overturning stream function (Fig. 1) which gives the general impression of the overall thermohaline meridional circulation, conforms to a simple scheme that has emerged from numerous computer simulations of differing complexity (e.g., England 1993; Toggweiler et al. 1989; Maier-Reimer et al. 1991; Wright and Stocker 1991; Fichefet et al. 1994; Rahmstorf 1994; Sakai and Peltier 1995; Manabe and Stouffer 1995). Today's forward or clockwise (as seen from the eastern boundary looking west through a vertical plane) gyre of the salinity conveyor occupies most of the ocean from the surface-subsurface layers to a depth of 3 km. This gyre conveys NADW and is thought of as the main wheel of the modern ocean climate. A much weaker abyssal reverse (counterclockwise) gyre conveys AABW (largely in the deepest layer, i.e., below 3 km). Its return (southward) flow joins the southward flow of NADW between 3 and 4 km. In addition, there is a weak, shallow, wind-driven reverse (counterclockwise) gyre in the mid to high latitudes, induced by Ekman convergence in the subtropics.

Note that although glacial NADW production is 8 Sv, much lower than today's value of 13 Sv, the intensity of the glacial conveyor is comparable to that of modern times. If we quantify the southward transport at 30°N, the Upper NADW outflow, or forward conveyor branch, does indeed rise to only 8 Sv (also see the inventory of the glacial water masses given by, e.g., Oppo and Lehman 1993 and Oppo et al. 1995). However, if we take into account the deep reverse branch comprising AABW and its mixture with lower NADW still produced in the central northern Atlantic (see below in the discussion of convection patterns), the total transport amounts to 12 Sv, which gives almost the same intensity of the deep water outflow as the present-day North Atlantic combined deep outflow. The intensity of this NADW-depleted conveyor, similar to present-day conveyors' intensity, seems to fit well the recent finding based on proxy data analysis (Yu et al. 1996). However, the modern overturning strength is too low. A slight cooling of the sea-surface to the southeast of Greenland and in the Nordic Seas may easily double the NADW production (see Sec. 5.2). This is because the sea-surface density taken as averages of summer and winter values is somewhat biased toward summer values, and therefore the convection is not deep enough to form the vigorous observed present-day overturning. Yet we may consider the numbers as a first approximation, keeping in mind that the glacial conveyor was probably relatively even weaker than our regional modeling indicates. Hence, we believe that the glacial conveyor was probably still noticeably weaker than the modern one, in contrast to what is suggested by Yu et al. (1996). The global circulation modeling with sea-surface cooled down by 2 °C to the east of Greenland between 60°

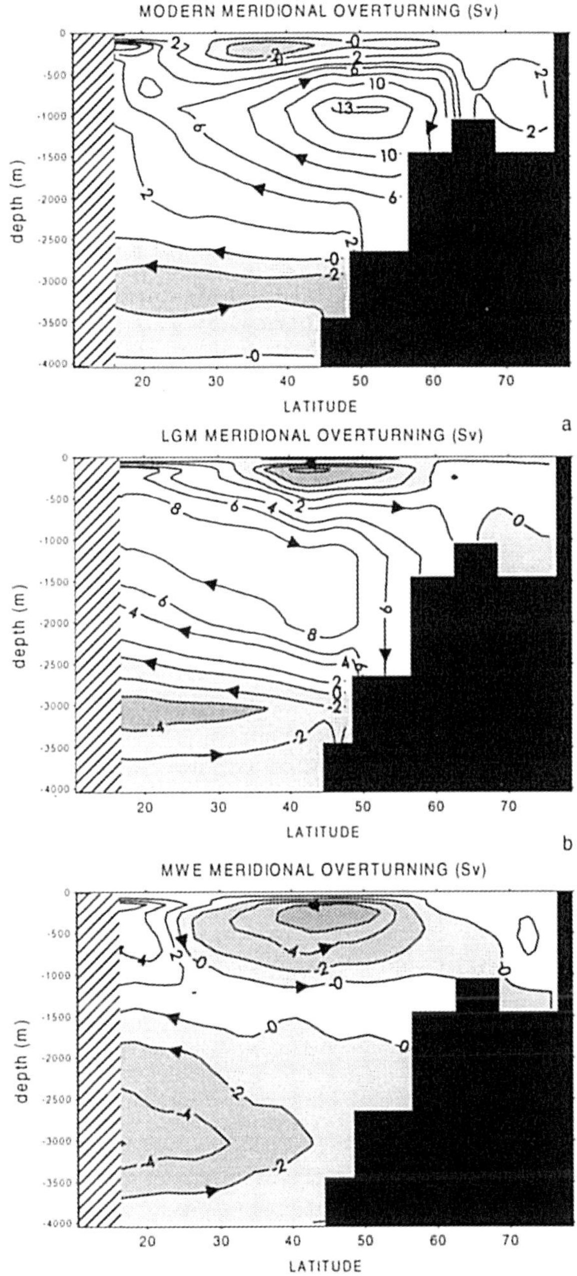

Fig. 1 a–c Overturning stream function showing total meridional transport in the North Atlantic (in Sv; 1Sv=10^6 m^3s^{-1}).
a Present-day overturning pattern.
b LGM.
c MWE. The patterns within the sponge layer at the southern boundary (see text) are masked. Areas of negative values are *shaded*. *Arrows* show direction of the transport. (After Seidov et al. 1996)

and 65°N gave 22 Sv NADW production (see below), which is probably too high; if one mediates the regional and global results there would be about 18 Sv, a number most appropriate for a coarse resolution study. This would give an LGM forward conveyor twice as weak but still comparable outflow which would fit the idea of still intensive glacial thermohaline circulation (Yu et al. 1996; Webb et al. 1997).

As we have already noted, the 3-D conveyor is essentially more complex than its 2-D image given by the total meridional overturning. The velocity fields are discussed in great detail in Seidov et al. (1996) and Seidov and Haupt (1997). Here we briefly overview how the glacial and MWE currents differ from the modern ones. The present-day, LGM, and MWE simulated velocity fields may be inspected in Seidov et al. (1996). The most noticeable feature is the deviation of the paleo-North Atlantic Drift from its modern northeastern path. Strong zonality of the subpolar front indicates a reduced supply of water that can be downwelled as NADW in the northern North Atlantic and the NGS. Moreover, the route of the return southward flow in the deep ocean changed radically, a robust feature emerging in all our glacial experiments regardless of complexity. In the eastern part of the basin, the incursion of AABW dominates the near-bed transport up to the Faeroe-Shetland Ridge, in agreement with the water mass contouring by Sarnthein et al. (1994). The most striking feature of the MWE currents in addition to the changes found at LGM is the reversal of the Norwegian Current and the inflow-outflow regime in the Norwegian-Greenland Seas (see Seidov et al. 1996 for details). Hence the redeposited sediment in the eastern mid to high latitudes might be of different origin. Today, sediment drifts there transport material largely from northeast to west and southwest. In contrast, during the LGM some sediment drifts could change direction of transport to redeposit the grains from south to north and northwest.

Today's southward deep return current is a deep ocean western boundary flow forming a strong countercurrent under the Gulf Stream. This western boundary current is well recognized as the most prominent feature of the thermohaline circulation (Stommel and Arons 1960). Though a noticeably weaker western boundary current still existed at the LGM and MWE, the descending branch returned to the southwestern basin largely as a broad zonal westward flow in the mid latitudes. The glacial countercurrent under the paleo-Gulf Stream was weaker, deeper, and farther eastward. At the LGM a southward deep ocean flow originated near the Rockall Plateau at depth of about 2 km, occurring in the eastern part of the basin, rather than in the western part as today. This particular feature of the computer model in the eastern North Atlantic is again in good agreement with the contouring by Sarnthein et al. (1994). Using the trajectory-tracing model below, we demonstrate that the simulated glacial deep water indeed moved along the eastern flank of the Mid-Atlantic Ridge. Therefore as water in the western deep Atlantic contained more AABW at the LGM, the deep and abyssal water during the glacial time was older than at present.

To understand the mechanism of change of deep water formation, one should consider changes in the convection regime. The modern deep convection sites are found mainly in the NGS, east and south of Iceland, and in the Labrador Sea. The present-day convection pattern in our calculations is consistent with the patterns given by Maier-Reimer et al. (1993), Toggweiler et al. (1989), and Rahmstorf (1995). Shallow convection starts in the Gulf Stream area and marks a further progression of subtropical water to high latitudes carried by the North Atlantic Current becoming deeper as it progresses northeastward and westward around the Greenland coast.

The major feature of glacial convection, as compared to its modern state, was the far weaker glacial convection in the NGS and the southward shift of the main convection sites. The maximum ventilation at the LGM occurred in the mid-North Atlantic between 50° and 60°N. This result is well supported by proxy data (Duplessy et al. 1988; Sarnthein et al. 1995). A southward shift of both the convection sites and the polar front led to a decrease in the northward transport of warm and salty water. Hence a positive feedback worked to establish the glacial mode of circulation. An even farther southward shift of the convection was limited by Ekman pumping in the subtropics which protected the major anticyclonic gyre from shrinking even more, i.e., providing a negative feedback to balance the positive one. The MWE convection sites were shifted further southward and the convection was very shallow, penetrating no deeper than 600 m in the central North Atlantic. There was no convection to the north of 50°N and the Nordic Seas were completely convection-free during that time slice (Seidov et al. 1996).

5.2
Sediment Accumulation Rates and Sediment Transport

Total sediment accumulation is governed by the requirement that sources and sinks balance at the sea-surface and be the same for all three time slices. The spatial distribution of sediment is however different, and reveals two distinctly different circulation modes between present, and at the LGM and MWE. Fig. 2 presents the sediment rates for the three chosen time slices. A very high accumulation rate associated with the Holocene/Modern mode of circulation is found in the vicinity of Iceland in the Irminger Basin, along the Reykjanes Ridge at both the south and north sides of Iceland, at the Rockall Plateau, and in the NGS (Fig. 2a). This is in agreement with the present-day concept of sediment trapping in these areas (McCave and Tucholke 1986; Bohrmann et al. 1990; Wold 1992). We find that the modeled sediment accumulation rates are smaller than those which have been measured. This is because of the low eolian sediment input and also because of the missing lateral input (Haupt 1995). The sediment is transported by the deep western southward boundary current with local maxima near Newfoundland and farther to the south in and near the Caribbean (McCave and Tucholke 1986). We note that Iceland and the Caribbean are represented by seamounts in the model bottom topography; hence the nonzero accumulation rates are artifacts at the exact positions of these islands.

As the glacial North Atlantic Current diverged from today's northeastern path, the accumulation rate in the Iceland and Irminger Basins dropped, and

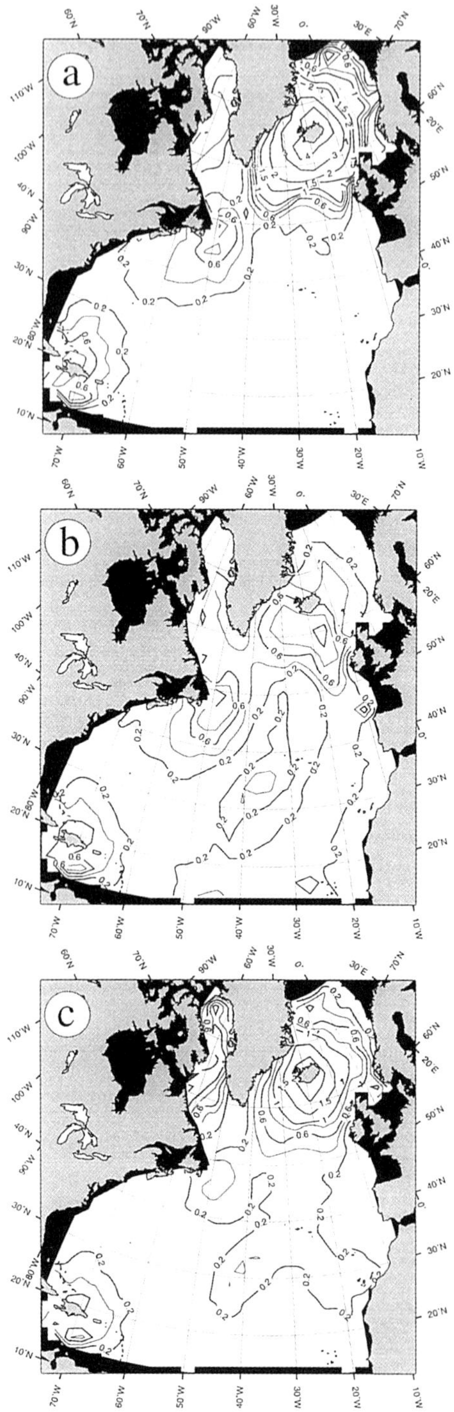

Fig. 2a–c Sedimentation rate (cm/1000 year) predicted by sediment transport model using the HM (**a**), the LGM (**b**) and the MWE (**c**) ocean circulation patterns from the OGCM (see text)

most of the sediment mass was spread over the abyssal valley in the Canary Basin (Fig. 2b). Note that although the sedimentation rate is far lower there than in the Iceland Basin, the sediment mass is roughly the same because bottom area in the latitude-longitude grid cells increases rapidly to the south. A relatively high glacial accumulation rate is found in the Newfoundland Basin. However, we do not see continuous southward sediment transport along the American coast, a signature of all our experiments based on the modern ocean surface climatology. Instead, a noticeable southward sediment transport is found at the eastern flank of the Mid-Atlantic Ridge.

Though an increased northward incursion of AABW at the LGM in the eastern North Atlantic is consistent with the contouring by Sarnthein et al. (1994), the conclusions about an increased reverse abyssal gyre cannot be attributed likewise to the western part of the basin. Indeed, the southward transport, though noticeably curtailed, still existed in the western deep ocean. Analysis of the velocity maps suggests (and trajectory analysis confirms; see below) that in the west the northward flow of AABW at LGM was probably the same as or weaker than today's. Simultaneously, the eastern flank of the glacial AABW intrusion (initially through the Vema Fracture Zone) was enhanced considerably. The exchange between the eastern and northern parts of the deep North Atlantic was stronger at the LGM than today.

As the meltwater North Atlantic Current diverged from today's northeastern path, the accumulation rate in the Iceland and Irminger Basins dropped and a significant part of the sediment mass was spread over the abyssal valley in the Canary Basin (Fig. 2c). Note that although the sedimentation rate is far lower there than in the Iceland Basin, the sediment mass is roughly the same because bottom area in the latitude/longitude grid cells increases rapidly to the south. The sedimentation rate around Iceland decreased by over a factor of two as compared with its modern value (cf. Fig. 2a and b). This is mainly because of the absence of ventilating convection both in the NGS and in the Irminger Sea. At the LGM, a relatively high glacial accumulation rate still emerges in the model in the Newfoundland Basin (Fig. 2b). However, we do not see continuous southward sediment transport along the North American coast, a signature of all our experiments based on the modern ocean surface climatology. Instead, a noticeable southward sediment transport is found at the eastern flank of the Mid-Atlantic Ridge. At the MWE this tendency is increased significantly. The accumulation rate in the Newfoundland Basin decreased in comparison with the LGM run, whereas deposition was stronger at the eastern flank of the Mid-Atlantic Ridge. The area between Newfoundland and the Caribbean is almost totally free of sediment deposition.

5.3 Trajectory tracing

Although most of the features common to the three sediment transport patterns (the HM, LGM and MWE) can be explained by comparing them with the circulation patterns, fundamental differences between glacial and interglacial ventilation and sedimentation regimes cannot be revealed using the velocity maps solely. The transport in a transit area of intense ventilation can differ principally depending on whether convection is taken into account or ignored. Let us hypothesize that a velocity field is not computed using a prognostic circulation model, but is instead diagnostically calculated from the ocean climatology, e.g., using data of Levitus (1982). These climatological fields were formed by all processes including convection. However, the convection is not present in the data directly. Moreover, the convection pattern cannot be reconstructed using these data without running a prognostic model. Hence direct impact of convection on the water motion is missed from diagnostic calculations based on these climatological data. In other words, diagnostic velocity would contain no information about convection because hydrostatic instability was removed from the processed climatological data. This means that the Lagrangian particles whose trajectories were calculated on the basis of the observed ocean climatology instead of a prognostic model would drift along the trajectories undisturbed by convection, and therefore these simulated trajectories would be wrong.

In our experiments, the particles illustrating the flow were deployed in different areas in the northern NA (Fig. 3 in Seidov and Haupt 1997). In each of the areas, about 30 particles started to travel through the Eulerian velocity fields in all three cases – the HM, MWE and LGM. In the following maps (Fig. 3 and Fig. 4) we employ two different techniques to show the trajectories. In Fig. 3 the trajectories are colored to show the depth of a particle. (convection sites are depicted by different shades of gray: the deeper the convection, the darker the shade). We use the sunlight spectrum colors, from dark red in the uppermost layer (< 100 m) to violet and black in the two deepest layers, to visualize vertical migration of the water parcels. In contrast, Fig. 4 shows the particles' pathways with both depth and elapsed time shown in small rectangles attached to the trajectories (only two pairs of the trajectories are repeated in the black and white Fig. 4). Although the model time in the trajectory-tracing calculations was over 500 years, only the tracks for the first 100-200 years of the elapsed time are shown in the maps to avoid confusion.

A striking feature of the trajectory map is the change in the glacial deep ocean circulation regime which is not so obvious from the velocity maps. This change is far more complex than a simple increase in the zonality of the surface current, a well-known feature of the LGM surface circulation (CLIMAP 1981; Ruddiman and McIntyre 1981; Kellogg 1980).

The deep water production at the LGM is found in the model only in the central part of the north central North Atlantic (Fig. 3), which is in agreement with the convection pattern in Fig. 4b. We note that water descends in the subpolar gyre in spite of upward motion induced by Ekman divergence. Hence thermohaline currents would drive water along isopycnals in the subsurface layers regardless of ventilating convection; i.e., ventilation of subsurface and intermediate water would occur regardless of deep convection. The pronounced convective chimney forms an intensive cyclonic circulation around a homogenized col-

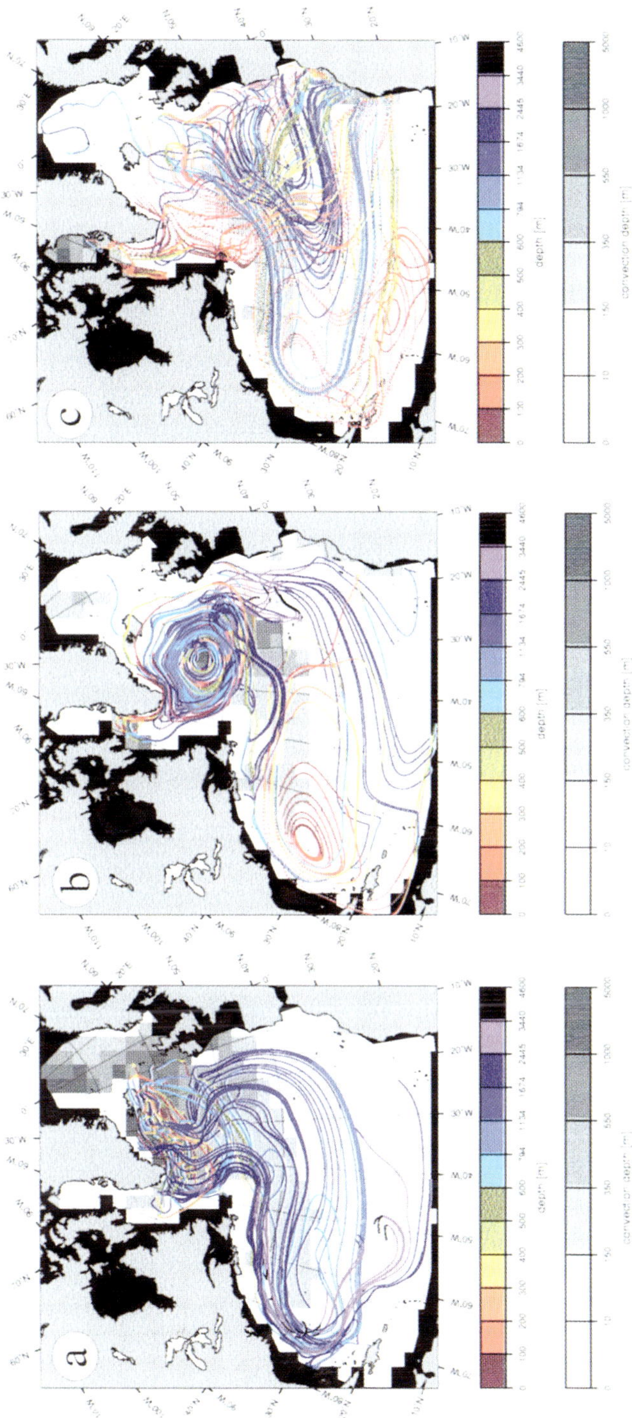

Fig. 3 a–c Trajectories of particles deployed at the HM in the central part of the northern North Atlantic. **a** Holocene/Modern. **b** LGM. **c** MWE. The convection depths are shown by *different shades of gray.* The first 100 years of the particles' history are shown. Depth is indicated by *colors* from the color palette; as the particle descends or upwells the color of the trajectory changes

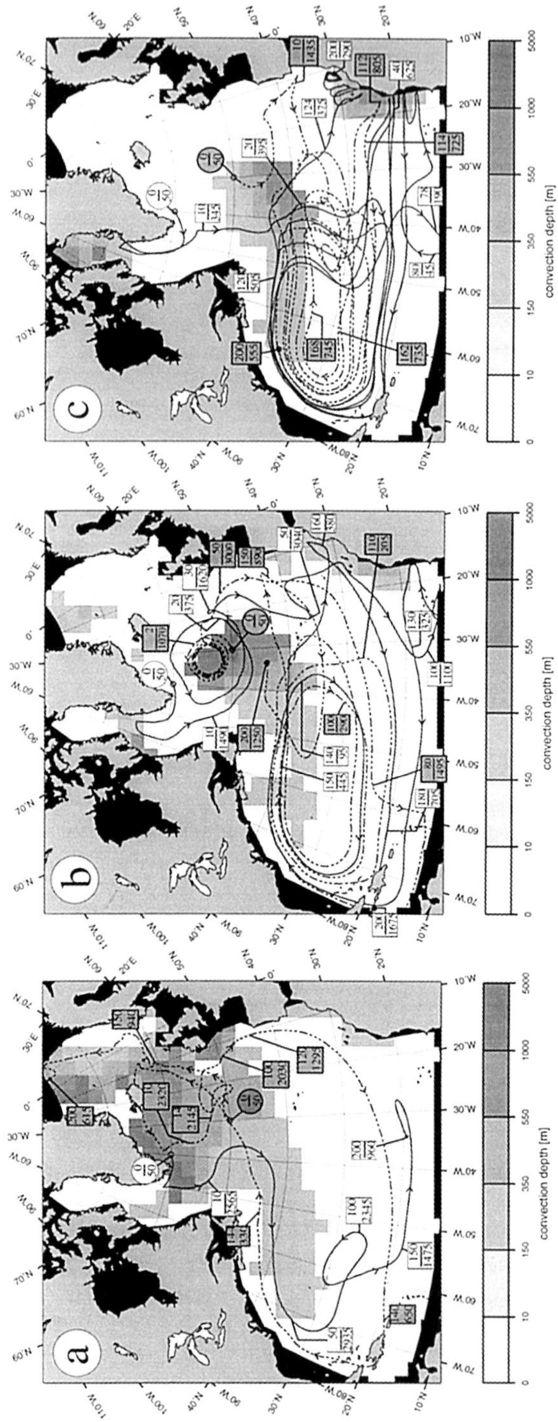

Fig. 4a–c Two hundred-year history of pairs of Lagrangian particles for HM (**a**), LGM (**b**), and MWE (**c**) from the assemblages shown in Fig. 3. *Small rectangles* show elapsed time and depth; *small circles* indicate starting points, the *arrows* show the direction of motion, and the *bullets* indicate the end points of the trajectories. One of the trajectories of each pair is presented by a *broken line*

umn of high-density water (Fig. 3b). As particles deployed at the surface during the LGM became trapped inside this column, they were propelled downward, forming most of the 8 Sv of glacial NADW (see above).

Fig. 3b indicates that there was some cold intermediate water produced in the NGS at the LGM. This water, flowing farther into the North Atlantic over the Greenland-Iceland sill, was not dense enough to subduct under the intermediate to deep water formed in the central part of the subpolar gyre. It mixed with the subpolar water and stayed at an intermediate depth flowing along the S-shaped route at that depth in the north central part of the basin (see in Plates 1d and 2b in Seidov and Haupt 1997). Most of the deep return flow occurred along the eastern slope of the Mid-Atlantic Ridge, though some water still contoured the American east coast. In contrast, water sinking in the NGS today is dense enough to descend even deeper after spilling over the sills into the North Atlantic. Together with the portion of NADW formed east of Greenland and in the Labrador Sea, this water travels southward in the western boundary current comprising most of the 13 Sv of simulated modern NADW outflow. Most of the present-day deep flow contours the American east coast, and a smaller portion is routed along the west slope of the Mid-Atlantic Ridge. We point out that the curtailment of the forward conveyor in the western part of the ocean was not complete. This implies that the conclusions based on analysis of the glacial proxy data assembled along the eastern meridional sections probably are not valid for the western part of the basin.

Fig. 3a and 3b compares present and the glacial trajectories' "spaghetti" in the subtropical anticyclonic gyre. The modern and LGM trajectory maps conform to the Luyten-Pedlosky-Stommel (LPS) ventilated thermocline theory (Luyten et al. 1983) and the computer experiment of Cox and Bryan (1984). The thermocline in the subtropics is maintained by Ekman pumping. Therefore as water circulates in the gyre, it descends and ventilates the thermocline. The water is brought up to subsurface layers in the western boundary along upward sloping isopycnals (orange-colored segments of the trajectories in the vicinity of the western boundary in Fig. 3). Flowing eastward, in the segment of western boundary current outflow, the water convects because of heat loss to the atmosphere (because the specified SST is colder than the outflowing subsurface water in this outflow zone). This shallow convection is shown in light gray in Fig. 3. The subsurface water parcels in this ventilation zone come in contact with the atmosphere. They then start to descend again to repeat the whole ventilation cycle. Those water parcels which flow northeastward below the convection depth do not contact the atmosphere again and therefore do not ventilate. Yet, at any given time, new water parcels descend from the surface in the ventilation zone. However, ventilation is restricted to the central and western parts of the gyre. In complete agreement with the LPS theory, there is a shadow zone attached to the eastern boundary, practically unreachable for ventilation. Modern elapsed times (Fig. 3a) for a single ventilation cycle in the gyre agree well with the estimates of Cox and Bryan (1984); short loops have an advection age of no more than 3 to 5 years before the water returns to the ventilation zone, whereas the longer paths in the subtropical gyre take more than 10 to 15 years to return to ventilation zone. It is obvious from the colors of the trajectories in Fig. 3 that the glacial thermocline is far more deeply ventilated than its modern analogue. There is evidence (Slowey and Curry 1992 1995) that the subtropical thermocline was indeed better ventilated during the last glacial period. As Fig. 4 reveals, our Lagrangian calculations agree well with these findings.

Figs. 3 and 4 indicate that the speed of the particles in the upper and intermediate layers was not lower at the LGM than today. The deep and abyssal glacial flows, though routed differently from today, were even stronger in the eastern part of the basin. This confirms the concept that the LGM conveyor was at least as intense as the present one (Yu et al. 1996; Boyle 1996). However we stress that, in contrast to a similar or even higher intensity meridional conveyor at intermediate depths, the forward conveyor was definitely weaker at the upper to intermediate depths.

The MWE trajectories indicate that at that time slice there was no ventilation of the deep ocean in the NA (Fig. 3c), although there was still quite intensive ventilation of the subtropical thermocline. The spaghetti form of the MWE trajectories indicate also that the Nordic Seas were rather isolated at that time. The selected trajectories illustrate the collapse of the forward conveyor and delineate the shallow and slow motion within the upper ocean layers.

6 Modeling of the Global Conveyor

6.1 Convection Regime

The global ocean conveyor is weaker than in the regional NA experiments. The MWE data, in the amount suitable for setting the data up on a regular grid, exist only for a small region in the northern North Atlantic and the Nordic Seas (Sarnthein et al. 1995). The LGM salinity data over the entire globe are rather speculative if compared to the North Atlantic, where they aresufficient to form a data set on a regular grid (Duplessy et al. 1988). On the other hand, the global simulations are far more advanced and physically consistent than any regional modeling because they give a continuous hydrodynamics of interconnected oceans without any need to introduce artificial sponge layers. Hence, it is a question of tradeoff when one considers advantages and disadvantages of global versus regional modeling based on limited data. One definite advantage of global simulations is the presence of all major sources of deep water. The convection patterns (Fig. 5) clearly delineate key differences in the dominant process of deep ocean ventilation in the northern and southern hemisphere at present and in the past. In Fig. 5 the convection depth (the depth to which the convection due to hydrostatic instability penetrates) is shown as vertical bars. The deeper the convection, the higher the bars (see legend in the diagrams). If multiplied by the area over which the convection occurred (here the surface of the grid cells), the depth

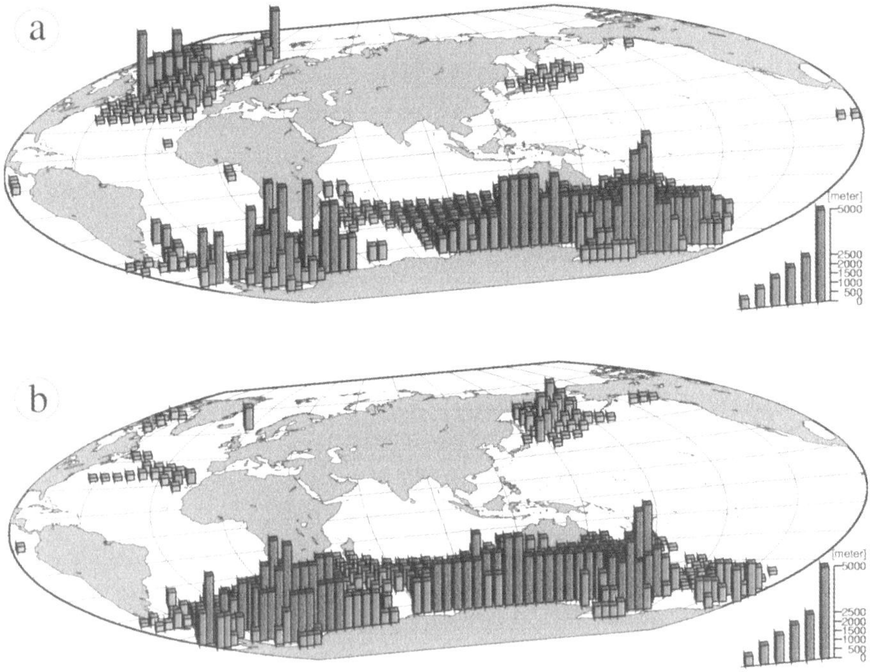

Fig. 5a, b. Diagrams of convection
a The HM
b The MWE. The *heights of the bars* are equal to the depth of convection

of convection would represent the volume of convectively mixed water. Fig. 5a indicates the dipole nature of deep ocean ventilation at present. The southern deep water sources around the Antarctic and particular in the Weddell and Ross Seas are counterbalanced by the northern NA source only, that is NADW balances AABW to form upper-to-deep ocean forward conveyor and deep-to-abyssal ocean reversed conveyor. Clearly, the LGM deep water production in the central NA, and reduced but still running convection in the Nordic Seas, led to somewhat curtailed but still running forward and abyssal reversed conveyors. The most important feature of the global HM and LGM forward conveyor is the coherent deep flow starting in the Nordic Seas and northern NA and flowing around the globe in the Antarctic Circumpolar Current with branches penetrating to the Indian and Pacific Oceans. Fig. 5b implies that the deep ocean circulation must have changed radically because at MWE the bipolar convection regime was replaced by a regime with only southern deep ventilation. The whole northern NA and the Nordic Seas were convection-free at the MWE.

Fig. 6 shows the present-day velocity vectors at 150 m (a) and 2500 m (b) depths. Figs. 7 and 8 depict the same for LGM and MWE, respectively. A comparison of the MWE and HM and LGM deep ocean currents (Figs. 6b, 7b, and 8b)

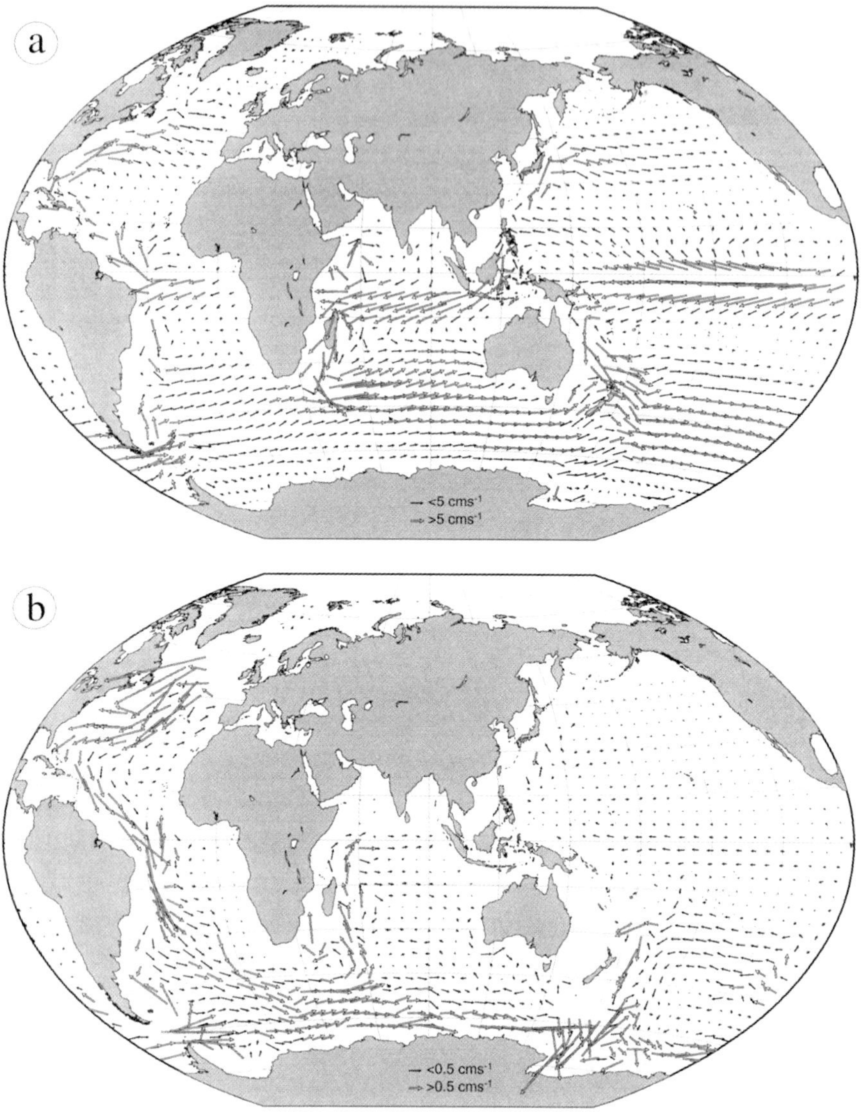

Fig. 6a, b Holocene/Modern velocity vectors at 150 (**a**) and 2500 m (**b**). Note different scales for the upper and deep ocean currents

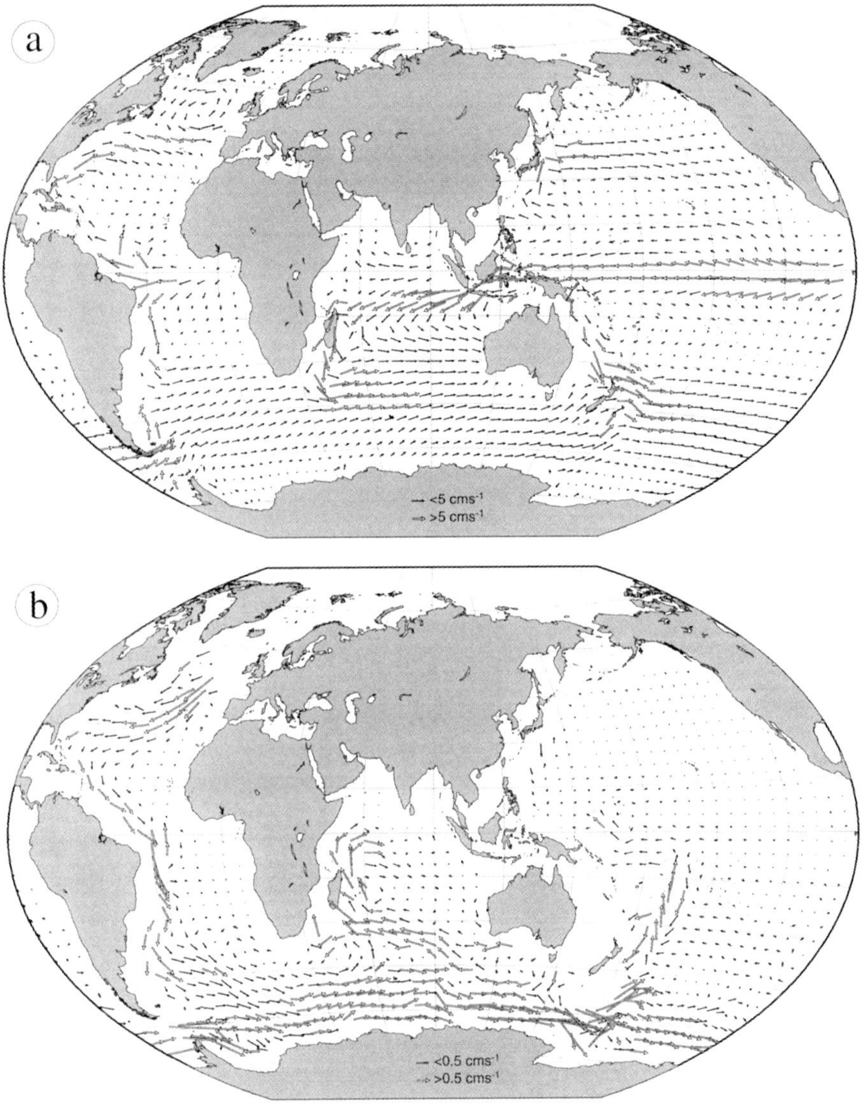

Fig. 7a, b As in Fig. 6 for LGM

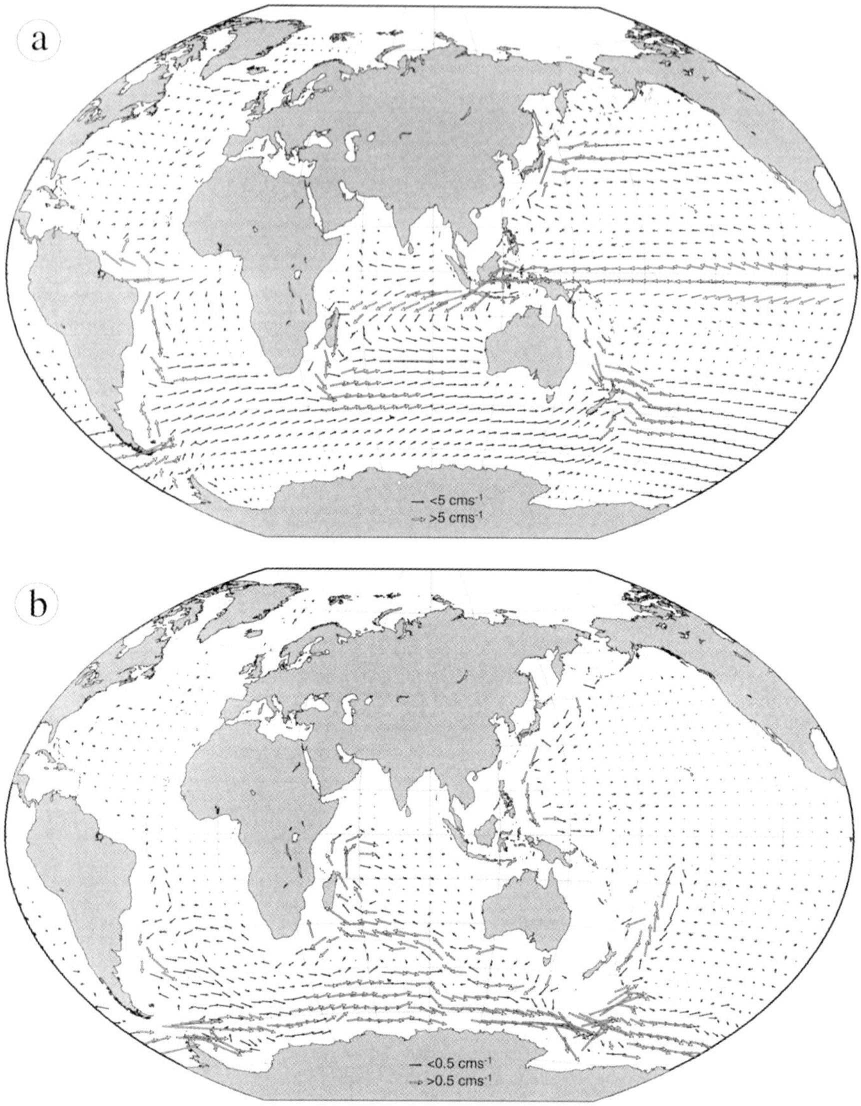

Fig. 8a, b As in Fig. 6 for MWE

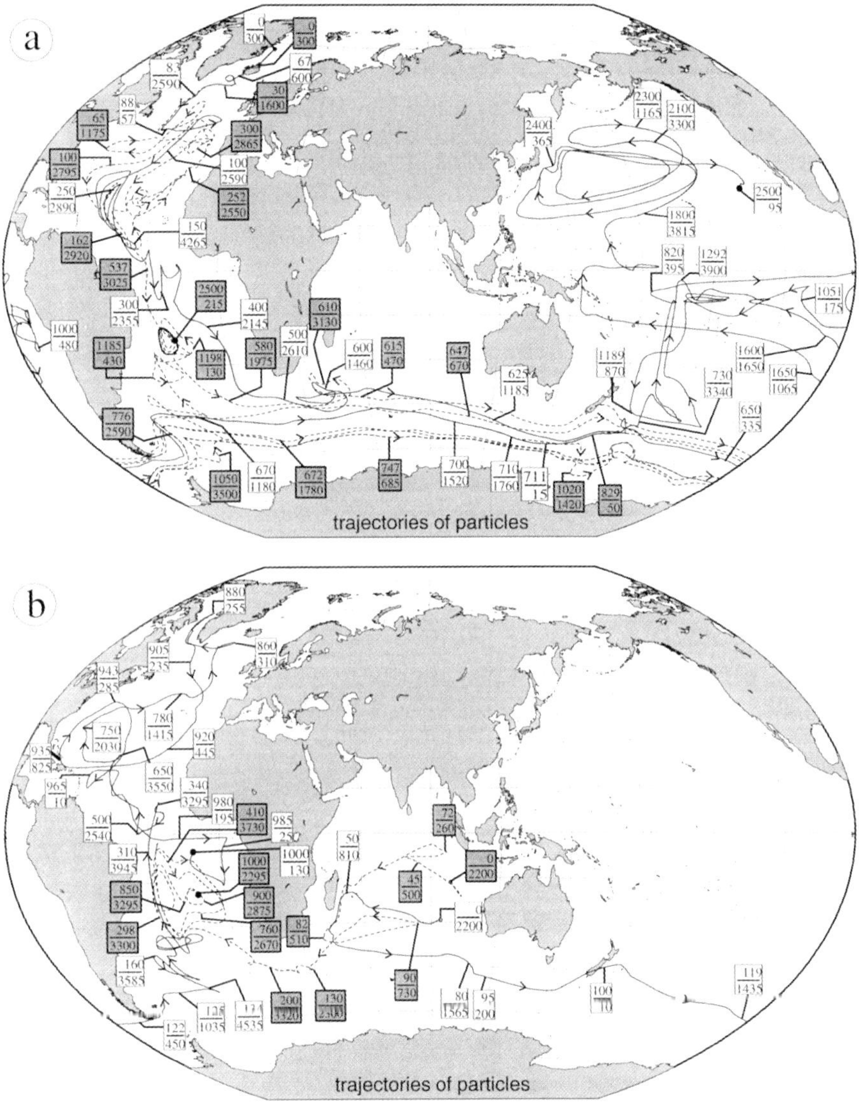

Fig. 9a, b Pairs of trajectories (one trajectory is shown as a *solid* and the other as a *dashed line*) visualizing the deep ocean conveyor
a Modern conveyor
b MWE. *Small rectangles* show time elapsed after deployment (*upper numbers*) and the depth at which the particle is found on the trajectory at that time (see Fig. 4).

indicates that the major change of the conveyor indeed took place at MWE, as the consideration of the convection regime implies. Here we concentrate on tracing the conveyor using the Lagrangian particles. As was mentioned above, these trajectories visualize the true three-dimensional water motion because they take into account both vertical motion and mixing in the convection chimneys. Since the vector maps suggest that the dramatic change of the conveyor took place at MWE rather than at LGM, we present the trajectories for the present-day and MWE time slices only. Figure 9a shows two pairs of trajectories calculated using the modern velocity field and Fig. 9b depicts the trajectories calculated using the MWE velocities. The particles to trace modern and MWE conveyor were deployed in different places. Two different sites were chosen because no deep convection has been found in the northern NA, whereas a site of deep convection exists southwest of Australia at both HM and MWE. However, only at MWE did particles deployed in the latter area progress westward and enter the central Atlantic. First, we briefly overview the OGCM results represented by velocity vectors and then emphasize how the semi-Lagrangian calculation enhances water transport analysis and understanding of the circulation change.

6.2 OGCM Results

The three-dimensional distribution of the horizontal currents is far more complex than Figs. 6–8 display. The deep inflow of AABW into the central and North Atlantic is masked in these figures. However, Figs. 6 and 8 give a clear impression of how different the deep and upper ocean flows are. The present-day deep flows emerge as a truly global feature. This consolidated current system justifies, to a certain extent, the term conveyor. The upper ocean circulation system, however, does not give such an unambiguous impression of a continuous flow system. The intensive subtropical gyres are indeed connected by their marginal extensions to form a system that might have been recognized as connected rings of a chain comprising the upper band of the conveyor. It should be noted that, because of the course resolution, we cannot hope to model the correct operation of the upper band. For example, the Agulhas retroflection, which would extend the leg of the Indian subtropical gyre into the South Atlantic, can only be resolved in an eddy-resolving simulation. Nevertheless, one may say that no coherence comparable to the deep ocean conveyor structure can be found in the upper layers. Moreover, the water traveling in the uppermost levels is strongly modified by short-term air-sea interactions. The time scale of such interactions is about 2 to 3 months, which is an order of magnitude shorter than the time needed for water to travel around the globe within this circulation system. Since the water characteristics would change on that short-term time scale, the question of their origin as the conveyors' water is therefore meaningless.

Because the structure of the glacial conveyor in the NA is not principally different from the modern one, the LGM conveyor is not illustrated here. Yet the glacial conveyor is characterized by a noticeable (here almost 60%) decrease in its intensity and the deep flows, especially in the NA, take different routes. Also, the NA conveyor branch became shallower than at present. However, as show in Seidov et al. (1996), the NA conveyor still existed at the LGM. Since there was a weaker modern conveyor in the cited study, the relative changes are larger here (see above). The present-day overturning rate (here we ended up with 23 Sv) is perhaps an overestimate. We tried to obtain the convection in the northern NA that would penetrate deep enough (deeper than 2 km) by cooling the sea-surface by 2 °C (see above). The course resolution employed in this study led to an overestimate of the convective mixing and to overturning somewhat stronger than in other coarse resolution studies. The mediating estimate of the glacial to present-day overturning ratio is perhaps within 50 to 70%.

The major changes took place at the MWE. Although the impact is tied to the NGS and the northern NA only, the whole deep conveyor to the eastern Australian coast is affected. The NADW production was completely switched off, and there is no deep southward flow in the western Atlantic. Moreover, the reversed deep ocean flow is found in this simulation to the north of a substantially curtailed deep ACC in the Indian and Atlantic sectors of the Southern Ocean. Although our calculations do not unambiguously indicate the reversal of the whole deep conveyor branch from the Atlantic to the northern Pacific, one may easily recognize that the deep flow has indeed an opposite direction over rather a long leg from the eastern Indian ocean to the northern NA. Additionally, a clearly seen deep southward-flowing western boundary current developed in the northern Pacific (Fig. 8b). This flow, which is absent from the present-day deep current system, is a signature of the possibly reversed Pacific branch of the global conveyor.

6.3 Trajectory Tracing

We have already emphasized above that the horizontal velocity vectors, even if inspected at each level, may be misleading because they do not show vertical motion. This is especially true in the areas of convection where water is mixed vertically and particles may be transferred to the deep ocean. Hence, the true 3-D motion would be essentially different from what the vector maps might have suggested. To illuminate the deep ocean ventilation and subsequent water transport by the conveyor, the particles were deployed in the areas where they can be transferred to the deep ocean within convection chimneys to trace the deep ocean conveyor leg.

Some of the particles deployed in the NGS and northern NA penetrated into the deep ocean and traveled southward in the western boundary current (Fig. 9a). Some of these particles pass further into the South Atlantic, although many remained trapped in the NA subtropcal gyre. Some of those which manage to travel to the South Atlantic turn backward. About 10 to 20% of all particles deployed in the shaded area in Fig. 5a reached the ACC and were transported further eastward in this current. Only 2 to 5% of these particles ever emerged in the northern Pacific. Hence, this study indicates that only a tiny fraction of the NADW can physically travel along the whole leg of the deep conveyor. However, it does not mean that the NADW water cannot reach the Pacific in significant amounts. One must consider the duration of the process which can be of several thousand years of a stable modern-like conveyor. Some additional calculations that combine the Lagrangian technique with incorporation of geochemical tracers are required to quantify the total amount of NADW reaching the northern Pacific at the slow rate indicated by our experiments (no more than 1-2 Sv in the model).

In spite of uncertainties in our present-day calculations due to a rather coarse horizontal resolution and a somewhat incomplete glacial and meltwater sea-surface climatology, the MWE curtailment of the conveyor and emergence of the reversed deep conveyor is a very robust feature. In many additional experiments with additional disturbances of MWE sea-surface conditions, a switched-off or even reversed NA conveyor was a permanent feature. All attempts to find a trajectory originating somewhere in the NA and continuing into at least the Indian Ocean have failed. On the contrary, the particles deployed at the surface to the southwest of Australia travel far into the South Atlantic. A tiny fraction of the deployed ensemble was found north of the equator. However, no particles managed to pass into the northern NA. Hence, although there was indeed a strong incursion of the AABW into the NA, it upwelled mostly to the south of 50°N, which agrees well with the results of Seidov et al. (1996), who show isopycnal outcrop to the south of 50°N. Such a southward-shifted density outcrop isolated the northern NA from the rest of the World Ocean. Hence, the main driving mechanism of the deep ocean circulation during MWE was restricted to the Southern Ocean.

7
Discussion and Conclusions

The primary task of our chapter is to demonstrate that the major meltwater events in the NA might have affected the deep ocean branch of the global conveyor very substantially and that these changes can be clearly and unambiguously traced in Lagrangian calculations. The characteristics of ocean circulation, such as the ventilation of the deep ocean, sedimentation transport, water parcel motion, meridional overturning and potential vorticity analysis, all indicate a weakening and some shallowing of the main conveyor at the LGM. These characteristics indicate a complete collapse after the very localized freshwater discharge at the MWE occurred. This may be instrumental for understanding and predicting global climate change on the basis of the top-analogue examples given by the major deglaciations. Our results also demonstrate, not only in an idealized sensitivity simulation, but based on numerous proxy data, that the deep ocean circulation is indeed sensitive to localized high-latitudinal forcing that might be able to destroy water convection in the World Ocean.

The Lagrangian calculations help to elaborate the true three-dimensional water motion and therefore are the only means by which conveyor modes can be genuinely visualized. Moreover, the trajectory-tracing technique may indicate whether specific parts of the global ocean are interconnected via the deep conveyor branches, or essentially isolated. For example, the northern NA during the MWE is characterized by very old nonventilated intermediate-to-deep water (Sarnthein et al. 1995). Our calculations may shed some additional light on this problem. If the MWE water in the NA was a mixture of the AABW, originating in the Weddell Sea and some water sinking in the eastern Indian ocean, it may explain extreme aging of the Atlantic water, stronger than it would be if only the AABW ventilated the NA areas.

A combined circulation/sedimentation/particle tracing modeling approach was employed to understand particular aspects of the glacial-interglacial change of the North Atlantic and World Ocean circulation which are difficult to address using single-component models. Primary among those are the ventilation regimes, including the characterization of vertical mixing in convective chimneys and the advective ages of ventilated water. Circulation studies might focus on the NADW outflow or deep ocean circulation routing. Most of these questions are traditionally addressed using geochemical tracers such as $\delta^{18}O$, $\delta^{13}C$, and $\Delta^{14}C$, which are extremely useful in ocean circulation studies and hence widely employed. However, the tracers alone cannot provide sufficient constraints over a simulated past circulation, as has been shown recently by LeGrand and Wunsch (1995). In their study they showed that there exist an infinite number of states that would satisfy a tracer distribution aimed at constraining the circulation, at least for the currently available proxy data sets.

Another problem arising in paleoceanographic investigations is the parallel analysis of both surface and benthic habitats and/or sediment transport features. Commonly, assumed features of a water flow thought to be suitable for explaining a particular distribution of proxy data are largely based on speculation. There is no guarantee that this hypothetical flow would satisfy hydrodynamic restrictions posed over the ocean by the wind stress, ocean geometry, bottom morphology, and sea-surface heat and fresh waterfluxes.

Based on only partly known SSS, recently corrected SST, and simulated glacial wind stress, our results largely conform to current ideas about ventilation and overturn in the North Atlantic at the height of the last glaciation and during the subsequent meltwater event near 13 500 ^{14}C years B.P. The sediment transport model output contains features that agree well with interpretation of sediment data in the northern North Atlantic. In Seidov and Haupt (1997) we discussed the LGM sedimentation pattern resulting from our simulation in comparison with the observed or inferred data of Cremer et al. (1993) and McCave and Tucholke (1986). It was pointed out that our simulations are in good qualitative agreement with their findings. Moreover, in our experiments, the sediment load in the vicinity of the North American coast is lower during the LGM than today, in agreement with the interpretation given by Boyle and Keigwin (1987) on the basis of nutrient concentration analysis. The Caribbean water at depths just above 2 km was nutrient-depleted during the glacial period. One interpretation is that there was a lower proportion of northern component water, which, in turn, means a weaker deep western boundary current and a stronger southern source water incursion.

In the eastern part of the NA, our sediment model agrees with data that indicate increased flow along the eastern flank of the Mid-Atlantic Ridge. Dowling and McCave (1993) and Robinson and McCave (1994) provide evidence that the Feni Drift was substantially enhanced at the last glacial maximum. Generally, the glacial sediment record indicates an enhanced Holocene bottom current driven by the Iceland-Greenland overflow characteristic of modern NADW production in the Norwegian-Greenland Seas.

Our trajectory-tracing model reveals very different fates of the water-volumes in the HM and LGM cases, once exposed to the air-sea interaction and then mixed downward in convective chimneys. NADW production in the center of the glacial North Atlantic is strongly supported by the proxy data (Duplessy et al. 1988 1991; Sarnthein et al. 1994 1995). In general, the most recent studies indicate enhanced upper NADW production and decreased lower NADW production (e.g., Oppo et al. 1995). However, there exists evidence that, despite decreased lower NADW production, the glacial conveyor was not, in total, weaker than today (Yu et al. 1996). This may appear contradictory to the fact that the glacial northward heat transport was substantially weaker at the LGM. Our trajectory analysis explains this seemingly controversial result. It shows that although the forward conveyor responsible for meridional heat transport became weaker, the abyssal reversed branch of the conveyor strength increased proportionally to allow the same rate of the southward transport of the tracers in the deep ocean found by Yu et al. (1996). Mediating the results of Seidov et al. (1996), Seidov and Haupt (1997) and the global conveyor simulations presented above, one may give the ratio of LGM to HM conveyor intensity as within 50–70% of modern magnitude.

Duplessy et al. (1988) argue that during glacial times most of the deep eastern North Atlantic was filled with southern source water which probably penetrated up to 45°N. Our experiments, deploying particles in the surface layers in the subpolar gyre and in the deep and abyssal areas in the subtropical gyre, are in complete agreement with this conclusion. Moreover, Michel et al. (1995) point to a steep gradient in $\delta^{13}C$ distribution northward of 30°N in the North Atlantic delineating the border between southern and northern source water at the LGM. We add that this border probably shifted farther to the north in the eastern part of the basin during that time period.

Moreover, we point out that the trajectory analysis is in agreement with the notion of a far better ventilated and deeper glacial thermocline in the subtropical North Atlantic (Slowey and Curry 1995). These authors provide indications that the glacial thermocline was shallower than today with its base raised by about 100 m. However, the water inside the glacial thermocline was up to 4 °C cooler than today. Our Fig. 3 confirms that the glacial thermocline was better ventilated and that there was an enhanced production of subtropical mode water. Deeper glacial ventilation, evident in Fig. 3b, implies that the base of thermocline had to rise upward because water pumped into the thermocline was more than 4 °C colder at the glacial ventilation points (westward boundary outflow region; compare Fig. 3a and b). To quantify this effect, the difference between today's and the LGM simulated temperature is shown in a meridional section of the temperature anomaly field in Fig. 8 in Seidov and Haupt (1997). In essence, Fig. 8 displays a penetration of cold anomalies deep into the thermocline. It is clear, however, that this penetration is limited to the upper 1 km and that the lower 500 m of this layer is more strongly cooled than the upper 500 m. This implies that the thermocline base has been raised. A comparison of the temperature profiles in the subtropical gyres indicates that on average the thermocline depth was about 200 m shallower than today, which differs from the estimates of Slowey and Curry (1995). There may be several reasons for this discrepancy. Firstly, the subtropical gyre is dangerously close to the sponge layer at the southern wall, which might have distorted the behavior of the thermocline base. Second, we use the annual mean surface forcing and, therefore, we are unable to simulate an extreme winter southward migration of the density outcrop intersecting the Ekman pumping within the subtropical gyre. Nevertheless, the general tendencies sketched in Fig. 1 of Slowey and Curry (1995, p. 717) are evident in our trajectory maps (Fig. 3b as compared to Fig. 3a). The particles dive deeper into the thermocline and stay there longer. The northern limit of the subtropical gyre, marked by the shallow convection along the North Atlantic Drift, is shifted noticeably southward.

As some findings have indicated (e.g., Sarnthein et al. 1995), the Nordic Seas were more isolated from the northern North Atlantic at the LGM than today. This is strongly supported by our Lagrangian calculations. As Fig. 7 displays, the water from the subtropics probably had a very limited chance to enter the Nordic Seas, largely because the outflow over the Iceland-Greenland Ridge was reduced during the LGM.

Based on a threefold numerical simulation of the North Atlantic circulation and sedimentation and twofold simulation of the World Ocean water transport we draw the following conclusions:

1. present-day and LGM meridional thermohaline circulation is characterized by a forward global deep ocean conveyor. At MWE the freshwater fluxes that maintain the "observed" high latitudinal sea-surface salinity appear to be sufficient to suppress this forward conveyor. A reversed deep ocean conveyor replaced the forward one on the whole distance from NA to the southeast Indian Ocean.
2. The convection regime is the most crucial process for the conveyor dynamics. At MWE the bimodal convection regime, with both NADW and AADW sources, was replaced by the regime with only southern deep water source driving the conveyor. As a result, a reversed conveyor emerged in the Indian-Atlantic sector of the Southern Ocean. The meltwater North Atlantic was essentially isolated from the other parts of the World Ocean.
3. Sedimentation in the North Atlantic is non-linearly coupled to the circulation modes associated with different surface climatology. The LGM and MWE sed-

imentation patterns differ fundamentally from the Holocene/Modern pattern in a way that cannot be expected a priori. The major differences indicated by sediment transport and trajectory-tracing models are summarized in following conclusions.

4. The results of sediment transport model suggest that there was almost no sediment deposition in the Gulf Stream area, and a lower rate than at present in the Caribbean during MWE.
5. The trajectory-tracing model indicates far stronger ventilation of the thermocline in the glacial low latitudes than is observed today. The trajectory-tracing model indicates complete absence of the deep ocean ventilation at the MWE. The model reveals disappearance of the deep western boundary current and westward contraction of the subtropical warm pool. This contraction leads to collapse of the eastern continuation of the North Atlantic drift and contributes to formation of the upper ocean reverse conveyor.
6. Concerning the use of numerical models as a principal source of sediment transport and Lagrangian calculations, it should be stressed that at high latitudes, the sediment transport and water-volume tracing models can only operate with the simulated rather than observed circulation patterns. The water transport calculations depend critically on knowledge about where the convection occurs and how deep it ventilates.

Acknowledgements

This study could not have been carried out without the efforts invested by the Kiel group into paleoreconstructions of the glacial-interglacial North Atlantic. We are especially grateful to Michael Sarnthein for inspiring us to study the meltwater oceanography using the reconstructions provided by the Kiel group and Karl Stattegger for the support and helpful comments. We deeply appreciate Avan Antia's help on correcting our English. The study was accomplished within the framework of the German Climate Program and supported by the Deutsche Forschungsgemeinschaft (DFG) and SFB313 of Kiel University.

References

Bard, E., F. Rostek, and C. Sonzogni, Interhemispheric synchrony of the last deglaciation inferred from alkenone palaeothermometry, Nature, 385, 707–710 1997.

Beck, J. W., J. Récy, F. Taylor, R. L. Edwards, and G. Cabioch, Abrrupt changes in early Holocene tropical sea-surface temperature derived from coral records, Nature, 385, 705–707 1997.

Beveridge, N. A. S., H. Elderfield, and N. J. Shackleton, Deep thermohaline circulation in the low-latitude Atlantic during the last glacial, Paleoceanography, 10, 643–660 1995.

Bitzer, K., and R. Pflug, DEPOD: A three-dimensional model for simulating clastic sedimentation and isostatic compensation in sedimentary basin, in Quantitative Dynamics Stratigraphy, edited by T. A. Cross, pp. 335–348, Prentice Hall, New York 1990.

Bohrmann, G., R. Henrich, and J. Thiede, Miocene to Quaternary paleoceanography in the northern North Atlantic: Variability in carbonate and biogenic opal accumulation, in Ge-

ological History of the Polar Oceans: Arctic Versus Antarctic, edited by U. Bleil and J. Thiede, pp. 647–675, Kluwer Acad., Norwell, Mass. 1990.
Bond, G. C., Climate and conveyor, Nature, 377, 383–384 1995.
Bond, G.et al., Evidence for massive discharges of icebergs into the North Atlantic Ocean during the last glacial period, Nature, 360, 245–249 1992.
Böning, C. W., and M. D. Cox, Particle dispersion and mixing of conservative properties in an eddy-resolving model, J. Phys. Oceanogr., 18, 320–338 1988.
Böning, C. W., R. Döscher, and R. G. Budich, Seasonal transport in the western North Atlantic: Experiments with an eddy-resolving model, J. Phys. Oceanogr., 21, 1271–1289 1991.
Boyle, E., Deep water distillation, Nature, 379, 679–680 1996.
Boyle, E. A., and L. D. Keigwin, North Atlantic thermohaline circulation during the past 20,000 years linked to high-latitude surface temperature, Nature, 330, 35–40 1987.
Boyle, E., and A. Weaver, Conveying past climates, Nature, 372, 41–42 1994.
Broecker, W., The great ocean conveyor, Oceanography, 4, 79-89 1991.
Broecker, W. S., and G. H. Denton, The role of ocean atmosphere reorganizations in glacial cycles, Geochim. Cosmochim. Acta, 53, 2465–2501 1989.
Bryan, F., High-latitude salinity effects and interhemispheric thermohaline circulations, Science, 323, 301–304 1986.
Bryan, F., Parameter sensitivity of primitive equation ocean general circulation models, J. Phys. Oceanogr., 17, 970–985 1987.
Bryan, F., and W. Holland, A high resolution simulation of the wind- and themohaline-driven circulation in the North Atlantic Ocean, in Parameterization of Small-Scale Processes, edited by P. Müller and D. Henderson, pp. 99-115, Hawaii Inst. of Geophys., Honolulu 1989.
Bryan, K., A numerical method for the study of the circulation of the world ocean, J. Comput. Phys., 4, 347–376 1969.
Climate: Long-Range Investigation Mapping and Prediction (CLIMAP) Project Members, Seasonal reconstructions of the Earth's surface at the last glacial maximum, Map and Chart Ser. MC-36, pp. 1–18, Geol. Soc. of Am., Boulder, Colo. 1981.
Colin de Verdière, A., Buoyancy driven planetary flows, J. Mar. Res., 46, 215–265 1988.
Cox, M. D., A primitive equation, 3-dimensional model of the ocean, Tech. Rep. No. 1, 250 pp., Ocean Group, Geophys. Fluid Dyn. Lab., Princeton, Univ., Princeton, N.J. 1984.
Cox, M., An idealized model of the world ocean, I, The global-scale water masses, J. Phys. Oceanogr. 19, 1730–1752 1989.
Cox, M., and K. Bryan, A numerical model of the ventilated thermocline, J. Phys. Oceanogr., 14, 674–687 1984.
Cremer, M., J.-C. Faugeres, F. E. Grousset, and E. Gonthier, Late Quaternary sediment flux on sedimentary drifts in the northeast Atlantic, Sediment. Geol., 82, 89–101 1993.
Dowling, L. M., and I. N. McCave, Sedimentation on the Feni drift and late Glacial bottom water production in the northern Rockall Trough, Sediment. Geol. 1993, 79-87 1993.
Drijfhout, S. S., E. Maier-Reimer, and U. Mikolajewicz, Tracing the conveyor belt in the Hamburg large-scale geostrophic ocean general circulation model, J. Geophys. Res., 101, 22,563–22,575 1996.
Duplessy, J.-C., and N. J. Shackleton, Response of global deep-water to Earth's climate change 135,000-107,000 years ago, Nature, 316, 500–507 1985.
Duplessy, J.-C., N. J. Shackleton, R. G. Fairbanks, L. Labeyrie, D. Oppo, and N. Kallel, Deepwater source variations during the last climatic cycle and their impact on the global deepwater circulation, Paleoceanography, 3, 343–360 1988.
Duplessy, J.-C., L. Labeyrie, A. Julliet-Lerclerc, J. Duprat, and M. Sarnthein, Surface salinity reconstruction of the North Atlantic Ocean during the last glacial maximum, Oceanol. Acta, 14, 311–324 1991.
Einsele, G., Sedimentary Basins; Evolution, Facies, and Sediment Budget, pp. 628, Springer-Verlag, New York 1992.

England, M. H., Representing global-scale water masses in ocean general circulation models, J. Phys. Oceanogr., 23, 1523–1552 1993.
Fichefet, T., and S. Hovine, The glacial ocean: A study with a zonally averaged, three-basin ocean circulation model, in Ice in Climate System, NATO ASI Ser., Ser. I, 12, edited by W. R. Peltier, pp. 433–458, Springer-Verlag, New York 1993.
Fichefet, T., S. Hovine, and J.-C. Duplessy, A model study of the Atlantic thermohaline circulation during the last glacial maximum, Nature, 372, 252–255 1994.
Goldschmidt, P., Accumulation rates of coarse-grained terrigenous sediment in the Norwegian-Greenland Sea: Signals of continental glaciation, Mar. Geol., 128, 137–151 1995.
Goldschmidt, P. M., S. Pfirmann, I. Wollenburg, and R. Henrich, Origin of sediment pellets from the Arctic seafloor. Sea ice or icebergs?, Deep Sea Res., 372, 252–255 1992.
Gordon, A., Interocean exchange of thermocline water, J. Geophys. Res., 91, 5037-5046 1986.
Gordon, A. L., S. E. Zebiak, and K. Bryan, Climate variability and the Atlantic Ocean, Eos Trans. AGU, 73, 161, 164–165 1992.
Guilderson, T. P., R. G. Fairbanks, and Rubenstone, J. L., Tropical Temperature Variations Since 20,000 Years Ago: Modulating Interhemispheric Climate Change, Science, 263, 663–664 1994.
Hasselmann, K., An ocean model for climate variability studies, Prog. in Oceanogr., 11, 69–92 1982.
Haupt, B. J., Numerische Modellierung der Sedimentation im nördlichen Nordatlantik, Ber. 54, pp. 1–129, Sonderforschungsbereich 313, Univ. Kiel, Kiel, Germany 1995.
Haupt, B. J., C. Schäfer-Neth, and K. Stattegger, Modeling sediment drifts: A coupled oceanic circulation-sedimentation model of the northern North Atlantic, Paleoceanography, 9, 897–916 1994.
Haupt, B. J., C. Schäfer-Neth, and K. Stattegger, Three-dimensional numerical modeling of late Quaternary paleoceanography and sedimentation in the northern North Atlantic, Geol. Rundsch., 84, 137–150 1995.
Honjo, S., Particle fluxes and modern sedimentation in the polar oceans, in Polar Oceanography, Part B, edited by W. O. Smith, pp. 687-739, Academic, San Diego, Calif. 1990.
Hsü, K. J., Physical Principles of Sedimentology, pp. 231, Springer-Verlag, New York 1989.
Keigwin, L. D., G. Jones, and S. J. Lehman, Deglacial meltwater discharge, North Atlantic deep circulation, and abrupt climate change, J. Geophys. Res., 96, 16811–16826 1991.
Kellogg, T. B., Paleoclimatology and paleo-oceanography of the Norwegian and Greenland Seas; Glacial-interglacial contrasts, Boreas, 9, 115-137 1980.
Killworth, P. D., Deep convection in the world ocean, Rev. Geophys., 21, 1–26 1983.
Kroopnick, P. M., The distribution of 13C and ΣCO2 in the world oceans, Deep Sea Res., 32, 57–84 1985.
Lautenschlager, M., and K. Herterich, Atmospheric response to ice age conditions – Climatology near the Earth's surface, J. Geophys. Res., 95, 22,547–22,557 1990.
LeGrand, P., and K. Wunsch, Constraints from paleotracer data on the North Atlantic circulation during the last glacial maximum, Paleoceanography, 10, 1011–1045 1995.
Lehman, S. J., and L. D. Keigwin, Sudden changes in North Atlantic circulation during the last deglaciation, Nature, 356, 757–762 1992.
Levitus, S., Climatological atlas of the world ocean, NOAA Prof. Pap., 13, 173 pp., U.S. Govt. Print. Off., Washington, D.C. 1982.
Levitus, S., and T. P. Boyer, World Ocean Atlas 1994, vol. 4, (Temperature; 117 pp.), NOAA Natl. Environ. Satell. Data and Inf. Ser., Washington, D.C. 1994.
Levitus, S., R. Burgett, and T. P. Boyer, World Ocean Atlas 1994, vol.3, (Salinity; 99 pp.) NOAA Natl. Environ. Satell. Data and Inf. Ser., Washington, D.C. 1994.
Lorenz, S., B. Grieger, P. Helbig, and K. Herterich, Investigating the sensivity of the atmospheric general circulation Model ECHAM 3 to paleoclimate boundary conditions, Geol. Rundsch., 85, 513–524 1996.

Luyten, J. R., J. Pedlosky, and H. Stommel, The ventilated thermocline, J. Phys. Oceanogr., 13, 292-309 1983.
Maier-Reimer, E., and U. Mikolajewicz, Experiments with an OGCM on the cause of the Younger Dryas, Rep. 39, 13 pp., Max-Plank-Inst. für Meteorol., Hamburg, Germany 1989.
Maier-Reimer, E., U. Mikolajewicz, and K. Hasselmann, On the sensitivity of the global ocean circulation to changes in the surface heat flux forcing , Rep. 68, 67 pp., Max-Plank-Inst. für Meteorol., Hamburg, Germany 1991.
Maier-Reimer, E., U. Mikolajewicz, and K. Hasselmann, Mean circulation of the Hamburg LSG OGCM and its sensitivity to the thermohaline surface forcing, J. Phys. Oceanogr., 23, 731–757 1993.
Manabe, S., and R. J. Stouffer, Two stable equilibria of a coupled ocean-atmospher model, J. Clim., 1, 841-866 1988.
Manabe, S., and R. J. Stouffer, Simulation of abrupt change induced by freshwater input to the North Atlantic Ocean, Nature, 378, 165–167 1995.
Marotzke, J., and J. Willebrand, Multiple equilibria of the global thermohaline circulation, J. Phys. Oceanogr., 21, 1372–1385 1991.
Maslin, M. A., N. J. Shackleton, and U. Pflaumann, Surface water temperature, salinity, and density changes in the northeast Atlantic during the last 45,000 years: Heinrich events, deep water formation, and climatic rebounds, Paleoceanography, 10, 527–544 1995.
McCartney, M. S., Recirculating components to the deep boundary current of the nothern North Atlantic, Prog. in Oceanogr., 29, 283–383 1992.
McCave, I. N., and B. E. Tucholke, Deep current controlled seimentation in the western North Atlantic, in The Geology of North America, vol. M. The Western North Atlantic Region, edited by P. R. Vogt and B. E. Tucholke, pp. 451–468, Geol. Soc. of Am., Boulder, Colo. 1986.
Michel, E., L. D. Labeyrie, J.-C. Duplessy, N. Gorfti, M. Labracherie, and J.-L. Turon, Could deep Subantarctic convection feed the world deep basins during the last glacial maximum?, Paleoceanography, 10, 927–942 1995.
Mikolajewicz, U., A meltwater induced collapse of the 'conveyor belt' themohaline circulation and its influence on the distribution of Δ14C and δ18O in the oceans, Rep. 189, 25 pp., Max-Plank-Inst. für Meteorologie, Hamburg, Germany 1996.
Miller, M. C., I. N. McCave, and P. D. Komar, Threshold of sedimentation under unidirectional currents, Sedimentol., 24, 507–528 1977.
Oppo, D. W., and S. J. Lehman, Mid-depth circulation of the subpolar north Atlantic during the last glacial maximum, Science, 259, 1148–1152 1993.
Oppo, D. W., M. E. Raymo, G. P. Lohmann, A. C. Mix, J. D. Wright, and W. L. Prell, A δ13C record of Upper North Atlantic Deep Water during the past 2.6 million years, Paleoceanogr., 10, 373–394 1995.
Pacanowski, R., K. Dixon, and A. Rosati, The GFDL modular ocean users guide, Tech. Rep. 2, Ocean Group, Geophys. Fluid Dyn. Lab., Princeton Univ., Princeton, N. Y. 1993.
Pfirmann, S., M. A. Lange, I. Wollenburg, and P. Schlosser, Sea ice characteristics and the role of sediment inclusions in deep-sea deposition: Arctic – Antarctic comparisons, in Geological History of the Polar Oceans: Arctic Versus Antarctic, edited by U. Bleil and J. Thiede, pp. 187–211, Kluwer Acad., Norwell, Mass. 1990.
Puls, W., Numerical simulation of bedform mechanics. Mitteilungen des Inst. für Meeresk. Univ. Hamburg, Hamburg, Germany, 147 pp. 1981.
Rahmstorf, S., Rapid climate transitions in a coupled ocean-atmosphere model, Nature, 372, 82–85 1994.
Rahmstorf, S., Bifurcations of the Atlantic thermohaline circulation in response to changes in the hydrological cycle, Nature, 378, 145–149 1995.
Robinson, S. G., and I. N. McCave, Orbital forcing of bottom-current enhanced sedimentation on Feni Drift, NE Atlantic, during the mid-Pleistocene, Paleoceanography, 9, 943–972 1994.

Ruddiman, W. F., and A. McIntyre, The mode and mechanism of the last deglaciation: Oceanic evidence., Quat. Res., 16, 125–134 1981.
Sakai, K., and W. R. Peltier, A simple model of the Atlantic thermohaline circulation: Internal and forced variability with paleoclimatological implications, J. Geophys. Res., 100, 13,455–13,479 1995.
Sarmiento, J. L., On the north and tropical Atlantic heat balance, J. Geophys. Res., 91, 11,677-11,689 1986.
Sarnthein, M., K. Winn, S. J. A. Jung, J.-C. Duplessy, L. Labeyrie, H. Erlenkeuser, and G. Ganssen, Changes in east Atlantic deepwater circulation over the last 30,000 years – eight time slice reconstructions, Paleoceanography, 9, 209–267 1994.
Sarnthein, M. et al., Variations in Atlantic Ocean paleoceanography, 50°-80°N: A time slice record of the last 30,000 years, Paleoceanography, 10, 1063–1094 1995.
Schmitz, W. J., Jr., On the interbasin-scale thermohaline circulation, Rev. Geophys., 33, 151–173 1995.
Schulz, H., Meeresoberflächentemperaturen im frühen Holozän 10,000 Jahre vor heute, Ph.D. dissertation, Univ. Kiel, Kiel, Germany 1994.
Seibold, E., and W. H. Berger, The Sea Floor; An Introduction to Marine Geology, 2nd ed., Springer-Verlag, New York 1993.
Seidov, D. G., Numerical modeling of the ocean circulation and paleocirculation, Mesozoic and Cenozoic Oceans, Geodyn. Ser., vol.15, edited by K. J. Hsu, pp. 11–26, AGU, Washington, D. C. 1986.
Seidov, D., An intermediate model for large-scale ocean circulation studies, Dyn. Atmos. Oceans, 25/1, 25–55 1996.
Seidov, D., and R. Prien, A coarse resolution North Atlantic ocean circulation model: An intercomparison study with a paleoceanographic example, Ann. Geophys., 14, 246–257 1996.
Seidov, D., and B. J. Haupt, Simulated ocean circulation and sediment transport in the North Atlantic during the last glacial maximum and today, Paleoceanography, 12, No. 2, 281–305 1997.
Seidov, D., M. Sarnthein, K. Stattegger, R. Prien, and M. Weinelt, North Atlantic ocean circulation during the last glacial maximum and subsequent meltwater event: A numerical model, J. Geophys. Res., 101, 16,305–16,332 1996.
Semtner, A. J., Finite difference formulation of a world ocean model, in Advanced Physical Oceanographic Modelling, edited by J. O'Brien, pp. 187-202, D. Reidel, Norwell, Mass. 1986.
Send, U., and J. Marshall, Integral effects of deep convection, J. Phys. Oceanogr., 25, 855–872 1995.
Shanks, A. L., and J. D. Trent, Marine snow: Sinking rates and potential role in vertical flux, Deep Sea Res., Part A, 27, 137–143 1980.
Shapiro, R., The use of linear filtering as a parameterization of atmospheric diffusion, J. Atmos. Sci., 28, 523–531 1971.
Slowey, N. C., and W. B. Curry, Enhanced ventilation of the North Atlantic subtropical gyre thermocline during the last glaciation, Nature, 358, 665–668 1992.
Slowey, N. C., and W. B. Curry, Glacial-interglacial differences in circulation and carbon cycling within the upper western North Atlantic, Paleoceanography, 10, 715–732 1995.
Stocker, T. F., The variable ocean, Nature, 367, 221–222 1994.
Stommel, H., and A. B. Arons, On the abyssal circulation of the world ocean, II, An idealized model of the circulation pattern and amplitude in the oceanic basins, Deep Sea Res., 6, 217–233 1960.
Stull, R. B., Transilient turbulence theory, I, The concept of eddy-mixing across finite distances, J. Atmos. Sci., 41, 3351–3367 1984.
Sündermann, J., North Sea Dynamics, Dynamics, edited by J. Sünderman and W. Lenz , 693 pp., Springer-Verlag, New York 1983.

Sündermann, J., and R. Klöcker, Sediment transport modeling with applications to the North Sea, in North Sea Dynamics, pp. 453–471, Springer-Verlag, New York 1983.
Syvitski, J. P. M., and T. M. C. Hughes, Delta 2: Delta progadation and basin filling, Comput. Geosci., 18, 839–897 1992.
Tetzlaff, D. N., and J. W. Harbaugh, Simulating Clastic Sedimentation, Van Nostrand Reinhold, New York 1989.
Toggweiler, J. R., K. Dixon, and K. Bryan, Simulations of radiocarbon in a coarse-resolution world ocean circulation model, 1, Steady state prebomb distribution, J. Geophys. Res., 94, 8217–8242 1989.
Weaver, A. J., and T. M. C. Hughes, Rapid interglacial climate fluctuations driven by North Atlantic ocean circulation, Nature, 367, 447–450 1994.
Weaver, A. J., and E. S. Sarachik, The role of mixed boundary conditions in numerical models of the ocean's climate, J. Phys. Oceanogr., 21, 1470–1492 1991.
Weaver, A. J., J. Marotzke, P. F. Cummins, and E. S. Sarachik, Stability and variability of the thermohaline circulation, J.. Phys. Oceanogr., 23, 39–60 1993.
Webb, R. S., D. H. Rind, Scott J. Lehman, R. J. Healy, and D. Sigman, Influence of ocean heat transport on the climate of the Last Glacial Maximum, Nature, 385, 695–699.
Weinelt, M., Veränderungen der Oberflächenzirkulation im Europäischen Nordmeer während der letzten 60.000 Jahre - Hinweise aus stabilen Isotopen, Ber. 41, pp. 1–106, Sonderforschungsbereich 313, Univ. Kiel, Kiel, Germany 1993.
Wold, C. N., Paleobathymetry and sediment accumulation in the northern North Atlantic and southern Greenland-Iceland-Norwegian Sea, Ph.D. thesis, Univ. of Kiel, Kiel, Germany 1992.
Wright, D., and T. F. Stocker, A zonally averaged ocean model for the thermohaline circulation, I, Model development and flow dynamics, J. Phys. Oceanogr., 21, 1713–1724 1991.
Yu, E.-F., R. Francois, and M. P. Bacon, Similar rates of modern and last-glacial ocean thermohaline circulation inferred from radiochemical data, Nature, 379, 689–694 1996.
Zanke, U., Zusammenhänge zwischen Strömung und Sedimenttransport, Teil 1, Berechnung des Sedimenttransports – allgemeiner Fall –, Mitt. Franzius Inst. Wasserbau Küsteningeneuerswesen Tech. Univ. Hannover, 47, 214–345 1978.
Zhang, S., C. Lin, and R. J. Greatbatch, A thermocline model for ocean-climate studies, J. Mar. Res., 50, 99–124 1992.

SEDLOB and PATLOB: Two Numerical Tools for Modeling Climatically-Forced Sediment and Water Volume Transport in Large Ocean Basins

B. J. Haupt, K. Stattegger and D. Seidov

1 Introduction

The processes of sediment erosion, transport and deposition in large ocean basins depend strongly on sediment input from various sources and on oceanic circulation patterns. Most sedimentation models developed during the last decade are designed for small basins and specific sediment input simulating alluvial or deltaic basin fill (cf. Bitzer and Pflug 1989; Cao and Lerche 1994; Paola et al. 1992; Slingerland et al. 1994; Syvitzki and Daughney 1992; Tetzlaff and Harbaugh 1989). In order to simulate large basins we need to couple an ocean general circulation model (OGCM) with an sedimentation model.

Oceanic thermohaline circulation is controlled mainly by the morphology of a basin and by climate. Given a specific steady-state oceanic circulation pattern from an OGCM with its temperature, salinity, velocity fields and convection depths, one can add sediment characterized by its physical properties to the circulating water volumes. The proper representation of important topographic features depends on the spatial resolution of the model basin.

With respect to sediments, a numerical model should allow simulation of:

1. sediment distribution patterns on the sea floor, especially accumulation and erosion of sediments integrated over time intervals long enough to represent the stratigraphic architecture; and
2. transport paths of water volumes and defined sediment particles from prescribed sources.

Two numerical models, SEDLOB (SEDimentation in Large Ocean Basins) and PATLOB (PArticle Tracing in Large Ocean Basins) were developed for this purpose. Especially PATLOB is a useful tool to address both sedimentation and deep ocean ventilation problems. In this chapter, the structure and most important algorithms of these models are described and applied to the modern North Atlantic. Earlier versions of the models can be found in Haupt (1995) and Haupt et al. (1994, 1995).

2
The Sediment Transport Model SEDLOB

SEDLOB mainly consists of two coupled submodels which are linked with each other (Fig. 1). The first submodel includes sediment transport in a 3–D water column (Bitzer and Pflug 1990; McCave and Gross 1991; Zanke 1977b), and the second two-dimensional (2-D) submodel simulates the near bottom processes in a 1-cm-thick layer (Puls 1981; Sündermann and Klöcker 1983; Zanke 1978). This 2-D boundary layer is always parallel to the bottom and continuously follows the dynamic changes of the topography. Although the model is used for the deep ocean, the 1-cm thickness is chosen for the bottom layer on the basis of experimental studies of flows in channels. These studies provide detailed information concerning the interdependence between temperature, salinity, viscosity, and velocity. Additionally, the bed and suspension transport are well documented in a set of equations for this 1-cm-thick layer.

2.1
Three-Dimensional Submodel of SEDLOB

The upper 3-D submodel of SEDLOB (Fig. 2) simulates the lateral inflow and outflow of particles from coastal sources as well as the inputs of eolian dust (worldwide approximately 60–360 million tons/year), and melting icebergs (worldwide approximately 100 Mt/year including glaciers) (Allen 1985; Goldschmidt et al. 1992; Möller 1986; Pickard and Emery 1988). Moreover, several biological processes such as dying plankton and fecal pellet production can act as sediment sources.

2.2
Two-Dimensional Submodel of SEDLOB

The 2-D submodel of SEDLOB simulates the exchange of sediment between the water column and the ocean bottom. Within this layer, which follows the bottom

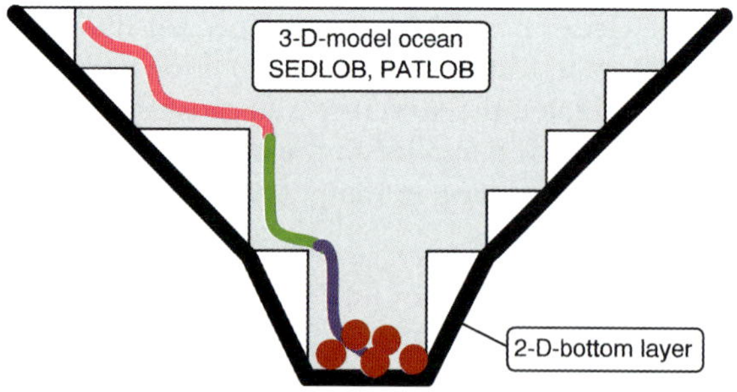

Fig. 1 Coupling of the two submodels

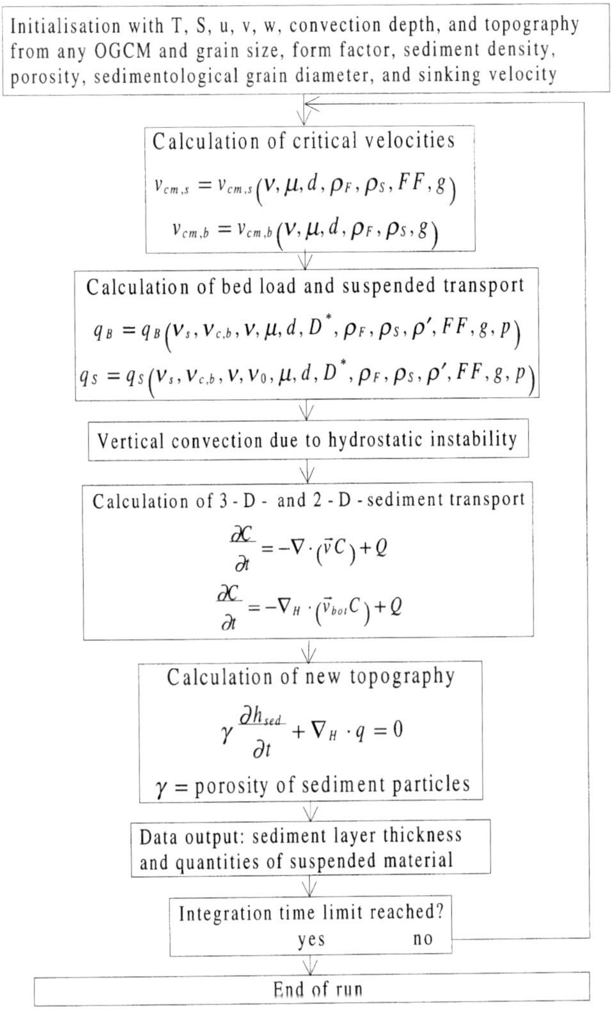

Fig. 2 Flow chart of SEDLOB

topography, erosion, transport (sliding, rolling, and skipping), and deposition of sediment is calculated based on the critical shear velocities of the bed load and the suspension load, the bottom slope, and bottom roughness (bottom friction) (Anderson and Humphrey 1989; Bogárdi 1974; Garde and Ranga Raju 1977; Hsü 1989; Puls 1981; Sündermann and Klöcker 1983; Zanke 1976; 1977a; 1978; 1982). The change in the bottom topography is calculated from the changing sediment content in the 1-cm-thick bottom layer (Krohn 1975; Sündermann and Klöcker 1983).

2.3 Modeled processes depending on scales

Since our knowledge about many geological and biological processes is very limited, we assume a medium grain size homogenous mixture of sediment (Bitzer and Pflug 1989; Sündermann and Klöcker 1983) and that suspended particles in the water column are transported by currents (Bitzer and Pflug 1989). In order to calculate the vertical transport, one should take into account not only the vertical velocity *w*, but also the settling velocity w_s of suspended material. In addition to the vertical transport by upwelling/sinking water, the settling velocity relative to the water motion is superimposed on the water motion to obtain the true particle transport. The settling velocity depends on grain size, density and kinematic viscosity from the surrounding water as well as particle density, form factor and sedimentological grain diameter, and gravitational acceleration (Gibbs et al. 1971; Gibbs 1985; McCave and Gross 1991; Zanke 1977b). It must be emphasized that the vertical velocities are spatially variable, and are not preset fixed values, as in many other models. This is significant, considering that transport and deposition mainly depend on the settling velocity, w_s, which normally exceeds the vertical velocity of the surrounding water (Fig. 3; McCave 1984). The movement of sediment is based on mechanical processes (Dietrich et al. 1975; Miller et al. 1977; Zanke 1982).

Tectonic processes are not considered here because our sedimentation reconstructions cover only geologically short time periods lasting from several hundreds to several thousands of years. During such time intervals, the overall ocean geometry may be considered stationary for the North Atlantic (Wold 1992). Tectonic subsidence or uplift is much slower than the expected sedimentation rate within a range from a few millimeters to several meters over a 1000-year timespan (Shaw and Hay 1989). As a consequence, tectonic movements and their effects on topography are negligible (Stephenson 1989).

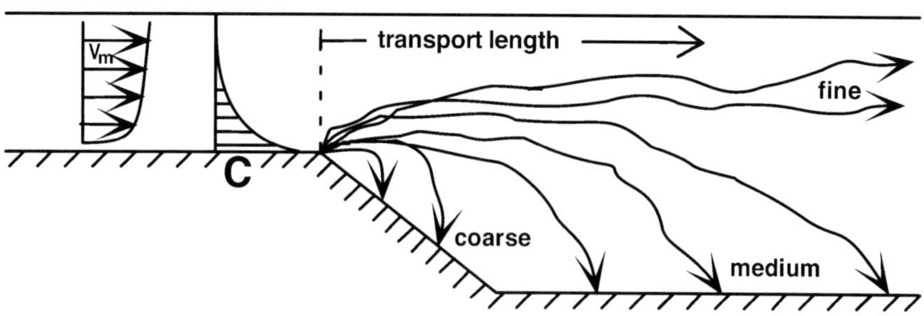

Fig. 3 Sketch of different transport length due to different grain size, and different settling velocity w_s. (After Middleton and Southland 1984)

2.4
Model Equations of SEDLOB

In the following section, the equations of the 3-D and 2-D submodels are listed separately. The symbols and units used are listed in the Appendix. For the sake of clarity, all equations are shown in Cartesian coordinates.

2.5
The Three-Dimensional Submodel of SEDLOB

The 3-D submodel of SEDLOB consists of a transport equation with a source term Q (Bryan 1969; Dietrich et al. 1975; Eppel 1977/78; Fahrbach et al. 1989; Gerdes 1988; Struve 1978; Tetzlaff and Harbaugh 1989)

$$\frac{\partial C}{\partial t} = -\nabla \cdot \left(\vec{v}C\right) + Q \tag{1}$$

and a continuity equation (conservation of mass) for an incompressible medium (and a continuity equation (conservation of mass) for an incompressible medium ($d\rho_F / dt = 0$) (Apel 1987; Bryan 1969; Fahrbach et al. 1989; Kurz 1977; Krauß 1973; LeBlond and Mysak 1978; Pond and Pickard 1986; Tetzlaff and Harbaugh 1989):

$$\nabla \cdot \vec{v} = \frac{\partial u}{\partial x} + \frac{\partial v}{\partial y} + \frac{\partial w}{\partial z}. \tag{2}$$

The hydrostatic equation gives the local pressure p (Bryan 1969; Cox 1984; Haupt 1990):

$$p(z) = p_{surf} + g\int_z^0 \rho_F \, dz. \tag{3}$$

The nonlinear equation of state is given by the UNESCO formula (UNESCO 1981; see also Millero and Poisson 1981)

$$\rho_F = \rho_F\left(T, S, p\right). \tag{4}$$

The settling velocity w_s of a single particle is calculated using the approximation given in Zanke (1977b)

$$w_s = w_s\left(\nu, \mu, d, \rho_F, \varphi_S, FF, g\right)$$
$$= \frac{12\nu}{d\left(2.7 - 2.3FF\right)} \left(\sqrt{1 + \left(0.21 \left(\frac{\left(\frac{\rho_S - \rho_F}{\rho_F}\right) g}{\nu^2} \right)^{\frac{1}{3}} \right)^3 \left(2.7 - 2.3FF\right)} - 1 \right). \tag{5}$$

The equation for dynamic viscosity μ

$$\mu = \mu(T, S, p) \tag{6}$$

is approximated by a polynom (Matthäus 1972). The total vertical velocity of w_g is the sum of the water velocity and the particle settling velocity:

$$w_g = w + w_s. \tag{7}$$

At the surface, the "rigid-lid" approximation is used:

$$w_{surf} = 0 \quad \text{for} \quad z = 0. \tag{8}$$

The "rigid-lid" approximation eliminates external gravity waves and allows for a longer time step (Δt) (Cox 1984; Haupt 1990; LeBlond and Mysak 1978). At lateral boundaries "no-flux" and "no-slip" boundary conditions are used:

$$u, v, C_n = 0. \tag{9}$$

No bottom friction is used, but rather a "free-slip" boundary condition is employed at the bottom:

$$\frac{\partial u}{\partial z}, \frac{\partial v}{\partial z} = 0. \tag{10}$$

The fluxes through the bottom and lateral boundaries are set to zero:

$$\frac{\partial T}{\partial n}, \frac{\partial S}{\partial n}, \frac{\partial C}{\partial n} = 0. \tag{11}$$

The vertical velocity w at the bottom is calculated using the continuity equation

$$w = -\left(u \frac{\partial H}{\partial x} + v \frac{\partial H}{\partial y} \right). \tag{12}$$

2.6 The Two-Dimensional Submodel of SEDLOB

In many aspects, the 2-D submodel of SEDLOB is similar to the 3-D submodel. The sediment transport at the bottom has the form

$$\frac{\partial C}{\partial t} = -\nabla_H \cdot \left(\vec{v}_{bot} C \right) + Q. \tag{13}$$

The submodel uses the same hydrostatic equation for the local pressure p [cf. Eq. (3)], the same set of nonlinear equations for density [cf. Eq. (4)] and viscosity [cf. Eq. (6)]. Similarly, the total vertical velocity, w_g, is the sum of the vertical velocity w [cf. Eq. (2)] and the settling velocity w_s of a single particle [cf. Eq. (5)]. Even though the 1-cm bottom boundary layer is quasi-2-D, this vertical velocity is needed for coupling both submodels.

The critical velocities for sediment transport are approximated by polynominal equations given in Zanke (1977a; Figs. 4, 5). One has to take into account

1. the critical velocities for starting bed load transport

$$v_{cm,b} = v_{cm,b}\left(v, \mu, d, \rho_F, \rho_S, g\right) = 2.8\left(\left(\frac{\rho_S - \rho_F}{\rho_F}\right) g\, d\right)^{0.5} + 14.7\frac{v}{d}c \quad \text{where} \quad c = 1; \tag{14}$$

2. the critical velocities for initiating of suspension load transport

$$v_{cm,s} = v_{cm,s}\left(v, \mu, d, \rho_F, \rho_S, FF, g\right) = 8.4\frac{12v}{d\left(2.7 - 2.3FF\right)}\left(\sqrt{1 + \left(0.21\left(\frac{\left(\frac{\rho_S - \rho_F}{\rho_F}\right) g}{v^2}\right)^{\frac{1}{3}}\right)^3 \left(2.7 - 2.3FF\right)} - 1\right); \tag{15}$$

3. the critical velocity for deposition

$$v_{cm,d} = v_{cm,d}\left(v, \mu, d, \rho_F, \rho_S, FF, g\right) = 3.93\frac{12v}{d\left(2.7 - 2.3FF\right)}\left(\sqrt{1 + \left(0.21\left(\frac{\left(\frac{\rho_S - \rho_F}{\rho_F}\right) g}{v^2}\right)^{\frac{1}{3}}\right)^3 \left(2.7 - 2.3FF\right)} - 1\right). \tag{16}$$

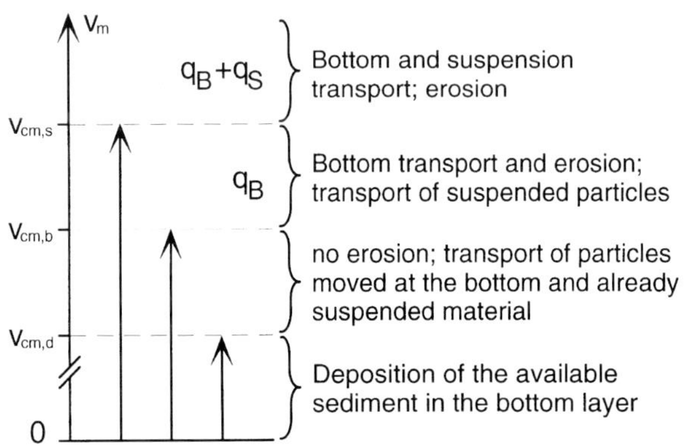

Fig. 4 Critical velocities for initiating bed load and suspension load transport (cf. Fig. 5)

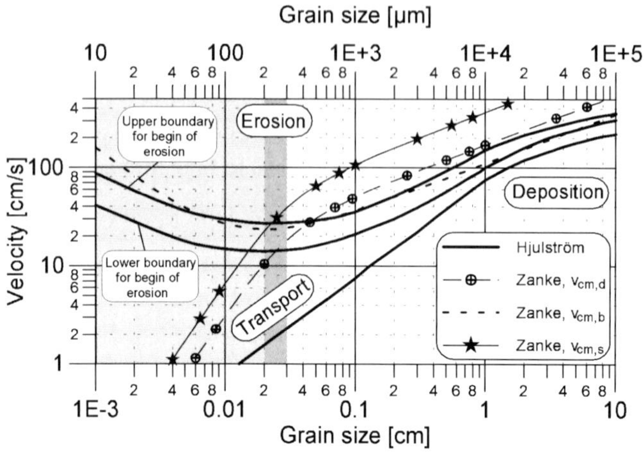

Fig. 5 Critical velocities for beginning bed load transport and suspension load transport The well-known Hjulström (1935) curve is given as reference

The bed load transport and the suspended load transport are calculated using their dependence on the reduced bottom velocity $\vec{v}_{bot}$ in the 1-cm-thick bottom layer, also called *Prandtl*'s boundary layer (Sündermann and Klöcker 1983; Zanke 1978). The formula for calculating the bed load transport is

$$
\begin{aligned}
q_B &= q_B\left(v_s, v_{c,b}, v, \mu, d, D^*, \rho_F, \rho_S, \rho', FF, g, p\right) \\
&= \frac{1}{p} 10^{-7} \left(\frac{v_s^2 - v_{c,b}^2}{w_s^2} D^{*2} \right)^2 v \quad \text{where} \quad D^* = \left(\frac{\rho' g}{v^2} \right)^{\frac{1}{3}} d \quad \text{and} \quad p \approx 0.7,
\end{aligned} \tag{17}
$$

and for calculating the suspension transport is

$$
\begin{aligned}
q_S &= q_S\left(v_s, v_{c,b}, v_{c,s}, v, v_0, \mu, d, D^*, \rho_F, \rho_S, \rho', FF, g, p\right) \\
&= 10^{-8} \frac{H}{h_1} \frac{\left(v_s^2 - v_{c,b}^2\right)\left(v_s^2 - v_{c,s}^2\right)}{w_s^4} D^{*4} \frac{1}{p} \left(\frac{v}{v_0 - v} \right)^{\frac{1}{4}}.
\end{aligned} \tag{18}
$$

The total sediment transport is computed by summing Eqs. (17) and (18):

$$
q = q_B + q_S. \tag{19}
$$

The relationship of critical velocities and sediment transport is summarized in Fig. 4.

A diagram of critical velocities versus grain size is displayed in Fig. 5.

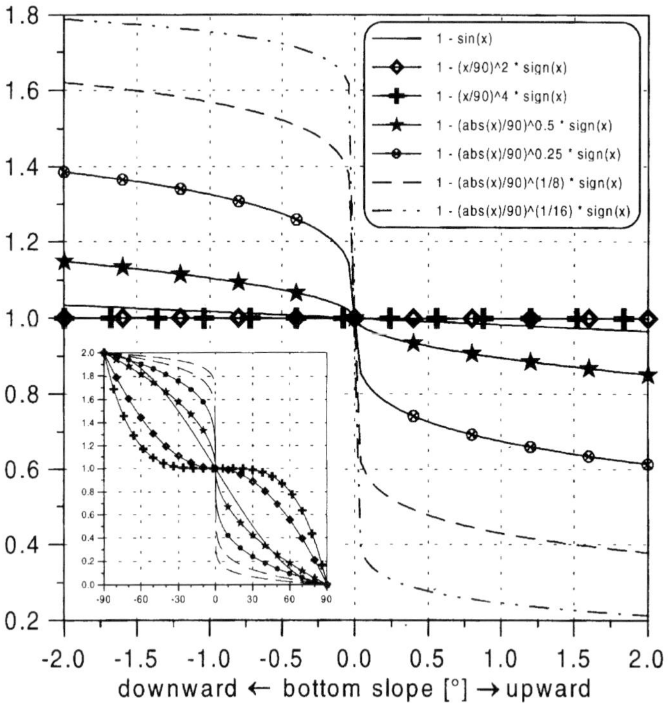

Fig. 6 Modification of sediment transport and critical velocities. The main figure shows the used functions in a range from –2° to 2°, while the lower left figure shows the range from –90° to 90°, and the upper right box gives the functions embedded in the model

The set of equations dealing with the critical velocities and the sediment transport are modified by the bottom slope. A downward flow leads to an increase in the transport capacities and a decrease in the critical velocities described above and vice versa. This modification is achieved by multiplying the transport velocities by an empirical function dependent on the bottom slope (Haupt 1995; Fig. 6).

The change in the bottom topography due to erosion and deposition is computed using the sediment continuity equation (Sündermann and Klöcker 1983; Tetzlaff and Harbaugh 1989):

$$\gamma \frac{\partial h_{sed}}{\partial t} + \nabla_H \cdot q = 0. \tag{20}$$

Sediment can be eroded or deposited according to Gross and Dade (1991) and Tetzlaff (1989)

$$\begin{aligned} &\frac{\partial h_{sed}}{\partial t} < 0 \quad \text{erosion} \\ &\frac{\partial h_{sed}}{\partial t} > 0 \quad \text{deposition.} \end{aligned} \tag{21}$$

This technique makes the simulation of the process of redistribution of the already deposited sediment possible (Frohlich and Matthews 1991). Sediment is neither eroded nor deposited when an equilibrium between the sediment transport and the sediment load exists. The equilibrium is checked at every time step in our model.

2.7 Coupling of SEDLOB's submodels

The coupling of both submodels facilitates the sediment exchange between suspension load in the 3-D water column and the bottom layer (Fig. 7). The 2-D bottom layer is initialized at every grid point (i, j) with the data of the 3-D submodel of SEDLOB. This is achieved by projecting the deepest "water grid point" of the 3-D submodel onto the bottom. Since the projected velocity may belong to different layers, the resulting 2-D velocity field is very inhomogeneous in areas where steep gradients in the bottom topography exist. To obtain a smoother flow, the velocity field is smoothed, using a moving average technique. In a large set of numerical experiments, it was found that smoothing with five to ten passes is sufficient to obtain an adequate velocity field in the bottom layer. The smoothed bottom velocity enables the model to run for more than 500 years

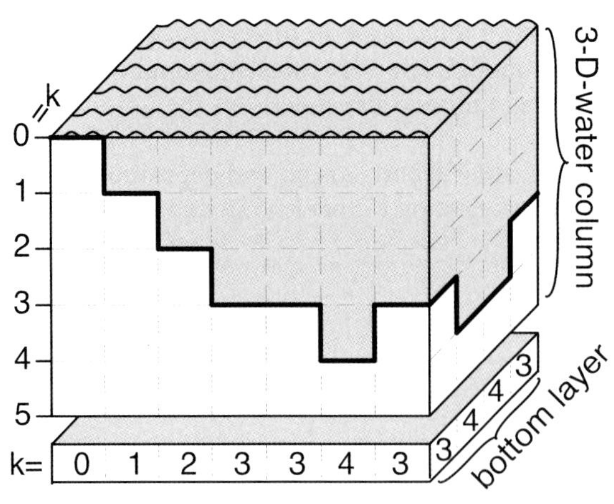

Fig. 7 Projection and coupling of both submodels

without producing ripples and spikes of sediment transport capacity in the adjacent grid points. Additional smoothing of other fields (bottom topography, sediment concentration near bottom, etc.) is therefore not required.

2.8 Numerical design

SEDLOB uses a staggered *Arakawa*-B-grid with a half grid distance's shift between *T–S* points and *u–v* points (Cox 1984; Mesinger and Arakawa 1976), and a Cartesian coordinate system with the vertical axis directed downward. The time integration is carried out using an "upstream differencing" scheme (Mesinger and Arakawa 1976; Struve 1978). This scheme is known to be rather effective (Dube et al. 1986), and may become essential for long-term integrations. It should be noted that this scheme, which is also called the donor cell, upward, or upwind scheme, is positive definite: positive values, like concentration, always remain positive during integration ["possitivity"; ($C_i^o \geq 0$ for all *i*) $\rightarrow$ ($C_i^N \geq 0$ for all *i* and *N*)] (Eppel 1977/78; Smolarkiewicz 1983). This is an important feature for mass transports, e.g., the transport of marine and eolian sediments, vapor, or gas in the atmosphere all are positive definite. Using second-order or higher-order integration accuracy schemes can introduce some difficulties because of a negative solution of the equation results (Smolarkiewicz 1983). Furthermore, a given disturbance is transported in the direction of physical advection and not, as in other discretization schemes, in the opposite direction (Mesinger and Arakawa 1976; Struve 1978). Furthermore, the "upstream differencing" scheme is mass conservative. All these requirements are only satisfied if the *Courant-Friedrichs-Levy* criterion is not violated (Eppel 1977/78; Mesinger and Arakawa 1976; Smolarkiewicz 1983; Struve 1978):

$$c_x \geq u\frac{\Delta t}{\Delta x}, c_y \geq v\frac{\Delta t}{\Delta y}, c_z \geq w_g\frac{\Delta t}{\Delta z}, \tag{22}$$

where c_x, c_y, c_z are Courant numbers ($c_{x,y,z} \leq 1$).Thus, the maximum time step in the model is:

$$\Delta t \leq \min\left(\frac{dx_{i,j}}{|u_{i,j,k}|}, \frac{dy_{i,j}}{|v_{i,j,k}|}, \frac{dz_k}{|w_{g\,i,j,k}|}\right). \tag{23}$$

Here $dx_{i,j}$, $dy_{i,j}$, dz_k denote the grid steps. The three velocity components are by $u_{i,j,k}$, $v_{i,j,k}$ and $w_{g_{i,j,k}}$.

Because a normal "upwind"-scheme has a strong implicit diffusion, we modify it in the 2-D submodel. There are different schemes to overcome the deficiencies of the simple upwind formulation. For example, the self-adjusting hybrid scheme (SASH), or the flux correction technique (FCT) offer better functionality while retaining the advantages of the upward scheme. However, the usefulness of these schemes is rather limited because of excessive computer time required. In addition, a positive definite solution is not guaranteed. Yet any emerging negative values are small enough to be neglected (Smolarkiewicz 1983; Struve 1978). An appropriate numerical scheme is essential for obtaining "sediment fronts", produced by sediment slumps, or local sediment clouds, etc. Smolarkiewicz (1983) introduces an "antidiffusion" with an "antidiffusion velocity" to keep the fronts sharp in spite of the high artificial diffusion inherent to the upwind schemes.

The numerical advection scheme is illustrated below for one-dimensional advection only. In a normal upwind scheme, two terms are in balance: the local changes in time, and the advective term. Smolarkiewicz (1983) adds another term with a small implicit diffusion at a low computational cost:

$$\underbrace{\frac{\partial C}{\partial t}+\frac{\partial}{\partial x}(uC)}_{\text{normal upwind scheme}} = \underbrace{\frac{\partial}{\partial x}\left(K_{impl}\frac{\partial C}{\partial x}\right)}_{\text{implicit diffusion}}, \quad \text{where} \quad K_{impl}=0.5\left(|u|\Delta x-\Delta t\, u^2\right). \tag{24}$$

Thus, for the normal upwind scheme in the 3-D submodel the following discretization on a staggered grid is chosen:

$$C_{n+1,i}=C_{n,i}-\left\{F\left(C_{n,i},C_{n,i+1},u_{n,i+\frac{1}{2}}\right)-F\left(C_{n,i-1},C_{n,i},u_{n,i-\frac{1}{2}}\right)\right\}, \tag{25}$$

where

$$F\left(C_i,C_{i+1},u\right)=\left[\left(u+|u|\right)C_i+\left(\left(u-|u|\right)C_{i+1}\right)\right]\frac{\Delta t}{2\Delta x}. \tag{26}$$

In the 2-D submodel we use the scheme of Smolarkiewicz (1983) with $\tilde{u}$ as antidiffusion velocity. The function F has the same form as in Eq. (26).

$$C_i^*=C_{n,i}-\left\{F\left(C_{n,i},C_{n,i+1},u_{n,i+\frac{1}{2}}\right)-F\left(C_{n,i-1},C_{n,i},u_{n,i-\frac{1}{2}}\right)\right\} \tag{27}$$

$$C_{n+1,i}=C_i^*-\left\{F\left(C_i^*,C_{i+1}^*,\tilde{u}_{n,i+\frac{1}{2}}\right)-F\left(C_{i-1}^*,C_i^*,\tilde{u}_{n,i-\frac{1}{2}}\right)\right\}, \tag{28}$$

where

$$\tilde{u}_{i+\frac{1}{2}}=\frac{\left(\left|u_{i+\frac{1}{2}}\right|\Delta x-\Delta t\, u_{i+\frac{1}{2}}^2\right)\left(C_{i+1}^*-C_i^*\right)}{\left(C_i^*+C_{i+1}^*+\varepsilon\right)\Delta x}. \tag{29}$$

ε is a small value (here 10^{-15}) to ensure $\tilde{u}=0$ when $C_{i+1}^*=C_i^*=0$.

A growth or decay of the initial signal can be obtained by scaling the antidiffusion velocity $\tilde{u}$ by a factor *Sc*:

$$\tilde{u}_{scale} = Sc\,\tilde{u} \tag{30}$$

The best result was achieved using a scaling factor of $1 \leq Sc \leq 1.08$ after Smolarkiewicz (1983). With $Sc = 0$ the above described scheme is identical to the normal upwind scheme without antidiffusion. Three experiments with different scaling factors are discussed below. These experiments were carried out to demonstrate how antidiffusion works and to check the program's overall performance. Starting from a given distribution of sediment in an anticyclone velocity field, the calculations are shown for one full rotation around the center (Fig. 8). The first experiment (I) has been carried out without antidiffusion $Sc = 0$, the second one (II) with a scaling factor $Sc = 1$, and the third one (III) with a factor $Sc = 1.08$ (Figs. 9, 10). With a scaling factor $Sc = 1$ we obtain results similar to those obtained by Smolarkiewicz (1983), i.e., best fit. The horizontal extent and intensity of the initial perturbation was preserved with good accuracy.

Without antidiffusion (experiment I) the experiment suffers from strong diffusion, which results in the signal being flattened and expanded horizontally. When the antidiffusion was overestimated, the initial perturbation was deformed and new maxima appeared (experiment III). It should be stressed that mass was conserved in all experiments.

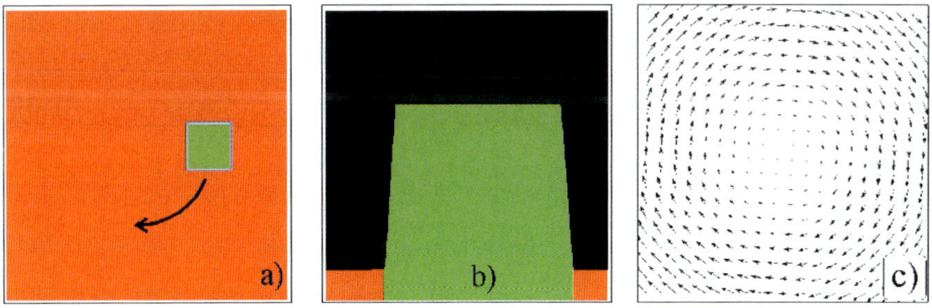

Fig. 8a–c Sediment distribution of the three experiments I, II, and III as a map view (**a**) and as a sectional view (**b**). The velocity field is circular (**c**)

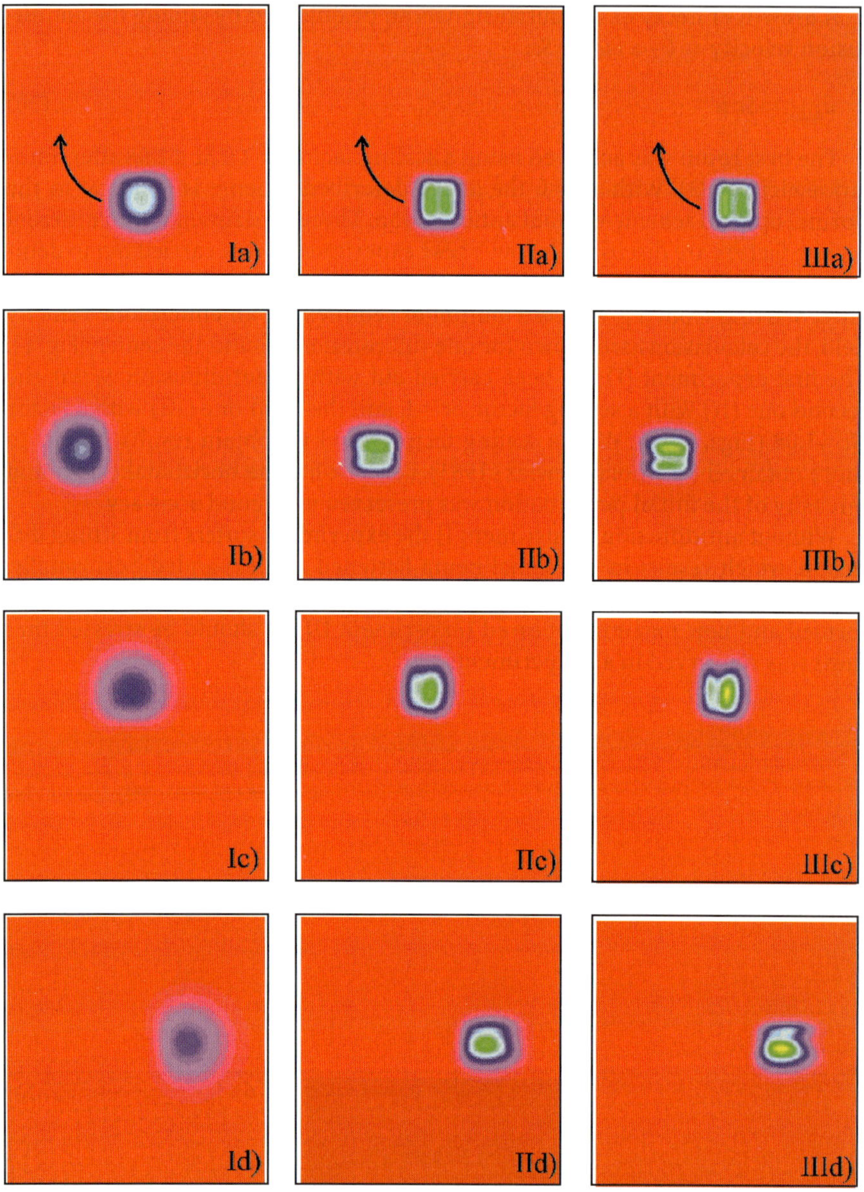

Fig. 9 Sediment distribution of the three experiments I, II, and III as a map view. Experiment I is calculated without antidiffusion ($Sc = 0$), the experiments II ($Sc = 1$), and III ($Sc = 1.08$) are calculated with antidiffusion (Smolarkiewicz 1983)

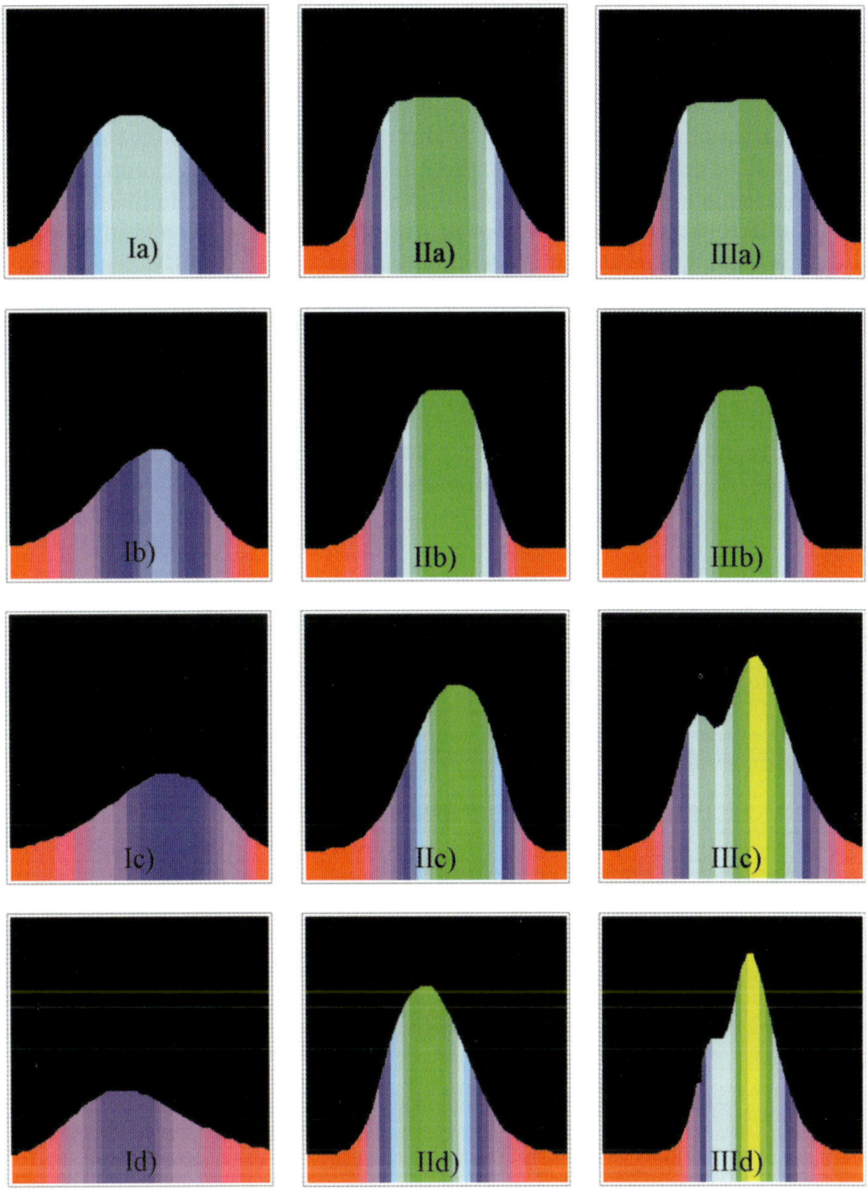

Fig. 10 Sediment distribution of the three experiments I, II, and III as a sectional view. Experiment I is calculated without antidiffusion ($Sc = 0$), the experiments II ($Sc = 1$), and III ($Sc = 1.08$) are calculated with antidiffusion (Smolarkiewicz 1983)

3
The particle-tracing model PATLOB

The semi-Lagrangian (hybrid Eulerian-Lagrangian approach) model PATLOB (Fig. 11) traces material parcels, e.g., water parcels, sediments, pollutants, natural or artificial organic material, etc. from their source area/origin until they are dissolved or deposited. This model is a useful tool to address both sedimentation and deep ocean ventilation problems. PATLOB was developed in order to use semi-Lagrangian calculations in combination with SEDLOB, which uses the output of the OGCM. Thus, the assumptions made for PATLOB are similar to those made for SEDLOB. Like SEDLOB, PATLOB mainly consists of two coupled submodels which are linked together. Hence, data flow between two submodels and the projection of data onto the 2-D 1-cm-thick bottom layer works identically in SEDLOB and PATLOB. Additionally, PATLOB takes the change in the bottom topography into account, which is calculated with SEDLOB. This is also relevant to particles which have a settling velocity.

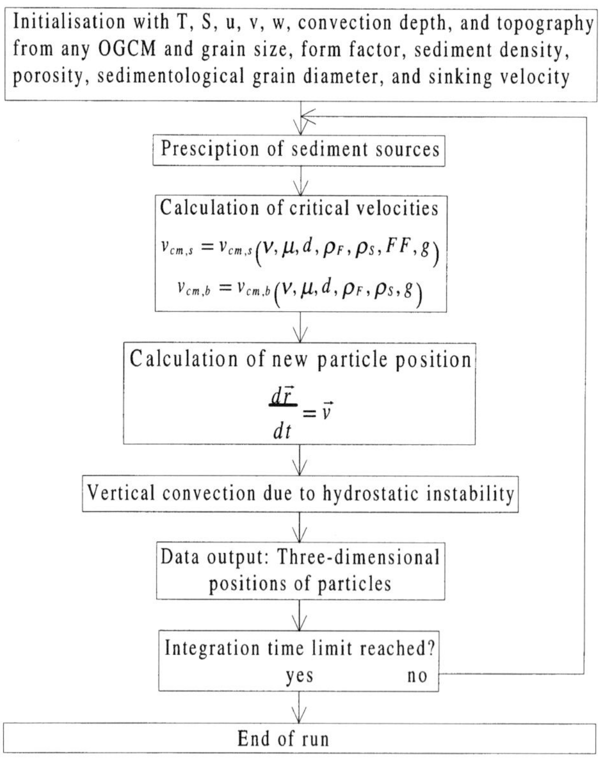

Fig. 11 Flow chart of PATLOB

3.1 Formulation of PATLOB

PATLOB uses the same approximation for the settling velocity, w_s, the critical shear velocities of bed load ($v_{cm,b}$) and suspension load ($v_{cm,s}$), and the critical velocity $v_{cm,d}$ for a final deposition of the parcels found in SEDLOB. Furthermore, the critical velocities are updated by the bottom slope.

The new location of every particle is calculated from the old position (Goldstein 1985; Kurz 1977):

$$\frac{d\vec{r}}{dt} = \vec{v}. \tag{31}$$

For the 3-D submodel the equation above is written in Cartesian coordinates as

$$\begin{aligned} r_{x,n+1} &= r_{x,n} + u_r \, \Delta t \\ r_{y,n+1} &= r_{y,n} + v_r \, \Delta t \\ r_{z,n+1} &= r_{z,n} + w_{g_r} \, \Delta t, \end{aligned} \tag{32}$$

and for the 2-D submodel as

$$\begin{aligned} r_{bot,x,n+1} &= r_{bot,x,n} + u_{bot,r} \, \Delta t \\ r_{bot,y,n+1} &= r_{bot,y,n} + v_{bot,r} \, \Delta t, \end{aligned} \tag{33}$$

where r is used for the location interpolated inside of the numerical grid and n for the time step. This means that the velocity components provided by the OGCM are interpolated to the current position of the Lagrangian particle from nearby points on the Eulerian numerical grid.

4 The vertical convection in SEDLOB and PATLOB

SEDLOB and PATLOB take into account vertical convection. This is an important feature because convection does not transport water and sediment in the same way as advection does. Tracers are advected by currents, whereas convection due to hydrostatic instability mixes water, sediments, and the tracers vertically in "turbulent" water columns or "chimneys" (Fig. 12). The convection due to hydrostatic instability determines the depth of vertical mixing in the ocean. In the model, the convection depth indicates how many layers participate in mixing, that is to which layer a particle entering such a turbulent water column is propelled within the chimney. We use different techniques to introduce this mixing into the two models. In SEDLOB, the sediment concentrations of vertical grid boxes affected by the convection depth are mixed to obtain a homogenous sediment distribution in every time step. In PATLOB, a parcel entering a chimney at the top is propelled downward to the base of the chimney and vice versa. This is equivalent to the reflection of every particle around the middle depth of a con-

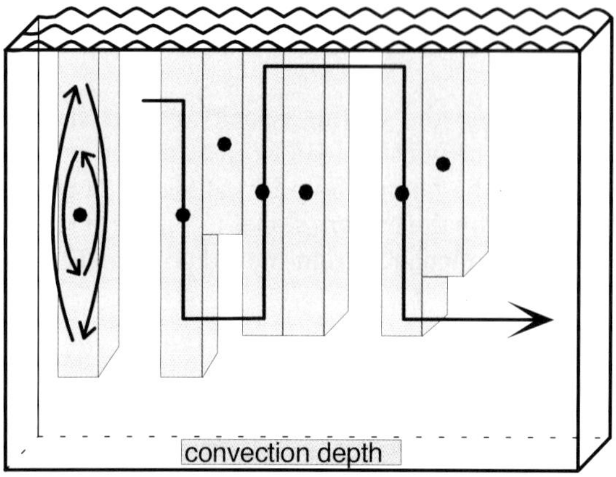

Fig. 12 Sketch of two particles entering an area with different vertical convection depths (Seidov and Haupt 1997)

vection site. If a particle enters a convection site, it is only mixed once. The vertical position remains unchanged if the particle enters an adjacent convection column having the same convection depth; in a case where the convection is deeper, the particle is brought to its new depth, either upward, or downward (Fig. 12). Additional details of this technique for incorporating convection into SEDLOB and PATLOB are given in Seidov and Haupt (1997).

5 Validation of SEDLOB and PATLOB

The models were originally developed and designed to study sedimentation processes in the North Atlantic and to be integrated with paleoceanographic modeling (Haupt 1995; Haupt et al. 1994, 1995). Therefore, the control experiments presented here concentrate on the modern North Atlantic. We review the results based on model runs with a 0.5°×0.5° horizontal resolution (95 grid points in both horizontal directions). For calculating the sediment transport through gateways and cross-sections, e.g. the Denmark Strait, the Iceland-Faeroe-Scotland-Ridge, or Barents Sea inflow and outflow, it is important to use a high vertical resolution which represents the topography in a realistic manner. Therefore, we use a model topography derived from the *ETOPO5* (1986) data set (Fig. 13) with 17 vertical layers which are 50, 50, 50, 50, 100, 100, 100, 250, 250, 250, 250, 500, 500, 500, 500, 500, and 1000 mthick. The maximum bottom slope in the direction of flow is less than 2.65°. The staircase-type bottom topography must never exceed 5° (Puls 1981), otherwise the turbulent bottom flow will detach from the seabed and the equation for sediment transport [Eq. (20)] will no longer be valid (Fig. 14). We use a spherical coordinate system in which the equator has been rotated up to 60°N along zero meridian in order to minimize the convergence of meridians in high latitudes.

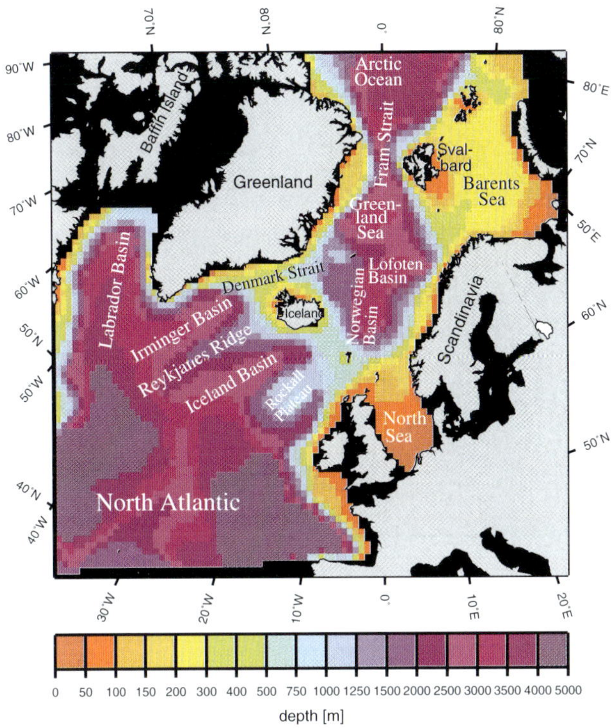

Fig. 13 Modern model bottom topography of the northern North Atlantic on a 0.5° × 0.5° grid with 17 unevenly spaced vertical layers

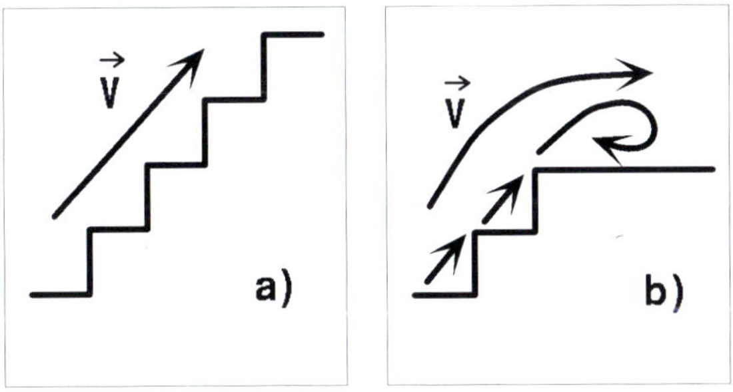

Fig. 14a, b Sketch of currents at the staircase bottom topography without a detached sediment transport (**a**). In the case of a detachment, the bottom transport is in the opposite direction (Puls 1981)

Furthermore, a high-resolution bottom topography is required for the better understanding of the influence that the additional sediment sources near Iceland and in the Greenland-Iceland-Norwegian Seas have on the sedimentation patterns in the northern North Atlantic.

5.1 Model initialization

As outlined above, in order to run the two models, one needs the output of an OGCM: temperature, salinity, velocity, and convection depths. A detailed de-

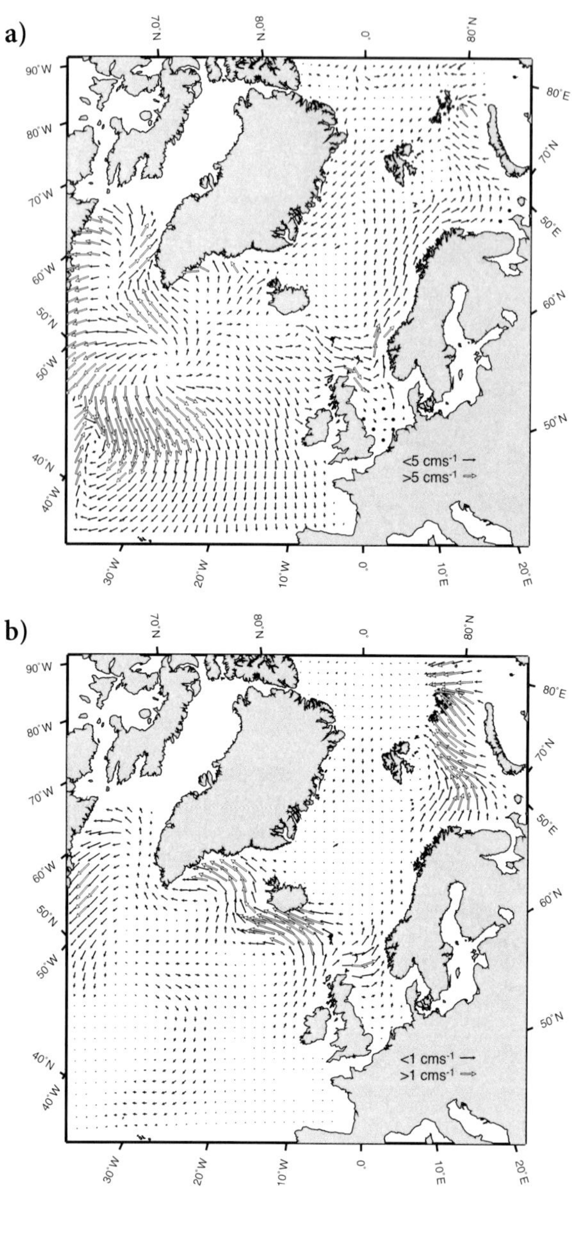

Fig. 15a, b Modern velocity: **a** at z=25 m and **b** at the bottom. The bottom velocity in the 1-cm-thick bottom is smoothed with a sliding average with ten passes

scription of the data fields is given in Haupt (1995). Here, we present two examples of the circulation pattern: the first one shows the circulation of the northern North Atlantic at 25 m depth (Fig. 15a), and the second one the bottom circulation resulting from the projection of the 3-D velocity field onto the bottom (Fig. 15b). The bottom velocity is smoothed with a ten-pass sliding average. Although

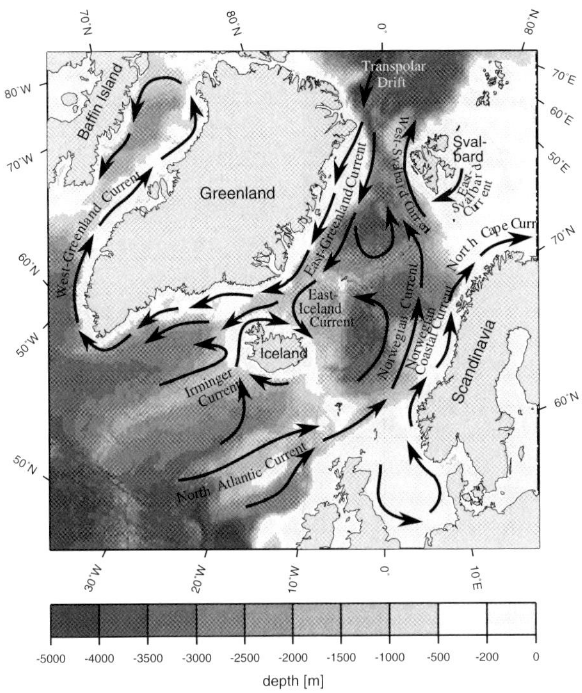

Fig. 16 The North Atlantic upper ocean circulation pattern

this experiment uses the closed boundary conditions with artificial walls at about 40°N, the model output from the OGCM shows the major currents around Iceland and in the Norwegian-Greenland Seas. These are mainly the West-Greenland Current, the East-Greenland Current, the outflow through the Denmark Strait, the Irminger Current, the North Atlantic Current, the Norwegian Current parallel to the Norwegian Coastal Current which enters the Barents Sea, and the West-Svalbard Current, or the Transpolar Drift (Fig. 16).

The sediment content of the 3-D water column in SEDLOB can be controlled in two different ways: sediment can be manually added to or taken from the water by sources and sinks, or alternatively, the exchange of sediment between the 3-D and the 2-D submodels is calculated automatically. Here, the external source of sediment is prescribed. Sediment sources and quantities inferred from the lateral river input and melting ice sheets are taken from Table 1 (Fig. 17). The lateral river sediment input is restricted to the grid points adjacent to the coastline. To limit the simulations to the case where sediment sources are known only at the lateral boundary, internal sources such as sediment derived from submarine fan deposition or from icebergs and fjords are not considered.

All our experiments were initialized with the same parameters (settling velocity of 0.05 cm s^{-1} = 43.2 m day^{-1} (Shanks and Trent 1980), density of sediment, grain size, and sedimentological grain diameter, form factor of sediment

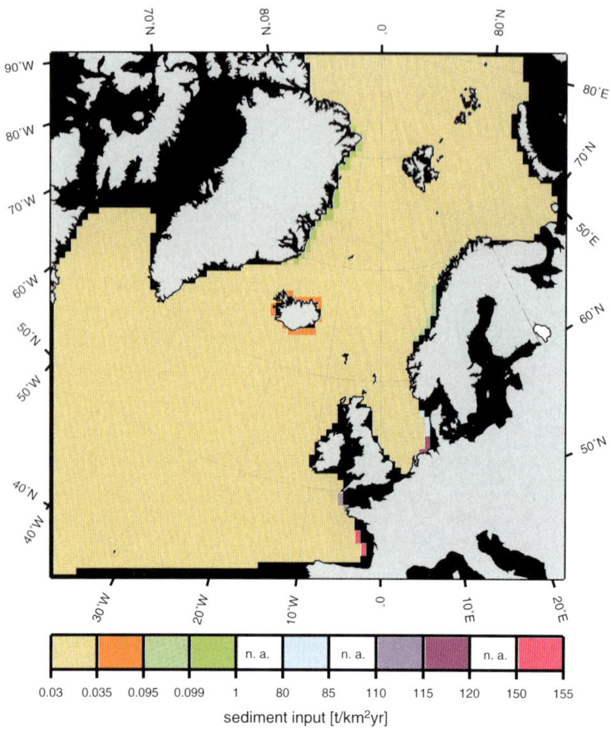

Fig. 17 Sediment sources prescribed as initial conditions (see Table 1). In the case of conversions the sediment density $\rho_S = 2.6\ cm^{-3}$ and the porosity $\gamma = 0.75$ were used (Zanke 1982)

Table 1 Survey of used sediment sources. In case of conversions the sediment density $\rho_S = 2.6\ g/cm^3$ and the porosity $\gamma = 0.75$ were used (Zanke 1982). The lower sediment input is used in case of a given range (Fig. 17)

Region	Area [$\times 10^6\ km^2$]	Sediment input [$t/km^2\ yr$]	Reference
northern North Atlantic	n. a.	3.156×10^{-2}	Honjo 1990 / Miller et al. 1977
East-Greenland	n. a.	6.96×10^{-2}	Enos 1991
Iceland	n. a.	6.1×10^{-2}–1.085×10^{-1}	Enos 1991
Norwegian coast	n. a.	6.06×10^{-2}–6.93×10^{-2}	Enos 1991
Elbe	0.13	84	Milliman / Syvitski 1992
Weser	0.038	33	Milliman / Syvitski 1992
Seine	0.065	114.2	Milliman / Syvitski 1992
Loire	0.155	150	Milliman / Syvitski 1992
southern England	n. a.	< 10	Einsele 1992

particles, and sediment porosity). In both experiments using the sedimentation model SEDLOB, we employed reduced critical velocities to initiate bed load $v_{cm,b}$ and suspension load $v_{cm,s}$. These were set to 0.002 cm s^{-1} and 0.02 cm s^{-1}, respectively. In control runs we figured out that a reduction is necessary to obtain realistic transports in the bottom layer in the deeper ocean basins. With these altered initial conditions the model was capable of eroding sediment when the critical velocities were weaker than the velocities predicted by the OGCM. Consequently, it produces more patchy sediment structures. Both simulations were run over 500 years. Unlike the OGCM, it is not possible to run SEDLOB into a steady-state condition. The forward time integration led to continuous changes of the bottom slope and therefore the critical velocities for initiating bed load and suspension load also changed. This is equivalent to the sediment availability which influences the maximum possible sediment concentration and transport in the fluid, depending on the bottom slope inclination.

6 Results and Discussion

In the first SEDLOB experiment (E1) only the eolian sediment input from the atmosphere was taken into account (Fig. 18a), whereas the second experiment (E2) added lateral sediment input from rivers and icebergs (Fig. 17). The sedimentation rate is given in cm/1000 years. The locations of the main sediment drifts south of Iceland and south of Greenland are well reproduced (Bohrmann et al. 1990; McCave and Tucholke 1986). However, the sedimentation rate is affected by the chosen distribution of the sediment supply. Both experiments show similar sedimentation patterns. They are mostly formed along the margins of the current axes. Differences can be found especially in regions where strongly selective river sediment input was added to the eolian sediment portion (e.g., in the Bay of Biscay, or in the eastern German Bight). Higher sedimentation rates occur also in the coastal areas, where currents run approximately parallel to the shore line. Here the water takes up a relatively low sediment input and accumulates it while moving along the coast. Whenever coastal currents depart from the coast to the open sea, the speed of the current slows down. Sediment transport capacity decreases and sediments are deposited. This phenomenon is found especially on continental slopes with a downward steepening bottom topography. An example of this is found off Southeast-Greenland where the East-Greenland Current flows through the Denmark Strait into the deep Irminger Basin. This feature can also be found to the south-southeast of Iceland and to the west of Lofoten, where the Norwegian Coastal Current turns to the east into the Norwegian Sea. Sediments are also deposited at higher rates on the Vøring Plateau. The higher sedimentation rates of these shelf areas are in good agreement with the recorded sedimentation rates from sediment cores.

The high sedimentation rate area in the northeasternmost part of the model area has a different origin. In both experiments, E1 (Fig. 18a) and E2 (Fig. 18b), the high sedimentation rates are due to the closed northern boundary and are

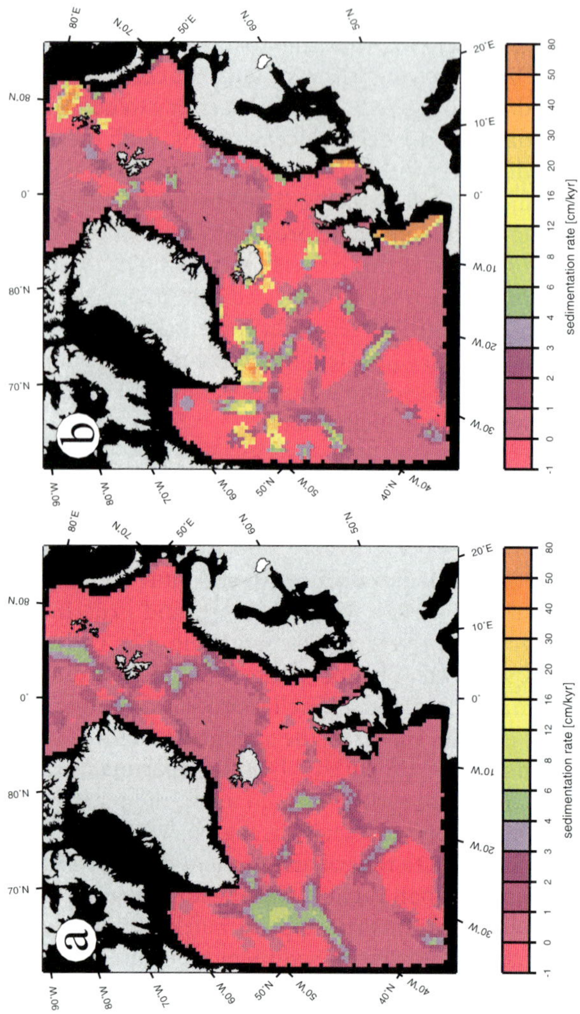

Fig. 18a, b Present-day sedimentation rate (cm/1000 years).
a Experiment E1: only the eolian sediment input from the atmosphere (1×10^{-13} g cm^{-2} $s^{-1} \approx 0.0864$ mg cm^{-2} day^{-1}) is considered (Honjo 1990; Miller et al. 1977). The critical velocities for starting of bed load $v_{cm,b}$ and for beginning of suspension load $v_{cm,s}$ are set to 0.002 cm s^{-1} respectively 0.02 cm s^{-1}.
b Experiment E2: additional sediment sources are applied; see Table 1 and Fig. 17

therefore an artifact. Figures 19 and 20 show the artificial "return flow" east of Franz Josef Land (northeastern Barents Sea). In Figs. 15a, b, 19, and 20, the sediment transport from North Scandinavia by the strong North Cape Current through the Barents Sea can be clearly seen. The southern model boundaries are responsible for artificially low sedimentation rate areas south of 50°N. In the OGCM the inflow of the North Atlantic Current is maintained using a southern sponge layer where the numerical solution is restored to modern climatology in a narrow latitudinal belt near this latitude (Seidov et al. 1996). Because of the lack of information concerning the sediment transport in the water column

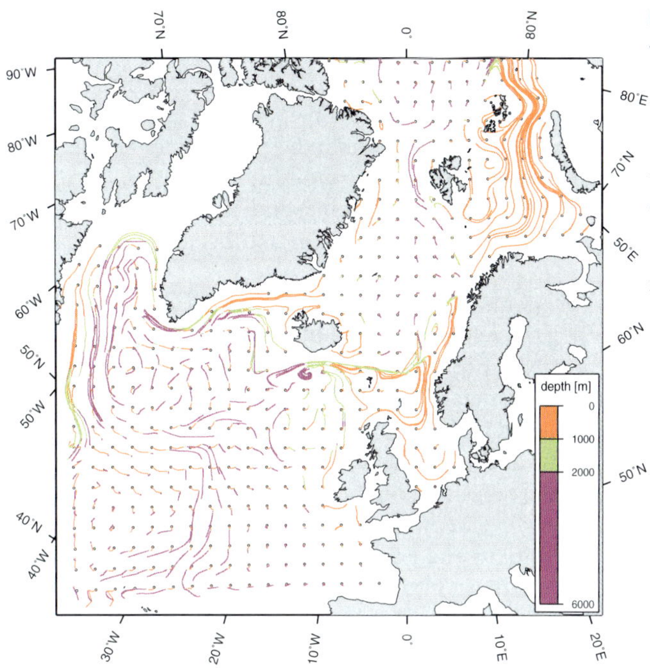

Fig. 19 Trajectories of 495 Lagrangian particles for the 3-D water motion tracing during 2 years. The small circles show the location where the particles were launched. The settling velocity is 0.05 cm s^{-1}. The depth is indicated by colors from the color palette

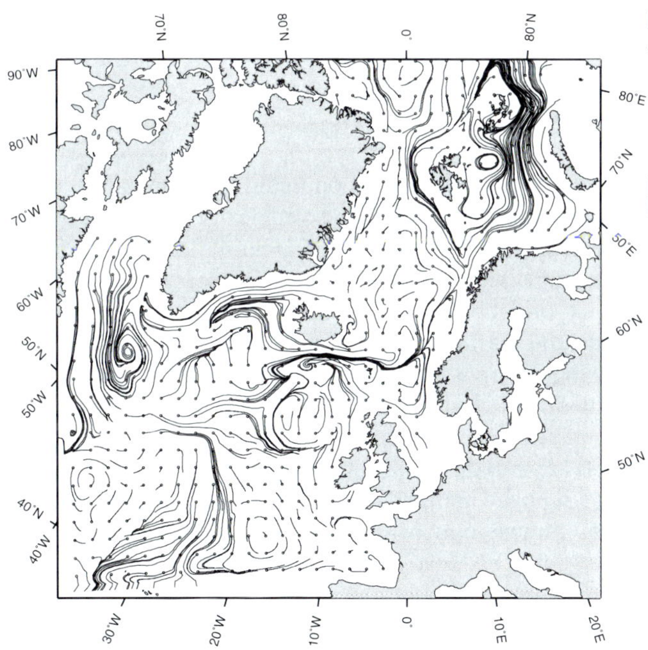

Fig. 20 Trajectories of 495 Lagrangian particles started in the bottom layer. The small circles show the location where the particles were launched. This experiment was run over 10 years

across these walls, we cannot arrive at reasonable sediment dynamics in this location.

The high sediment input to the English and French shelf areas from southern England and the French rivers Seine and Loire is responsible for the observed high sedimentation rates (Einsele 1992). Contrarily, the runoff from the German rivers Elbe and Weser is not limited to the river mouths. Although most of the sediment is deposited in the eastern German Bight, some is still transported to and deposited in the Norwegian Channel and on the southern Vøring Plateau (63°N, 5°–7°W).

A comparison of the experiments E1 and E2 shows that small local changes in the sediment supply can affect the sediment distribution in remote areas. Theses changes are clearly seen in the transport through gateways and along vertical cross sections. Figure 21a (experiment E1) and Fig. 21b (experiment E2) show the transport in the water column and at the bottom in tons km^{-2} year^{-1}. Both experiments indicate that most of the sediment transport occurs in the bottom layer. The transport through the cross section increases with the increase of additional lateral sediment supply everywhere except for two locations. Whether the decrease found between Svalbard and Franz Josef Land is a real feature or an artifact created by the closed northern boundary cannot be answered with this model. In Experiment E2 (Fig. 21b), the bottom transport over the Iceland-Faeroe-Ridge changes from a southward sediment transport (Fig. 21a) to a northward transport. In this region, the currents are generally parallel to the Iceland-Faeroe-Ridge (Figs. 15b, 20). In a set of several supplementary experiments we discovered that a small shift of this cross section to the north or to the south resulted in a change in the transport direction.

7 Conclusions

Integrated numerical models of oceanic circulation, sedimentation, and tracing water volumes lead to a better understanding of the complexity of interactions in the climatically forced ocean-sediment system. Two 3-D numerical models, SEDLOB and PATLOB, were developed to reconstruct the sedimentary history of the North Atlantic. Both models consist of a 3-D submodel for the water column and a 2-D bottom layer to model the specific features of near-bottom process motion. The models were tested using different horizontal and vertical resolutions in the North and northeastern Atlantic. High resolution experiments aimed at the simulation of detailed features of sediment flux through gateways and cross sections were discussed. The models are initialized using the output of an OGCM and different sediment sources. We supplied material by vertical eolian input from the atmosphere and from lateral sediment input by rivers and ice. In all experiments, the critical velocities for movement of bed load and suspension load were reduced to arrive at more realistic transports at the bottom in ocean basins. The employed polynominal equations for the sediment transport and the critical velocities were modified by an empirical function to introduce the dependence on bottom slope.

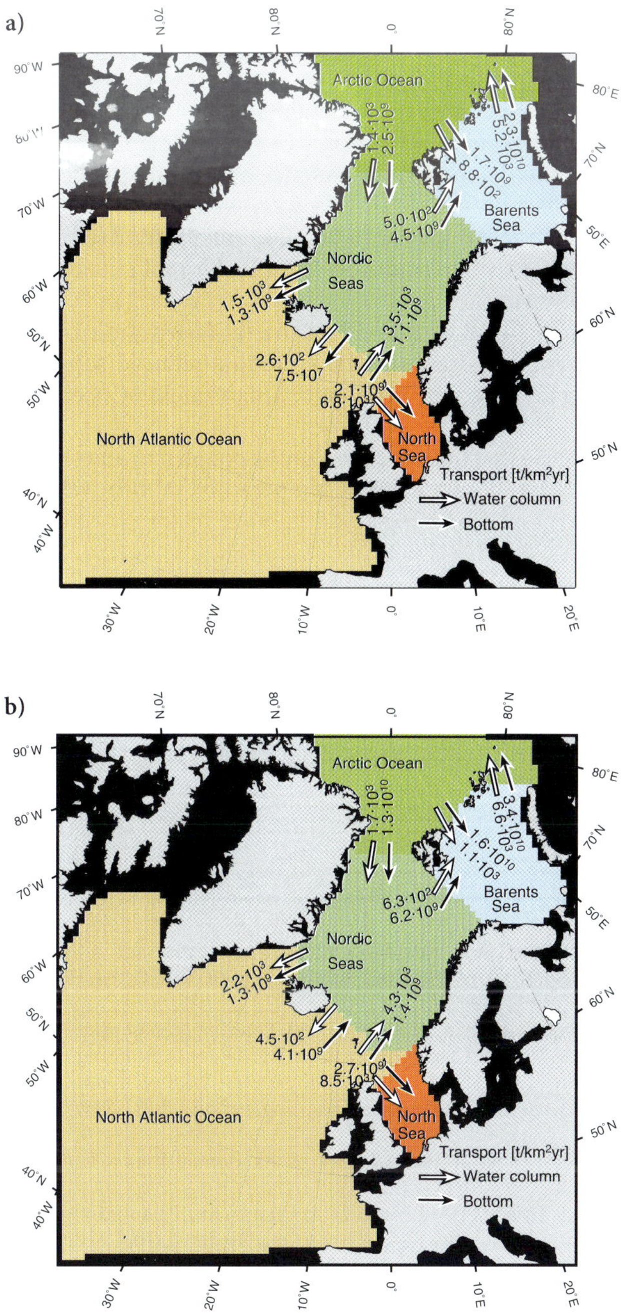

Fig. 21a, b Calculated transport rates for the three-dimentional water column and for the bottom layer: experiment E1 (**a**) and experiment E2 (**b**)

The simulated sediment distribution fits well the observed location of the main sediment drifts south of Iceland and Greenland. Additional lateral sediment input did not change the regional distribution of the high sedimentation areas. However, these changes in the input did affect the sedimentation rates and the transport through the cross sections. The increased sediment transport was predicted by the models both in the main body of water, and within the bottom layer. In comparison to the distribution within the water column the transport in the bottom layer showed a weaker response to the addition of lateral sediment sources. The calculation of cross section transport is a valuable tool in mass balancing through oceanic gateways.

In both experiments with SEDLOB we found that coastal downward currents led to reduced bottom current velocities and therefore to a reduced sediment transport capacity. This effect is especially pronounced in the areas of steep bottom gradients.

Finally, we want to emphasize that both models may be coupled to any OGCM which provides the adequate input data fields of temperature, salinity, velocity and convection depth.

Acknowledgments

This research was supported by the Deutsche Forschungsgemeinschaft (DFG) and SFB313 of Kiel University. We appreciate Avan Antia's und Derek Dreger's help on correcting our English. We thank Johannes Wendebourg for his very useful comments and suggestions which led to improvement of the manuscript.

References

Allen, J. R. L. (1985) Principles of Physical Sedimentology, George Allen & Unwin, London, 272 pp

Anderson, R. S., and N. F. Humphrey (1989) Interaction of Transport Processes in the Evolution of Arid Landscapes, Quantitative Dynamic Stratigraphy, ed. T. A. Cross, Prentice Hall, Englewood Cliffs, New Jersey, 349–361

Apel, J. R. (1987) Principles of Ocean Physics, Academic Press, London, 38, 634 pp

Bitzer, K., and R. Pflug (1989) DEPOD: A Three-Dimensional Model for Simulating Clastic Sedimentation and Isostatic Compensation in Sedimentary Basins, Quantitative Dynamic Stratigraphy, ed. T. A. Cross, Prentice Hall, Englewood Cliffs, New Jersey, 335–348

Bogárdi, J. L. (1974) Sediment Transport in Alluvial Streams, Akadémiai Kiado, Budapest, 826 pp

Bohrmann, G., R. Henrich, and J. Thiede (1990) Miocene to Quaternary Paleoceanography in the Northern North Atlantic: Variability in Carbonate and Biogenic Opal Accumulation, Geological History of the Polar Oceans: Arctic versus Antarctic, ed. U. Bleil and J. Thiede, Kluwer Academic Publishers, Netherland, 647–675

Bryan, K. (1969) A Numerical Method for the Study of the Circulation of the World Ocean, Journal of Computational Physics, 4, 347–376

Cao, S., and I. Lerche (1994) A Quantitative Model of Dynamical Sediment Deposition and Erosion in Three Dimensions, Computer & Geosiences, 20, 4, 635–663

Cox, M. D. (1984) A Primitive Equation, 3-Dimensional Model of the Ocean, GFDL Ocean Group Technical Report No. 1, Geophysical Fluid Dynamics Laboratory/NOOA, Princeton University
Dietrich, G., K. Kalle, W. Krauß, and G. Siedler (1975) Allgemeine Meereskunde - Eine Einführung in die Ozeanographie, Gebrüder Borntraeger, Berlin, third edition, 593 pp
Dube, S. K., M. E. Luther, and J. J. O'Brien (1986:) Documentation of the FSU Indian Ocean Model, Cooperative Indian Ocean Modelling Project Technical Report, Prepared at Mesoscale Air-Sea Interaction Group, The Florida State University, Tallahassee, FL 32306, 87 pp
Einsele, G. (1992) Sedimentary Basins; Evolution, Facies, and Sediment Budget, Springer-Verlag, Berlin, 628 pp
Enos, P. (1991) Sedimentary Parameters for Computer Modeling, Sedimentary Modeling: Computer Simulation and Methods for Improved Parameter Definition, ed. E. K. Franseen, W. L. Watney, C. G. St. C. Kendall and W. Ross, Kansas Geological Survey, 233, 63–98
Eppel, D. P. (1977/78) Numerische Simulation mesoskaliger Phänomene I., Skriptum zur Vorlesung - gehalten im Institut für Meereskunde an der Universität Kiel, Forschungszentrum Geesthacht GmbH - GKSS, Geesthacht, 105 pp
ETOPO5 (1986) Digital Relief of the Surface of the Earth, National Geophysical Data Center, Boulder, Colorado
Fahrbach, E., H. Franz, G. Gust, M. Hantel, J. Meincke, P. Müller, M. Rhein, W. Roether, and J. Willebrand (1989) Numerical Data and Functional Relationships in Science and Technology - Oceanography, Landolt–Börnstein, New Series Springer-Verlag, Berlin, Heidelberg, New York, London, Paris, Tokyo, Hong Kong, 3c, 398 pp
Frohlich, C., and R. K. Matthews (1991) Strata-Various: A Flexible Fortran Program for Dynamic Forward Modeling of Stratigraphy, Sedimentary Modeling: Computer Simulation and Methods for Improved Parameter Definition, ed. E. K. Franseen, W. L. Watney, C. G. St. C. Kendall and W. Ross , Kansas Geological Survey, 233, 449–461, 524 pp
Garde, R. J., and K. G. Ranga Raju (1977) Mechanics of Sediment Transportation and Alluvial Stream Problems, Wiley Eastern Limited, New Delhi, Bangalore, Bombay, 483 pp
Gerdes, R. (1988) Die Rolle der Dichtediffusion in numerischen Modellen der Nordatlantischen Zirkulation, Berichte aus dem Institut für Meereskunde an der Christian-Albrechts-Universität Kiel, 179, 176 pp
Gibbs, R. J., M. D. Matthews, and D. A. Link (1971) The Relationship between Sphere Size and Settling Velocity, Journal of Sedimentary Petrology, 41, No. 1, 7–18
Gibbs, R. J. (1985) Settling Velocity, Diameter, and Density for Flocs of Illite, Kaolinite, and Montmorillonite, Journal of Sedimentary Petrology, 55, 65–68
Goldschmidt, P. M., S. L. Pfirmann, I. Wollenburg, and R. Henrich (1992) Origin of Sediment Pellets from the Arctic Seafloor: Sea Ice or Icebergs?, Deep-Sea Research, 39, 539–565
Goldstein, H. (1985) Klassiche Mechanik, AULA-Verlag, Wiesbaden, 8. edition, 443 pp
Gross, T. F., and W. D. Dade (1991) Suspended Sediment Storm Modeling, Marine Geology, 99, 343–360
Haupt, B. J. (1990) Fluß-Gradient-Beziehung in wirbelauflösenden Modellen, Diplomarbeit, Institut für Meereskunde Kiel, Universität Kiel, 97 pp
Haupt, B. J., Chr. Schäfer–Neth, and K. Stattegger (1992) Towards Modelling the Paleocirculation and Sedimentation of the Northern North Atlantic, Fourth International Conference on Paleoceanography, Kiel, 137–138
Haupt, B. J., Chr. Schäfer-Neth, and K. Stattegger (1994) Modelling Sediment Drifts; A Coupled Oceanic Circulation-Sedimentation Model of the Northern North Atlantic, Paleoceanography, 9/6, 897–916
Haupt, B. J., Chr. Schäfer-Neth, and K. Stattegger (1995) 3-D Numerical Modelling of Late Quaternary Paleoceanography and Sedimentation in the Northern North Atlantic, Geologische Rundschau, 84, 137–150

Hjulström, F. (1935) Studies of the Morphological Activity of Rivers as Illustrated by the River Fyris, Bulletin of the Geological Institution of the University of Upsala, Upsala
Honjo, S. (1990) Particle Fluxes and Modern Sedimentation in the Polar Oceans, Polar Oceanography, Part B, ed. W. O. Smith Jr., Academic Press, Boston, Massachusetts, 687–739
Hsü, K. J. (1989) Physical Principles of Sedimentology, Springer-Verlag, Berlin, 231 pp
Krauß, W. (1973) Methoden und Ergebnisse der Theoretischen Ozeanographie/Methods and Results of Theoretical Oceanography – Dynamics of the Homogeneous and the Qasihomogeneous Ocean, Volume I, Gebrüder Borntraeger, Berlin, 302 pp
Krohn, J. (1975) Ein mathematisches Modell des großräumigen gezeitenbedingten Sedimenttransports mit Anwendung auf die Nordsee, Diplomarbeit, Institut für Meereskunde Hamburg
Kurz, M. (1977) Leitfäden für die Ausbildung im Deutschen Wetterdienst – Synoptische Meteorologie, Selbstverlag des Deutschen Wetterdienstes, Offenbach am Main, 8, 127 pp
LeBlond, P. H., and L. A. Mysak (1978) Waves in the Ocean, Elsevier Scientific Publishing Company, Amsterdam, Oxford, New York, 602 pp
Matthäus, D. J. (1972) Die Viskosität des Meerwassers, Beiträge zur Meereskunde, 29, 93–107
McCave, I. N. (1984) Erosion, Transport and Deposition of Fine–Grained Marine Sediments, Fine-Grained Sediments: Deep-Water Processes and Facies, ed. D. A. V. Stow and D. J. W. Piper, 35–69
McCave, I. N., and B. E. Tucholke (1986) Deep Current Controlled Sedimentation in the Western North Atlantic, The Geology of North America, ed. P. R. Vogt und B. E. Tucholke, Geological Society of America, Boulder, Colorado, M, 451–468
McCave, I. N., and T. F. Gross (1991) In-Situ Measurements of Particle Settling Velocity in Deep See, Marine Geology, 99, 403–411
Mesinger, F., and A. Arakawa (1976) Numerical Methods Used in Atmospheric Models, GARP Publication Series, 17/1, 64 pp
Middleton, G. V., and J. B. Southard (1984) Mechanics of Sediment Movement, Society of Economic Paleontologists and Mineralogists, USA, second edition, 400 pp
Miller, M. C., I. N. McCave, and P. D. Komar (1977) Threshold of Sediment Motion under Unidirectional Currents, Sedimentology, 24, 507–527
Millero, F. J., and A. Poisson, (1981) International One-Atmosphere Equation of State of Sea-Water, Deep Sea Research, 28A, 625–629
Möller, F. (1986) Einführung in die Meteorologie, Band 1: Meteorologische Elementarphänomene, B. I.-Hochschul-Taschenbücher, Band 276, Bibliographisches Institut Mannheim, Wien, Zürich, 222 pp
Paola, C., P. L. Heller, and C. L. Angevine (1992) The Large Scale Dynamics of Grain-Size Variation in Alluvial Basins, 1: Theory, Basin Research, 4, 73–90
Pickard, G. L., and W. J. Emery (1988) Descriptive Physical Oceanography – An Introduction, Pergamon Press, Oxford, 4. erweiterte Auflage, 249 pp
Pond, S., and G. L. Pickard (1986) Introductory Dynamical Oceanography, Second Edition, Pergamon Press, Oxford, 329 pp
Puls, W. (1981) Numerical Simulation of Bedform Mechanics, Mitteilungen des Instituts für Meereskunde der Universität Hamburg, Eigenverlag des Instituts für Meereskunde der Universität Hamburg, 147 Seiten
Seidov, D., M. Sarnthein, K. Stattegger, R. Prien, and M. Weinelt (1996) North Atlantic ocean circulation during the Last Glacial Maximum and subsequent meltwater event: A numerical model, J. Geophys. Res., 101, 16,305–16,332
Seidov, D., and B. J. Haupt (1997) Simulated ocean circulation and sediment transport in the North Atlantic during the last glacial maximum and today, Paleoceanography, 12, No. 2, 281–305
Shanks, A. L., and J. D. Trent (1980) Marine Snow: Sinking Rates and Potential Role in Vertical Flux, Deep-Sea Research, 27A, 137–143

Shaw, C. A., and W. W. Hay (1989) Mass-Balanced Paleographic Maps: Modeling Program and Results, Quantitative Dynamic Stratigraphy, ed. T. A. Cross, Prentice Hall, Englewood Cliffs, New Jersey, 277–291

Slingerland, R., J. W. Harbaugh, and K. P. Furlong (1994) Simulating Clastic Sedimentary Basins, Prentice Hall, Englewood Cliffs, New Jersey, 220 pp

Smolarkiewicz, P. K. (1983) A Simple Definite Advection Scheme with Small Implicit Diffusion, American Meteorological Society, 111, 479–486

Stephenson, R. (1989) Beyond First-Order Thermal Subsidence Models for Sedimentary Basins?, Quantitative Dynamic Stratigraphy, ed. T. A. Cross, Prentice Hall, Englewood Cliffs, New Jersey, 113–125

Struve, S. (1978) Transport und Vermischung einer passiven Beimengung in einem Medium mit einem vorgegebenen Geschwindigkeitsfeld, Berichte aus dem Institut für Meereskunde an der Christian-Albrechts-Universität Kiel, 57, 28 pp

Sündermann, J., and R. Klöcker (1983) Sediment Transport Modelling with Applications to the North Sea, North Sea Dynamics, ed. J. Sündermann and W. Lenz, Springer-Verlag, Berlin, Heidelberg, New York, 453–471

Syvitski, J. P. M., and S. Daughney (1992) Delta2: Delta Progradation and Basin Filling, Computer & Geosciences, 18/7, 839–897

Tetzlaff, D. M. (1989) Limits to the Predictive Ability of Dynamic Models that Simulate Clastic Sedimentation, Quantitative Dynamics Stratigraphy, ed. T. A. Cross, Prentice Hall, Englewood Cliffs, New Jersey, 55–66

Tetzlaff, D. M., and J. W. Harbaugh (1989) Simulating Clastic Sedimentation, Van Nostrand Reinhold, New York, 202 pp

UNESCO (1981) Tenth Report of the Joint Panel on Oceanographic Tables and Standards, UNESCO Technical Papers in Marine Science, 36, 24 pp

Wold, C. N. (1992) Paleobathymetry and Sediment Accumulation in the Northern North Atlantic and Southern Greenland-Iceland-Norwegian Sea, Dissertation, Mathematisch-Naturwissenschaftliche Fakultät der Universität Kiel, 255 pp

Zanke, U. (1976) Über die Naturähnlichkeit von Geschiebeversuchen bei einer Gewässersohle mit Transportkörpern, Mitteilungen des Franzius-Instituts für Wasserbau und Küsteningenieurswesen der Technischen Universität Hannover, 44, 289–324

Zanke, U. (1977a) Neuer Ansatz zur Berechnung des Transportbeginns von Sedimenten unter Strömungseinfluß, Mitteilungen des Franzius-Instituts für Wasserbau und Küsteningenieurswesen der Technischen Universität Hannover, 46, 156–178

Zanke, U. (1977b) Berechnung von Sinkgeschwindigkeiten von Sedimenten, Mitteilungen des Franzius-Instituts für Wasserbau und Küsteningenieurswesen der Technischen Universität Hannover, 46, 230–245

Zanke, U. (1978) Zusammenhänge zwischen Strömung und Sedimenttransport; Teil 1: Berechnung des Sedimenttransportes – allgemeiner Fall –, Mitteilungen des Franzius-Instituts für Wasserbau und Küsteningenieurswesen der Technischen Universität Hannover, 47, 214–345

Zanke, U. (1982) Grundlagen der Sedimentbewegung, Springer-Verlag, Berlin, Heidelberg, New York, 402 pp

Appendix

Symbols and definitions			
c_x, c_y, c_z	=	Courant numbers	–
C	=	sediment concentration in the fluid	g cm^3
C_n	=	sediment concentration at time step n	g cm^3
d	=	grain size	cm
$d_{xi,j}, d_{yi,j}, d_{zk}$	=	grid spacing	cm
D^*	=	sedimentological grain diameter $\left(\frac{\rho' g}{\nu^2}\right)^{\frac{1}{3}} d$	–
FF	=	form factor	–
g	=	gravitational acceleration	g cm^{-2} s^{-1}
h_{sed}	=	change of bottom topography due to erosion, transport, and deposition	cm
H	=	water depth	cm
i, j, k	=	zonal, meridional, and vertical grid indices	–
n	=	time step, $n=1, 2, 3,...$	–
p, p_{surf}	=	pressure, surface pressure	g cm^{-1} s^{-1}
q	=	total sediment transport	cm^3 cm^{-1} s^{-1}
q_B, q_S	=	bed load and suspension load transport	cm^3 cm^{-1} s^{-1}
Q	=	sediment source term	cm^3 cm^{-1} s^{-2}
$r_{x,n}, r_{y,n}, r_{z,n}, r_{bot,x,n}, r_{bot,y,n}$	=	zonal, meridional, and vertical location inside of the numerical grid at time step n	cm
$\vec{r}$	=	location of a particle	cm
t, t_n, t_0	=	time, time at time step n, starting time	s
T	=	temperature	°C
S	=	salinity	‰
u, v, w	=	zonal, meridional, and vertical velocity components	cm s^{-1}
u_{bot}, v_{bot}	=	reduced zonal and meridional bottom velocity components	cm s^{-1}
$\vec{v}, \vec{v}_{bot}$	=	three-dimensional and two-dimensional velocity vector	cm s^{-1}
w_s, w_g	=	settling velocity, total vertical velocity	cm s^{-1}
x, y, z	=	zonal, meridional, and vertical component of the grid	cm
Δt	=	time step	s
$\Delta x, \Delta y, \Delta z$	=	zonal, meridional, and vertical grid spacing	cm
γ	=	sediment porosity	–
∇, ∇_H	=	three- and two-dimensional Laplacian operator	cm^{-1}
μ	=	dynamic viscosity of seawater	g cm^{-1} s^{-1}
ν	=	kinematic viscosity of seawater	cm^2 s^{-1}
φ	=	bottom slope	°

Symbols and definitions			
ρ_F, ρ_S	=	density of sea water and sediment	g cm^{-3}
ρ'	=	relative density $\frac{\rho_S - \rho_F}{\rho_F}$	–

Digital Database of Modern Evaporites and their Predicted Distribution Based on Results from an Atmospheric General Circulation Model Simulation

C. N. Wold, G. J. Schwartz, C. Morrill

1 Introduction

Evaporites are probably the most significant of climatically sensitive sediments, because they form in only regions where the rate of evaporation greatly exceeds the rate of rainfall plus runoff. Evaporites, coals, carbonates, tillites, and thick clastic deposits are among some of the climate-sensitive sediments that have been used as qualitative and indirect evidence for paleoclimatic conditions.

Several authors have used AGCMs and semi-quantitative methods to explain the distribution of climatically sensitive sediments for the geologic record. However, only a few models have been based on first principles. Briggs and Pollack (1967) developed a model to simulate the spatial characteristics of the Late Silurian Salina Formation of the Michigan Basin. Their model had two dimensions in the horizontal plane, i.e. a Cartesian coordinate map view of the basin. The rate of evaporation from the surface of the brine was prescribed. They assumed that water was seeping into the basin radially and were able to simulate a "bull's-eye" pattern of evaporites similar to those observed in the Salina Formation.

Pollard and Schulz (1994) developed a proxy formation model for evaporites that calculated the rate of evaporation over a hypothetical, subaqueous evaporite basin. In general, a PFM simulates the physical, chemical, and/or biological conditions necessary for the formation of a climatically sensitive sediment (Wold and DeConto, in press). Pollard and Schulz (1994) applied the PFM for evaporites to the Triassic. The paleoclimate was simulated with GENESIS (Pollard and Thompson 1995; Thompson and Pollard 1995a; b) using a 4.5° × 7.5° grid for the atmosphere and 2° × 2° grid for the land surface. The experiment of Pollard and Schulz (1994) did not predict Carnian (225 Ma; Late Triassic) evaporites in the proto-Atlantic rift. Wold et al. (1994) explained this by the fact that the rift was not resolved on either the 4.5° × 7.5° or 2° × 2° grids. Wold et al. (1994) increased the resolution to better resolve the rift valley and mountain flanks. Using the results of the Late Triassic climate simulation (Hay et al. 1994; Wilson et al. 1994) Wold et al. (1994) then ran the PFM for evaporites and found much better agreement with the observed data.

Wold and DeConto (in press) used the PFM for evaporites (Pollard and Schulz 1994) to predict the potential locations of Late Cretaceous evaporite basins

GENESIS (2.0) was run for 35 years to simulate a mid-Campanian (80 Ma) climate with a high-stand of sea level. The PFM was driven by monthly-mean meteorological data from GENESIS. Wold and DeConto (in press) found that the locations of all compiled evaporites whose age was known to be either Campanian or Senonian were correctly predicted by the PFM. However, the Late Cretaceous evaporites in northeast Thailand were not predicted by the PFM using mean annual climate forcing from the GENESIS (2.0) simulation.

The evaporite database presented here was used to test the PFM for evaporites and to help validate the present-day GENESIS (2.0) climate simulation. We anticipate that the digital evaporite database will also be of interest to the hydrologic community.

2 Evaporite Database

Lefond (1969) outlined the location and age of the majority of salt deposits on Earth. He included small-scale maps showing the locations of most of these, as well as descriptions of those not shown on his maps. Lefond (1969) also included chemical analyses for many of the deposits. He attempted to show or mention the location of all known solar salt production operations that existed up to 1969. Lefond (1969) was the primary source of information used to compile the solar salt operations, included in the new database. Gordon (1975) showed general trends in the distribution of evaporites by latitude for the Phanerozoic. The resolution of his data was limited to 10° latitudinal bands. He noted that the present-day distribution of evaporites is centered about the subtropical high-pressure zones near 30°N and 30°S. Gordon (1975) also noted that very few evaporites occur along the low-pressure equatorial zone. Cogley (1991) published a 1° × 1° global data set of hydrographically significant terrains. Two of Cogley's (1991) terrains that were relative to the present compilation included SLTW (any type of salt water) and SFLT (salt flats). SLTW included oceans, estuaries, lagoons and saline lakes on land. SFLT was any type of intermittently dry evaporite deposit on land including non-marine saline pans, saline mudflats, marginal marine sabkhas, and salinas.

2.1 Non-Marine Evaporites

Non-marine evaporites in this compilation include saline pans and saline mudflats (Fig. 1B), and saline lakes (Fig. 1C). The largest perennial saline lakes where evaporites are precipitating include Lake Chad in northern Africa, Lake Balkash in Russia and the Dead Sea in Israel. A large portion of the dissolved salt in saline lakes is often derived from the dissolution of older evaporite deposits by inflowing rivers and groundwater. Saline pans are normally dry, but occasionally become flooded. The water then evaporates and salt is deposited. Examples of two very large saline pans include Salar de Uyuni in Bolivia and Lake Eyre in Aus-

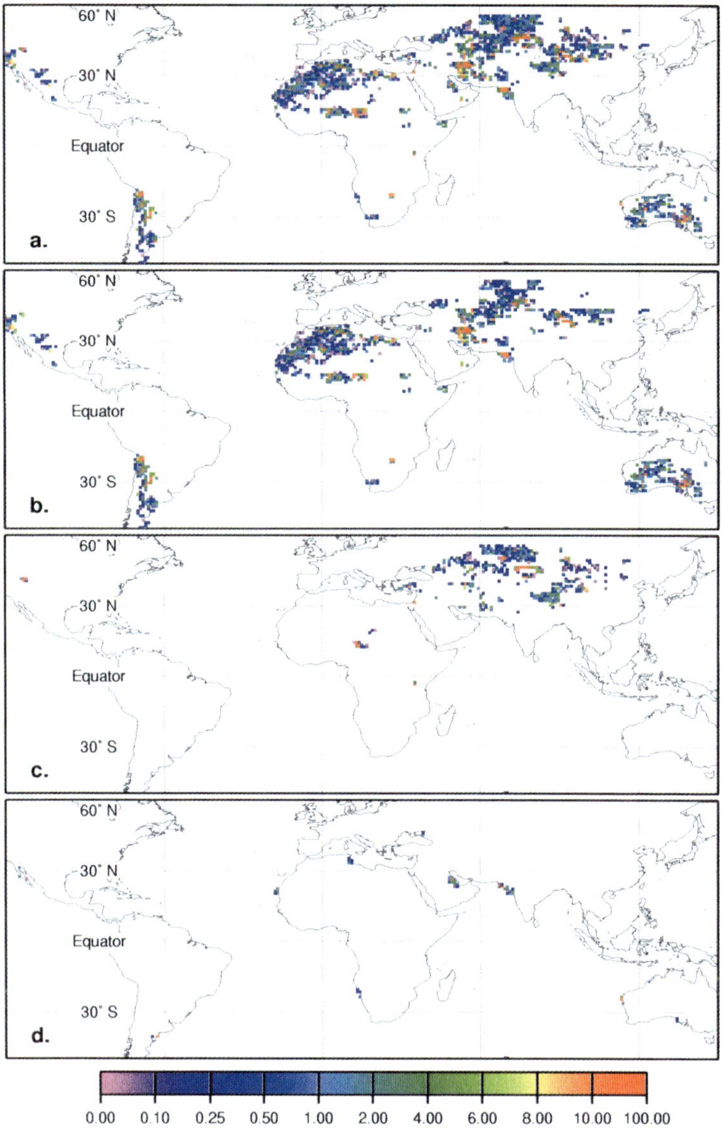

Fig. 1a The 1° × 1° distribution of modern evaporites shown as *colored squares* representing the percent coverage of 1° × 1° grid cells by evaporites. The present coastlines are shown. The map is a Mollweide Projection centered on the equator and 0° longitude. The grid spacing between lines of latitude is 30° and 60° between the meridians. **b** 1° × 1° distribution of saline pans and saline mud flats. The evaporites in Australia could also be called saline lakes. **c** 1° × 1° distribution of saline lakes. **d** 1° × 1° distribution of marginal marine sabkhas and salinas. There is some overlap between the 1° × 1° grid cells shown in these diagrams because at the resolution of a 1° × 1° latitude-longitude grid, the area of each grid cell is approximately 10000 km^2. This in some grid cells there is more than one type of evaporite deposit present.

tralia. Saline mudflats may be isolated or occur along the margins of saline lakes and saline pans. The amount of inflow to a saline lake can vary by large amounts annually. If inflow is decreased or stops for a number of years, shallower saline lakes may dry up to become saline pans or saline mud flats (Smoot and Lowenstein 1991).

2.2 Marginal Marine Evaporites

Marginal marine settings where evaporites form may be divided into two groups, salinas and sabkhas (Fig. 1D). Evaporites form subaqueously in salinas and in soils above the water table in sabkhas. One of the best examples of a modern salina is Lake MacLeod (Logan 1987) on the coast of western Australia. Large regions along the coast of the United Arab Emirates and in Abu Dhabi are covered by sabkhas (Kinsman 1969; Warren and Kendall 1985). Salinas and sabkhas form at or below sea level and their primary source of water is the ocean. They may be separated from the sea by a permeable barrier or form on the coastal plain just above the tidal zone, however they are not affected by tidal changes. Salinas may form in tectonic depressions or behind coastal barriers. Sabkhas are thinner deposits that are more likely to be flooded or subsequently eroded. Much greater volumes of salt are likely to accumulate and be preserved in salinas than in sabkhas (Handford 1991).

2.3 Solar Salt Operations

Solar salt operations (Fig. 2) are man-made facilities that use solar radiation to evaporite seawater or naturally occurring brines. Their purpose is to concentrate the brine for commercial extraction of various salts. The present compilation does not include facilities where the brine is heated artificially to extract salt. Solar salt operations are principally used to extract halite (rock salt) from seawater for human and animal consumption.

The distribution of 659 solar salt operations that existed prior to and during 1969, are shown in Fig. 2. The solar salt operations do not extend as far towards the poles as the naturally occurring evaporites (Fig. 1A). There is a greater concentration of solar salt operations near the equator than the naturally-occurring evaporites. This is because humans are able to extract salt from seawater during seasonally dry months of the year. Naturally occurring evaporites (Fig. 1A) do not form at these tropical latitudes along coastlines, because they would probably be washed away during the rainy months.

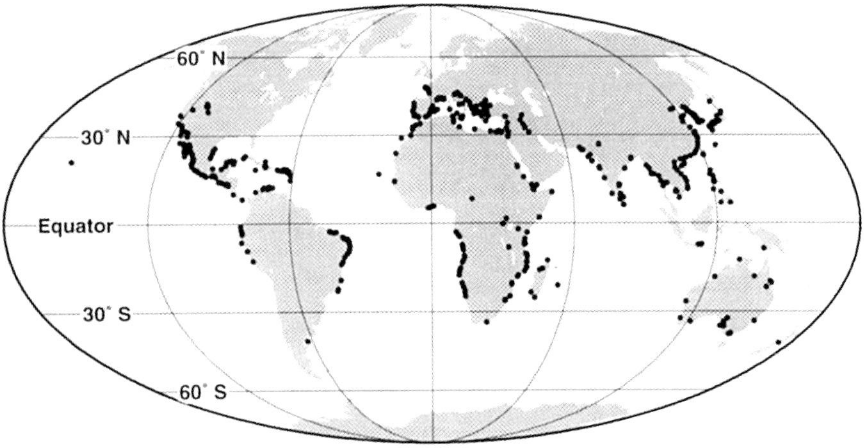

Fig. 2 There are 659 solar salt operations in the database. They were digitized as points, and shown here as *black dots*.

3 Analysis of the Modern Evaporite Database

From Fig. 3A, it is apparent that naturally occurring evaporites (excluding solar salt operations) are distributed with some degree of symmetry about 30°N and 30°S latitude (e.g. Gordon 1975). We analyzed the data to find the mean latitude of evaporites in each hemisphere. The Northern and Southern Hemispheres were treated as separate data sets. The saline lakes in Antarctica are small (total area of about15 km^2; Appendix 3) and anomalous (perennial ice cover) with respect to all other evaporites and were not included in the mean latitude calculations.

We calculated the arithmetic mean latitude of evaporites

$$\bar{x} = \sum_{i=1}^{N} x_i ,$$

where N was the total number of latitude points and x_i was a single latitude value. The actual number of points within each 10° band was directly proportional to the total area of evaporites within that 10° band. The mean latitude of evaporites in the Northern Hemisphere was 33.5°N

and the mean latitude in the Southern Hemisphere was 27.6°S. The standard deviation (σ_x) was calculated from:

$$\sigma_x = \sqrt{\frac{1}{N-1} \sum \left(x_i - \bar{x} \right)^2}$$

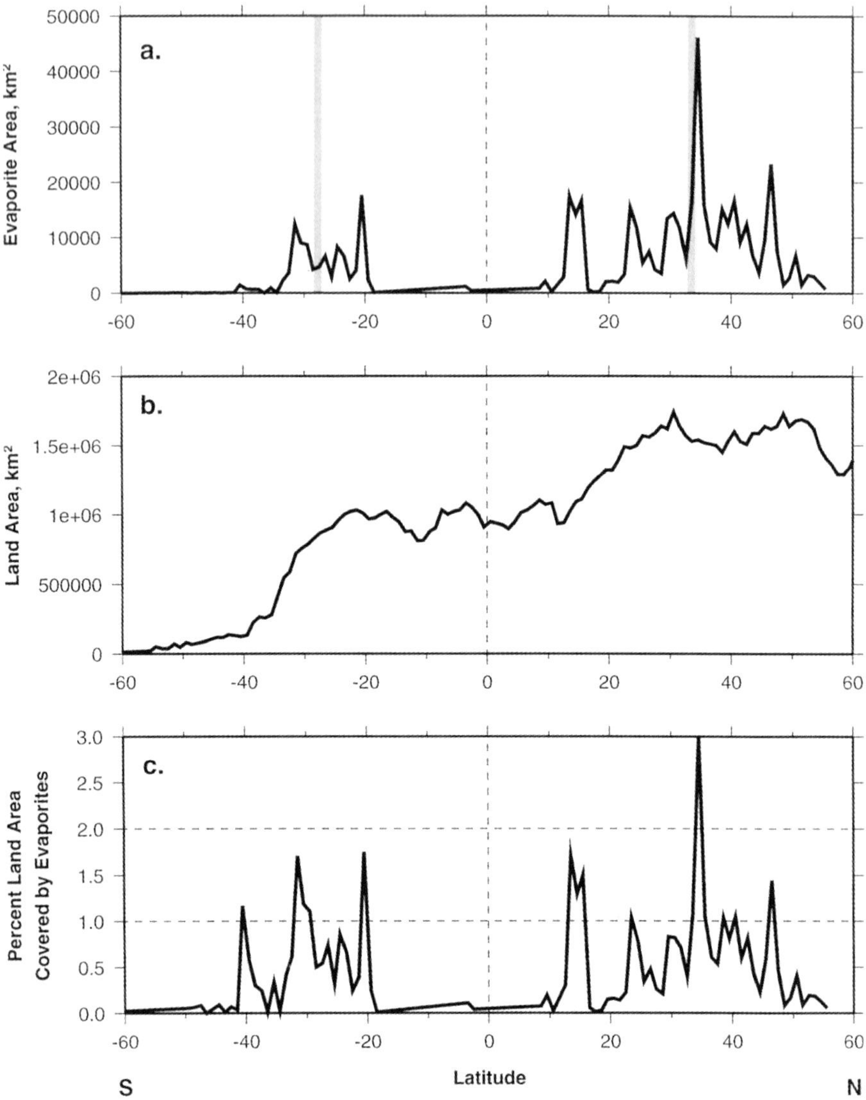

Fig. 3a Total land area (km^2) covered by naturally occurring evaporites at the present time. The *gray-shaded area* shows the standard deviation of evaporites with respect to latitude. The *white line in the middle of each gray area* is the mean latitude of evaporites.
b Total present-day land area (km^2) in every 1° interval of latitude from 60°S to 60°N.
c Percent of present-day land area covered by evaporites. *Horizontal dashed lines* were drawn at 1 and 2% of total land area covered by evaporites. The *vertical-dashed line in the center of each diagram* is the location of the equator

and was 11.3° in the Northern Hemisphere and 7° in the Southern Hemisphere. The mean latitude of Northern Hemisphere evaporites is 3.4° north of 30° N and the Southern Hemisphere evaporites are 2.4° north of 30°S latitude. The fact that these are not centered on 30° latitude probably reflects the annual average northward shift of the meteorological "thermal" equator (e.g. Barry and Chorley 1992).

It is apparent from any map of the Earth, that the majority of present-day land area is in the Northern Hemisphere. This is shown graphically in Fig. 3B. The distribution of land area explains to a first approximation, the larger observed area of evaporites in the Northern Hemisphere (Fig. 3A). When we divide the area covered by evaporites by the total land area, we can plot the percent of total land area covered by evaporites (Fig. 3C). This suggests that if the land were more evenly distributed between the Northern and Southern Hemispheres, then the evaporites would also be more evenly distributed.

At the latitudes where evaporites occur, they typically occupy about 1% of the land area (Fig. 3C). In the Southern Hemisphere evaporites reach more than 1% of land area near 40°, 30°, and 20°S latitude. In the Northern Hemisphere evaporites peak over 1% at 15°, 25°, 35°, 40°, and 46°N latitude. The differences between the left and right sides of Fig. 3C are due to differences in climate, the land-sea distribution, and the availability of closed drainage basins or land areas with no surface runoff.

The majority of Phanerozoic evaporites were formed in basins along continental margins with restricted marine connections. Examples include the Late Miocene Mediterranean, Aptian South Atlantic, Jurassic Gulf of Mexico and North Atlantic, Permian Zechstein, and Silurian Michigan Basins. It is interesting to note that based on the geologic record, the present-day evaporites most likely to be preserved millions of years from now are the marginal marine salinas and sabkhas and that these are centered near 30°N and 30°S latitude (Fig. 1D). Although the marginal marine evaporites (Fig. 1D) represent a very small portion of modern evaporites (Fig. 1A), they do closely approximate the mean latitudes (33.5°N and 27.6°S) of evaporite distribution at the present time.

4 The GENESIS AGCM

The GENESIS (Global ENvironmental and Ecological Simulation of Interactive Systems) version 2.0 AGCM was developed at the National Center for Atmospheric Research (NCAR; Pollard and Thompson 1995; Thompson and Pollard 1995a; b). GENESIS (2.0) includes an AGCM derived from GENESIS version 1.02b that was a highly modified version of the NCAR Community Climate Model version 1 (CCM1). The AGCM includes a diurnal cycle with solar-radiation calculations every 1.5 h and uses multi-layer randomly overlapped clouds. The infrared radiative effects of CO_2, CH_4, N_2O, and chlorofluorocarbons are all treated explicitly. The AGCM resolutions are a horizontal grid of 3.75° x 3.75°, and 18 levels in the vertical. GENESIS (2.0) also has a Land-Surface-Transfer

Model (LSX; Pollard and Thompson 1995) that accounts for the physical effects of vegetation. The land surface grid has a resolution of 2° × 2°. Up to two vegetation layers (trees and grass) can be specified for each 2° × 2° grid cell. Within LSX is a soil model with the same horizontal resolution and 6 layers in the vertical extending to a depth of 4.25 m. Heat is diffused linearly, soil moisture nonlinearly, and ice is predicted for each soil layer. LSX also includes a snow model with three layers. Inland lakes, snow cover, and the land-sea distribution can all be represented as fractional data on the 2° × 2° grid cells. A 50 m mixed-layer slab ocean model captures the oceans thermal capacity and transports heat poleward. GENESIS also includes a three-layer dynamic and thermodynamic sea ice model.

We used the results from GENESIS (2.0) simulation of present-day climate to drive the PFM for evaporites. The results from the GENESIS (2.0) simulation were stored as data files on the mass-storage system at the National Center for Atmospheric Research (Boulder, Colorado). These files were then later retrieved and used to run the PFM for evaporites.

5 The Proxy Formation Model for Evaporites and its Prediction of the Modern Evaporite Distribution

The PFM for evaporites (Pollard and Schulz 1994) consists of a $1 \times 1\ m^2$ column of water that is 50 m deep and assumed to be well mixed. The PFM is initialized with a salt-content sufficient to precipitate gypsum or halite. Positive values of evaporation minus precipitation (E-P) over the brine surface indicate the potential for precipitation of evaporite minerals. The PFM for evaporites is driven by monthly-mean meteorological forcing fields including: (1) atmospheric precipitation; (2) surface air temperature; (3) upward sensible heat flux; (4) surface-incident solar radiation; (5) air density; (6) surface-incident infrared radiation; (7) wind speed; and (8) relative humidity over the brine surface. The reader is referred to Pollard and Schulz (1994) for a more detailed description of the PFM for evaporites.

When seawater evaporates completely, a typical sequence of minerals precipitates. Calcite ($CaCO_3$) precipitates when the brine has been concentrated to approximately three times (105) the normal salinity (35) of seawater. Gypsum ($CaSO_4 \cdot 2H_2O$) precipitates at 5 times (175) the normal concentration of seawater. Then halite (NaCl) begins to precipitate at 11 times (385) the normal concentration of seawater. Finally, the potassium salts precipitate at about 100 times the normal concentration of seawater (Anonymous 1985; Warren 1988). These are the four major evaporite minerals that precipitate from seawater. There are a host of other minerals that precipitate at different concentrations of the brine depending on the original composition and subsequent mixing with other waters.

The output from the PFM is the rate of evaporation from water at the Earth's surface. Two simulations were performed using the PFM, one where the brine was initialized with a salinity of 175 (equivalent to 1175 kg/m^3 = 1000 kg H_2O + 175 kg total dissolved salts), to approximate the onset of gypsum precipitation, and the other with an initial salinity of 385 (equivalent to 1385 kg/m^3 = 1000 kg H_2O + 385 kg total dissolved salts) to approximate the onset of halite precipitation (Warren 1988). The rate of total atmospheric precipitation (mm/day) calculated by GENESIS (2.0) was subtracted from the rate of evaporation in each $2° \times 2°$ grid cell as calculated by the proxy formation model. Because this model depends entirely on AGCM results, and knows nothing about surface drainage patterns, it can only predict the potential for evaporite precipitation from a brine. Positive values for E-P indicated the potential for gypsum precipitation (Fig. 4) when the proxy formation model was initialized with a salinity of 175. Similarly, positive values of E-P indicated potential for halite precipitation (Fig. 5) when the proxy formation model was initialized with a salinity of 385.

Figure 4 shows the results of the PFM initialized for gypsum (S=175). The contours in Fig. 4A show where the model predicts the potential for gypsum to precipitate based on mean annual climate (12 months of forcing). The greatest values of E-P on the contours indicate regions with the highest potential. The PFM predicts the potential for the evaporite to precipitate based only on climate, the other major factor, land surface morphology, is not considered. The PFM predicted almost all of the observed evaporites between 30°N and 30°S latitude using 12 months of climate forcing (Fig. 4A). However, the model did not do a very good job predicting those evaporites that lie north of 30°N or south of 30°S latitude.

It then occurred to us that because the majority of modern evaporites occur in closed basins or in areas with no surface runoff, we should investigate seasonal patterns of E-P. This makes sense because even if the annual average of E-P is too small to indicate the potential for evaporite precipitation, the fact that those conditions occur over a closed drainage area means that the water will eventually evaporate and potentially precipitate evaporite minerals. Fig. 4B is the PFM driven by three months of forcing for June, July, and August (JJA; Northern Hemisphere summer) from the GENESIS (2.0) simulation. In this simulation, where E-P indicates the potential for gypsum formation in the Northern Hemisphere, we find that all of the Northern Hemisphere evaporites were predicted except for those farthest south in Mexico and those farthest north in North America (Nevada and Utah, USA). The Mexican evaporites are predicted by the PFM driven by 12 months of climate data. Those in Nevada and Utah are not predicted, probably because of the inability of the AGCM to adequately resolve mountains to the west resulting in a decreased rainshadow-effect and too high precipitation over this area.

The PFM for evaporites initialized for gypsum and driven by Southern Hemisphere summer (DJF; December, January, February) 3-month climate is shown in Fig. 4C. Note the strong seasonality of the PFM prediction in western and southern South America. None of these evaporites were predicted for the JJA forcing (Fig. 4B), and all of the South American evaporites were predicted in the DJF case (Fig. 4C). The greatest potential for gypsum precipitation during the

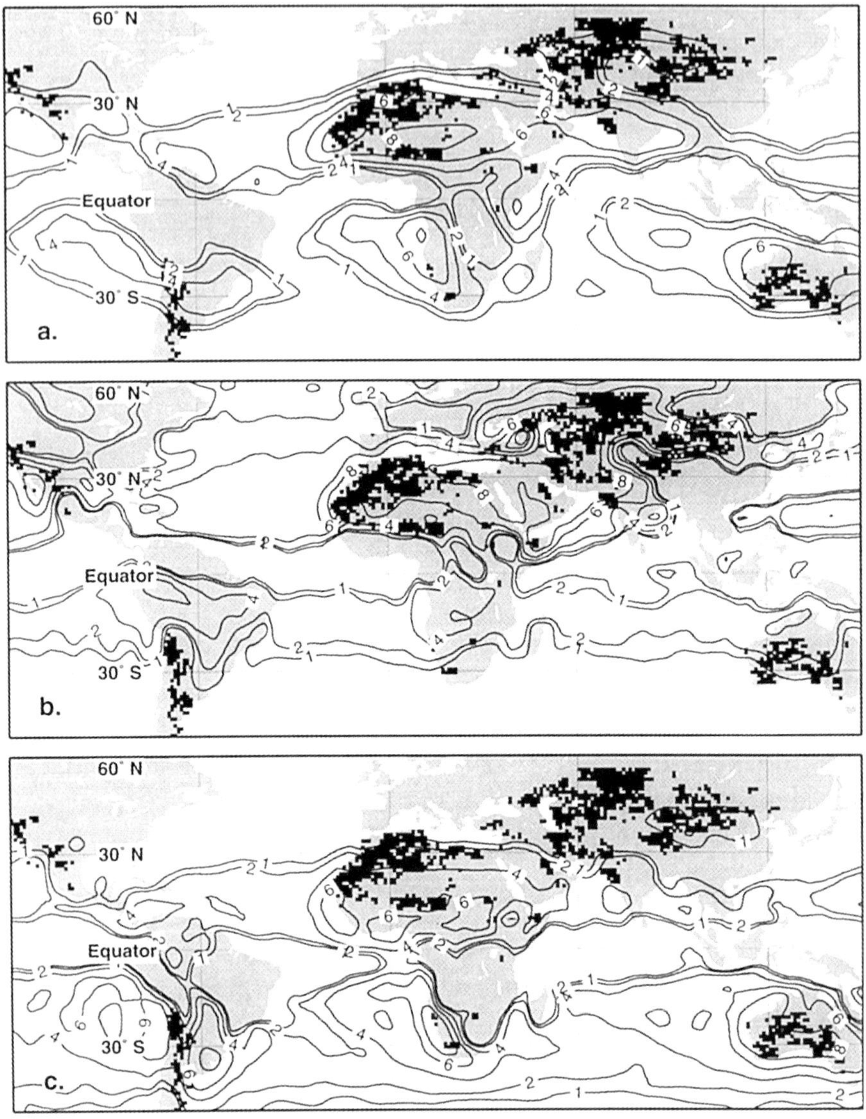

Fig. 4a Average annual rate of evaporation predicted by the PFM initialized for gypsum (initial salinity = 175) minus the average annual rate of total atmospheric precipitation (E-P) calculated by GENESIS (2.0). The contours range from 0 to 8 mm-day, e.g. only positive values of E-P are shown. The most positive values of E-P are over regions with the greatest potential for gypsum deposition. The compiled evaporites are shown as in Fig. 1A.
b PFM prediction of modern evaporites for the Northern Hemisphere summer (June, July, August).
c PFM prediction of modern evaporites for the Southern Hemisphere summer (December, January, February)

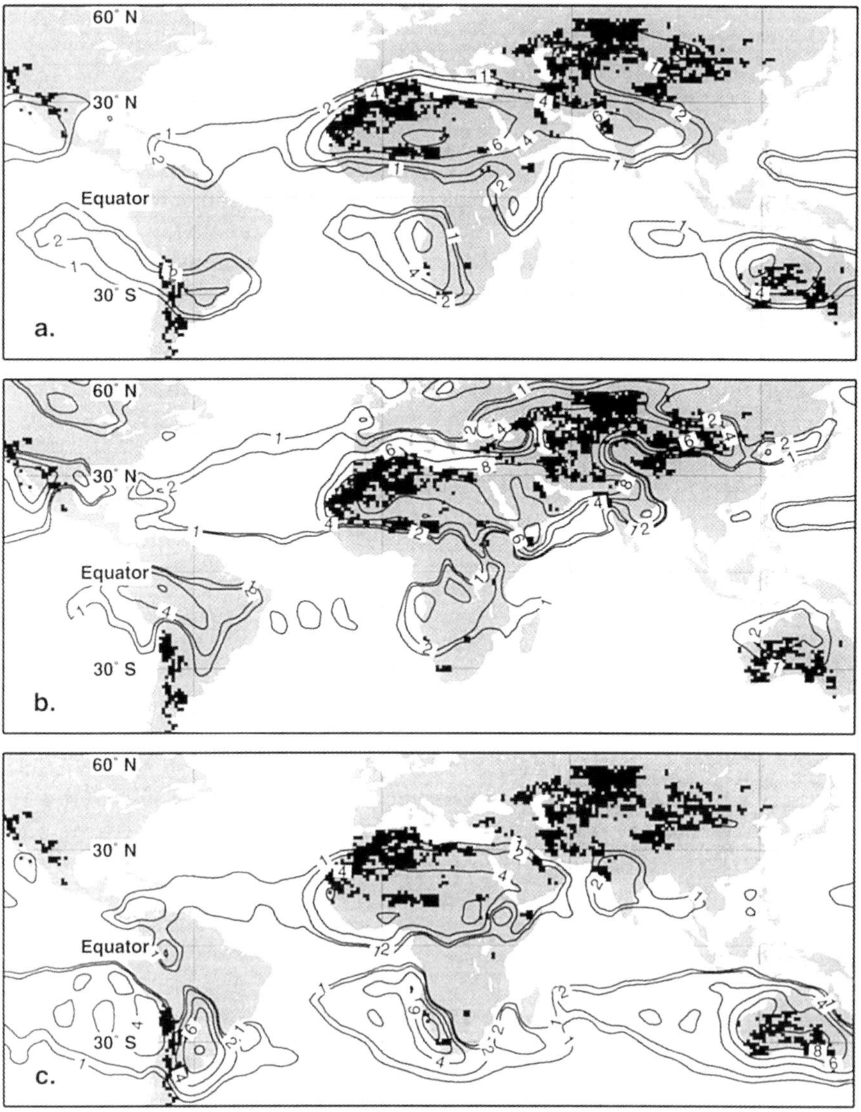

Fig. 5 Same as Fig. 4, but the PFM was initialized for halite (initial salinity = 385).
a Mean annual climate forcing (12-month).
b Northern Hemisphere summer (JJA).
c Southern Hemisphere summer (DJF)

Southern Hemisphere summer occurs over southwest Africa where saline pans and marginal marine sabkhas and salinas occur, and over western and central Australia where all types of modern evaporites occur. All of the evaporites in Africa were predicted by the PFM driven by 12-month forcing (Fig. 4A).

The PFM for evaporites initialized for halite (S=385) presents a more focused picture of E-P than when the model is initialized for gypsum. This is because it takes more energy to evaporate a brine to the point where halite will start to precipitate than it does for gypsum. The result in the PFM prediction is that the geographic distribution of positive E-P contours will focus more on those regions having the very highest (climatic) potential for evaporite formation (Fig. 5A). The PFM for evaporites initialized for halite and run for the Northern Hemisphere summer (Fig. 5B) predicts a smaller region of potential evaporite formation than the PFM did when initialized for gypsum (Fig. 4B) and the pattern of E-P contours more closely matches the observed evaporite distribution. Saline lakes in west central North America (Nevada and Utah, USA) and on the Himalayan Plateau are not predicted, again due to incomplete resolution of mountain ranges by the AGCM.

6 Discussion

The new database significantly expands the regions covered by GGHYDRO (Cogley 1991). This is primarily due to the different method that we used to compile the new database. Cogley (1991) used a 1° × 1° overlay with 100 points within each 1° × 1° grid cell. To calculate the percent coverage of salt, Cogley (1991) counted only those features that were located directly below one of the 100 points. Thus, smaller salt-accumulating bodies that were located between adjacent points were not counted (J. G. Cogley 1995; pers. comm.). In the present compilation, however, all salt-accumulating bodies that could be identified on the 1:1,000,000 scale maps (AMS 1934–1965) were digitized. The result is a much broader distribution of present-day evaporites (Fig. 1). The Cogley (1991) data (Fig. 6) is primarily a subset of the new database.

To compare the new evaporite database (Fig. 1) with GGHYDRO (Fig. 6), we plotted the SFLT field, and filtered the SLTW field to exclude coastal points and oceans. The most notable locations where GGHYDRO indicated evaporites and the new database did not, were in Brazil, South Africa, central North Africa, and near the Black Sea. Some of the differences could be because SLTW, may refer to any water that is more saline than freshwater, but is not necessarily precipitating evaporites. The Brazilian locations between 15°N and 15°S latitude in Fig. 2 can be thrown out because these are covered by the Amazon rainforest. There are a few grid cells in South Africa, central North Africa and around the Black Sea (Fig. 6) that were not included in the new database. We searched the AMS (1934-1965) maps for these locations, but found no indication of evaporites. Otherwise, the Cogley (1991) database representing salt flats (SFLT) and salt water on land (subset of SLTW) is a subset of the new database.

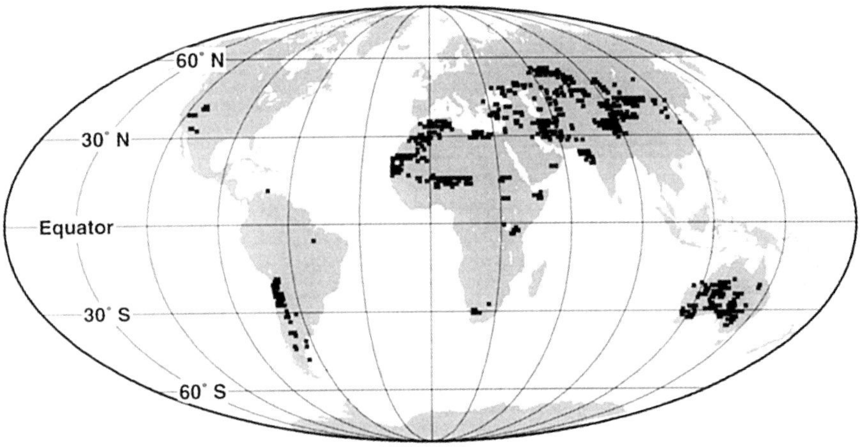

Fig. 6 The distribution of two subsets of the Cogley (1991) GGHYDRO database. Cogley's (1991) SFLT (salt flats) and SLTW (salt water) filtered to remove oceans and inland seas

7
Conclusions

The new database of modern evaporites consists of digitized polygons, points, and fractional coverage data representing both naturally occurring and man-made (solar salt operations) evaporites. The new database is a significant improvement over the previous evaporite database (Cogley 1991) because it more accurately represents the actual distribution of evaporites, and subdivides them into: (1) non-marine saline pans and saline mudflats; (2) non-marine saline lakes; and (3) marginal marine sabkhas and salinas. The mean latitude of evaporites in the Northern Hemisphere is 33.5°N with a standard deviation of 11.3°. The mean latitude of evaporites in the Southern Hemisphere is 27.6°S with a standard deviation of 7.0°. This represents a shift of the mean latitude of evaporites by approximately 3° north of 30° latitude in both hemispheres. This is most likely due to a meteorological thermal equator that lies a few degrees north of the geographic equator.

The majority of large, Phanerozoic evaporite deposits were formed in basins along continental margins with restricted marine connections. The present-day evaporites most likely to be preserved in the geologic record are the marginal marine evaporites centered near 30°N and 30°S latitude. Although the marginal marine evaporites represent a very small portion of the total present-day evaporites, their distribution closely approximates the mean latitudes of all modern evaporites.

Most of the observed evaporites between 30°N and 30°S latitude were predicted by the proxy formation model using 12-month (mean annual) forcing from the GENESIS (2.0) simulation of present-day climate.

Acknowledgements

We thank colleagues from the Climate Change Research Section, Climate and Global Dynamics Division of the National Center for Atmospheric Research (NCAR) for their encouragement and support. This research was supported in part by an appointment to the Global Change Distinguished Postdoctoral Fellowships Program sponsored by the U.S. Department of Energy, Office of Health and Environmental Research, and administered by the Oak Ridge Institute for Science and Education. Office space, computer, and travel support were kindly provided by NCAR. All of the illustrations were prepared by CNW using GMT version 2.1.4 (Wessel and Smith 1991).

Appendix I: Compilation Methods

An important source of information for the present compilation was Lefond (1969). His book contained information on the largest saline pans, saline mud flats, salt lakes, marginal marine sabkhas, and salinas, as well as information on solar salt and brine extraction facilities. He also outlined the geological history and locations of the largest pre-Holocene salt deposits. Using Lefond (1969) as a guide, saline pans, saline mud flats, salt lakes, marginal marine sabkhas, and salinas were digitized from the 1:1000000 scale maps of the world published by the U. S. Army Mapping Service (AMS; later called the Defense Mapping Agency) in the International Map of the World (AMS 1934-1965). Many smaller salt-forming basins were apparent and/or labeled on the 1:1000000 scale maps and were also digitized.

The naturally occurring evaporite deposits were digitized as polygons and the solar salt operations were digitized as points. The solar salt operations were either described or shown on much smaller scale maps in Lefond (1969). Solar salt operations in Mexico were digitized from a 1:5000000 scale map shown on a Bipolar Oblique Conic Conformal Projection. All the other solar salt operations were digitized from the 1:10000000 scale GEBCO (1984) maps shown on a Mercator projection. Some solar salt operations that could not be located on the GEBCO (1984) maps were found in the Times Atlas of the World (Bartholomew 1955–1959). The locations of naturally-occurring evaporites were digitized from the International Map of the World (AMS 1934–1965). There were three map projections used for the "International Map of the World," a Transverse Mercator, International Polyconic and Lambert Conic Conformal Projection. The digitized polygons were classified according to one of three categories: (1) non-marine saline pans and mudflats; (2) non-marine saline lakes; and (3) marginal marine sabkhas and salinas. The total number of digitized polygons was approximately 3934 saline pans and saline mud flats, 1176 saline lakes, and 200 marginal marine sabkhas and salinas, for a total of 5310 digitized polygons. There were 659 solar salt operations included as points in the database. This represents a minimum estimate of the total number of solar salt operations that exist at the present time.

Zerzan (1989) published a FORTRAN program (overlap) to calculate the area of two overlapping, 2-dimensional polygons. The algorithm was converted to the C-programming language, by J. Kleypas (NCAR, Boulder, CO) and was revised by CNW to calculate the area of overlap between a digitized polygon and the present 1° × 1° grid of latitude and longitude.

Appendix II: Regional Aspects of the Modern Evaporite Distribution

The majority of modern evaporite forming environments in the Americas are saline pans and saline mud flats (Fig. 1B, 1C). These occur primarily in the rain shadow and intermontane basins along the western margin of the north-south-trending continental divide formed by the Rocky and Andes Mountains. The largest concentration of evaporites is in South America (Fig. 1). There are a large number of saline pans and mud flats in Argentina, Bolivia, and Chile. These occur predominantly in the rain shadow along the eastern side of the Andes. This is an example of an alluvial fan-saline pan subenvironment, where saline pans and mud flats are surrounded by alluvial fans extending from mountain flanks (Smoot and Lowenstein 1991). The northernmost salt-forming body in the Americas, is the Great Salt Lake in Utah (Fig. 1C), where there are saline pans, mud flats and solar salt operations surrounding a perennial saline lake. There are a few marginal marine salinas around the Gulf of California and on the coast of the Gulf of Mexico (Fig. 1D). There are also marginal marine sabkhas on the southeastern margin of Argentina between the Colorado and Parana Rivers (Fig. 1D). In Central America, the Caribbean and Brazil, salt is produced by the solar evaporation of seawater (Fig. 2).

All of the present-day salt-producing areas in Europe are solar salt operations (Fig. 2). The majority of modern evaporites in Africa, are saline pans and mud flats in the northwestern part of the continent (Fig. 1B). These evaporites have accumulated in local depressions between eolian dunes and are fed by ephemeral streams. The primary source of salt for human consumption south of the equator in Africa, comes from widely distributed solar salt operations (Fig. 2). The largest evaporite-accumulating areas on the Arabian subcontinent are marginal marine sabkhas in the United Arab Emirates and Abu Dhabi on the southwest side of the Persian Gulf (Fig. 1D). There are a few saline pans and mud flats in Syria and northern Iraq. There are some saline lakes in the intermontane basins in Turkey (between the Mediterranean and Black Seas) and a few sabkhas adjacent to the Black Sea. The depression around the northern part of the Caspian Sea is covered with salt lakes, saline pans, and saline mud flats. The mountainous region south of the Caspian depression has widely distributed saline pans and saline mud flats (Fig. 1B).

In China and the region of the former Soviet Union, there are a large number of saline lakes (Fig. 1C). This was the only region covered by the AMS (1934-1965) maps that explicitly differentiated between saline lakes and saline pans. The existence of so many saline lakes is probably due in large part, to dissolution by groundwater and rivers of pre-Holocene evaporite deposits. The former Soviet Union contains some of the world's largest halite deposits (Lefond 1969).

There is also a great concentration of saline pans and saline mud flats (Fig. 1B). This is due to a mean-annual deficit of rainfall relative to the mean-annual rate of evaporation, and to the fact that the basins are closed or have no surface drainage (e.g. Fig. 3.1 in Smoot and Lowenstein 1991). The northernmost evaporite basins in Eurasia are found at 55.5°N in Siberia (Fig. 1). The highest saline lakes in the world (approximately 5560 m above sea level) are found on the northern side of the Himalayan Plateau (34.5°N, 79.5°E).

In Australia, non-marine saline lakes, saline pans and saline mud flats are all generally referred to as salt lakes on the AMS (1934-1965) maps. However, these may range from perennial saline lakes to saline pans or saline mud flats. Thus, we had no justification to subdivide the non-marine evaporites in Australia. The Australian evaporite basins are shown in Fig. 1.

The mapping of saline lakes in Antarctica is not complete (Burton 1981), and only small lakes and ponds remain to be compiled. Antarctica was not included in the International Map of the World (AMS 1934-1965), so we had to search the literature for information on saline lakes in Antarctica. The surface area and locations of Antarctic saline lakes was taken from the literature. These values were added up to produce the fractional coverage of four, 1° × 1° grid cells. This compilation does not include any digitized polygons for saline lakes in Antarctica.

We found descriptions of several saline lakes in Antarctica whose salinity was high enough to at least precipitate gypsum (greater than 175). Almost all of the saline lakes in Antarctica have perennial ice-cover (Wharton et al. 1993). The best-known exception is Don Juan Pond in South Victoria Land. It has the highest known salinity among the saline lakes in Antarctica, and has been observed to remain ice-free during Antarctic winter temperatures down to –50°C (Chinn 1993). The oldest (greater than 6 ka) saline lakes occur in the dry valleys of South Victoria Land (Burton 1981). Don Juan Pond, Lake Vanda, and Lake Bonney from this region, were included in two 1° × 1° grid cells (centered at 161.5°E, 77.5°S, and at 162.5°E, 77.5°S). These lakes have a salinity at least ten times normal seawater, and Lake Bonney is known to precipitate halite and gypsum (Simmons et al. 1993). Five lakes from the Vestfold Hills, Lakes Dingle, Stinear, Deep, Club, and Organic (Burton 1981; Matsumoto et al. 1983; Bird et al. 1991), were included in one grid cell (78.5°E, 68.5°S). We found three lakes from the Syowa Coast (Lützow-Holm Bay) that were saline enough to precipitate gypsum. These were Lakes Hunazoko, Suribati, and Itiziku (Tominaga 1981; Matsumoto et al. 1983) and were included in one grid cell (39.5°E, 69.5°S).

Appendix III: Availability of Data

The data presented in this paper, as well as these and additional diagrams are described and available to download through the Internet (World Wide Web) at the following location:

http://www.odsn.de/climate

References

AMS, 1934-1965, International Map of the World, Scale = 1:1,000,000, U. S. Army Mapping Service.

Anonymous, 1985, World Survey of Potash Resources: London, The British Sulfur Corporation Limited, 144 p.

Bartholomew, J., 1955–1959, The Times Atlas of the World Mid-Century Edition: Boston, Houghton Mifflin Company.

Bird, M. I., Chivas, A. R., and Radnell, C. J., 1991, Sedimentological and stable-isotope evolution of lakes in the Vestfold Hills, Antarctica: Palaeogeography, Palaeoclimatology, Palaeoecology, v. 84, p. 109–130.

Blackett, P. M. S., 1961, Comparisons of ancient climate with the ancient latitude deduced from rock magnetic measurements: Proceedings of the Royal Society of London, v. A263, p. 1–30.

Briden, J. C., and Irving, E., 1964, Palaeolatitude spectra of sedimentary palaeoclimatic indicators, in Nairn, A. E. M., ed., Problems in Palaeoclimatology: New York, Interscience, p. 199–224.

Briggs, L. I., and Pollack, H. N., 1967, Digital model of evaporite sedimentation: Science, v. 155, p. 453–456.

Burton, H. R., 1981, Chemistry, physics and evolution of Antarctic saline lakes: Hydrobiologia, v. 82, p. 339–362.

Chinn, T. J., 1993, Physical hydrology of the Dry Valley lakes, in Green, W. J., and Friedmann, E. I., eds., Physical and Biogeochemical Processes in Antarctic Lakes: Antarctic Research Series, p. 1–51.

Cogley, J. G., 1991, GGHYDRO - Global Hydrographic Data Release 2.0: Trent University, Trent Climate Note 91-1.

Crowley, T. J., and North, G. R., 1991, Paleoclimatology, Oxford Monographs on Geology and Geophysics No. 18: New York, Oxford University Press, 339 p.

Drewry, G. E., Ramsay, A. T. S., and Smith, A. G., 1974, Climatically controlled sediments, the geomagnetic field, and trade wind belts in Phanerozoic time: Journal of Geology, v. 82, p. 531–553.

GEBCO, 1986, General Bathymetric Chart of the Oceans: Canadian Hydrographic Service, scale 1:10000000.

Gordon, W. A., 1975, Distribution by latitude of Phanerozoic evaporite deposits: The Journal of Geology, v. 83, p. 671–684.

Gyllenhall, E. D, Engberts, C. J., Markwick, P. J., Smith, L. H., and Patzkowsky, M. E., 1991, The Fujita-Ziegler model: a new semi-quantitative technique for estimating paleoclimate from paleogeographic maps: Palaeogeography, Palaeoclimatology, Palaeoecology, v. 86, p. 41–66.

Handford, C. R., 1991, Marginal marine halite: Sabkhas and salinas, in Melvin, J. L., ed., Evaporites, Petroleum and Mineral Resources: Developments in Sedimentology: New York, Elsevier, p. 1–68.

Hay, W. W., Thompson, S. L., Pollard, D., Wilson, K. M., and Wold, C. N., 1994, Results of a climate model for Triassic Pangaea: Zentralblatt für Geologie und Paläontologie, v. 1, no. 11/12, p. 1253–1265.

Hay, W. W., and Wold, C. N., in press, The Role of Mountains and Plateaus in a Triassic Climate Model, in Tectonic Boundary Conditions for Climate Reconstructions, Crowley, T. and Burke, K. (Eds.), Oxford Monographs on Geology and Geophysics.

Irving, E., and Briden, J. C., 1962, Palaeolatitude of evaporite deposits: Nature, v. 196, p. 425–428.

Kinsman, D. J. J., 1969, Modes of formation, sedimentary associations, and diagnostic features of shallow-water and supratidal evaporites: Bulletin of the American Association of Petroleum Geologists, v. 53, p. 830–840.

Lefond, S. J., 1969, Handbook of World Salt Resources: New York, Plenum Press, 384 p.

Logan, B. W., 1987, The MacLeod Evaporite Basin, Western Australia - Holocene Environments, Sediments and Geological Evolution, Memoirs of the American Association of Petroleum Geologists: Tulsa, American Association of Petroleum Geologists, 140 p.

Matsumoto, G., Torri, T., and Hanya, T., 1983, Stenols and phytol in lake sediments from the Syowa and Vestfold Oases in the Antarctic: Geochemical Journal, v. 17, p. 1–8.

Parrish, J. T., Ziegler, A. M., and Scotese, C. R., 1982, Rainfall patterns and the distribution of coals and evaporites in the Mesozoic and Cenozoic: Palaeogeography, Palaeoclimatology, Palaeoecology, v. 40, p. 67–101.

Patzkowsky, M. E., Smith, L. H., Markwick, P. J., Engberts, C. J., and Gyllenhaal, E. D., 1991, Application of the Fujita-Ziegler paleoclimate model: Early Permian and Late Cretaceous examples: Palaeogeography, Palaeoclimatology, Palaeoecology, v. 86, p. 67–85.

Pollard, D., and Schulz, M., 1994, A model for the potential locations of Triassic evaporite basins driven by paleoclimatic GCM simulations: Global and Planetary Change, v. 9, p. 233–249.

Pollard, D., and Thompson, S. L., 1995, Use of a land-surface-transfer scheme (LSX) in a global climate model: the response to doubling stomatal resistance: Global and Planetary Change, v. 10, p. 129–161.

Simmons, G. M., Jr., Vestal, J. R., and Wharton, R. A., Jr., 1993, Environmental regulators of microbal activity in continental Antarctic lakes, in Green, W. J., and Friedmann, E. I., eds., Physical and Biogeochemical Processes in Antarctic Lakes: Antarctic Research Series, p. 165–195.

Smoot, J. P., and Lowenstein, T. K., 1991, Depositional environments of non-marine evaporites, in Melvin, J. L., ed., Evaporites, Petroleum and Mineral Resources: Developments in Sedimentology: New York, Elsevier, p. 189–348.

Taylor, J. R., 1982, An Introduction to Error Analysis, A Series of Books in Physics: Mill Valley, University Science Books, 270 p.

Thompson, S. L., and Pollard, D., 1995a, A global climate model (GENESIS) with a land-surface-transfer scheme (LSX). Part I: Present climate simulation: Journal of Climate, v. 8, p. 732–761.

Thompson, S. L., and Pollard, D., 1995b, A global climate model (GENESIS) with a land surface-transfer scheme (LSX). Part II: CO2 sensitivity: Journal of Climate, v. 8, p. 1104-1121.

Tominaga, H., and Fukui, F., 1981, Saline lakes at Syowa Oasis, Antarctica: Hydrobiologia, v. 82, p. 375–389.

Warren, J. K., 1988, Evaporite Sedimentology: New York, Prentice-Hall, 285 p.

Warren, J. K., and Kendall, C. G. St., 1985, Comparison of sequences formed in marine sabkha (subaerial) and salina (subaqueous) settings - modern and ancient: Bulletin of the American Association of Petroleum Geologists, v. 69, p. 1013–1023.

Wessel, P., and Smith, W. H. F., 1991, Free software helps map and display data: EOS - Transactions of the American Geophysical Union, v. 72, p. 441.

Wharton, R. A., Jr., McKay, C. P., Clow, G. D., and Andersen, D. T., 1993, Perennial ice covers and their influence on Antarctic lake ecosystems, in Green, W. J., and Friedmann, E. I., eds., Physical and Biogeochemical Processes in Antarctic Lakes: Antarctic Research Series, p. 53–70.

Wilson, K. M., Pollard, D., Hay, W. W., Thompson, S. L., and Wold, C. N., 1994, General circulation model simulations of Triassic climates: preliminary results, in Klein, G. d. V., ed., Pangaea: Paleoclimate, Tectonics, and Sedimentation During Accretion, Zenith, and Breakup of a Supercontinent: Geological Society of America Special Paper 288, p. 91–116.

Wold, C. N., and DeConto, R. M., in press, Proxy formation model used to predict the locations of Late Cretaceous evaporites. in Barrera, E., and Johnson, C., eds., The Evolution of Cretaceous Ocean/Climate Systems, Geological Society of America Special Publication.

Zerzan, J. M., 1989, Overlap: A FORTRAN program for rapidly evaluating the area of overlap between two polygons: Computers & Geosciences, v. 15, p. 1109–1114.

Ziegler, A. M., Barret, S. F., and Scotese, C. R., 1981, Palaeoclimate, sedimentation and continental accretion: Royal Society of London Philosophical Transactions, ser. A, v. 301, p. 253–264.

Ziegler, A. M., Raymond, A. L., Gierlowski, T. C., Horrell, M. A., Rowley, D. B., and Lottes, A. L., 1987, Coal, climate and terrestrial productivity: the present and early Cretaceous compared, in: Scott, A. C., ed., Coal and Coal-bearing Strata: Recent Advances: London, Geological Society Special Publication No. 32, p. 25–49.

A Coupled Physical-Chemical-Biological Model for the Western Baltic

K. Fennel and T. Neumann

1 Introduction

The shallow and strongly stratified Baltic Sea, with only narrow channels to the North Sea is very sensitive to anthropogenic influences. Increased nutrient inputs during the last decades have led to a significant eutrophication especially in the nearshore regions (Nehring 1992; Nehring and Matthäus 1991; Wulff 1990). Indications of changes in the phytoplankton biomass and species composition have been reported (Schulz and Kaiser 1986; Wulff et al. 1990).

The Baltic Sea is characterized by a high variability of physical, chemical and biological parameters on mesoscale and other spatial and temporal scales. Natural variations, mainly caused by fluctuations of the hydrographical conditions, overlie longterm changes. Stratification, topography, vertical mixing, mesoscale circulation patterns, external forcing as well as chemical-biological transformations between the various inorganic and organic compounds are controlling factors for changes in the total amount and distribution of nutrients and plankton. Complex non-linear interactions exist between the compartments of the ecosystem. A better understanding of the interactions of physical transport processes and chemical-biological dynamics is required. A coupled chemical-biological and circulation model is used for the western Baltic to study the effects of variable physical forces on the dynamics of nutrients and plankton (Fennel and Neumann 1996). The most important compartments of the system are included so that the model is able to grasp the interactions of mesoscale transport processes and chemical-biological transformations.

Model simulations over 3 months were performed for two years, 1994 and 1995, with realistic external forcing to investigate interannual and regional differences in the dynamics of nutrients and plankton. In addition, model experiments with different horizontal resolution were carried out to investigate the influence of grid resolution.

2
The Coupled Model

We use the Modular Ocean Model (MOM 1) which is a version of the primitive equation ocean model of Bryan and Cox (GFDL-model). The model area covers the southwestern Baltic Sea (Fig. 1) with a horizontal grid scale of 1 nautical mile, and a vertical resolution of 2 m for the first 12 layers and gradually increasing for deeper layers with a total of 22 layers. The model area is closed by solid walls. This model is able to produce mesoscale current patterns which evolve in response to the wind (see for example Fennel and Seifert 1995). The horizontal turbulent viscosities and diffusivities are constant. For the vertical mixing we use the Richardson-number model of Pacanowski and Philander (1981). The numeric values of the mixing parameters are given in Table 1.The model is forced by realistic wind and solar radiation measured at the weather station Arkona and the IOW mast-station at the Darss sill.

The chemical-biological dynamics is introduced by a simple model which describes the behaviour of nutrients and plankton with four state variables: limiting nutrient (N), phytoplankton (P), zooplankton (Z), and detritus (D) (Fennel 1995). The chemical-biological model is incorporated into the circulation model by means of advection-diffusion-equations for the state variables which are considered as non-conservative tracers. The equations for the chemical-biological state variables read explicitly:

$$\frac{\partial N}{\partial t} + v \cdot \nabla N - A\Delta N = -r\frac{N^2}{\alpha^2 + N^2}P + L_{PN}\left(P - P_0\right) + L_{ZN}\left(Z - Z_0\right) - L_{DN}D$$

$$\frac{\partial P}{\partial t} + v \cdot \nabla P + W_{SP}\frac{\partial P}{\partial z} - A\Delta P = r\frac{N^2}{\alpha^2 + N^2}P - \beta\left(1 - \exp\left(-I^2P^2\right)\right)Z - L_P\left(P - P_0\right)$$

$$\frac{\partial D}{\partial t} + v \cdot \nabla Z - A\Delta Z = \beta\left(1 - \exp\left(-I^2P^2\right)\right)Z - L_Z\left(Z - Z_0\right)$$

$$\frac{\partial D}{\partial t} + v \cdot \nabla D + W_{SD}\frac{\partial D}{\partial z} - A\Delta D = L_{PD}\left(P - P_0\right) + L_{ZD}\left(Z - Z_0\right) - L_{DN}D.$$

Here w_{SD} and w_{SP} are the sinking speeds of dead and live particles, respectively. We assume that dead particles sink at a constant rate. Since the sinking of phytoplankton occurs event-like rather than a steady process (see for example Bodungen et al. 1981), we assume a variable sinking velocity for phytoplankton, following the approach of Stigebrandt and Wulff (1987).

The role of light is included in the growth rate r. We adopt the PI formula of Steele (1962). The photosynthetically active radiation is controlled by the light attenuation due to water and self-shading of phytoplankton.

The loss rates L_{PN}, L_{ZN}, L_{PD} and L_{ZD} specify how much of the loss of phytoplankton (L_P) and zooplankton (L_Z) is transferred to nutrients N and detritus D. The rate L_{DN} prescribes the recycling of detritus to nutrients. The background

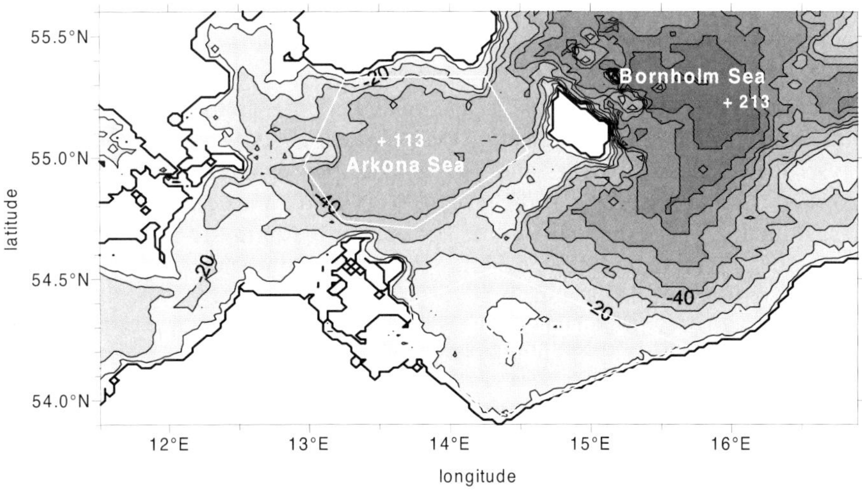

Fig. 1 The model area is the southwestern Baltic. The monitoring stations *113* and *213* are indicated

Table 1 Physical and biological parameters

A_H	Horizontal eddy diffusivity	10E+5 $cm^2\,s^{-1}$
A_M	Horizontal eddy viscosity	10E+5 $cm^2\,s^{-1}$
FRMXT	Maximal vertical diffusivity	50 $cm^2\,s^{-1}$
FRMXU	Maximal vertical viscosity	50 $cm^2\,s^{-1}$
BVDC	Background vertical diffusivity	0.00134 $cm^2\,s^{-1}$
BVVC	Background vertical viscosity	0.0134 $cm^2\,s^{-1}$
a	Half-saturation constant	0.3
r_{max}	Maximal growth rate	1 day^{-1}
β	Maximal grasing rate	0.5 day^{-1}
I	Ivlev constant	1.1
L_{PN}	Loss P - N	0.02 day^{-1}
L_{PD}	Loss P - D	0.02 day^{-1}
L_{ZN}	Loss Z - N	0.01 day^{-1}
L_{ZD}	Loss Z - D	0.02 day^{-1}
L_{DN}	Loss D - N	0.02 day^{-1}
w_{SD}	Sinking velocity for D	500 cm day^{-1}
w_{SP}	Sinking velocity for P	120 P^2 cm day^{-1}
P_0	Background concentration of P	0.5 mmol N m^{-3}
Z_0	Background concentration of Z	0.5 mmol N m^{-3}

values P_0 and Z_0 are the winter concentrations which serve also as initial concentrations. The numeric values for the biological parameters are given in Table 1.

The solid walls bordering the model area affect the circulation patterns and influence the chemical-biological processes nearby. Winds parallel to the walls cause upwelling of nutrients and increase the phytoplankton concentrations. However, the spatial extent of these patterns is small. It is described by the first baroclinic Rossby-radius, which does not exceed 10 km in this area. We assume no significant influence on the Arkona Sea and the Pomeranian Bight, especially during the short model period.

3 Results

3.1 Interannual and Regional Variability of the Chemical-Biological Variables

Simulations for 1994 and 1995 were carried out to investigate the interannual and regional variations of chemical-biological variables according to the external forces. The simulations start after a spinup period of 5 days for the circulation model on 1 March. The initial distributions of temperature and salinity were computed from climatological data and are identical for both model runs. The initial concentrations of chemical-biological variables were set at 0.5 mmol N/m^3 for plankton and detritus. For nutrients a winter concentration of 5 mmol N/m^3 was assumed, following Nehring (1991). Both model runs differing only in external forcing, i.e., wind stress, atmospheric temperature and solar radiation. In response to the wind, a rich spectrum of mesoscale current patterns is generated which transport and redistribute material.

Time series of the phytoplankton variable, converted to mg chlorophyll/m^3, are shown in Figs. 2 and 3 for the locations of the monitoring stations 113 (Arkona Sea) and 213 (Bornholm Sea). The spring bloom can be seen in both years with high concentrations of chlorophyll, followed by a period with nearly stationary concentrations. Only short and sporadic fluctuations in phytoplankton values occur after the decline of the bloom.

The spring bloom event can be regarded as adjustment of the pelagic system to the changing physical state of water column. According to Sverdrup (1953) the bloom starts when the mixed layer is shallower than the critical depth. Generally, this criterion is not fulfilled before the formation of thermocline, which is mainly controlled by meteorological forcing and density stratification. The high phytoplankton concentrations during the bloom start to decrease when the limiting nutrient is exhausted. Afterwards, a nearly stationary biochemical cycle exists in the stratified water column as shown during May in Figs. 2 and 3. There is a balance between phytoplankton sources (growing after nutrient uptake) and sinks (losses by metabolism, grazing by zooplankton and sinking). Nutrient injections from deeper parts of the water column can disturb this balance and in-

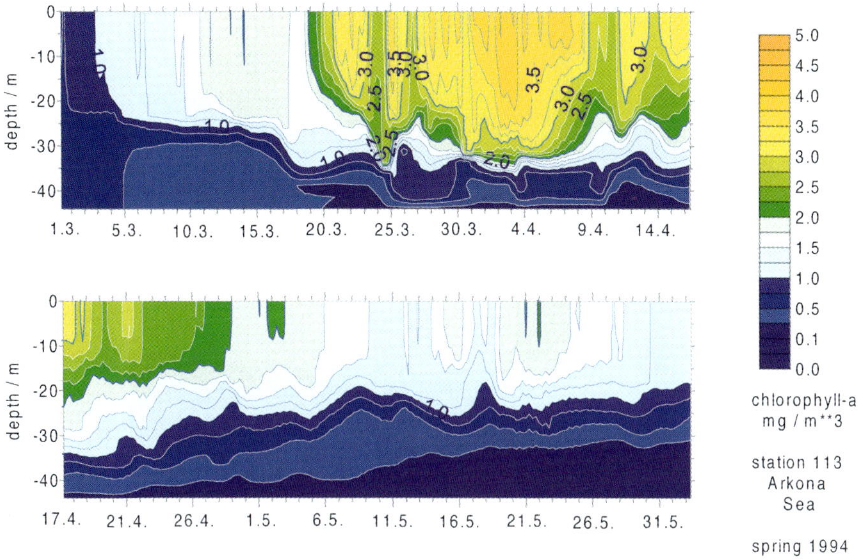

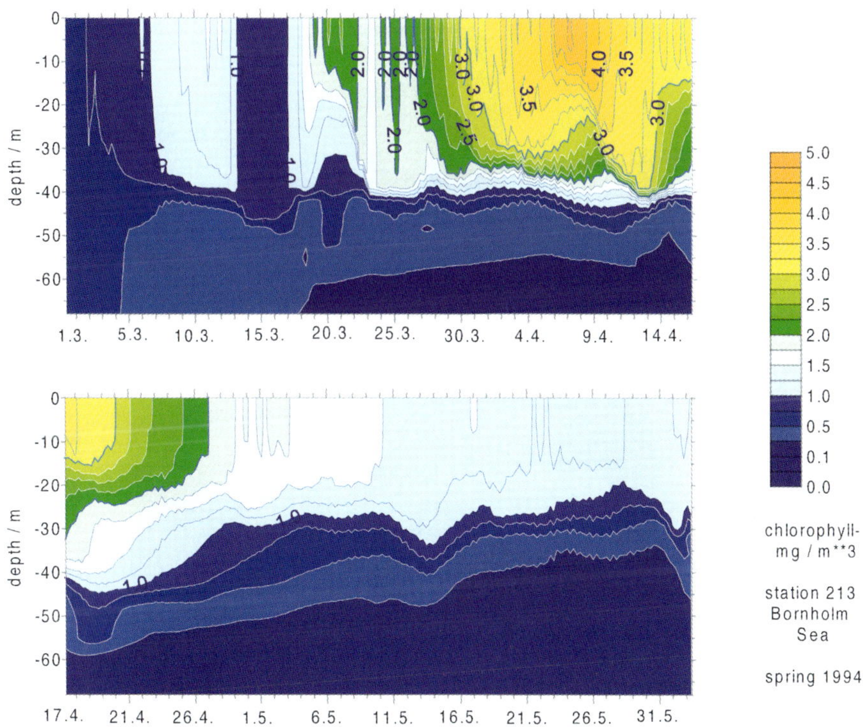

Fig. 2 Time series of model phytoplankton (vertical profiles) at the stations *113* and *213* in 1994

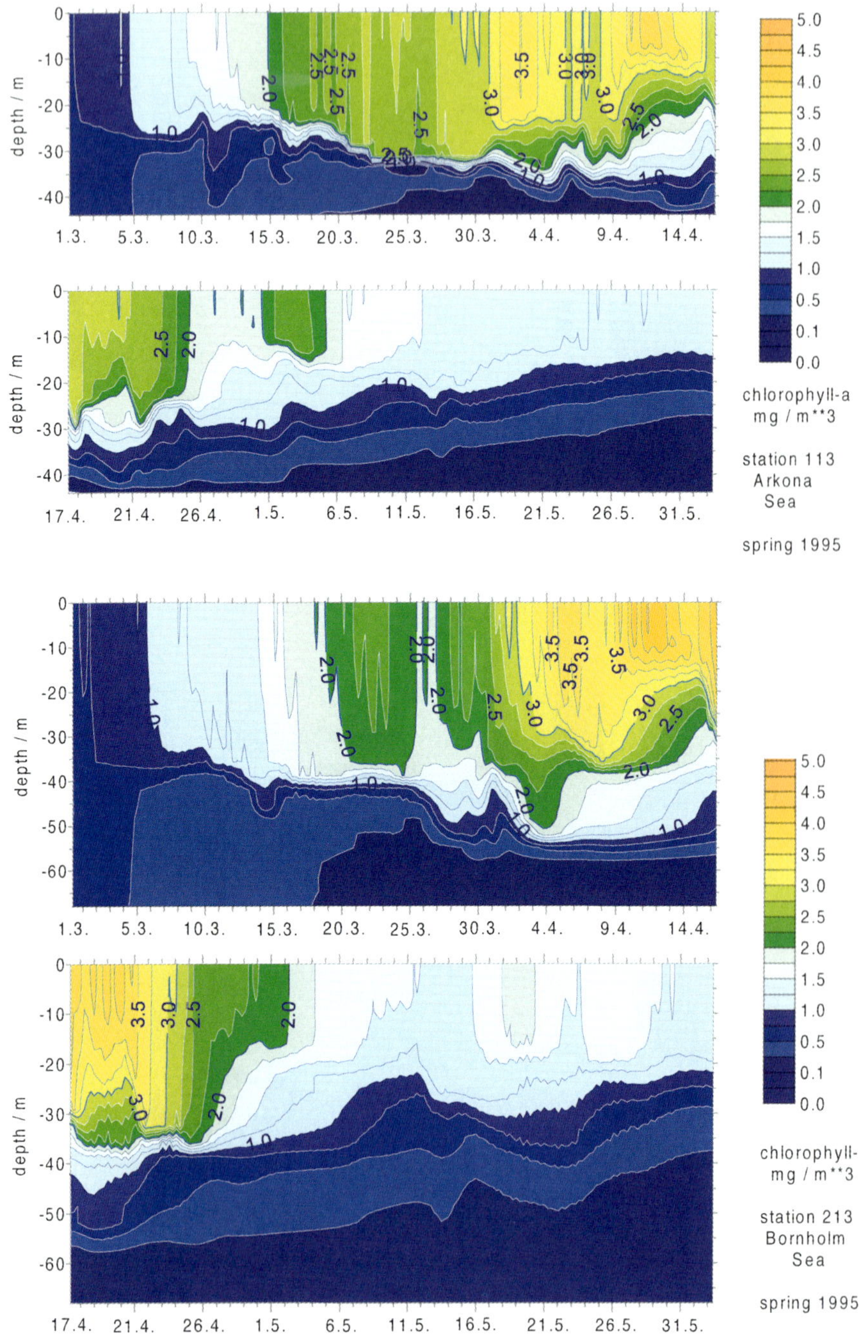

Fig. 3 Time series of model phytoplankton (vertical profiles) at the stations *113* and *213* in 1995

crease the concentrations of phytoplankton for a few days, e.g. 21.5.1994 at station 113 and 18.5.1995 at station 213.

Timing, duration and maximum values of the blooms differ regionally and interannually. There is a regional delay in timing of the bloom. It starts in the shallow regions of the model area like Greifswalder Bodden and nearshore regions, where the phytoplankton cannot leave the euphotic zone by sinking and hence the formation of thermocline is not necessary to satisfy the Sverdrup criterion. Later, in response to solar heating, a thermocline develops in the deeper basins and the concentrations increase there first in the Pomeranian Bight, the Mecklenburg Bight, the Arkona Sea and finally in the Bornholm Sea. In order to illustrate the differences in the commencement of the bloom a few selected horizontal distributions of chlorophyll are shown in Fig. 4. Comparing the time series of station 113 and 213 in 1994 (Fig. 2), it can be seen that the maximum concentrations occur 5 days later at station 213. This regional delay was observed in the Baltic and explained by the hydrographical differences of the regions (Kaiser and Schulz 1978).

Comparing the model runs of 1994 and 1995, we also find that the maximum values are reached in 1995 about 1 week later than 1994. The reason has to be the variable meteorological forcing, because it is the only difference in both model runs. Especially the wind stress is different in both years. It influences the depth of the mixed layer and explains the interannual differences. In 1994 moderate winds not stronger than 10 m/s from the end of March to the beginning of April led to a relatively shallow wind-mixed layer which supported the growth of phytoplankton. Maximum concentrations of phytoplankton were reached by 31 of

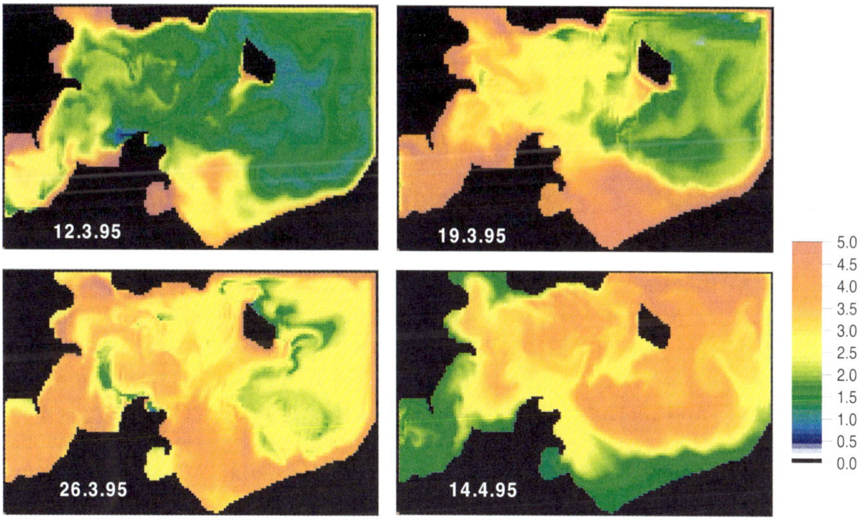

Fig. 4 Selected horizontal distributions of model phytoplankton. The uppermost level is shown in mg chlorophyll m^{-3} units

March. During the same time period in 1995, the winds were stronger, with speed exceeding 10 m/s, and therefore a deeper mixed layer existed until 9 of April which reduced the growth of the phytoplankton. Later, the winds decreased and an exponential growth of the phytoplankton started. During April, May and June of 1994, the frequency of strong winds was higher than in the same period in 1995. Nutrients are exhausted in the euphotic zone during this time and strong winds are an effective mechanism for transporting nutrients from deeper regions of the water column into the euphotic zone, e.g. 21.5.1994 at station 113 (Fig. 2).

3.2 The Influence of the Grid Resolution on Tracer Concentration

The values of chemical-biological variables in the model are averaged over grid boxes of 1 nautical mile × 1 nautical mile × 2 m size. Processes on smaller scales than grid size are not resolved explicitly, e.g. the chemical-biological nutrient uptake or the physical turbulence. These sub-grid processes are described via parametrizations by processes having spatial scales comparable to the grid resolution or larger.

There are also parts of the circulation pattern with spatial dimensions in the range of grid size, e.g. mesoscale circulation. The choice of grid width determines which parts are resolved explicitly and which are subgrid processes and treated by parametrizations. Because of the non-linear coupling of chemical-biological dynamics and circulation the question arises: how are the chemical-biological interactions influenced by changes of the model resolution?

Model experiments with the same parameters and forcing, but different horizontal resolution, were performed. We compare two simulations: a fine scale model run with the horizontal grid width of 1 nautical mile and a coarse scale model run with twice the horizontal grid width and twice the horizontal mixing coefficients. The doubling of mixing coefficients is motivated by its linear dependence on gridsize in the stability criterion of grid-Reynolds number

$$A_H > \frac{u_{\max} \Delta x}{2} ,$$

where A_H is the horizontal mixing coefficient, $u_{\max}$ is the maximal advection velocity and Δx the horizontal gridscale.

The circulation pattern shows different features in subregions of the model area in relation to the resolution. The fine scale model produces, for instance, typical currents with spatial extent between 5 and 10 nautical miles in the Arkona Sea (Figs. 5 and 7). In the Pomeranian Bight the velocity field is more or less homogeneous. The coarse scale model is not able to resolve the currents in the Arkona Sea properly (Fig. 6), i.e., those features became partially subgrid process by the doubling of the horizontal grid width.

Distribution patterns of the chemical-biological variables do roughly agree in both models. Nevertheless, there are differences comparing, for instance, Figs. 5

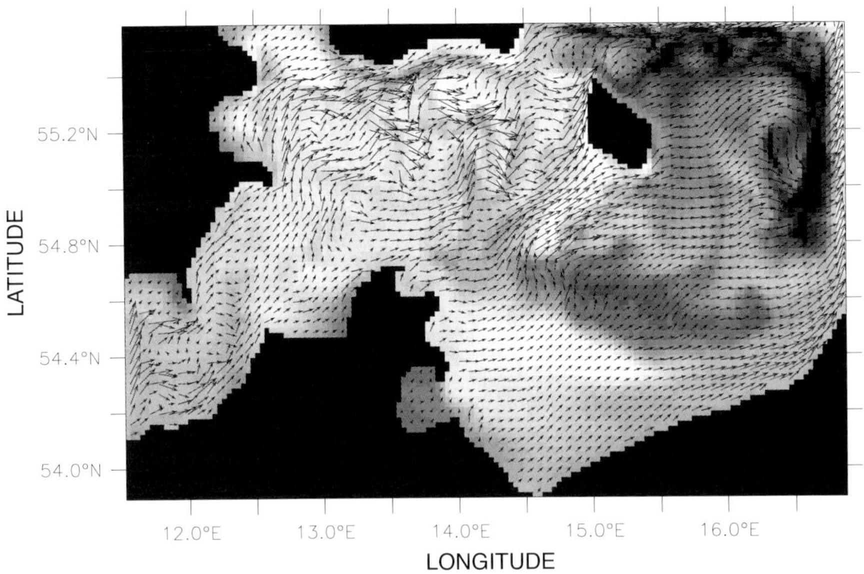

Fig. 5 Relative phytoplankton distribution and velocity field in the uppermost model level of the fine-scale model at day 29 (18.4.95). Only every 2nd velocity vector is drawn

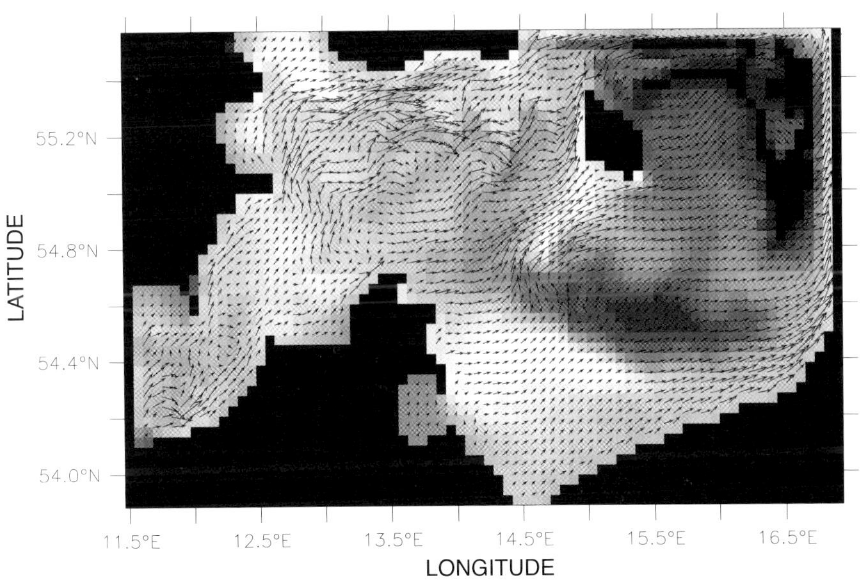

Fig. 6 Relative phytoplankton distribution and velocity field in the uppermost model level of the coarse-scale model at day 29 (18.4.95). Every velocity vector is drawn

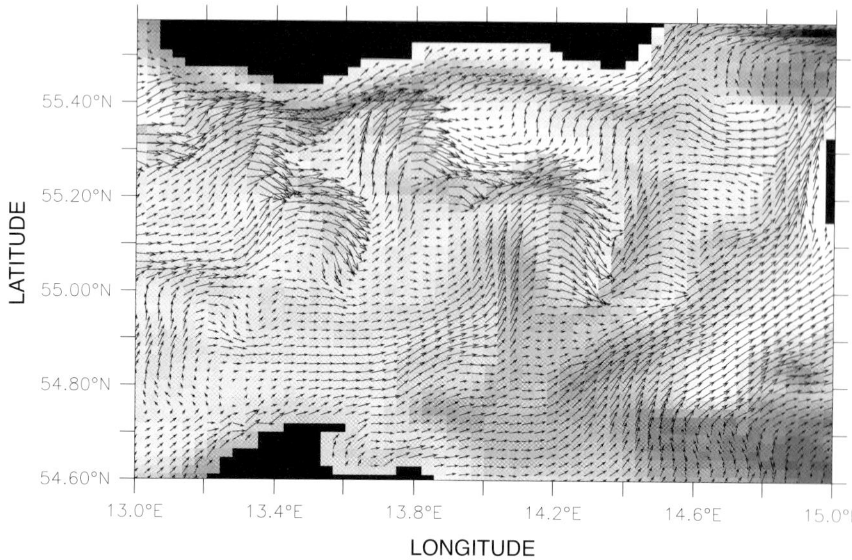

Fig 7 Detail of Fig. 5 showing currents in the Arkona Sea. Every velocity vector is drawn

and 6, which show the distributions of phytoplankton at model day 29. In the Arkona Sea especially, the structures and filaments in the fine-scale model are not reproduced in the coarser-scale model. The nearly unstructured distributions in the Pomeranian Bight are not effected much.

Averages of the chemical-biological variables over the upper 10 m of water column in the Arkona Sea and the Pomeranian Bight were computed to investigate the influence of resolution quantitatively (Fig. 8). The horizontal extent of the areas used for averaging is indicated in Fig. 1. In the Arkona Sea the values show remarkable deviations. The difference of phytoplankton concentration reaches values of 20% during the spring bloom, the nutrient concentration 30%. The average nutrient concentrations converge after the decline of the bloom. The difference of phytoplankton values decreases but the zooplankton concentrations show continuous deviations of 20%. In contrast, the values of the Pomeranian Bight agree well in both models.

A higher consumption of nutrients and production of phytoplankton occurs in the Arkona Sea during the spring bloom in the fine scale model run. The currents in the Arkona Sea cause a higher advective transport of the tracers in the fine-scale model run and make more nutrients available. An increased availability of nutrients supports higher growth rates of phytoplankton, which, in turn, decrease nutrient and increase phytoplankton concentrations.

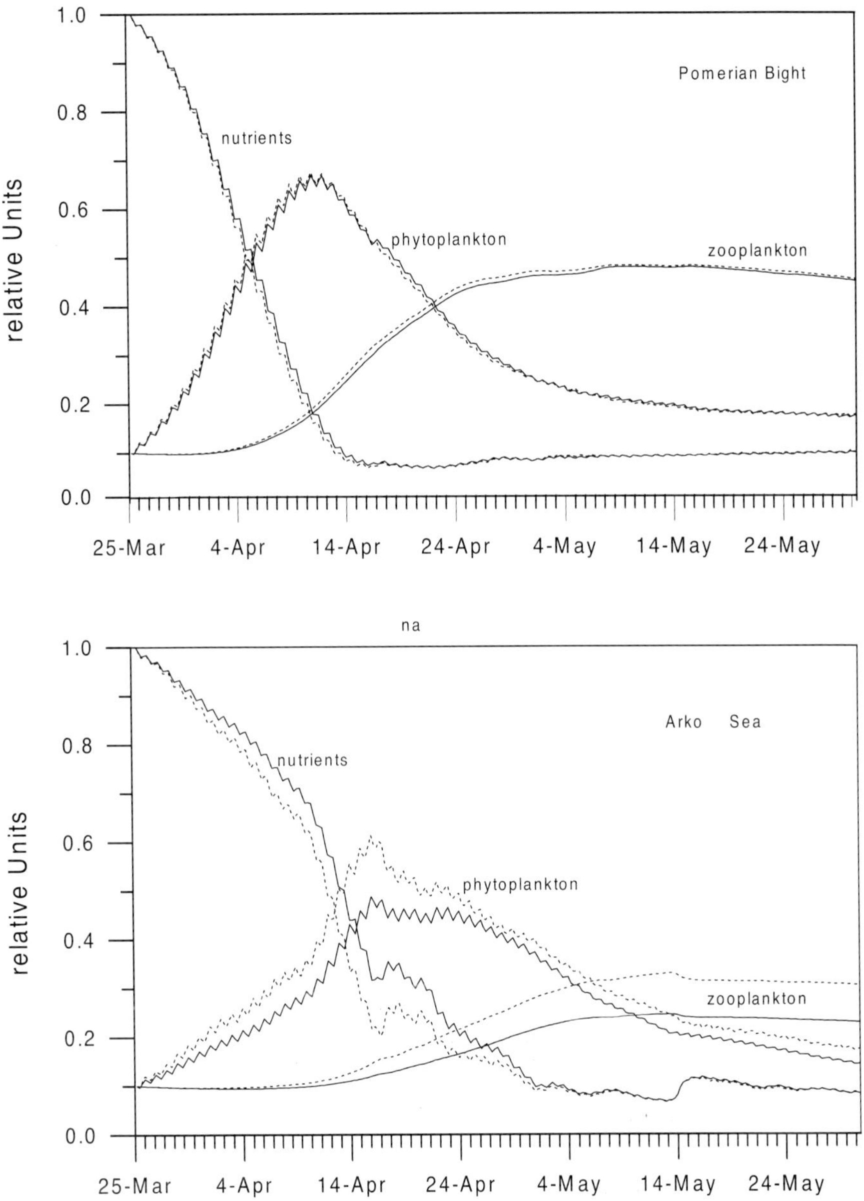

Fig. 8 Mean values for chemical-biological variables in two sub-regions of the model area. The *dashed lines* correspond to the fine-scale model run and the *solid lines* to the coarse-scale run, respectively

4 Discussion

A variety of mesoscale patterns in nutrient and plankton distributions are generated by the external forcing, and these correspond qualitatively with the satellite images of chlorophyll distributions (for instance in Siegel et al. 1996). The course of nutrient and plankton dynamics is well described by the model. Interannual differences are due to the variations in meteorological forcing. Regional differences are due to bottom bathymetry and background stratification.

The resolution of mesoscale advective processes is important for the chemical-biological dynamics. A change in grid size has a strong influence where spatial scales of advective processes are comparable to grid width. Processes which are not resolved after increasing the grid size still influence the chemical-biological dynamics. If the grid is fine enough to resolve the advective processes properly, then the differences of both results are negligible. Increasing grid scale filters out parts of the mesoscale variability and may cause significant deviations of the concentration and distribution of the chemical-biological state variables.

Acknowledgements

The authors thank Prof. Tom Osborn for revising the language. The investigations were partly supported by the project TRUMP of the Bundesministerium für Forschung und Technologie (project number: 03F0105B).

Institut für Ostseeforschung Contribution Number 245.

References

Bodungen vB, Bröckel vK, Smetacek V, Zeitzschel B (1981) Growth and sedimentation of the phytoplankton spring bloom in the Bornholm Sea (Baltic Sea). Kieler Meeresforsch Sonderh 5:49–60

Fennel W (1995) A model of the yearly cycle of nutrients and plankton in the Baltic Sea. J Mar Syst 6:313–329

Fennel W, Neumann T (1996) The mesoscale variability of nutrients and plankton as seen in a coupled model. DHZ 48 Nr. 1:49–71

Fennel W, Seifert T (1995) Kelvin wave controlled upwelling in the western Baltic. J Mar Syst 6:289–300

Kaiser W, Schulz S (1978) On the causes for the differences in space and time of the commencement of the phytoplankton bloom in the Baltic. Kieler Meeresforsch Sonderh 4:161–170

Nehring D (1992) Hydrografisch-chemische Langzeitveränderungen und Eutrophierung der Ostsee. Wasser Boden 44:632–638

Nehring D, Matthäus W (1991) Die hydrografisch-chemischen Bedingungen in der westlichen Ostsee im Jahre 1991. Deutsche Hydrogr Zeitung 44:217–238

Pacanowski RC, Philander SGH (1981) Parametrization of the vertical mixing in numerical models of the trophic ocean. J Phys Oceanogr 11:1443–1451

Schulz S, Kaiser W (1986) Increasing trends in plankton variables in the Baltic Sea – a further sign of eutrophication? OPHELIA Suppl 4:249–257

Siegel H, Gerth M, Schmidt T (1996) Water exchange in the Pomeranian Bight investigated by satellite data and shipborne measurements. Cont Shelf Res 16 No 14:1793–1817

Stigebrandt A, Wulff F (1987) A model for the nutrients and oxygen in the Baltic proper. J Mar Res 45:729–759

Sverdrup HU (1953) On the conditions for the vernal blooming of phytoplankton. J Cont Int Explor Mer 18:287–295

Wulff F, Stigebrandt A, Rahm L (1990) Nutrient dynamics in the Baltic Sea. AMBIO 19 No 3:126–133

Wulff A, Ærtebjerg G, Nicolaus G, Niemi A, Ciszewiski P, Schulz S, Kaiser W (1986) The changing pelagic ecosystem of the Baltic Sea. OPHELIA Suppl 4:299–319

Steele J (1962) Environmental control of photosynthesis in the sea. Limnol Oceanogr 7:137–150

Field and Numerical Studies of Near-Bed Aggregate Dynamics

W. Ritzrau and H. Fohrmann

1 Introduction

The small-scale distribution and the physical and biological modification of particulate matter influence near bottom transport processes of particles and consequently their incorporation in the sediment. Early on in sedimentology, major focus was set on the size distribution of particles above the sea bed and in the sediments. The vertical distribution of different particle size classes and its relationship to the hydrodynamic regime was empirically investigated (Rouse 1937). Only recently have biologists become interested in the water mass above the sediment-water interface. However, the dynamics of particles, their interaction and modification within the last few meters above the seafloor (the benthic boundary layer – BBL) is in many terms poorly understood.

Mc Cave (1984) proposed a more or less even distribution of the mass of the various existing size classes of particles in the deep ocean and identified a hierarchy in the encounter processes controlling particle interaction close to the seafloor. Differential settling, the scavenging of small particles by larger faster-sinking particles (Stolzenbach 1993) is the dominant mechanism of particle interaction, followed by turbulent encounter and Brownian motion. However, the dominance of either of these processes is strongly dependent on the hydrodynamic regime and the prevailing size distribution of particles.

The aggregation theory of particles was mainly applied to processes in the upper ocean, for example to explain sedimentation events of phytoplankton from the euphotic zone (Jackson 1990). Hill and Nowell (1990) applied the mathematical formulations of particle encounter in the context of particles settling through the water column. In well-controlled laboratory experiments they verified and improved the aggregation theory for different particles in the sub- and super Kolmogorov size range. Only recently, they applied the aggregation theory to the boundary layers of shelf environments (Hill and Nowell 1995). Following the modification of a given particle community over time, they showed in numerical experiments that various processes have to be taken into account. The general size distribution found in the ocean is not necessarily altered by particle interaction. One basic assumption of their layer-averaged model is the absence of strong vertical concentration gradients of particles in the simulated layer, an assumption which is only valid for the upper bottom nepheloid layer. However, strong gradients of particle concentration and composition have been observed in the benthic boundary layer close to the seafloor (Thomsen et al. 1994, 1995; Townsend et al. 1992; Ritzrau 1996). To interpret these field observations and to answer the question whether aggregate formation and disintegration influence not only the vertical distribution of the concentration of suspended matter but also the aggregate size distribution would require an expanded multiple-layer model with a fine scale resolution close to the seafloor.

A different approach was proposed by Boudreau (1997), who developed a theory to describe the dynamics of exchange processes across the sediment-water interface. His analytical model comes to one exact solution for the vertical distribution for a single particle size class and a defined hydrodynamic regime. This approach is comparable to the classical description of the suspension distribution close to the seafloor (Rouse 1937). Here, the ratio of the settling velocity (w_S) of a distinct particle size class over the product of the friction velocity times the van Karman's constant (w_S/u^*k) determines the shape of the exponential concentration distribution. The combination of the Rouse equation defining the near-bed concentration distribution and the Karman-Prandtl equation (the "law of the wall") describing the velocity distribution above the seafloor allows an estimation of the horizontal flux of distinct particle classes (Muschenheim 1987). This approach led to the idea of hydrodynamic sorting of particles with different settling velocities. Consequently, distinctly different fluxes of particles with different densities and hence settling velocities should occur close to the seafloor. Thus, particle interactions, changing the size and property distribution of aggregates continuously, would need to be considered as well.

The grain size distribution in the sediments at the seafloor, which is frequently resolved down to the submicron size class, does not necessarily tell how these particles arrived, let alone describe the mode of their transport in suspension. Yet the understanding of the dynamics of particle interactions and modification within the benthic boundary layer is a key parameter to interpreting the sedimentary record in terms of sediment origin and transport paths. Residence times of particles can be extremely prolonged, since, for example, rebound particles can stay in suspension without any contact with the sediment (Walsh et al. 1988). Resuspended particles may also enter the dynamic modification process and be redistributed as suspension over wide distances.

During the past few years, the benthic boundary layer in shallow areas (Ritzrau and Graf 1992) as well as at continental margins (Thomsen et al. 1994, 1995; Ritzrau 1996; Ritzrau and Thomsen 1997; Townsend 1992) has been identified as a distinct layer of water close to the seafloor. Here, not only the composition but also the microbiological modification of particles differs significantly from the major portion of the water column. In the BBL the concentration of particulate organic carbon is significantly higher (Fig. 1) and the particles tend to be richer in organic matter than in the water column above (Thomsen and Graf 1995). The data presented in Fig. 1 combine samples from 21 stations at the East Greenland continental margin over a wide depth range from 136 to 2808 m

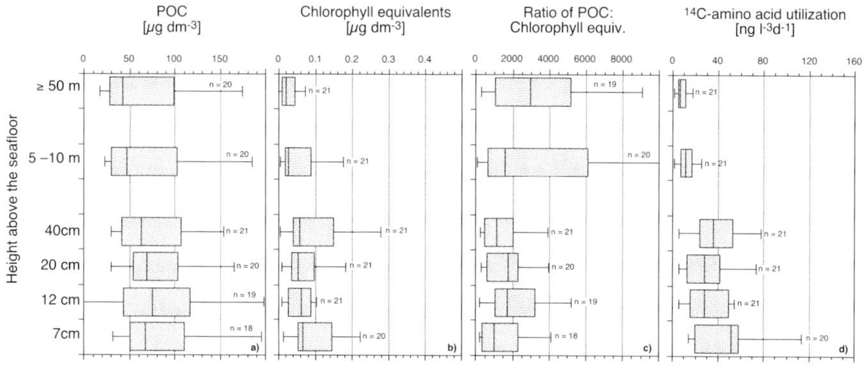

Fig. 1a–d Vertical distribution of particulate organic carbon **(a)**, chlorophyll equivalents **(b)**, ratio of POC to chlorophyll equivalents **(c)** and ^{14}C-utilisation **(d)** in the benthic boundary layer (7, 12, 20 and 40 cm height above seafloor) and in the IWC (≥ 50 and 5–10 m hight above seafloor). The presented data represent 21 stations at the East Greenland continental margin over a depth range from 136 to 2808 m taken in 1993 and 1994 during the RV Polarstern cruises ARK IX 2 and 3 and ARK X/I, respectively. Samples were taken in the North East Water Polynya between 77° and 81°N and between 5° and 16°W from May through July 1993 and on a slope normal transect at 75°N in July 1994. Lines within the boxes indicate the median of the data. Upper and lower limits of the boxes represent 25 and 75%, respectively; and the limits of the bar ends, 10 and 90%, respectively, of the data of each presented parameter. The box plots represent 118 samples for POC **(a)**, 125 for chlorophyll equivalents **(b)**, 89 for the ratio of POC to chlorophyll equivalents **(c)** and 125 for the microbial activity **(d)**. The abundances of data analysed for each parameter and height are listed next to the box plot.

water depth (see legend of Fig. 1 for details). Increases in microbial biomass after resuspension events have been observed in the laboratory (Wainright 1987; 1990) as well as in the field (Ritzrau and Graf 1992). Microbial utilisation of labile organic substrates (^{14}C amino acids) appears to be significantly faster (Fig. 1; Ritzrau 1996), and microbial production is significantly higher compared to the water column above (W. Ritzrau, unpubl. data). Although no statistically significant differences of the four displayed parameters between the sampling heights in the BBL were detectable (Fig. 1a–d, Kruskal-Wallis test $p \geq 0.15$), there seems to be a trend within the distribution of concentrations and microbial rates in the BBL. A hyperbolic distribution pattern of the medians with higher values in close vicinity to the seafloor and further above the bed, and a decrease in concentrations or rates in between. Similarly, the contribution of phytogenic material to the total particulate organic matter (ratio of POC to chlorophyll equivalents, Fig. 1c), i.e. the composition of aggregates changed also with height above the seafloor. Low values at 7 and 40 cm above the seafloor suggest that aggregates with potentially higher food value accumulate in these layers. These patterns suggest that hydrodynamic sorting of particles or aggregates with different composition and properties may occur over vertical scales < 0.5 m. Since the distribution of some parameters like POC (Fig. 1a) or also total suspended matter (data not shown) is rather homogeneous, a methodological artefact for the dis-

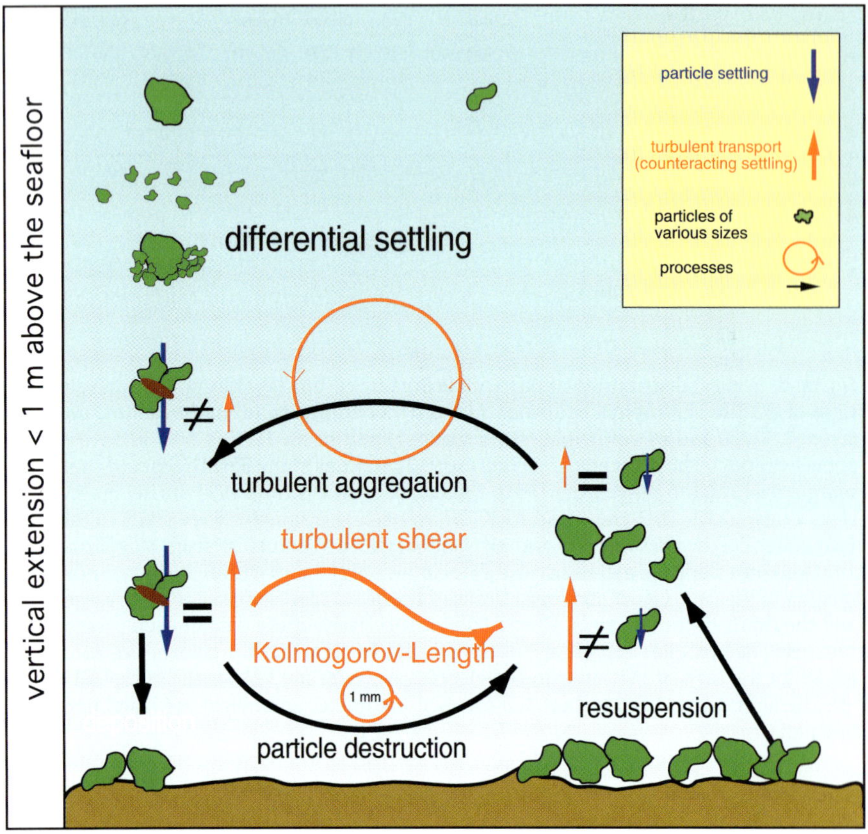

Fig. 2 Conceptual model of processes influencing aggregate dynamics in the benthic boundary layer (After Graf and Ritzrau 1997). Processes indicated were numerically implemented in the model. The hydrodynamic regime is described by using simple standard equations (See text for details).

tribution of microbial activity and the ratio of POC to chlorophyll equivalent can be excluded. These empirical findings lead to a conceptual model which schematically incorporates the various processes which influence the particle distribution in the benthic boundary layer (Fig. 2). Since most particles in the BBL occur as aggregates, i.e. composites of various smaller components, this model distinguishes areas of aggregate production due to aggregation or differential settling and areas of aggregate disintegration. If these interactive processes cause an increase or decrease of aggregates in one distinct size class, they cause a decrease or increase in the other classes, respectively. For example, aggregate disintegration results in smaller fragments (smaller particles), which are consequently redistributed within the BBL by the hydrodynamic regime. Thus, not only settling of aggregates but also aggregate interaction and disintegration determine the concentration distribution of suspended matter close to the seafloor. However, the vertical extension of these processes is unknown and may be smaller or larger than the range of sampling heights in the BBL displayed in Fig. 1a–d; but it is most likely linked to the hydrodynamic regime. The dynamics of these aggregation-disaggregation cycles are determined by the composition and the properties of the involved aggregates and may be different for different aggregate types. The crude field measurements of bulk parameters, like the concentration of total suspended matter or particulate organic carbon, do not allow for such discrimination of aggregate properties, which are required to evaluate the processes of interest. In contrast, microscopic investigation of the aggregate images and the resulting size class distribution for different heights above the seafloor support the idea of hydrodynamical sorting in the BBL. It has to be kept in mind, how-ever, that the observed preserved particles only potentially represent fragments of aggregates *in situ*. Sample handling most likely destroys their original configuration.

Moreover, since the hydrodynamic regime is very dynamic close to the seafloor, bulk measurements of particle composition do not allow for discrimination of the various aggregate sizes and properties. This issue has been approached by applying optical methods in the field. Aggregate size distribution, abundance and settling velocity were studied in various environments by analysing photographic images of particles taken *in situ* (Milligan 1995; Asper 1986, 1987; Chen et al. 1994; Rathmeyer and Wefer 1996). In a similar attempt, using an *in situ* video imaging system combined with a bottom water sampler, particles in the BBL were observed at the East Greenland continental margin (Thomsen and Ritzrau 1996). This device videotaped particles at defined distances close to the seabed. The recorded video sequences allowed not only the size but also the behaviour and speed of aggregates in the flow to be estimated.

Based on this experimental evidence and clues, and in contrast to the approaches described so far, a numerical multilayer 1-D diffusion-advection model was developed to investigate particle or aggregate dynamics in the BBL, under inclusion of effects of particle interactions on the particle distribution in the BBL. Initially, the basic setup of the presented model used distinct particle size classes. With inclusion of particle interactions, aggregates are formed or destroyed. However, for simplicity, we prefer to use the term aggregate to describe the model results. A few case studies are included to display the strength and sensitivity of this model.

The model presented includes the particular hydrodynamic regime of the benthic boundary layer and provides a continuous distribution of various distinct aggregate size classes from the seafloor into the water column, as suggested by Newberger and Caldwell (1981). The features of the different aggregate classes and the settings of the hydrodynamic regime can be changed easily. The advantage of the numerical simulation over the analytical solution is that the model provides not only the final distribution pattern for selected aggregate classes, but also the temporal development of their distribution pattern. The model allows this approach either for a specific height above the seafloor or for the entire benthic boundary layer.

Implementing aggregate interaction like differential settling, turbulent aggregation and/or aggregate disaggregation as a simple reaction term in the diffusion-advection equation at each time step describes the natural process rather straightforwardly. At each time step, the newly formed aggregate is redistributed according to its "new" properties. Thus, the influence of aggregate interaction on the continuous concentration distribution of various aggregate size classes becomes transparent.

2
Model Description

The used numerical model consists of 85 layers with logarithmic grid spacing, ranging from 1 mm at the lower boundary at the sediment-water interface to 1660 m at the upper boundary. Due to the specific interest in particle dynamics in the BBL, particular emphasis was set on a high model resolution in the vicinity of the seafloor (58 model layers for 0–1 m above the sediment). The model is mass conservative. The large vertical extension (0–5600 m) eliminates effects at the upper boundary of the model.

3
The Basic Diffusion-Advection Equation

The distribution of particles in the BBL is predicted using a 1-D-advection-diffusion model described by the basic equation

$$\frac{\partial C_{i,j}(z)}{\partial t}=\frac{\partial}{\partial z}\left(A_z\frac{\partial C_{i,j}(z)}{\partial z}\right)-w_{s\,i,j}\frac{\partial C_{i,j}(z)}{\partial z}+R_{i,j}(z) \tag{1}$$

with: $C_{i,j}$ (z) the concentration ($g\,l^{-1}$) of a particle fraction i or j at height z above the sediment (m). Turbulent transport is implemented as a diffusion analogue process, which distributes aggregates upwards and downwards determined by the concentration gradient in the BBL. The eddy-diffusion coefficient [A(z) in ($m^2\,s^{-1}$)] describes the intensity of this mixing process, i.e. the higher A(z) the faster the turbulent transport in the BBL. The settling velocity $w_{s\,i,j}$ ($m\,day^{-1}$) describes the different aggregate size fractions and is the leading coefficient for the advective term in Eq. (1). The reaction term $R_{i,j}(z)$ allows processes such as aggregate interactions or aggregate modification as a function of z to be included. In the model, turbulent aggregation, differential settling and aggregate disaggregation are used to investigate aggregate interaction (see below). At every time step of the numerical simulation, aggregate interaction is computed as a function of depth and aggregate concentration. Subsequently, the produced or destroyed aggregates are added or subtracted [Eq. (13)] to or from the concentration of the aggregate size classes. At the next time step they will be redistributed again by the diffusion-advection routine.

4
Relating the Diffusion-Advection Equation to the Hydrodynamic Regime

The hydrodynamic regime is defined according to the law of the wall. For each experiment the velocity profile in the BBL is defined by the current velocity at 100 cm above the seafloor (u_{100}), by "no slip" at the sediment-water interface and by the sediment roughness height (z_o). Depending on the mean velocity $\left(\bar{u}\right)$ the friction velocity u_*, which is required in the following, can be calculated by:

$$\frac{\bar{u}}{u_*} = \frac{1}{k} \ln \left| \frac{z - z_o}{z_o} \right|. \quad (2)$$

According to Newberger and Caldwell (1981), two approaches have been used to describe the distribution of A(z) in the BBL. In one case, the coefficient $A_c(z)$ was set to a constant value for all model layers (Table 1, experiment 1). The subscript *c* indicates a constant diffusion coefficient over the total model height. However, their results suggest that A(z) should be linearly proportional to the friction velocity and the height above the seafloor. Thus, the hydrodynamic regime is taken into account and the formulation for A(z) is

$$A(z) = u_* z k, \quad (3)$$

following the standard textbooks on boundary layer hydrodynamics.

Table 1a Ranges for variables used in the numerical experiments without particle interaction included. In all experiments, aggregates were distributed among three aggregate size classes of 0.000066, 0.000128 and 0.000234 m in diameter for the according settling velocities of 30, 50 and 80 m day^{-1} (Gibbs 1985). In all nine cases, the initial conditions of each size class were 50 g m^{-3} distributed homogeneously from 0–1 m above the seafloor amounting to a total mass of 50 g in each aggregate size class. In all experiments z_o was set at 0.001 (m).

Hydrodynamic parameter	Hydrodynamic regime 1	Hydrodynamic regime 2
u_{100} (m s^{-1})	0.1	0.5
$A_{1c,1d}(z)$ (m^2 s^{-1})	0.0–0.042	0.0–1.250
u_* (m s^{-1})	0.006	0.03
Height of the boundary layer (m)	23.7	118.7
Dissipation energy, ε, (m^3 s^{-2})	9.0×10^{-11}–4.6×10^{-5}	1.1×10^{-8}–0.0058
Kolmogorov length, λ (z), (m)	0.0006–0.0161	0.00018–0.0048

Table 1b Summary of the parametrisation of the numerical experiments.

Parametrization of A(z)	Reaction term	Experiment number		
		Hydro-dynamic regime 1	Hydro-dynamic regime 2	
$A_c(z)$ = constant	–	1		
$A_{lc}(z)=u^*kz$ and for $z > h$ $A_{lc}(z) = A(h)$	–	2	3	Diffusion advection alone
	–	4	5	
$A_{ld}(z) = u^*kz$ and for $z \geq h$ $A_{ld}(z) = h/z^* A(h)$	Σ differential settling + turbulent aggregation	6	7	Particle interaction included
	Σ (differential settling + turbulent aggregation) – – dissaggregation	8	9	

In this expression k is the von Karman's constant (k = 0.41). The upper limit of the BBL is characterised by the boundary layer thickness (h) according to Smith (1977),

$$h = \frac{0.4u_*}{f} \tag{4}$$

with the Coriolis parameter (f) set to 10^{-4} s^{-1} (Hill and Nowell 1995). In the first set of numerical experiments the influence of A(z) was investigated. Table 1 summarises the parameters used in these experiments. In experiments 2 and 3, $A_{lc}(z)$ was kept constant above the height of the boundary layer,

$$A_{lc}(z) = A(h) \text{ for } z > h. \tag{5}$$

The subscript *lc* indicates a linear increase of A(z) in the bottom boundary layer and a constant value for z > h.

However, above the boundary, no vertical velocity gradient persists and thus fluid shear as the potential force of turbulence generation is greatly reduced. Thus, in experiments 4 and 5 we applied a modified parameterisation of A(z), describing a height-dependent decrease in the intensity of turbulent diffusion above the BBL. The dimensionless ratio of the height of the boundary layer over the distance from the seafloor determines the rate of this decrease and in this way includes the hydrodynamic regime in this indirectly.

$$A_{ld}(z) = \frac{h}{z} A(h) \text{ for } z > h. \tag{6}$$

Here the subscript *ld* indicates a linear increase in the BBL and a decrease above.

According to Gust (1989), the height-dependent distribution of the energy dissipation rate (ε) in the boundary layer is given by

$$\varepsilon(z) = \frac{u_*^{\,3}}{kz}, \tag{7}$$

closely related to the energy dissipation is the size distribution of the turbulent eddies in the boundary. The Kolmogorov length (λ) describes the smallest possible eddy size and is calculated as

$$\lambda\,(z) = \left(\frac{\nu^3}{\varepsilon_z}\right)^{0.25}, \tag{8}$$

with ν as kinematic viscosity. In the presented model λ(z) is used to parameterise aggregate disaggregation.

5
Aggregate Interaction

Generally, three mechanisms, Brownian motion, turbulent aggregation and differential settling, determine physicochemical aggregate production in the BBL (McCave 1984). All three processes are strongly dependent on size distribution, abundance of aggregates and in the case of turbulent aggregation on the hydrodynamic regime. In the diffusion-advection equation [Eq. (1)] aggregate size classes are defined using the settling velocity. The aggregate size of each class is derived using the relation of settling velocity and aggregate diameter $w_s = 1.73\ D^{0.78}$ (Gibbs 1985; see legend of Table 1). Assuming spherical shape and a density of 2.65 kg dm^{-3}, the mass ($M_{i,j}$) of aggregates of each size class is determined. At each time step aggregate abundances [$N_j(z)$] are calculated by dividing $C_j(z)$ by the aggregate mass M_j. Whether and how much aggregates interact strongly depends on aggregate encounter probabilities. The significance of the aggregate encounter processes changes with distance to the sediment surface (McCave 1984). Depending on the concentration distribution and the hydrodynamic regime, aggregate interaction becomes a function of height above the seafloor (z) in the presented numerical approach. In the water column and the upper BBL, differential settling dominates aggregate interaction, a process which describes the scavenging of small suspended aggregates by larger faster-sinking aggregates (Hill and Nowell 1990; Hill 1992). The ensuing encounter rate (K_{ds}) is proportional to the square sum of diameters, to the absolute difference of settling velocities of the involved aggregates and to aggregate abundance. It takes the form (e.g. Hill and Nowell 1995):

$$K_{ds}(z) = \frac{\pi}{4}\left(d_i + d_j\right)^2 \left|w_{s\,i} - w_{s\,j}\right| N_i\,N_j, \tag{9}$$

where the subscript ds denotes differential settling.

By increasing turbulent shear towards the seafloor, particle aggregation due to turbulent encounter becomes more important and can dominate the aggregate interactions. For aggregates much smaller than the Kolmogorov scale, the turbulent encounter rate of aggregates is calculated as

$$K_{ts}(z) = 0.16\,\alpha \left(d_i + d_j \right)^3 N_i N_j \sqrt{\frac{\varepsilon(z)}{\nu}}, \tag{10}$$

where the subscript ts denotes turbulent shear. The strong influence of the aggregate diameter is obvious. For aggregates similar to and larger than the Kolmogorov length, a different formulation was suggested by Hill et al. (1992):

$$K_{ts}(z) = 1.08\,\alpha \left(d_i + d_j \right)^2 \left[\varepsilon(z) \left(d_i + d_j \right) \right]^{1/3} N_i N_j . \tag{11}$$

Thus, if the aggregate size is similar to or greater than Kolmogorov length $[\lambda(z)]$, Eq. (11) is used in the model. If the aggregates are smaller than $\lambda(z)$, turbulent aggregation is calculated according to Eq. (10).

The model allows turbulent aggregate formation within one distinct size class as well as between different size classes. In contrast, differential settling is only possible between classes of different sizes. The settling velocities of the aggregate, i.e. size classes, are defined as a multiple of the smallest class ($2^{i,j}$). Thus, turbulent aggregate production within one size class results in a aggregate of the next largest size class ($2^{i,j+1}$). Consequently, the produced mass is transferred from the class $2^{i,j}$ to the class $2^{i,j+1}$. In the case of differential settling, the generated mass is added to the size class of the larger involved aggregate. In the case of turbulent aggregation of the largest size class, the resulting aggregates are removed from this class and transferred to an extra aggregate size class. This size class acts as an aggregate sink which accepts aggregates but does not return them to the smaller size classes.

The final production of a new aggregate is strongly dependent on the probability of aggregates to stick together after the encounter. This probability is the so-called stickiness (α), a dimensionless factor ranging from 0–1 (Alldredge and Mc Gillivary 1991). For $\alpha = 1$, each encounter results in a new aggregate, whereas for $\alpha = 0.1$, only every tenth encounter produces a new aggregate. Values of stickiness for various aggregate types have been investigated empirically (Alldredge and Mc Gillivary 1991) and range for natural aggregates between 0.3 and 0.7. In the model the stickiness can be chosen freely. However, for simplicity, in the presented case studies the stickiness was set to unity.

6
Aggregate Disaggregation

Strong gradients of current velocity lead to turbulence and increasing shear stress. The resulting relative motion between an aggregate and the fluid is largest close to the seafloor. Thus, it is most likely that aggregates are destroyed in this area. Following the arguments of Hill and Nowell (1995), we use their formulation of an aggregate disaggregation coefficient $g_{i,j}(z)$. This coefficient is mainly determined by the rate of turbulent energy dissipation [$\varepsilon(z)$] and relates the aggregate size with the Kolmogorov length λ (z), the smallest possible eddy size. Aggregate splitting and aggregate erosion is not distinguished in this model.

$$g_{i,j}(z) = 0.1\frac{\beta\,\lambda(z)}{kh}\left(\frac{\varepsilon\,(z)}{\nu}\right)^{0.5}\left(\frac{d_i}{\lambda(z)}\right)^{2}. \tag{12}$$

According to Smith (1977), the coefficient β is set to 15, a reasonable value for shelf environments.

The loss of mass in the aggregate size classes i or j is calculated by multiplying the disaggregation coefficient $g_{i,j}(z)$ with aggregate abundance ($N_{i,j}$) and the according aggregate mass ($M_{i,j}$).

The sum of aggregate production (turbulent aggregation and differential settling) and aggregate disintegration for each aggregate size class [Eq. (13)] is added to the diffusion-advection equation (Eq. 1) at each time step.

$$R_i(z) = \sum_i\left(K_{ts_z} N_{i_z} N_{j_z} M_i + K_{ds_z} N_{i_z} N_{j_z} M_i\right) - g_{i_z} N_{i_z} M_i. \tag{13}$$

If this sum is negative, the resulting mass is added to the next smallest size class, assuming that the aggregate splits into two fragments of the same size. If the sum is positive, the resulting mass is added to the next largest size class. For the largest class, if turbulent aggregation dominates over disaggregation, the resulting mass is transferred to the aggregate sink. In the case of a negative sum for the smallest size class, disaggregation dominates aggregate production, the resulting mass stays in this fraction. Introduction or loss of mass due to sediment erosion and deposition is not considered in the present state of the model.

7
Results and Discussion

The first set of numerical experiments was designed to analyse the influence of the eddy diffusion coefficient A(z) on the vertical distribution of the three aggregate size classes without aggregate interactions (Table 1b, exps. 1–5). The eddy diffusion coefficient A(z) describes the distribution of the intensity of turbulent transport in the benthic boundary layer and above. Experiment 1 was run with a height-independent constant $A_c(z)$ of 0.001 m^2s^{-1} (Boudreau 1997). In the other four experiments the diffusion coefficient in the BBL $A_{lc}(z)$ and $A_{ld}(z)$ was linearly proportional to the height above the seabed and the friction velocity u_* [Eq. (3)], and in this way related to the hydrodynamic regime. In experiments 2 and 3 $A_{lc}(z)$ was constant above the BBL, whereas in experiments 4 and 5 $A_{ld}(z)$ decreases with distance above the BBL. Two different hydrodynamical regimes represented by current velocities of u_{100} of 0.1 (regime 1) and 0.5 ms^{-1} (regime 2) were simulated. The first velocity of 0.1 m s^{-1} is a reasonable value for shelf areas, whereas the latter describes a more vigorous regime. The results of the experiments 1–5 are summarised in Fig. 3a–f.

In all five numerical experiments the initial homogeneous concentration distributions of the three aggregate size classes with settling velocities of 30, 50 and 80 m day^{-1} were redistributed and for all experiments steady state was reached after a simulation time of 22 h. Significant differences in concentration distributions for all three aggregate size classes evolved between the constant $A_c(z)$ of 0.0001 m^2s^{-1} (Fig. 3, dotted line) and the height dependent A(z) [$A_{lc}(z)$ dashed and $A_{ld}(z)$

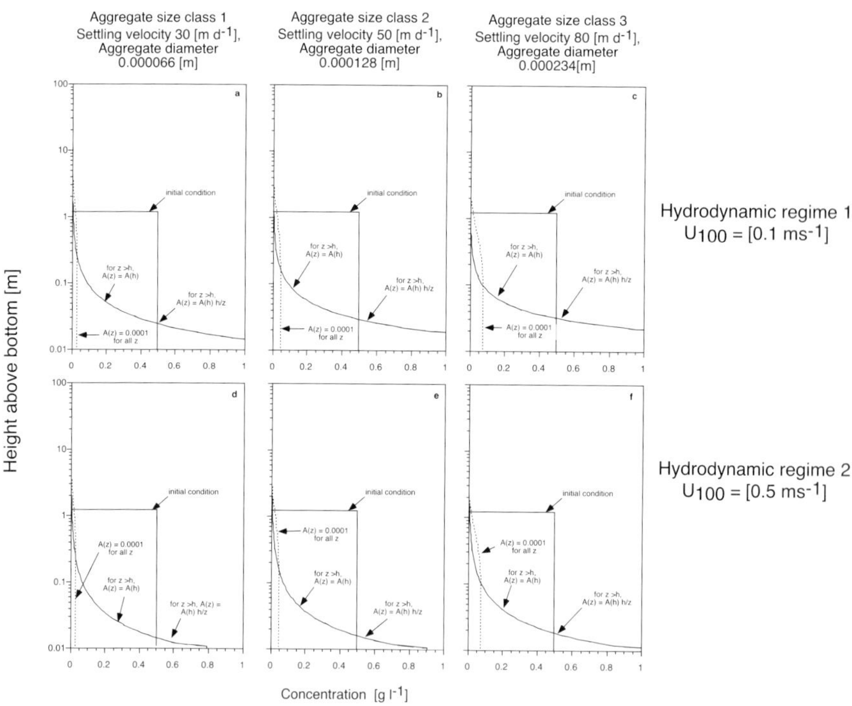

Fig. 3 The influence of the parameterisation of the eddy diffusion coefficient [A(z)] on the vertical concentration distribution of the three studied particle sizes (settling velocities 30, 50 and 80 m day^{-1}, respectively) was investigated in experiments 1–5 (Table 1). **a–c** represent simulations with $u_{100} = 0.1$ m s^{-1}, **d–f** with $u_{100} = 0.5$ m day^{-1}. The three different approaches of the parameterisation of the eddy diffusion coefficient displayed are: $A_c(z)$ = constant over all model layers (dotted line) ; in the cases of $A_{lc,ld}(z) = u_*kz$ we distinguish for $z > h$ between $A_{lc}(z) = A(h)$ (dashed line) and $A_{ld}(z) = A(h)\ h/z$ (solid line) (See text for details).

solid line]. For all three size classes the constant $A_c(z)$ produced a rather homogeneous distribution of concentrations to about 1 m above the seafloor, but significantly lower than the initial conditions. Smaller aggregates (settling velocity of 30 m day^{-1}) diffused further into the water column, whereas the larger and heavier aggregates (settling velocities of 50 and 80 m day^{-1}) stayed closer to the seafloor.

In contrast, both height-dependent, $A_{lc}(z)$ and $A_{ld}(z)$ parameterisations (Table 1, experiments 2–5) generated logarithmic concentration profiles of all aggregate size classes and both hydrodynamic regimes (u_{100} of 0.1 and 0.5 m s^{-1}, Fig. 3a–f). In all four experiments (Table 1, experiments 2–5) the concentration distributions of the observed aggregate size classes were the same for the two approaches of the parameterisations of eddy diffusion coefficients $A_{lc}(z)$ and $A_{ld}(z)$ (Fig. 3b–f). However, preliminary results of simulations with strong discontinuities of the diffusion coefficient with rapidly decreasing values above the upper limit of the boundary displayed accumulations of small aggregates at the limit of the boundary layer. Mass accumulation especially of small aggregates at pycnoclines, is a well-known phenomenon (McCave 1986). In the model, the sharp gradient of the diffusion coefficient at the upper limit of the BBL could influence the aggregate behaviour like a strong density gradient in the field. Therefore, we decided to use a modified diffusion coefficient with decreasing values above the height of the boundary layer [Eq. (6)] in all experiments including aggregate interaction.

The magnitude of the diffusion coefficient strongly influences the vertical transport distance of all aggregate classes. Comparable to the Rouse parameter (u^*k/w_s), in the present model the interaction of the diffusion coefficient and the settling velocity w_s determines the concentration profile of each aggregate size class. Accordingly, at a u_{100} of 0.5 m s^{-1}, more material of the largest aggregate class (Fig. 3c, w_s 80 m day^{-1}) was transported further into the water column, compared to the case of u_{100} of 0.1 m s^{-1} (Fig. 3d). Consequently, a smaller amount of this matter accumulated close to the seafloor. Similar to the results of Newberger and Caldwell (1981), the logarithmic concentration distributions in experiments 2–5, verify our proposed approach to investigate near-bed aggregate distributions with a simple diffusion-advection equation. The profiles resembled the expected distributions of distinct particles suggested by Rouse (1937).

For the two hydrodynamic regimes, the concentration distribution for the three aggregate size classes differed significantly (Fig. 4a, b). In the faster flow (Fig. 4b, $u_{100} = 0.5$ m s^{-1}), concentrations near to the seafloor were lower than in the slower flow (Fig. 4a, $u_{100} = 0.1$ m s^{-1}) and the maximal vertical transport distance was larger. In the faster flow, concentrations of the three classes were similar above 0.1 m above the seafloor, in the slower flow they were similar above 0.5 m. Thus, in the faster flow ($u_{100} = 0.5$ m s^{-1}), each size class contributed between 0.3 and 0.8% of the total initially added suspended matter at distinct heights above the seafloor, within the last 100 m above the seafloor (Fig. 4d). The relative contribution of each aggregate size class to the total amount of suspended matter at a distinct depth was nearly similar for all three classes.

In the more tranquil hydrodynamic regime 1 ($u_{100} = 0.1$ m s^{-1}) a single size class controlled the grain size distribution at different heights above the seafloor (Fig.

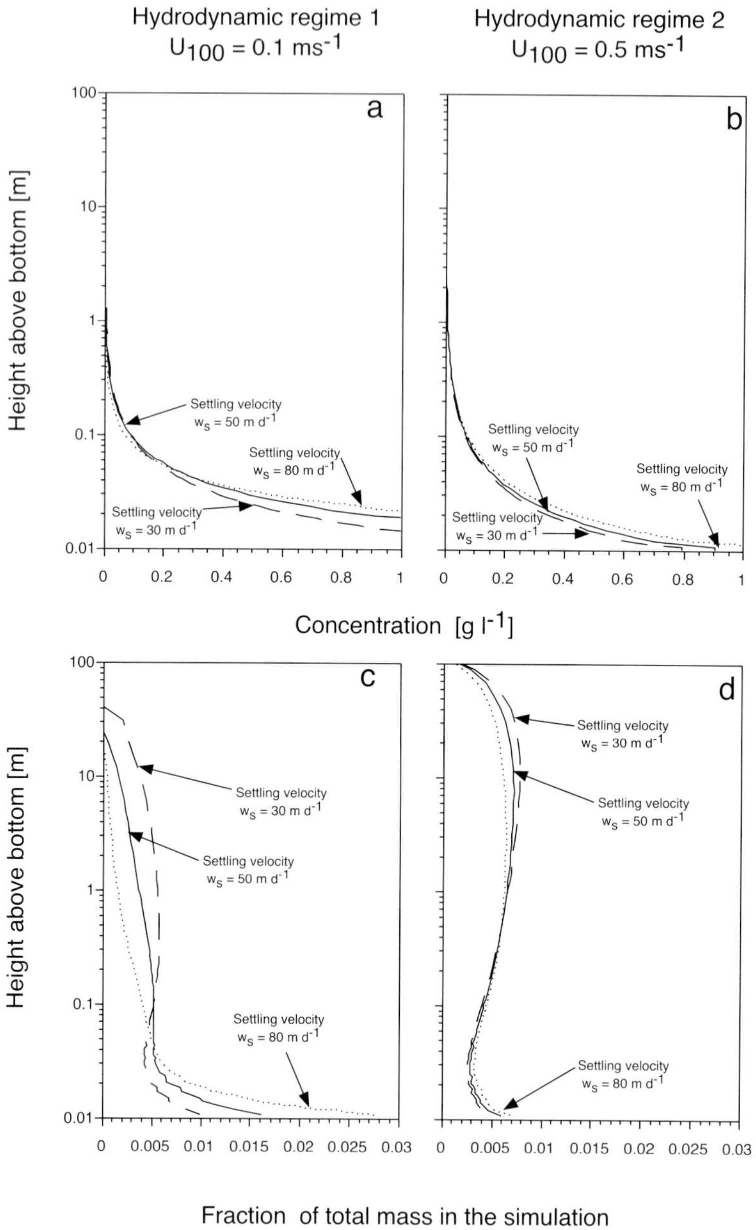

Fig. 4a–d Logarithmic concentration distribution of all three studied aggregate size classes with settling velocities 30 (dashed line), 50 (solid line) and 80 m day^{-1} (dotted line) in a diffusion advection simulation for the hydrodynamic regime 1 with $u_{100} = 0.1$ m s^{-1} **(a)** and hydrodynamic regime 2 with $u_{100} = 0.5$ m s^{-1} **(b)**. Fraction of the initially added total mass (sum of all three size classes, ca. 150 g) of each aggregate size class for the hydrodynamic regime 1 with $u_{100} = 0.1$ m s^{-1} **(c)** and regime 2 with $u_{100} = 0.5$ m s^{-1} **(d)**.

4c). The largest size class accumulated the highest concentration close to the seafloor, whereas above 0.25 m above the seafloor the mass distribution was dominated by small aggregates (>0.5% of the total initial concentration). The medium size class with a settling velocity of 50 m day^{-1} was most important between 0.03 and 0.25 m above the seafloor. Therefore, for the slower current velocities the simple 1-D-diffusion-advection model supports the idea of hydrodynamic sorting, similar to that proposed by Muschenheim (1987), even though his approach is based on the combination of the empirical Rouse and the Karman-Prandtl equation.

8 Aggregate Interaction

All experiments including particle interaction were also run with two different hydrodynamic regimes (Table 1, experiments 6–9). In Fig. 5a–f, concentration distributions of the three simulated aggregate size classes are summarised. Due to their wide range of concentrations, a log-log form was chosen for the presentation of the vertical concentration distribution up to 100 m above the seafloor. Similar to the previous simulations the current velocity at 1 m above the seafloor (u_{100}) was set at 0.1 m s^{-1} (experiments 6 and 7) and at 0.5 m s^{-1} (experiments 8 and 9). The results of these experiments were compared with the standard diffusion-advection runs (experiments 4 and 5). Compared to the standard runs of the simple diffusion-advection approach (dotted line), aggregate interaction produced significantly different concentration profiles (solid and dashed lines) after 22 h of simulation time. The advantage of the experiments with aggregate interaction included over the simple diffusion advection approach is that in these experiments the initial mass is freely exchanged between the aggregate size classes, a situation similar to processes in the field and in contrast to the simpler approach without aggregate interaction, where the final concentration distribution of a size class for a defined hydrodynamic regime is only dependent on the initial mass of this class. In experiments 4 and 5, without aggregate interaction, all three size classes account for one third of the initial mass. In all experiments with aggregate interaction significantly different mass transfer between the three size classes and the aggregate sink was observed (Table 2). Thus, aggregate interaction most likely affects the vertical concentration distribution of the different interacting aggregate size classes.

9 Turbulent Aggregation and Differential Settling

In experiments 6 and 7 aggregate formation from turbulent aggregation and differential settling was analysed. Either process moves aggregates from one size class to the next largest size class. Asproposed by McCave (1984), under the calmer hydrodynamic conditions differential settling dominates over turbulent aggregation over the total vertical extension of the model (model results not shown here).

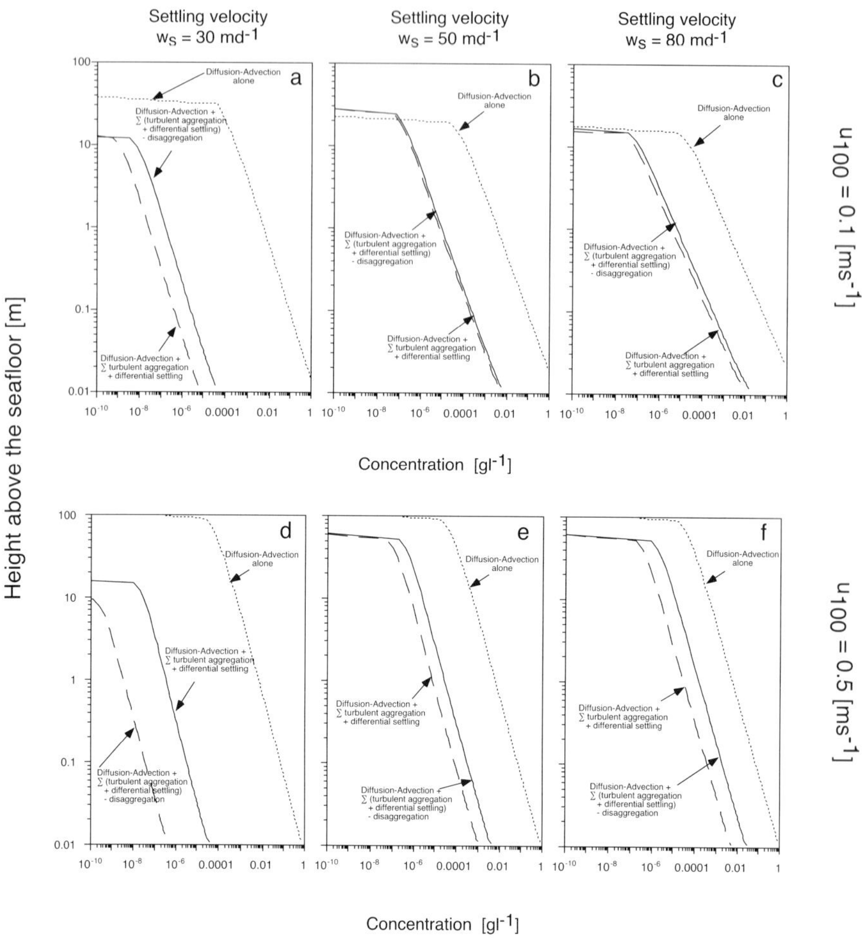

Fig. 5a–f Logarithmic concentration distribution of all three studied aggregate size classes for two different hydrodynamic regimes. **a–c** Regime 1 with $u_{100} = 0.1\ \mathrm{m\,s^{-1}}$ and **d–f** regime 2 with $u_{100} = 0.5\ \mathrm{m\,s^{-1}}$. Results are shown for aggregates with settling velocity of 30 **(a+d)**, 50 **(b+e)** and 80 m d^{-1} **(c+f)**. For these aggregate classes concentrations distributions are presented for the diffusion-advection model (dotted line, experiments 4 and 5), the model including aggregation (solid line, experiments 6 and 7) and the model including aggregation and disaggregation (dashed line, experiments 8 and 9).

As hypothesised, due to the transfer of mass to larger size classes, the mass concentration of each of the three aggregate size classes decreased over time, resulting in the presented profiles. In the case of the calmer hydrodynamic regime 1 ($u_{100} = 0.1\ \mathrm{m\,s^{-1}}$, experiment 6, Fig. 5a, dashed line), the smallest size class displays a profile with concentrations 5 orders of magnitude lower than those in the diffusion-advection standard run (experiment 4, Fig. 5a dotted line). The concentrations of the other two aggregate classes with settling velocities of 50 and 80 m day^{-1} were about 3 and 2 orders of magnitude lower than in the standard

Table 2 Final percent distribution of the initial mass into the distinct aggregate size classes for the numerical experiments 4–9. Experiments 4 and 5 describe a diffusion-advection approach without aggregate interaction. Experiments 6 and 7 include aggregate production due to turbulent aggregation and differential settling. Experiments 8 and 9 included aggregate production and desintegration. See Table 1a and b for detailed description of the experiments.

	Experiment	Aggregate size class 1, diameter 0.000066 (m) settling velocity 300 m day^{-1}	Aggregate size class 2, diameter 0.000128 (m) settling velocity 50 m day^{-1}	Aggregate size class 3, diameter 0.000234 (m) settling velocity 80 m day^{-1}	Aggregate sink, diameter ≥ 0.000234 (m)
Hydro-dynamic regime 1	4	33.3	33.3	33.3	–
	6	0.0002	0.075	0.093	99.83
	8	0.001	0.094	0.158	99.75
Hydro-dynamic regime 2	5	33.3	33.3	33.3	–
	7	0.00002	0.041	0.166	99.79
	9	0.0015	0.152	0.878	98.97

run (Fig. 5b, c), respectively. Aggregate production and the following redistribution by diffusion and advection strongly influences the relative contribution of each size class compared to the total mass at different heights above the seafloor. In contrast to the standard run, where the smallest size class dominates the total suspended matter above 0.25 m above the seafloor (Fig. 6a), particle interaction practically removes the smallest size class (Fig. 6b). Here, the smallest aggregate class accounted for less than 0.0002% of the initial mass. Only aggregates with settling velocities of 50 and 80 m day^{-1} contributed to the total mass of suspended matter (Fig. 6b). For the hydrodynamic regime 1, the medium size class (settling velocity of 50 m day^{-1}) dominated the suspended matter above 0.08 m above the seafloor, when aggregate disintegration was excluded. In this experiment aggregate formation transferred more than 99.8% into the aggregate sink class (Table 2).

In the hydrodynamic regime 2 with the higher current velocity of u_{100} of 0.5 m s^{-1}, the overall trend was similar (Fig. 5d–f). In this case, the higher eddy diffusion coefficient $A_{ld}(z)$ distributed aggregates further into the water column. Thus, less mass of each size class accumulated in close proximity to the seafloor, but, in contrast to the hydrodynamic regime 1, the larger aggregate size class dominated over the whole BBL (Fig. 6e). Similarly to the calmer hydrodynamic regime, significant differences evolved between simulation with particle interaction (experiment 7) and the standard runs. Between these two experiments the concentrations of suspended matter differed by various orders of magnitude and with aggregate interaction included (experiment 7), the three size classes

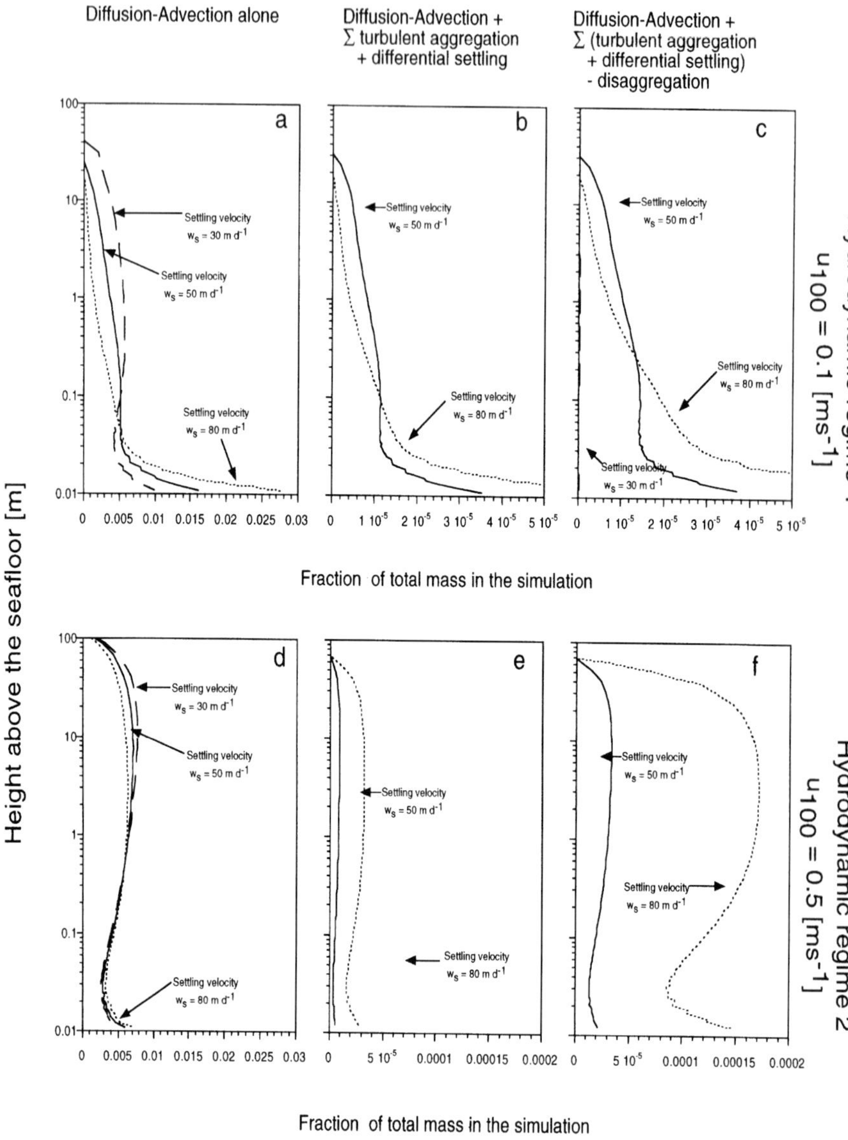

Fig. 6 Fraction of the initially added mass (sum of all three size classes, ca. 150 g) of each aggregate size class for two different hydrodynamic regimes. **a–c** Regime 1 with $u_{100} = 0.1\,\mathrm{m\,s^{-1}}$ and **d–f** regime 2 with $u_{100} = 0.5\,\mathrm{m\,s^{-1}}$. Results are shown for the diffusion-advection model alone (**a+d**, experiments 4 and 5), the model including aggregation (b+e, experiments 6 & 7) and the model including aggregation and disaggregation (**c+f**, experiments 8 and 9). For these cases aggregate mass fractions are plotted for aggregates with settling velocity of 30 (dashed line), 50 (solid line) and 80 $\mathrm{m\,day^{-1}}$ (dotted line). (NOTE different scales for **ad**, **bc** and **ef**).

contained less than 1% of the initial mass (Table 2). Again, more than 99% of the initial mass ended up in the aggregate sink term. Within the processes of aggregate interaction, turbulent aggregation became more important than differential settling for the more vigorous hydrodynamic regime 2.

10 Aggregate disaggregation

In a first attempt to describe the near-bed particle dynamics more completely, an aggregate disaggregation term was included into the diffusion-advection equation. Similarly to the aggregate production terms [Eqs. (9–11)], aggregate disaggregation [Eq. (12)] is a function of the hydrodynamic regime and mainly related to the ratio of the aggregate diameter and the Kolmogorov length. The amount of mass transfer between the various aggregate size classes is calculated as the sum of the mass of aggregate production minus the mass of aggregate disaggregation [Eq. (13)]. If this sum is positive, the difference is transferred to the next largest size class. If this sum is negative, i.e. aggregate disintegration dominates the aggregate interaction, the difference is transferred to the next smallest size class. Consequently, the vertical distribution of aggregate interactions will be dominated by aggregate production or aggregate disintegration as a function of the hydrodynamic regime.

The results of the numerical simulations suggest that disaggregation influenced the vertical concentration profiles and the size class distribution of the three aggregate classes close to the seafloor. Compared to the simulation without disaggregation, higher concentrations of each size class were observed when aggregate disintegration was included (Fig. 5a–f). For the calmer hydrodynamic regime 1 this influence was prominent only for the smallest size class, where aggregate disaggregation results in ten times higher concentrations than with turbulent aggregation and differential settling alone (Fig. 5a). In experiment 8 in the calmer hydrodynamic regime 1, without disaggregation, the smallest aggregate size class was unimportant, accounting for less than 0.00002% (Fig. 6b, Table 2), whereas with aggregate disintegration this aggregate class contains five times more of the initial mass (0.001%; Fig. 6c, Table 2). Only minor differences between the experiments with and without disaggregation were observed in the two larger aggregate size classes of settling velocities of 50 and 80 m day^{-1} (Fig. 5b, c). Even though the disaggregation term resulted in increased concentrations of small aggregates compared to the simulation without aggregate disintegration, the majority of the initial mass (>99.7%) was found in the aggregate sink class.

For the more vigorous hydrodynamic regime, the effect of particle disaggregation and thus transfer of mass from a larger to a smaller size class is important for all size classes (Fig. 5d–f). The most striking effects of particle disaggregation turn out in the relative composition of the total suspended mass (Fig. 6). In contrast to the simulation without disaggregation, the smallest aggregate class contributed about 75 times more to the total mass when disaggregation is included (Fig. 5f, Table 2). The largest class of aggregates ($w_s = 80$ m day^{-1}) contain about six times more of the initial mass the medium size aggregates ($w_s = 50$ m day^{-1}; Table 2). In the rougher hydrodynamic regime 2, hydrodynamic sorting did not occur, the whole BBL was dominated by the largest aggregate class. One interesting feature was the distinct c-shaped vertical distribution pattern of the relative mass fractions below 15 cm above the sediment (Fig. 6d–f). This shape occurred only in the stronger hydrodynamic regime 2 and was more pronounced if aggregate interaction was included. The vertical distribution patterns suggest, by linking particle distribution and interaction with the current regime, hydrodynamic sorting of different aggregate size classes can be simulated.

11 Relevance of Model Results to Field Studies

Only little information on the size distribution of aggregates is available with high resolution in the benthic boundary layer. Using an in situ particle camera at 1–2 m above the seafloor in a tidal inlet near Texel revealed that aggregate dynamics were strongly related to the tidal cycle (Chen et al. 1994). For the bottom water the aggregate size distribution over a tidal cycle was dominated by aggregate sizes between 93×10^{-6} and 379×10^{-6} m (93–373 μm). In this study, the mean current velocity in the inlet was more than 1 m s^{-1}, applying the same methods in the Elbe estuary, variations of mean particle sizes between 150×10^{-6} and 350×10^{-6} m were observed (Chen and Eisma 1995). The authors relate size distribution of aggregates to various parameters describing the composition of aggregates (e.g. organic content). Prediction of small-scale variations in the distribution of aggregate sizes due to hydrodynamic sorting has not been attempted. Our simulated size distribution patterns support the idea of hydrodynamic sorting in the BBL. Even though hydrodynamic sorting of aggregates could not be proved by the observed distribution of bulk measurements, this process was proposed to explain elevated microbial activities in the BBL (Ritzrau 1996). The numerical simulations presented here suggest that hydrodynamic sorting occurs even under a very simple approach of parameterisation of aggregate interaction.

12 Vertical distribution of aggregate size classes and benthic organisms

The distribution of aggregate size classes has major implication for feeding strategies of suspension-feeding benthic animals. If the aggregate formation due to turbulent aggregation and differential settling dominates the aggregate dynamics close to the seafloor, it is very likely that small-sized classes of aggregates ($< 10^{-6}$ m) may not be present and hence not be available to benthic organisms; but also the opposite may occur. Close to the seafloor, aggregates are subjected to the strongest shear, so that aggregate disintegration may produce small fragments which then become available for suspension feeders. Further on, the size distribution of aggregates in the BBL may provide information to interpret the vertical distribution patterns of epibenthic vagile animals (Brandt 1996; Childress et al. 1989; Angel 1990). These organisms would be predicted to live and feed at heights where their preferred aggregate size spectrum exists or accumulates.

In the presented experiments the different physicochemical properties of the aggregates sizes were not taken into account. For simplicity and in contrast to the available knowledge (e.g. Alldredge and Mc Gillivary 1991) the stickiness, a factor influencing the probability of aggregate formation, was set to unity. Especially the effect of various different aggregate sources may have strong effects on the stickiness and consequently on the near-bed aggregate dynamics. Further on, resuspension of fine-grained clay particles will influence the aggregation behaviour of the suspension (Muschenheim et al. 1990). Likewise, increasing the relative amount of freshly settled, most likely senescent phytoplankton cells (Riebesell 1992) will result in the abundance of transparent exopolymere particles (TEP; Alldredge et al. 1993; Passow et al. 1996) close to the seafloor. Their presence will also alter the aggregation but also disaggregation rates in the hydrodynamically energetic BBL. Thus, experiments with varying size-dependent stickiness values would simulate aggregate dynamics more realistically.

13 Conclusions

The model potentially improves the understanding of aggregate dynamics close to the seafloor. In contrast to the layer-averaging model of Hill and Nowell (1995), the presented results suggest that aggregate interaction influences both concentration profiles of the three investigated aggregate size classes and their overall size distribution close to the seafloor. Thus, it supports the idea of hydrodynamical sorting, a process which cannot be resolved by measuring bulk parameters of aggregate composition alone. Hydrodynamical sorting was observed in a more tranquil hydrodynamic regime 1 ($u_{100} = 0.1$ m s^{-1}). In the more vigorous hydrodynamic regime 2 ($u_{100} = 0.5$ m s^{-1}), a c-shaped distribution was obtained near the bed, which was more pronounced when aggregate interaction was included. Aggregate interaction strongly influenced the concentration distribution of different aggregate size classes. Especially the smallest aggregate class was strongly reduced when aggregates were allowed to interact. However, when aggregate interaction is taken into account, both aggregate formation (turbulent aggregation and differential settling) and disaggregation have to be calculated. Without disaggregation, the aggregate distribution was unrealistically dominated by the largest aggregate class.

The major goal was to introduce a new tool to investigate the near-bed aggregate dynamics using a simple diffusion advection approach including aggregate interactions. However, further experiments should focus on lower current velocities and varying roughness heights. The presented results suggest that the model has to be extended to more interacting aggregate size classes. Future experiments will explore the effects of aggregate properties, like size-dependent stickiness. Finally, interaction or exchange with the seafloor has to be included. Aggregate loss due to deposition and aggregation introduction due to sediment erosion have been identified as essential processes for the aggregate dynamics in the benthic boundary layer. Nevertheless, the presented model appears to be a rugged tool to study the major processes of near-bed aggregate dynamics for a wide variety of aggregate properties and hydrodynamic regimes.

Acknowledgements

This work was funded by the German Science Foundation and is publication number 328 of the Sonderforschungsbereich 313 at the University of Kiel. We thank M. Schlüter and S. Pesant for valuable discussions during the preparation of the manuscript. Comments of G. Gust were very helpful and improved the manuscript significantly.

Parameter List

$C_{ij}(z)$	=	concentration of fraction i,j	$(g\ l^{-1})$
z	=	height above the sediment	(m)
z_o	=	roughness height	(m)
$A_c(z)$	=	Eddy-diffusion coefficient constant over entire model height	$(m^2\ s^{-1})$
$A_{lc}(z)$	=	Eddy-diffusion coefficient, linearly increasing in the BBL, constant above the boundary layer	$(m^2\ s^{-1})$
$A_{ld}(z)$	=	Eddy-diffusion coefficient, linearly increasing in the BBL, decreasing above the boundary layer	$(m^2\ s^{-1})$
$w_{si,j}$	=	settling velocity of fraction i,j	$(m\ d^{-1})$
$K_{ts}(z)$	=	encounter rate, turbulent shear,	$(m^3\ s^{-1})$,
$K_{df}(z)$	=	encounter rate, differential settling	$(m^3\ s^{-1})$
$N_{ij}(z)$	=	particle abundance of fraction i,j at layer z	$(N\ cm^{-3})$
$d_{i,j}$	=	particle diameter of fraction i,j	(m)
$\varepsilon(z)$	=	dissipation energy	$(m^2\ s^{-3})$
ν	=	kinematic viscosity	$(m^2\ s^{-1})$
$R_{i,j}(z)$	=	reaction term of fraction i,j	
$M_{i,j}$	=	particle mass of fraction i,j	$(kg\ particle^{-1})$
$u(z)$	=	current velocity	$(m\ s^{-1})$
u_{100}	=	current velocity at 100 cm above the seafloor	$(m\ s^{-1})$
u_*	=	friction velocity	$(m\ s^{-1})$
t	=	time	(s)
k	=	von Karman's constant	(dimensionless)
α	=	particle stickiness; 0–1	(dimensionless)
$\lambda(z)$	=	Kolmogorov length	(m)

References

Alldredge AL, Mc Gillivary P (1991) The attachment probabilities of marine snow and their implications for particle coagulations in the ocean. Deep-Sea Res 38: 431–443

Alldredge AL, Passow U, Logan BE (1993) The abundance and significance of a class of large, transparent organic particles in the ocean. Deep-Sea Res 40: 1131–1140

Angel MV (1990) Life in the benthic boundary layer: connections to the mid-water and sea floor. Phil Trans R Soc Lond A 331: 15–28

Asper VL (1986). Accelerated settling of particulate matter by marine snow aggregates. Woods Hole Oceanographic Institution 86-12.

Asper VL (1987) Measuring the flux and sinking speed of marine snow aggregates. Deep-Sea Res 34: 1–17

Boudreau BP (1997) A mathemetical model for coupled sediment-suspended particle exchange. J Mar Syst 11(3/4): 279–303
Brandt A, Barthel D (1995) An improved supra- and epibenthic sledge for catching Peracarida (Crustacea, Malacostraca). Ophelia 43: 15–23
Chen S, Eisma D (1995) Fractal geometry of in situ flocs in the estuarine and coastal environments. Neth J Sea Res 32: 173–182
Chen S, Eisma D, Kalf J (1994) In situ size distribution of suspended matter during the tidal cycle in the Elbe estuary. Neth J Sea Res 32: 37–48
Childress JJ, Gluck DL, Carny RS, Gowing MM (1989) Bentho-pelagic biomass distribution and oxygen consumption in a deep-sea benthic boundary layer dominated by gelatinous organisms. Limnol Oceanogr 34(5): 913–930
Gibbs RJ (1985) Estuarine flocs: their size, settling velocity and density. J Geophys Res 90: 3249–3251
Graf G, Ritzrau W (1997) Biologische Aspekte sedimentologischer Prozesse. Geowissenschaften 15: 14–17
Gust G (1989) The benthic boundary layer. In: J. Sündermann (ed) Oceanography. Springer-Verlag, Berlin, p 345–398
Hill PS (1992) Reconciling aggregation theory with observed vertical fluxes following phytoplankton blooms. J Geophys Res 97: 2295–2308
Hill PS, Nowell ARM (1990) The potential role of large, fast sinking particles in clearing nepheloid layers. Phil Trans R Soc Lond A 331: 103–117
Hill PS, Nowell ARM (1995) Comparison of two models of aggregation in continental-shelf bottom boundary layers. J Geophys Res 100: 22, 749–22, 763
Hill PS, Nowell ARM, Jumars PA (1992) Encounter rate by turbulent shear of particles similar in diameter to the Kolmogorov scale. J Mar Res 50: 643–668
Jackson GA (1990) A model of the formulation of marine algal flocs by physical coagulation processes. Deep-Sea Res 37: 1197–1211
Logan BE, Wilkinson DB (1990) Fractal geometry of marine snow and other biological aggregates. Limnol Oceanogr 35: 130–136
McCave IN (1984) Size spectra and aggregation of suspended particles in the deep ocean. Deep-Sea Res 31: 329–352
McCave IN (1986) Local and global aspects of the bottom nepheloid layers in the world ocean. Neth J Sea Res 20: 167–181
Milligan TG (1995) An examination of the settling behaviour of a flocculated suspension. Neth J Sea Res 33: 163–171
Muschenheim DK (1987) The role of hydrodynamic sorting of seston in the nutrition or a benthic suspension feeder, Spio setosa (Polychaeta: Spionidae). Biological Oceanography 4: 265–289
Muschenheim DK, Kepkay PE, Kranck K (1989) Microbial growth in turbulent suspension and its relation to marine aggregate formation. Neth J Sea Res 23: 283–292
Newberger PA, Caldwell DR (1981) Mixing and the bottom nepheloid layer. Mar Geol 41: 321–336
Passow U, Alldredge AL, Logan BE (1994) The role of particulate carbohydrate exudates in the flocculation of diatom blooms. Deep-Sea Res 41: 335–357
Ratmeyer V, Wefer G (1996) A high resolution camera system (ParCa) for imaging particles in the ocean: system design and results from profiles and a three month deployment. J Mar Res 54: 589–603
Riebesell U (1992) The formation of large marine snow and its sustained residence in surface waters. Limnol Oceanogr 37: 63–76
Ritzrau W (1996) Microbial activity in the benthic boundary layer (BBL): Small scale distribution and its relationship to the hydrodynamic regime. Sea Res 36: 171–180
Ritzrau W, Graf G (1992) Increase of microbial biomass in the benthic turbidity zone of Kiel Bight, Baltic Sea, after resuspension by a storm event. Limnol Oceanogr 37: 1081–1086

Rouse H (1937) Modern conception of mechanics of turbulence. Trans Am Soc Civ Eng 102: 436–505

Smith JD (1977) Modelling of sediment transport on continental shelves. In: D. E. Goldberg, McCave, I. N., O'Brien, J. J. and Steele, J. H. (ed) The Sea. Vol. 6, Wiley Interscience, New York, p 539–577

Stolzenbach KD (1993) Scavenging of small particles by fast sinking porous aggregates. Deep-Sea Res 40: 359–369

Thomsen L, Ritzrau W (1996) Aggregate studies in the benthic boundary layer at a continental margin. J Sea Res 36: 143–146

Thomsen L, Graf G (1995) Benthic boundary layer characteristics of the continental margin of the western Barent Sea. Oceanologica Acta 6: 597–607

Thomsen L, Graf G, Martens V, Steen E (1994) An instrument for sampling water from the bottom nepheloid layer. Cont Shelf Res 14: 871–882

Thomsen L, Graf G, von Juterzenka K, Witte U (1995) An in situ experiment to investigate the depletion of seston above an interface feeder field on the continental slope of the western Barents Sea. Mar Ecol Prog Ser 123: 295–300

Townsend DW, Mayer LM, Dortch Q, Spinrad RW (1992) Vertical structure and biological activity in the nepheloid layer of the Gulf of Maine. Cont Shelf Res 12: 367–387

Wainright SC (1987) Stimulation of heterotrophic microplankton production by resuspended marine sediment. Science 238: 1710–1711

Wainright SC (1990) Sediment-to-water fluxes of particulate material and microbes by resuspension and their contribution to the planktonic food web. Mar Ecol Prog Ser 62: 271–281

Walsh I, Fischer K, Murray D, Dymond J (1988) Evidence for resuspension for rebound particles from near bottom sediment traps. Deep-Sea Res 35: 59–70

Possibilities of Cause-and-Effect Simulation in the Study of System Climate-Ocean-Sediments

S. V. Sirotinskaya

1 Introduction

The mathematical simulation of cause-and-effect relations (CE relations) in the system climate-ocean-sediments considered in this chapter is based on the ideas of the English logician J.S.Mill [1], proposed in the 19th century. He made the first known attempt to apply tools of logic to the detection of CE relations between phenomena described by propositions. However he considered only the simplest cases, only one cause and one effect. Unlike Mill's investigations, the i.e. cause-and-effect simulation (CE simulation) suggested here uses the concepts of plurality and interaction which have arisen as a result of the study of complicated geological objects. Additionally, tools of formal logic, modern mathematical logic [2,6] and computer processing of data [3,4] are also employed for the simulation and detection of CE relations.

CE simulation is most suitable for the analysis of non-numerical data (conventional maps, texts). As numerical data can be converted into qualitative data, this approach is also applicable to the analysis of heterogeneous data. In the case of only numerical data, its use without preliminary classification methods as developed in multivariate mathematical statistics is not likely to be reasonable.

CE simulation is intended for the study of complicated natural systems to which the system climate-ocean-sediments also belongs. With respect to this system it is assumed that: (1) all three components of the system are compound, consisting of more simple ones (e.g., climate includes the seasonal distribution of rain and snow, type of atmospheric circulation; the ocean is characterized by salinity, temperature, currents; sediments are represented by granulometric composition, velocity of sedimentation); (2) any simple sedimentation phenomenon (a property of sediments) may be considered as an effect of some climatic and (or) oceanic phenomena which are possible causes; (3) some sediment properties are perhaps convergent, that is, the same property might have been arisen under different combinations of climatic and oceanic phenomena. Further, with the term causes, the term climatic (oceanic) factors is also used.

As has been proved [3], the existence of CE relations between phenomena is expressed by the presence in a data file of certain logical dependencies, which are represented by Boolean algebra functions, and to reveal them, the shape at the functions must be given. To this end CE simulation is performed. Data processing consists of revealing the logical dependencies which coincide with theoretical functions derived by the simulation of different types of CE relations. As a result, an appropriate function is chosen and the formula expressing the CE relation in the system under study, e.g., in the system climate-ocean-sediments, is constructed. Before data processing, data formalization is fulfilled.

2 Data Formalization

CE simulation is based on mathematical logic including as main elements the notions proposition and predicate (proposition related to objects). Thus, to use these tools, available data have to be in the shape of propositions concerning attributes of natural phenomena. Depending on the kinds of data, there are different ways for such representation, as shown in Table 1.

Table 1. Data formalization

Kind of data	Concepts under formalization		Manner of data conversion
	Attribute	Phenomenon	
Conventional map	Unit of a map legend (f.i.,"sand")	Subdivision of a map legend (f.i. "Litological composition")	Construction of additional subdivision of a map legend by rearrangement of some units or(rand) by using the map content (e.g. construction of subdivision "Faults directions"
Isoline map	Values interval of numerical property, demostrated by map (e.g. Salinity is from 7.6 to 11.2%	Numerical properties corresponding to one isoline map (f.i., "Salinity of superficial layer")	a. Free division of properties values into intervals b. Division of properties values into intervals by means of multivariante mathematical statistic (methods: classification, regionalization) c. Selection of values intervals on the basis of Boolean algebra
Table	Values interval of numerical property	Numerical property, characterized by one column (line) of table	The same, as for isoline maps
Text	Separate proposition	Group of propositions	Construction of propositions based on text meaning and professional knowledge

Another important point in data formalization is the choice of object. This is determined by a concrete problem and the entire manner of investigating. In the case of the system climate-ocean-sediments, two main types of standard object groups are possible, stations or observation times.

After composing a predicate set and choosing standard objects, a binary table containing predicates values is constructed. Predicate values in such tables (e.g., Tables 2 and 4) are assigned by the following way: if a proposition about the ex-

Table 2. Example of a the case of the preliminary division or typification of quantitative properties

Objects	Index number of attributes (predicates)[a]											
	1	2	3	4	5	6	7	8	9	10	11	12
Station 1	1	1				1		1		1		1
Station 3	1		1			1		1		1		
Station 4	1	1					1	1			1	
Station 2		1			1			1				1
Station 5				1		1			1			1
Station 6		1		1					1			1
Station 7		1		1				1				1

[a] Columns of the table correspond to some predicates. They may be, e.g., the following: 1 – Content of sulphur in the bottom sediments of the station i, i=1,...,7, is above value c; 2,3,4 – Average annual salinity for the station i, i=1,...7, is 7–9‰ (2), >9–11‰ (3), >11‰ (4); 5,6,7 – Average annual temperature of the water for the station i, i=1,...7, is 4–6 °C(5), >6–8 °C(6), >8°C (7); 8,9 – Average annual content of the oxygen in the superficial water of the station i, i=1,...7, is 6–7‰ (8), 7–8‰ (9); 10,11,12 – Faults directions for the station i, i=1,...,7, are northeast 40–70°(10), northeast 0–20° (11), northwest 320–350° (12)

Table 3. Example of a case without preliminary division or typification of quantitative propertie

Objects	Index number of attributes (predicates)[a]							
	1	2	3	4	5	6	7	8
1961	1	7.7	10.0	7.1	13.0	890	80	300
1983	1	8.4	8.7	7.5	3.7	350	60	140
1994	1	8.1	10.4	5.9	8.6	680	90	530
1972		8.2	9.4	7.9	7.5	640	140	390
1965		7.9	8.0	7.4	3.8	800	70	670
1986		7.4	8.3	7.0	15.6	270	150	270
1977		7.1	8.4	8.2	5.6	540	120	440

[a] Column 1 includes values of a predicate which corresponds to some sediments property, e.g., it may be the following: in the year i, i=1,...,7, content of sulphur in the bottom sediments is above value c; b) Columns 2 – 8 characterize some oceanic and climatic phenomena. In particular, they may be the following ones: 2 – average salinity (‰); 3 – average temperature (°C); 4 – average content of oxygen (‰); 5 – average content of phosphate (kg/l); 6 – maximum area of ice (hundreds of km^2); 7 – zonal type of atmospheric circulation (number of days); 8 – sun radiance (number of h)

Table 4. Example of a binary table constructed by transformation of Table 3 on the basis of mathematical logic

Objects	Index number of attributes (predicates)[a]							
	1	2	3	4	5	6	7	8
1961	1	1	1	1	1	1	1	1
1983	1	1	1	1	1	1	1	1
1994	1	1	1	1	1	1	1	1
1972		1	1		1	1		1
1965		1		1	1	1	1	
1986				1				1
1977					1	1		1

[a] Index numbers of attributes (predicates) correspond to index numbers of properties (phenomena) from Table 3 and columns include values of the following predicates: 1 – In the year i, i=1,...,7, average content of sulphur in the bottom sediments is above value c; 2 – In the year i average salinity is from 7.7 to 8.4ä; 3 – In the year i average temperature is from 8.7 to 10.4 °C ; 4 – In the year i average content of oxygen is from 5.9 to 7.5ä; 5 – In the year i average content of phosphate is from 3.7 to 13.0 kg/l; 6 – In the year i maximum area of ice is from 350 to 890 hundreds of km^2; 7 – In the year i the number of days with a zonal type of atmospheric circulation is from 60 to 90; 8 – In the year i the number of hours with the sun radiance is from 140 to 530

istence of a certain attribute for the object under consideration corresponds to reality, a predicate accepts the value true (symbol 1); otherwise, it takes the value false (symbol 0 or blank). For example, the proposition: content of sulphur is over value c can be true for one station and false for an other, and the table column corresponding to this proposition contains the symbol 1 in the first case and has a blank in the second (e.g., stations 1, 2 in Table 2).

Any attribute related to the sediments is considered as a possible effect of the oceanic and/or climatic factors, that is, as a function of the latter. Thus, only one column with predicate values corresponding to the sediment attributes under study is included in a data table (e.g., column 1 in Tables 2, 3, 4). In accordance with these values, any standard object belongs to one of two groups: the positive or negative. For example, if the proposition is: Content of sulphur in sediments is over c, then the positive group consists of objects with the sulphur content over value c and the negative one includes objects with the sulphur content under value c. The division of objects into positive and negative is needed if data are quantitative or heterogeneous (both quantitative and qualitative). If they are entirely qualitative, the solution of some problems can be performed without the presence of negative objects.

Binary tables are not very suitable for computing. Thus, a final step is data formalization is the transformation of binary tables into sets of positive integers:

$$J_{il}, \ldots, J_{it}, \ldots J_{is_i}, \; i = 1, \ldots, m, \tag{1}$$

where j_{it}, t=1,..., s_i, are indexes of s_i predicates, whose values in the i-th table line are 1 (blank). For example, in the case of the value 1 the first line of the Table 2. can be represented as the following set: 1, 2, 6, 8, 12.

In the case of blanks, the whole of Table 4 is expressed by four sets: 1, 4, 7; 1, 3, 8; 1, 2, 3, 5, 6, 7; 1, 2, 3, 4, 7, including the index numbers of blanks from the four last lines of the table correspondingly.

3
Theoretical Principal of Cause-and-Effect Simulation

Apparently, different properties of sediments may have different relations to oceanic and climatic phenomena. In particular, some properties may arise under the influence of only climatic or only oceanic factors, and the others may be connected with both climatic and oceanic ones. In the latter case, climatic and oceanic factors either have or have no cause-and-effect dependencies between them. Additionally, any climatic and oceanic phenomenon may be expressed either by separate independent attributes or by a complex of interrelated attributes. In addition, types of CE dependencies can vary. Thus, one can formulate diverse hypotheses about CE relations in the system climate-ocean-sediments, each of them including propositions of three following groups: I. Connections between sediments and groups of oceanic and climatic phenomena; II. Connections of climatic and oceanic phenomena between them; III. Relation of mapped or observed attributes to a relevant phenomenon.

Besides the natural language, these propositions are expressed by formulas of mathematical logic. Taking one formula from each group, one can compose a formula set expressing a certain hypothesis on CE relations in the system climate-ocean-sediments (Tables 5–7).Depending on the propositions of groups I and II, these hypotheses have been divided into five main types A, B, C, D, E. Using, in addition, the different variants of proposition II (variants II.1 – II.3 for hypotheses B and C) and proposition III (variants III.1 – III.4 for hypotheses A, B, C and two of them for hypotheses D, E), it is possible to construct versions of hypothesis A, 12 versions each of hypotheses B and C, and 2 versions each of hypotheses D and E.

Hypotheses described by Tables 5–7 include implicitly the assumption that all oceanic (climatic) phenomena are connected with climatic (oceanic) phenomena and sediments by the same type of CE relations. However, it is clear that more complicated cases are also possible (Fig. 1). In particular, in the case of the hypotheses AB and C* two sets of climatic phenomena (Climate^1, Climat^2) have relations different from oceanic phenomena and sediments. In the case of the hypotheses AC and B*, the same is true for two sets of oceanic phenomena (Ocean^1, Ocean^2). Complicated hypotheses usually represent combinations of the main hypotheses expressed by Tables 5–7 (e.g., AB, AC) or are, in fact, some of their modifications (f.i., B*, C*).

Any formula set describing some version of the hypotheses A, B, C, D, E can be transformed into a Boolean algebraic equation, which is properly a mathematical model of CE relations for the given hypothesis. By means of this equation, a type of logical dependency between attributes (more exactly, between the variable predicates corresponding to attributes) is described. The equation

Table 5. Mathematical simulation of hypothesis A

Hypothesis type	Propositions and corresponding formulas[a]
A 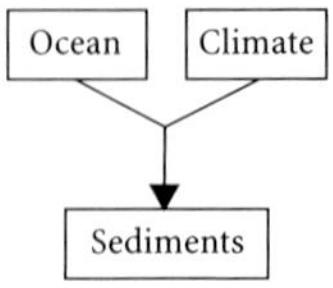	I. Sediments property S (f.i., lithological composition) is an effect of the interrelation between climatic (C) and oceanic (O) phenomena $S \sim C \,\&\, O$ (1)
	II. Oceanic and climatic phenomena have no CE-relations between them $O \not\rightarrow C \quad C \not\rightarrow O.$ (2)
	III.1. Both oceanic (O) and climatic (C) phenomena are expressed by separate independent attributes $O \sim o_1 \,\mathrm{v} \ldots \mathrm{v}\, o_j \,\mathrm{v} \ldots \mathrm{v}\, o_r,\ C \sim c_1 \,\mathrm{v} \ldots \mathrm{v}\, c_t \,\mathrm{v} \ldots \mathrm{v}\, c_k,$ (3) where $o_1, \ldots, o_j, \ldots, o_r$ and $c_1, \ldots, c_t, \ldots, c_k$ are sets of attributes relating to oceanic and climatic phenomena correspondingly.
	III.2. Both O and C are expressed by complexes of interrelated attributes $O \sim o_1 \& \ldots \&\, o_j \,\& \ldots \& o_r,\ C \sim c_1 \& \ldots \&\, c_t \,\& \ldots \& c_k,$ (4)
	III.3. O and C are expressed correspondingly by separate and interrelated attributes $O \sim o_1 \mathrm{v} \ldots \mathrm{v}\, o_j \,\mathrm{v} \ldots \mathrm{v} o_r,\ C \sim c_1 \& \ldots \&\, c_t \,\& \ldots \& c_k.$ (5)
	III.4. O and C are expressed correspondingly by interrelated and separate attributes $O \sim o_1 \& \ldots \&\, o_j \,\& \ldots \& o_r,\ C \sim c_1 \mathrm{v} \ldots \mathrm{v}\, c_t \,\mathrm{v} \ldots \mathrm{v} c_k.$ (6)

[a] Symbols & and v in the formulas of the table denote the logical operations of conjunction and disjunction correspondingly.

Table 6. Mathematical simulation of hypotheses B and C

Hypothesis type	Propositions and corresponding formulas
B Climate ↓ Ocean ↓ Sediments	I. Sediments property S is an effect of the interaction between some oceanic phenomena $O_1 \ldots O_s \ldots O_n$ $S \sim O_1 \,\&\, \ldots \,\&\, O_s \,\&\, \ldots \,\&\, O_n$ (7) II.1. A separate oceanic phenomenon is an effect a separate climatic one $O_s \sim C_p$ (8) II.2. A separate oceanic phenomenon is an effect of the interaction between some climatic phenomena $C_1 \ldots C_p \ldots C_m$ $O_s \sim C_1 \,\&\, \ldots \,\&\, C_p \,\&\, \ldots \,\&\, C_m$ (9) II.3. A separate oceanic phenomenon (f.i., the high salinity) is an effect of different, not interrelated climatic phenomena $O_s \sim C_1 \vee \ldots \vee C_p \vee \ldots \vee C_m$ (10) III. The same variants III.1, III.2, III.3, III.4 as for the hypothesis A
C Ocean ↓ Climate ↓ Sediments	Propositions and formulas for the hypothesis C are similar to these ones for the hypothesis B. They are derived from the model B by exchange of the oceanic phenomena to climatic ones, and vice versa

Table 7. Mathematical simulation of hypotheses D and E

Hypothesis type	Propositions and corresponding formulas
D Climate Ocean ↓ Sediments	A special case of hypothesis B. Hypothesis D differs from the latter one by the proposition of climate group II, which is the following: there are no CE relations between climatic and oceanic phenomena, that is $O \not\rightarrow C$ (11)
E Ocean Climate ↓ Sediments	A special case of hypothesis C. Hypothesis E differs from the latter one by the proposition of group II, which is the following: there are no CE relations between oceanic and climatic phenomena, that is $C \not\rightarrow O$ (12)

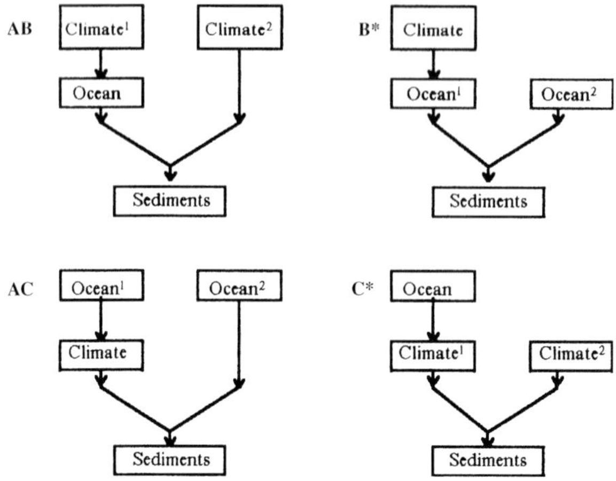

Fig. 1. Types of complicated hypotheses

structure is a certain sum of logical products (disjunctive normal form or DNF) and also a certain product of logical sums (conjunctive normal form or CNF) being equivalent to the first one. Besides, models of some hypotheses are expressed by a simple conjunction which, however, may be considered as a special case of DNF.

In connection with the variety of models, a problem arises: whether it is possible to identify in data files all types of CE relations. To the present, this problem has not been solved completely; but at least most versions of all main models A, B, C, D, E can be distinguished. In particular, they differ by the occurrence and number of logical variables corresponding to attributes from different phenomenon groups. This property allows to be criteria of deduce for testing the compatibility between data and models of CE relations (Table 8).

Most models absent in Table 8 have not been investigated. Some of them are trivial. In particular, the hypothesis C is described by the same equations as the hypothesis B, but CNF1 and CNF2 change places. The hypotheses D and E may be considered as special cases of the hypotheses B and C.

4 Data Processing

To start the data processing, the following information has to be input:

a) Data file T, which consists of m sets including index numbers of variables with values 1 for i-th object (observation), i=1 ,..., m.
b) File T^+ including m_1 sets of T which correspond to value 1 of the function f (to m_1 positive objects with respect to a sediments property), and file T– consisting of m_2 sets of T which correspond to the value 0 of the function f (to m_2 negative objects relatively to the sediment property).
c) Sequence of n variables, and minimum and maximum index numbers of its r+l subsequences characterizing r oceanic and l climatic phenomena.

Table 8. Criteria of compatibility between data and models of type A and B

Model type	Boolean equations and corresponding criteria[a]
A, the version consisting of propositions I, II, III.1	CNF: $f = (o_1 \vee ... \vee o_j \vee ... \vee o_r) \,\&\, (c_1 \vee ... \vee c_t \vee ... \vee c_k)$ (13) DNF: $f = o_1 c_1 \vee ... \vee o_j c_t \vee ... \vee o_r c_k$ (14) Here $o_1, ..., o_j, ..., o_r$ and $c_1, ..., c_t, ..., c_k$ are logical variables characterizing oceanic and climatic attributes correspondingly; f - Boolean function characterizing the sediments property. Criteria (1). Dependencies between sediments property and other attributes, revealed in data file (e.g., Tables 2, 4), correspond to the given model, if the function f has the following features: some disjunctions (logical sums) in the CNF of the function f include only variables of the set $o_1, ..., o_j, ..., o_r$, some consist only of variables $c_1, ..., c_t, ..., c_k$. (2) Each conjunction (a logical product) in the DNF of the function f consists of variables belonging to both these sets.
A, the version consisting of propositions I, II, III.2	Criteria: 1 Function f is expressed by one conjunction 2. This conjunction consists of variables belonging to both sets $o_1, ..., o_j, ..., o_r$ and $c_1, ..., c_t, ..., c_k$.
B, the version consisting of propositions I, II.1, III.1	CNF1: $f = (o'_1 \vee ,..., \vee o'_r) \,\&\, (o''_1 \vee ,..., \vee o''_m)$ (15) CNF2: $f = (c'_1 \vee ,..., \vee c'_k) \,\&\, (c''_1 \vee ,..., \vee c''_n)$, (16) where superscripts of variables are indexes of two oceanic and two climatic phenomena O′, O″, C′, C″. Criteria. Dependencies between the sediments property and other attributes in data file correspond to the given model, if the function f has the following features: (1) it can be expressed in the shape of two CNF, one of them consisting only of oceanic variables and the other including only climatic ones; (2) the number of disjunctions in CNF1 is equal to the number of this in the CNF2; (3) each disjunction in both CNF includes variables of one group characterizing one oceanic (for CNF1) and one climatic (for CNF2) phenomena; (4) there is the relation (8) from the Table 6 between disjunctions of the CNF1 and CNF2 (12)
B, the version consisting of propositions I, II.2, III.1	Criteria. A set of criteria for this version differs from this one for the previous version by the criteria (2) and (4). In particular, a number of disjunctions in the CNF1 is less than a number of these ones in the CNF2; disjunctions in the CNF1 and CNF2 have the relation (9), instead of (8), between them.

[a] Equations describe here the simplest cases (two phenomena of each kind).

The following results should be obtained: to select such variables among n analyzable ones which are arguments of the function f; to verify hypotheses about types of CE relations, that is, to test the compatibility between mathematical models of different hypotheses and available dependencies in the data file; to construct the function f formula describing the revealed CE relation.

Depending on the peculiarities of a problem under study and available information (conventional maps, isoline maps, tables, texts or some combination of them), different combinations of algorithms are used in data computing. Three main algorithms of these combinations are parts of the mathematical logic methods published before [3]. They are based on the mathematical theory developed for the computer projection [6]. Two of them are intended for the detection and analysis of conjunctive and disjunctive logical dependencies in data files. The third is the optimization algorithm. It belongs to a class of algorithms developed in connection with a problem of the Boolean functions minimization and is employed for the construction of minimum DNF of Boolean functions.

Further more, by a description of data processing, the mentioned algorithms are grouped into the following three procedures.

Procedure I. This includes the detection of disjunctive logical dependencies, minimization of Boolean functions expressing them, dependencies analysis;

Procedure II. Besides some additional operations, this includes the Boolean functions' minimization and dependencies analysis;

Procedure III. This procedure consists of the detection of conjunctive logical dependencies and Boolean functions' minimization.

A general way of data processing for most variants of the main hypotheses, investigated by now consists of the following ten steps.

1. Procedure I is applied k+l times to the file T^+. By this means, some combinations of variables V_s, $s=1,\ldots,q$, $q \leq k+l$, consisting of possible arguments of function f, are detected within each of k+l subsequence. Subsequences including only one variable, are also taken into account.
2. Variables which have not entered the combinations V_s are removed from the initial file T.
3. Each combination V_s is replaced by a new variable p_s which is the disjunction of all variables from V_s.
4. Hypotheses D and E are verified. To do this, the procedure III is applied separately to the part T^o of the file T^+, which includes only oceanic variables p^o_j, $j \leq k$, and to the part T^c of file T^+, which includes only climatic variables p^c_t, $t \leq l$. As a result, the following cases are possible: (1) function f cannot be expressed either as a conjunction of variables from T^o, or as a conjunction of variables from T^c. Then step 6 for a test of hypothesis A is realized; (2) function f is expressed only as a conjunction of variables from the file T^o. This means that the hypothesis D is true. Then step 8 is carried out; (3) function f is expressed only as a conjunction of variables from the file T^c. This means that hypothesis E is true. Then step 8 is carried out; (4) function f can be expressed both as a conjunction of variables belonging to both the file T^o and the file T^c. Then step 5 is realized.
5. Hypotheses B and C are verified. Here, the following cases are possible: (1) each variable from the conjunction relating to the file T^o is identical to a certain variable from the conjunction relating to the file T^c, and vice versa. It means that the versions I, II.1, III.1 of the hypotheses B or C are true, as the hypotheses B and C are not distinguishable between them. Further, step 8 is realized; (2) each variable from the conjunction relating to the file T^o can be represented in its turn as a function of variables from the conjunction relating to the file T^c, and a function type is also the conjunction. This means that

the versions I, II.2, III.1 of hypothesis B are correct. Then step 8 is realized; (3) each variable from the conjunction relating to the file T^c can be represented in its turn as a conjunctive function of variables from the conjunction relating to the file T^o. Then the versions I,II.2, III.1 of hypothesis C are true, and step 8 is fulfilled; (4) not all variables from the conjunction relating either to the file T^o or to the file T^c are submitted to the conditions of items (2) or (3). This means that some complicated hypothesis is true, and step 8 is realized.

6. Hypothesis A is verified. To do this, the files T^o and T^c are united in one file T^*. Then Procedure III is applied to T^*. If the function f can be expressed as a conjunction of variables from T^*, then the hypothesis A is accepted, and step 8 is realized. If there is no such dependence in T^*, the hypothesis A is rejected.
7. Conclusions as to the absence of CE relations between the phenomena under study are drawn.
8. Steps 1–7 are applied to the file T– instead of the file T^+, and then step 9 is fulfilled.
9. By the Morgan rule, the results of steps 1–7 and 8 are united. The variables which are not arguments of the function f are removed from data files (from T^* for the hypotheses A and from T^o, T^c for the others). Then the procedures II, III and also some additional operations which are different for the different hypotheses are applied to these files. As a result, a formula (DNF) describing the resultant type of CE relation for the problem under study is constructed.
10. End of data processing.

5 Possible Application

There were no real data to carry out demonstrative investigation of the system climate-ocean-sediments, and even data in Tables 2–4 are imaginary. Thus, only general trends of the system study are outlined here.

1. Spatial CE analysis. This trend is the most suitable for thematic investigations and study of phenomena having a low velocity of change. While using the spatial CE analysis, data relates to stations, profiles, areas or to other spatially distributed objects (e.g., Table 2) and the time is assumed to be fixed. As a result, regularities in the spatial distribution of a certain sediment properties are found, in particular the climatic-oceanic settings causing this property are revealed. Clearly, the knowledge of such settings may be helpful in paleoclimatic investigations. Dependencies revealed by CE analysis, are described by formulas. In the case of spatial CE analysis, they can also be expressed by maps.

2. Time CE analysis. Obviously, the time CE analyses is mainly appropriate for the study of phenomena with a high velocity of change. In this case, a data file includes time moments (years) of observations as objects (e.g., Tables 3, 4). It is clear that all observations must be related to one fixed territory (the same station, region and so on). If several territories are considered, several data files

are processed separately and the results are united. A possible application of time CE analysis is the detection of natural regularities in time and the identification of situations which deviate from them. Knowledge of factor combinations forming such situations may help in prediction and prevention of dangerous ecological events.

6 Conclusions

The methods of CE simulation were originally suggested for the purposes of prediction and retrospective analysis in mining geology. There were numerous cases of their application related mainly to gold and tin mineralization. The theoretical foundations have been developed for the two-component systems: mineralization property – geological environment.

This chapter presents the foundations of CE simulation for three-components systems such as the system climate-ocean-sediments. These investigations are under way. Only two cases have been studied to the present: one is the system basement-sedimentary cover-mineralization type for the region Copet-Dag (Turkmenia) and the other is the system deeper earth layers-upper earth layers-lead and copper mineralization for the territory Rudni Altai [5].

Available software [4] is intended for the study of two-component CE relations. Nevertheless, some combinations of programs which are parts of these systems can also be applied to the analysis of three-components CE relations.

In this chapter, the simulation of three-component CE relations is illustrated by the example of the system climate-ocean-sediments. Obviously, the same approach may be applied to the study of other such systems, e.g., the systems crustal processes-ocean-sediments and climate-ocean-biota.

Acknowledgements

Thanks are due to Dr. C.Tipper, University of Freiburg, Geological Institute, for critical review and helpful comments.

References

Mill J.S. (1914) System of syllogistic and inductive logic. Moscow (in Russian).
Novikov P.S. (1973) Elements of mathematical logic. Nauka Press, Moscow (in Russian).
Sirotinskaya S.V. (1986) Logical methods for analysis of geological information. Nedra Press, Moscow (in Russian).
Sirotinskaya S.V. (1994) Strategy of deposits search and their evaluationon the basis of computer technologies of qualitative data processing. Sci. de la Terre, Ser. Inf., 32, 353–360, Nancy.
Sirotinskaya S.V., Vedjaeva I.V. (1994) Stereometallogenic analysis of copper and lead ineralization of Rudni Altai on the base of mathetical logic. Otechestvennaya geologiya, 2, 23 – 31, Moscow (in Russian).
Yablonski S.V. (1979). Introduction to discrete mathematics. Nauka Press, Moscow (in Russian).

Topic B:
Trends and Periodicity in the Sedimentary Record as a Response to Environmental Changes

Climate Cycles and Climate Transitions as a Response to Astronomical and CO_2 Forcings

A. Paul and W. H. Berger

1 Introduction

The natural variability of climate occurs on a wide range of time scales, ranging from decades to millions of years. The time scales of importance to the ice ages of the late Cenozoic are of the order of 10 to 1000 ka. On these time scales, the so-called astronomical theory of paleoclimates and in particular its Milankovitch version have become the almost universally accepted basis of research (Berger 1992). In its traditional form, the astronomical theory states that changes in high-latitude summer insolation cause the waxing and waning of the continental ice sheets. This was first quantitatively formulated by Milankovitch (Berger 1988; Milankovitch 1995), who calculated how the incoming radiation at the top of the atmosphere varies as a function of latitude and the orbital parameters e, ϵ and $e \sin \tilde{\omega}$, where e denotes the eccentricity of the Earth's orbit around the Sun, ϵ is the obliquity and $e \sin \tilde{\omega}$ is the precessional effect, with $\tilde{\omega}$ being the longitude of the perihelion as measured from the moving equinox. Each of the orbital parameters can be expressed as a quasi-periodic function of time. According to the astronomical theory, low summer insolation could prevent the winter snow from melting. The high albedo of snow- and ice-covered areas would initiate a further cooling of the Earth. Eventually, this positive feedback would result in the build-up of the continental ice sheets.

The success of the astronomical theory is mainly due to two reasons. (1) Climatic time series as recorded in deep-sea sediments exhibit power spectra that show peaks corresponding to climate cycles with frequencies near 1/19, 1/23 and 1/41 ka^{-1}. The first two frequencies correspond to the precessional effect, with the frequency of the precessional angle $\tilde{\omega}$ 1/22 ka^{-1} split into two by the eccentricity frequency, whose mean is about 1/100 ka^{-1} (the most important terms in the series expansion of eccentricity have frequencies of about 1/404, 1/95, 1/124, 1/99 and 1/131 ka^{-1}, in decreasing order of amplitude – (Berger and Loutre 1991). The third frequency corresponds to obliquity. (2) Climate models forced at the astronomical frequencies produce changes in global ice volume in fair agreement with the deep-sea records.

Oxygen isotope data from deep-sea sedimentary cores are commonly taken as a direct proxy for global ice volume. It is assumed that water mass effects and, in the case of low-latitude planktonic records, temperature effects are small. Long records extending back through the time of the onset of Northern Hemisphere glaciation at about 2.5 Ma before present (B. P.) (Raymo 1994) and before show that marked changes in the character of the ice age oscillations occurred in the late Pliocene as well as in the mid-Pleistocene. Impressive records of this type stem from North Atlantic Deep Sea Drilling Program (DSDP) site 607 (Ruddiman et al. 1989), Ocean Drilling Program (ODP) site 677 in the Panama Basin (Shackleton et al. 1990), ODP site 659 in the subtropical eastern Atlantic (Tiedemann et al. 1994) and ODP site 806 on the Ontong Java Plateau (Berger et al. 1994; Berger et al. 1995; Bickert et al. 1997). In Fig. 13a, we show the benthic $\delta^{18}O$ record of ODP site 806B. As evidenced in this and similar cores, the ice volume oscillations in the early Pliocene were of low amplitude and high frequency. Then a major transition occurred from a warmer to a cooler climate which, subsequent to about 2.5 Ma B. P., was dominated by glacial cycles of a larger amplitude with a period near 40 ka, obviously related to obliquity variations. Data from a number of deep-sea cores indicate that another transition led to an even cooler climate witnessing the build-up of unusually large ice sheets at about 900 ka B. P. (Maasch 1988) and the onset of high-amplitude climate cycles with near-100-ka period at about 650 ka B. P. (Berger and Wefer 1992; Mudelsee and Schulz 1997). This latter multiple transition phenomenon is referred to as the mid-Pleistocene climate transition (Berger and Wefer 1992; Mudelsee and Stattegger 1997).

The climate transitions in the late Pliocene and mid-Pleistocene, as well as the climate cycle with near-100-ka period, pose a twofold problem to the astronomical theory. (1) The rhythmic variations in insolation have not changed in character over the entire Cenozoic (Berger et al. 1992; Laskar et al. 1993). (2) The cycle with near-100-ka period is the dominant feature of nearly all climatic time series of the late Pleistocene, but eccentricity-induced changes in insolation at about this period are smaller than 1 %. Hence, in order to explain the observed behaviour, the climate system must have active internal degrees of freedom, or there must be additional forcing mechanisms, or both.

In this chapter we consider the effect of active internal degrees of freedom such as bedrock depression and uplift, and additional forcing mechanisms such as the accretion of meteoroids and interplanetary dust particles (IDP) or a slow tectonic decrease in atmospheric carbon dioxide concentration. To this end we use a paleoclimate model (PCM) of the late Cenozoic climate based on the work of Saltzman and Verbitsky (1993). This model is globally averaged and zero-dimensional. It is an attempt to express the main processes which are believed to determine paleoclimatic change on long time scales ranging over millions of years, in a minimum form. We add two physical factors to it: the elevation-temperature feedback and inclination forcing.

2
The Inclination-Accretion Hypothesis

The astronomical theory of paleoclimates in its traditional form was challenged recently by Muller and MacDonald (1995). They put forward the hypothesis that the 100-ka glacial cycle is based on changes in orbital inclination rather than eccentricity. Orbital inclination i is the angle between the Earth's orbital plane and a reference plane. A useful choice of a reference plane is the invariable plane: the plane perpendicular to the total angular momentum vector of the solar system that is very close to Jupiter's orbital plane. With respect to the invariable plane, the orbital inclination shows a nearly sinusoidal cycle with a period of about 100 ka (Muller and MacDonald 1995). Since it has no direct effect on insolation, a mechanism must be found that could link glacial cycles to orbital inclination. One such mechanism could be the accretion of meteoroids and interplanetary dust particles (IDP).

Muller and MacDonald (1997b) have made a spectral reanalysis of four oxygen isotope records: from ODP sites 607 (Ruddiman et al. 1989) and 806 (Berger et al. 1994), the SPECMAP stack (Imbrie et al. 1984; Imbrie et al. 1992) and the stack by Hays et al. (1976). Muller and MacDonald note that in previous spectral analyses, the Blackman-Tukey method was commonly used. This method obtained the power spectral amplitudes by first computing the autocorrelation function and then performing a fast Fourier transform (FFT). With the lag parameter set to one-third (the usual value), narrow peaks are artificially broadened by a factor of 3. If the lag parameter is set to two thirds or greater, or if a direct FFT method is used, a very narrow and unsplit 100-ka peak appears. In fact, the width of this peak comes close to the width of a peak that results from calculating the spectral power of a pure 100-ka sine wave of the same duration as the oxygen isotope record (Muller and MacDonald 1997a).

The narrow width of the 100-ka peak in the oxygen isotope records strongly suggests that the 100-ka glacial cycle is astronomically driven, but Muller and MacDonald doubt that it is based on changes in insolation as the Milankovitch theory states. This is mainly for two reasons. (1) The only orbital parameter that can cause changes in insolation with a period of about 100 ka is eccentricity; but eccentricity-induced changes in insolation are very small and need to be amplified by the climate system. If the climate system were a free oscillator or a relaxation oscillator that could amplify the small input signal, the output signal would probably be a glacial cycle with a broad spectral peak, due to friction and phase instability. Such a broad peak would be at variance with the observed power spectra of oxygen isotope records. (2) The power spectrum of eccentricity does not show a single prominent peak with a period of about 100 ka, but several peaks, the five largest having periods near 95, 99, 124, 131 and 404 ka. In the spectra of Late Pleistocene oxygen isotope data, the peak with a period near 100 ka is unsplit, and there is only a very small peak with a period near 400 ka (but this peak is larger if the whole Pleistocene is considered – Berger 1992). If the climate system were a forced oscillator, then a split peak in the input signal would yield a split peak in the output signal. Hence, if there were an unsplit peak with a period near 100 ka in the response of such a climate system, then eccentricity could not be the driving force, but orbital inclination could.

Muller and MacDonald not only argue that the 100-ka glacial cycle is based on changes in orbital inclination, they also propose that the mid-Pleistocene climate transition might have been caused by an increase in the amount of meteoroids or dust (Muller 1994; Muller and MacDonald 1997a). In a study of helium-3 in oceanic sediments, Farley (1995) indeed reported a dramatic rise of the IDP accretion rate at about 1 Ma B. P. Since the coincident onset of both the major 100-ka glacial cycle and a higher IDP flux could hardly be fortuitous and could not be attributed to the regular changes in insolation, it might indicate a causal relationship.

There are a few additional pieces of evidence that support the inclination-accretion hypothesis. Farley and Patterson (1995) detected a 100-ka cycle of helium-3 in oceanic sediments (where the primary source for this isotope is accretion of IDP). A narrow ring of IDP extending only 2° from the invariable plane was observed by the Infrared Astronomical Satellite (IRAS; Sykes 1988). The number of noctilucent clouds (mesospheric clouds associated with the effects of high meteors and high-altitude dust) shows a strong peak within about a day of the date when the Earth passes through the invariable plane, supporting the contention that accretion increases significantly at such a date (Foyle and Haurwitz 1966; Thomas 1991).

We add that Farley and Patterson themselves consider the possibility that 100-ka climate effects could control the observed helium-3 flux, by focusing of IDP to particular regions of the seafloor; but they prefer an interpretation in which the 100-ka cycle in IDP accretion is the cause of helium-3 variability. In a subsequent study, Marcantonio et al. (1996) compare ^{3}He and ^{230}Th in two deep-sea cores from the equatorial Pacific Ocean and conclude that the helium-3 variability is indeed due to IDP focusing. They find a temporal pattern that points to a causal relationship between climate cycles and cycles of reorganization of near-bottom oceanic currents.

3 The Paleoclimate Model

As described in detail in Saltzman and Verbitsky (1992; 1993), the “full PCM” is composed of four time-dependent equations governing slow-response variables and three diagnostic equations for fast-response variables. The slow-response variables are total ice mass of the Northern Hemisphere ice sheets excluding the Greenland ice sheet Ψ, mean bedrock depression below the undepressed topography D (due to each of n individual ice sheets of assumed equal mass Ψ/n), atmospheric carbon dioxide concentration μ and mean ocean temperature θ. The fast-response variables are summer mean temperature at high latitudes of the Northern Hemisphere τ, mean elevation of the undepressed bedrock above sea level Z and an ice-calving catastrophe function C, adapted by Saltzman and Verbitsky (1992) for their zero-dimensional model from the one-dimensional model of Pollard (1982, 1983). A "reduced PCM" is obtained if the influence of atmospheric CO_2 and mean ocean temperature on ice mass is neglected. It is composed of only two time-dependent equations.

Here we merely give the equation governing total ice mass Ψ,

$$\frac{d\Psi}{dt} = \phi_1 - \phi_2\tau - \phi_3\Psi - nC, \tag{1}$$

and the equation for summer mean temperature τ,

$$\tau = \tilde{\tau} + \frac{B}{\tilde{\mu}}(\mu - \tilde{\mu}) + k_\theta(\theta - \tilde{\theta}^\star) + k_\Psi(\Psi - \tilde{\Psi}^\star) + k_h h' + k_R R' + k_I I'. \tag{2}$$

We adopt the convention of Saltzman and Verbitsky (1993) and denote the climatic ground state solely determined by longer-term tectonic processes by a wavy bar, and a constant representative of the late Pleistocene by a star. The variables h', R' and I' denote ice sheet height above present sea level, high-latitude summer insolation at $\phi_0 = 65°$ N and accretion of meteoroids and IDP, all expressed as departures from the climatic ground state. The coefficients ϕ_1, ϕ_2 and ϕ_3 are rate constants and the coefficients B, k_θ, k_Ψ, k_h, k_R, and k_I are sensitivity constants. All model parameters are as specified in Saltzman and Verbitsky (1993), except for those listed in Table 1. Equation (2) contains two features that are new to the PCM: the terms $k_h h'$ and $k_I I'$, which represent the elevation-temperature feedback and the inclination forcing.

The ice sheet height above present sea level is

$$h = (Z - Z_0) + H - D, \tag{3}$$

where Z_0 is the value of Z with respect to present ice mass and H is the ice sheet thickness. Considering the dry adiabatic lapse rate of temperature

Table 1 Selected model parameters

	Units	Cases 1–3	Case 4
ϕ_0	–	65° N	65° N
$\alpha_0^\star$	10^{15} kg a^{-1}	2.0	2.3
Z_0	m	300	596
k_R	°C (W m^2)$^{-1}$	0.1	0.107
k_h	K m^{-1}	-0.0065	–
k_I	K	-2.0	-2.0
ϵ_2^{-1}	ka	30.0	3.0
β_2	10^{-3} a^{-1}	–	6.12
β_5	10^{-3} ppmv(°C a)$^{-1}$	–	5.9
γ_0	10^{-4} °C a^{-1}	–	4.29
δ	–	1°	1°

$k_h = -6.5\,\mathrm{K\ km}^{-1}$, an increase in h should lead to a decrease in local surface temperature and a further increase in h. This is the elevation-temperature feedback. We must add that the non-local effects of an increase in h on surface temperature can be more complex than captured by Eq. (2) – see, e.g., Barron (1985); Paul (1996).

We test two simple paramterizations for the effect of changes in orbital inclination on temperature. In the first parameterization, we assume that increases in temperature are directly proportional to increases in orbital inclination, such that $I = i$, $I^\star = i^\star$ and $I' = i - i^\star$. In the second parameterization, we assume that decreases in temperature are proportional to the time the Earth spends in the ring of meteoroids and IDP during a year. The inclination parameter becomes

$$I = \begin{cases} (2/\pi)\arcsin(\delta/i), & \text{if } i > \delta \\ 1, & \text{if } i \le \delta \end{cases}. \tag{4}$$

In the derivation of Eq. (4), we take the ring of dust to be a thin disk of infinite horizontal extent, with a homogenous density inside, but sharp upper and lower boundaries. For simplicity we assume that this disk lies in the invariable plane. The orbit of the Earth is taken to be circular. We introduce a new parameter δ that is related to width of the disk: if the inclination of the Earth's orbit is low, then $i \le \delta$ and the Earth spends the entire year in the dust, if it is high, then $i > \delta$ and the Earth passes through the dust only twice per year (Muller 1994). The angles i and δ are both assumed to be small.

4
A Note on Numerical Methods

For the time discretization of the system of equations that constitutes the PCM, we use a semi-implicit scheme which we solve by a fixed-point iteration method (Deblonde and Peltier 1991). The ice sheet calving term is treated explicitly. To improve the convergence rate for this scheme we employ a relaxation factor (Press et al. 1992). The time step is adapted automatically and ranges between 10 and 1000 a.

The spectral analysis that is used to compare the model results and the δ^{18}O records is based on the Lomb-Scargle Fourier transform for unevenly spaced time series in combination with a Welch-Overlapped-Segment-Averaging procedure, as implemented in the SPECTRUM program (Schulz and Stattegger 1997). The advantage of SPECTRUM is that any interpolation of the time series is avoided which otherwise leads to an underestimation of high frequency components. Typically deep-sea records constitute unevenly spaced time series, while model output can be made evenly spaced. Two sets of parameters are used: in the first the number of segments is one ($n_{50} = 1$) and a rectangular window is chosen, while in the second the number of segments is three ($n_{50} = 3$) and a Hanning window is chosen. The first set of parameters is in the spirit of Muller and Macdonald (1997a, b), who advocate a spectral method that maximizes the frequency resolution or, equivalently, minimizes the resolution bandwidth. The second set of parameters means that the time series is split into n_{50} segments that overlap each other by 50 %. In order to suppress spurious peaks and side lobes, each segment is multiplied by a Hanning window. Then the n_{50} windowed segments are Fourier transformed separately. Averaging the resulting raw spectra yields a consistent estimate of an autospectrum with an increased resolution bandwidth, but a decreased standard error in the power spectral amplitudes. We refer to the power spectra corresponding to the two sets of parameters as the raw and the smoothed spectrum. In all figures that represent spectral analyses, we use a logarithmic decibel (dB) scale:

$$\mathrm{dB} = 10 \log_{10} G_{\mathrm{xx}}(f), \qquad (5)$$

where $G_{\mathrm{xx}}(f)$ denotes the power spectral amplitude. A horizontal line marks the 6 dB-bandwith, and a vertical line marks the standard error (see, e.g., Fig. 2).

5 The 100 000-Year Glacial Cycle in the Late Pleistocene

In our first series of experiments, we assume that to a first order the waxing and waning of the continental ice sheets is not affected by variations in atmospheric carbon dioxide concentration and mean ocean temperature (see Table 2, cases 1–3). Thus in Eq. (2) we set $k_\theta = 0$ and $B = 0$ and obtain the "reduced PCM", a pure ice sheet-bedrock model in the spirit of, e.g., Birchfield et al. (1981), Pollard (1982, 1983) and Deblonde and Peltier (1991).

We focus on the time period between 600 ka B. P. and the last deglaciation, after the onset of the 100-ka glacial cycle. In Fig. 1, we present three δ^{18}O records, which we wish to compare with our model results: the SPECMAP stack (Imbrie et al. 1984; Imbrie et al. 1992) and the benthic δ^{18}O records from ODP sites 659 (Tiedemann et al. 1994) and 806B (Berger et al. 1995). The SPECMAP time scale has been tuned to the precessional and obliquity cycles. For the data from ODP sites 659 and 806B, we use the following chronology: in order to minimize any bias by using a time scale that is tuned to the astronomical frequencies, we adopt the untuned time scale developed by Raymo (1997). This time scale uses three radiometric age constraints: the midpoints of the last two deglaciations (Terminations I and II) and the midpoint of stage 19, with constant sedimentation rates inbetween. The ages of Terminations I and II are assigned to be 13.5 and 128 ka B. P., and the midpoint of stage 19 is taken to be 772 ka old, 6 ka younger than the Brunhes-Matuyama boundary radiometrically dated at 778 ± 3.5 ka by Tauxe et al. (1996). In the time domain, looking at the ages of the remaining terminations, we find from Fig. 1 that the agreement between the tuned SPECMAP time series and the untuned time series is rather good for ODP site 659, but mediocre for ODP site 806B. In the frequency

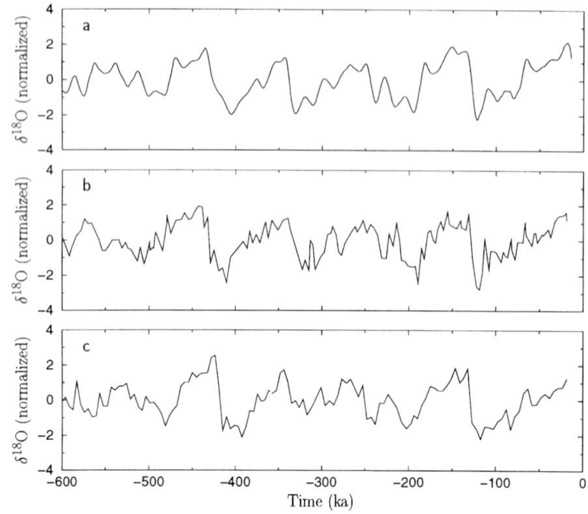

Fig. 1a–c Time series of **a** the SPECMAP stack, **b** the benthic $\delta^{18}O$ record of ODP site 659 and **c** the benthic $\delta^{18}O$ record of ODP site 806B. All data sets are normalized to have zero mean and unit variance. *Positive values* indicate glacials (large ice mass) and *negative values* indicate interglacials (small ice mass). For the data from ODP sites 659 and 806B the untuned time scale developed by Raymo (1997) is adopted

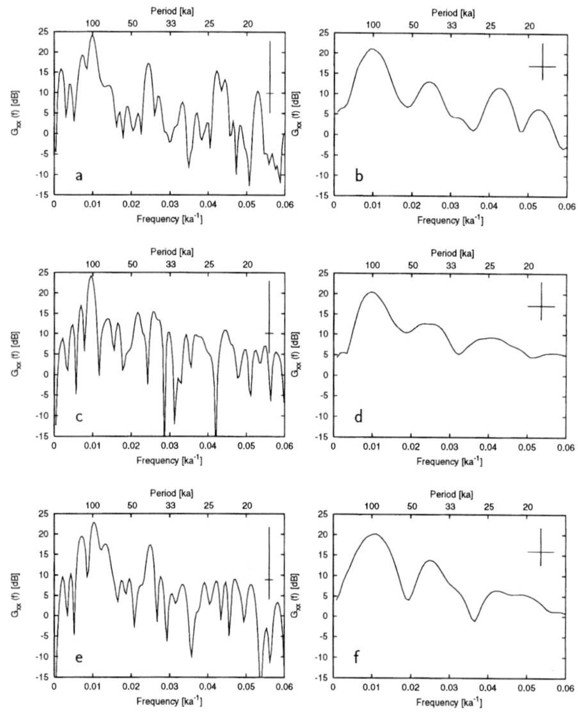

Fig. 2a–f Spectral analysis of **a** and **b** the SPECMAP, **c** and **d** the ODP 659 and **e** and **f** the ODP 806B $\delta^{18}O$ time series. The raw spectra are given on *the left-hand side* and the smoothed spectra are given on *the right-hand side*. Note that the power spectral amplitude $G_{xx}(f)$ is plotted on a logarithmic scale

domain, a different picture emerges. Figure 2 shows the corresponding raw and smoothed power spectra. The dominant 100-ka peak is clearly visible in all smoothed spectra. Furthermore, the obliquity peak with a period of about 41 ka is present in the SPECMAP stack and, slightly shifted towards a shorter period, in the record from ODP site 806B. In the same two time series the precessional peaks with periods near 23 and 19 ka are well separated. In the record from ODP site 659, the obliquity as well as the precessional peaks are smeared out. Our point is here that the raw spectra have large standard errors, which do not allow to regard the separate peaks with a period near 100 ka as significant, neither in the tuned nor in the untuned time series. As in the work of Saltzman and Verbitsky, we use the SPECMAP stack as our target for the late Pleistocene.

Figure 3 presents the July insolation at $\phi_0 = 65°$ N, along with the eccentricity (Berger 1978). Figure 4 gives the corresponding raw and smoothed spectra. In the first two experiments we neglect the temperature-elevation feedback and ignore any inclination forcing by setting $k_{\mathrm{h}} = 0$ and $k_{\mathrm{I}} = 0$ (see Table 2, case 1). Thus, the July insolation at 65° N is the only input to the PCM. The task is now to find a set of parameter values for which the output of the PCM is a good approximation to the changes in global ice volume (or equivalently, global ice mass) as recorded in the deep sea sediments. This means that the low 100-ka peak in the input signal must be transformed into a large 100-ka peak in the output signal.

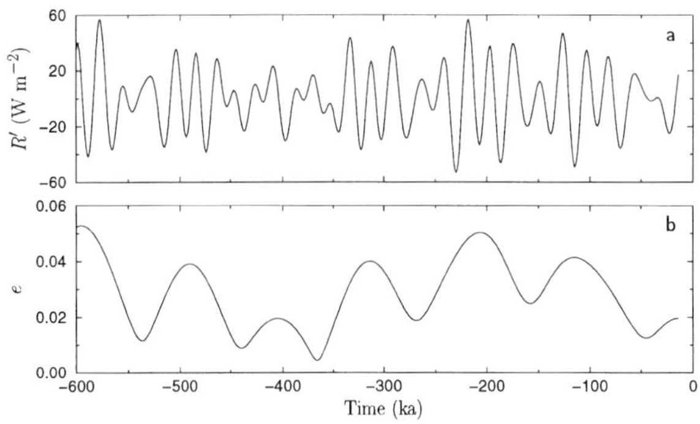

Fig. 3a, b Time series of **a** July insolation at 65° N and **b** eccentricity (Berger 1978)

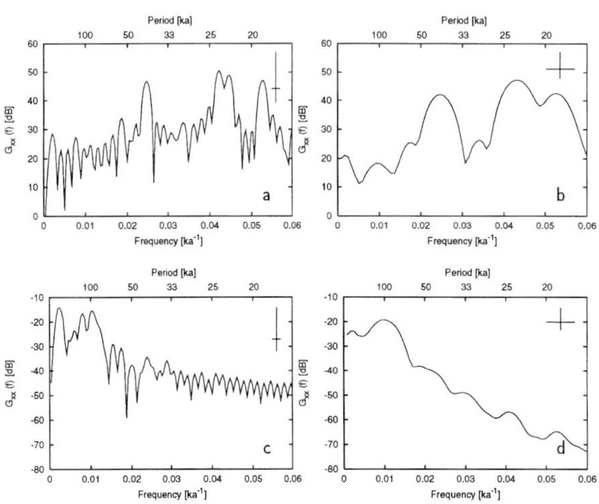

Fig. 4a–d Spectral analysis of **a** and **b** the insolation and **c** and **d** the eccentricity time series shown in Fig. 3. The raw spectra are given on *the left-hand side* and the smoothed spectra are given on *the right-hand side*. Note that the power spectral amplitude $G_{\mathrm{xx}}(f)$ is plotted on a logarithmic scale and for different ranges

Table 2 Summary of the numerical experiments

Case	Experiment
1 Standard	a No ice calving b With ice calving
2 Elevation-temperature effect included	a No ice calving b With ice calving
3 With inclination forcing	a Directly proportional to inclination b Proportional to inclination parameter c Proportional to inclination parameter, with insolation forcing d Proportional to inclination parameter, with insolation forcing and ice calving e Proportional to inclination parameter, with insolation forcing and ice calving, eccentricity set to present-day value
4 Global carbon cycle and heat storage in the ocean included	a With insolation forcing and ice calving b Combined insolation and inclination forcing, with ice calving c Combined insolation and inclination forcing, with ice calving and eccentricity set to present-day value

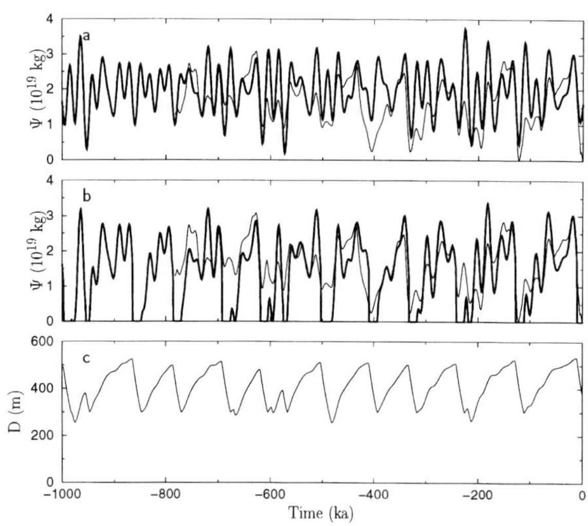

Fig. 5a–c Case 1 time-dependent solutions of the "reduced PCM". **a** Ice mass Ψ with the ice-calving mechanism suppressed ($C = 0$). **b** Ice mass Ψ and **c** bedrock depression D with the ice-calving mechanism activated ($C \neq 0$). The *thin curve* is the SPECMAP $\delta^{18}O$ record, scaled in terms of ice mass

The parameter values that we use in case 1 are taken from Saltzman and Verbitsky (1993), except that we specify a long time constant for bedrock depression, $\epsilon_2^{-1} = 30\,\text{ka}$ (see Table 1, cases 1–3). We start from the initial conditions $\Psi(0) = 2 \times 10^{19}\,\text{kg}$ and $D(0) = 0$ at 3 Ma BP. Without the ice-calving mechanism ($C = 0$), the PCM generates changes in ice mass which already show some resemblance to the $\delta^{18}O$ records (Fig. 5a). There is considerable power in

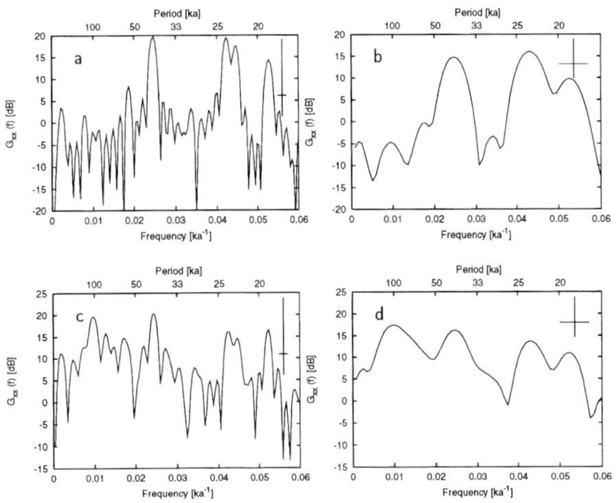

Fig. 6a–d Spectral analysis of the case 1 time-dependent solutions shown in Fig. 5. **a** and **b** With the ice-calving mechanism suppressed ($C = 0$). **c** and **d** With the ice-calving mechanism activated ($C \neq 0$). The raw spectra are given on *the left-hand side* and the smoothed spectra are given on *the right-hand side*. Note that the power spectral amplitude $G_{xx}(f)$ is plotted on a logarithmic scale and for different ranges

the precessional and obliquity bands (Fig. 6b); but the later terminations are incomplete, and the peak near a period of about 100 ka is only very low.

We note that in the case $B = k_\theta = k_h = C = 0$, the PCM becomes a linear forced oscillator and Equation (1) can be integrated analytically to yield

$$\Psi(t) = \Psi_0 + \int_0^t [\alpha_0 - \phi_2(k_R R' + k_I I')] \exp[-\alpha_3(t - t')]\, dt, \tag{6}$$

where

$$\alpha_0 = \alpha_0^\star, \;\; \alpha_0^\star = \phi_1 - \phi_2\big(\tilde{\tau}^\star - k_\Psi \tilde{\Psi}^\star\big) \text{ and } \alpha_3 = \phi_3 + \phi_2 k_\Psi .$$

With the ice-calving mechanism ($C \neq 0$), the PCM results for bedrock depression D and ice mass ψ shown in Figs. 5b and c are similar to those obtained by Saltzman and Verbitsky (1992). They exhibit complete terminations and an oscillation of a period close to 100 ka. In the raw spectrum, the 100-ka peak is of a width comparable to that in the raw spectra of the δ^{18}O records (Fig. 6c). In the smoothed spectrum, the amplitude is now maximum at the 100-ka period, and the relative amplitudes at the precession and obliquity frequencies are similar to the relative amplitudes of the SPECMAP time series (Fig. 6d). However, the relative amplitude at the 1/100 ka^{-1} frequency is still to small.

The ice-calving mechanism introduces an instability into the "reduced PCM" that is conditional, activated only by perturbations that become greater than a finite, non-zero threshold value. In the absence of Milankovitch forcing, stochastic or high-frequency periodic forcing can cause such perturbations. An instability of this type is characteristic of a relaxation oscillator. To the extent that stochastic or high-frequency periodic forcing can be considered as being generated internally by the climate system, the oscillation of a period near-100 ka is an example of a "quasi-free" oscillation. For comparison, the

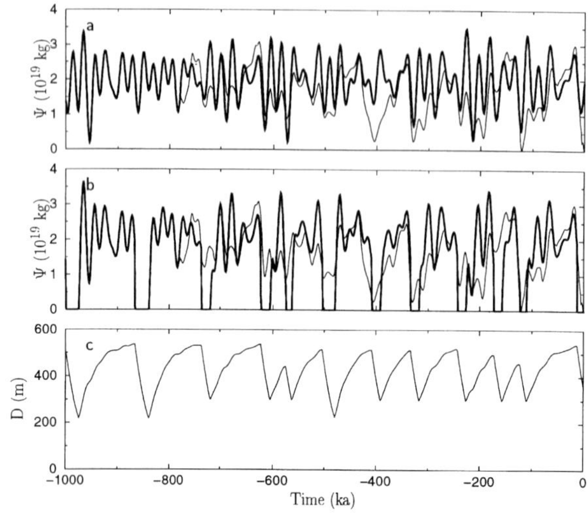

Fig. 7a–c Case 2 time-dependent solutions of the reduced PCM – elevation-temperature feedback included. **a** Ice mass Ψ with the ice-calving mechanism suppressed ($C = 0$). **b** Ice mass Ψ and **c** bedrock depression D with the ice-calving mechanism activated ($C \neq 0$). The *thin curve* is the SPECMAP $\delta^{18}O$ record, scaled in terms of ice mass

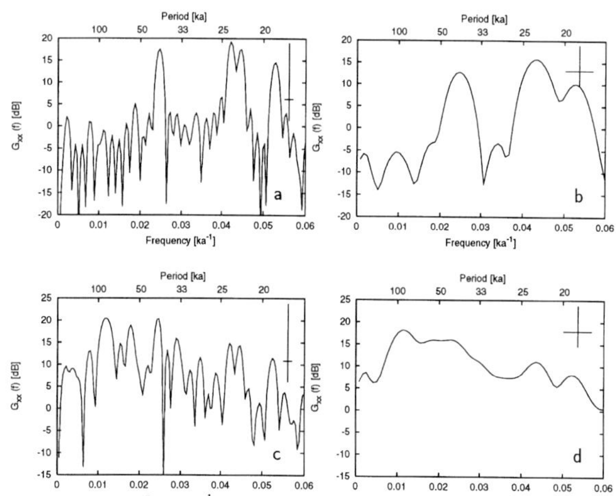

Fig. 8a–d Spectral analysis of the case 2 time-dependent solutions shown in Fig. 7 – elevation-temperature feedback included. **a** and **b** With the ice-calving mechanism suppressed ($C = 0$). **c** and **d** With the ice-calving mechanism activated ($C \neq 0$). The raw spectra are given on *the left-hand side* and the smoothed spectra are given on *the right-hand side*. Note that the power spectral amplitude $G_{xx}(f)$ is plotted on a logarithmic scale and for different ranges

instability built into the Milankovitch template of Berger et al. (1994) is unconditional, realizable for any small perturbations. Hence, the Milankovitch template is a free oscillator that admits free oscillations even in the absence of "internally generated noise" (Paul and Berger 1997).

Case 2 includes the elevation-temperature feedback ($k_h = -6.5\,\text{K km}^{-1}$). In the PCM, an increase in ice-sheet height above sea level causes a decrease in surface temperature and a corresponding decrease in the ablation rate, which constitutes a positive feedback. This additional feedback does not result in a substantially better simulation of the SPECMAP $\delta^{18}O$ record (Figs. 7a and c), nor does it lead to a further increase of spectral power in the 100-ka band. However, Fig. 8a shows that the 100-ka peak and the obliquity peak merge to give one broad peak.

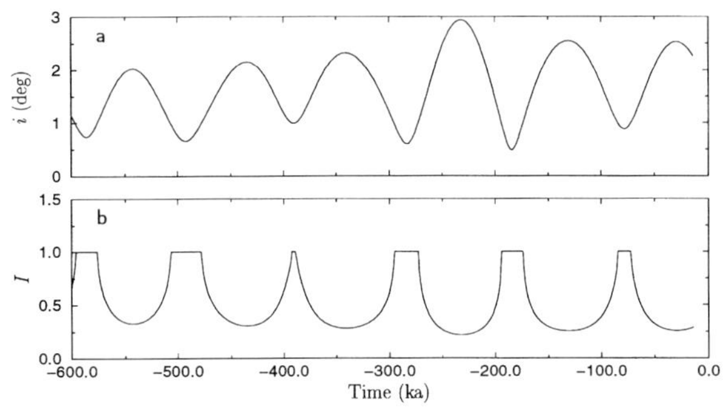

Fig. 9a, b Time series of **a** orbital inclination (Muller and MacDonald 1995) and **b** inclination parameter [as given by Eq. (4)]

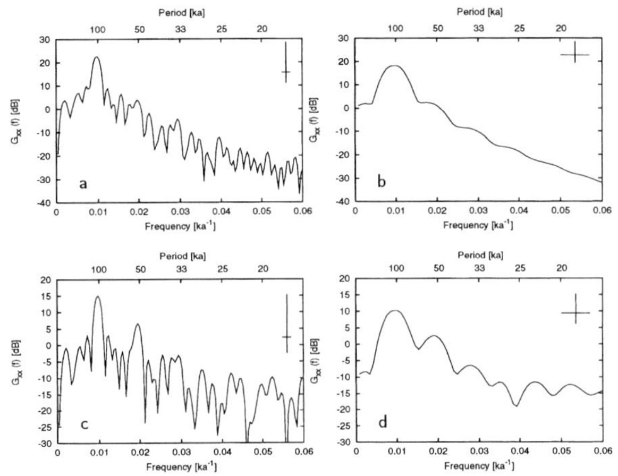

Fig. 10a–d Spectral analysis of **a** and **b** the orbital inclination and **c** and **d** the inclination parameter time series shown in Fig. 9. The raw spectra are given on *the left-hand side* and the smoothed spectra are given on *the right-hand side.* Note that the power spectral amplitude $G_{XX}(f)$ is plotted on a logarithmic scale and for different ranges

In case 3 we again neglect the elevation-temperature feedback and force the PCM by changes in orbital inclination (Fig. 9a). First we ignore changes in insolation and assume that increases in temperature are directly proportional to increases in orbital inclination. Thus we set $I' = i - i^*$. The resulting changes in ice mass Ψ follow the trend of the SPECMAP $\delta^{18}O$ curve, but the timing of the terminations is poor (Fig. 11a). The ice mass curve actually reflects an orbital inclination cycle with a period of the order of 1 Ma. We then take changes in temperature to be proportional to the inclination parameter I as defined in Eq. (4). Hence, we set $I' = I - I^*$. The time series shown in Fig. 9b displays a number of sharp edges that generate the high-frequency background visible in Figs. 10c and d. These sharp edges could be smoothed out if the ring of dust were taken to be a disk with a Gaussian-like density distribution in the vertical. The corresponding changes in ice mass are now more asymmetric, but the simulated time series still lags the SPECMAP time series (Fig. 11b). This phase lag can be reduced if the damping constant α_3 is set to the much larger value $1/100\,\text{ka}^{-1}$ (at the same time, the positive feedback constant α_0 must be changed from 2 to $0.2\,\text{kg a}^{-1}$ in order to preserve the equilibrium ice mass $\alpha_0/\alpha_3 = 2 \times 10^{19}\,\text{kg}$.).

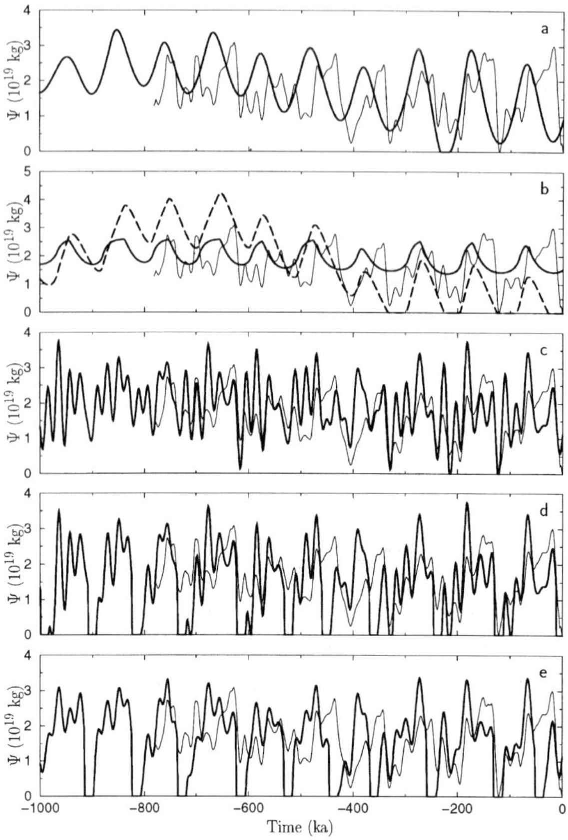

Fig. 11a–e Case 3 time-dependent solutions of the "reduced PCM" – with inclination forcing. The *thin curve* is the SPECMAP $\delta^{18}O$ record, scaled in terms of ice mass. **a** Forcing directly proportional to inclination ($I' = i - i^{\star}$). **b** Forcing proportional to inclination parameter ($I' = I - I^{\star}$). The *solid curve* refers to the case $\alpha_0 = 2.0 \times 10^{15}$ kg a^{-1} and $\alpha_3 = 1.0 \times 10^{-4}$ a^{-1}. The *dashed curve* refers to the case $\alpha_0 = 0.2 \times 10^{15}$ kg a^{-1} and $\alpha_3 = 0.1 \times 10^{-4}$ a^{-1}. **c** With $I' = I - I^{\star}$, including insolation forcing. **d** With $I' = I - I^{\star}$, including insolation forcing and ice-sheet calving. **e** With $I' = I - I^{\star}$, including insolation forcing and ice-sheet calving, eccentricity set to present-day value

Since we set $k_I = -2\,°C$, the time series for the inclination parameter I shown in Fig. 9b implies a cooling of the high-latitude surface temperature τ by about 1.4 °C. This is slightly larger than the cooling of the global temperature by 0.5–1.0 °C known from volcanic eruptions. Thus, for a reasonable value of the sensitivity parameter k_I inclination forcing alone can produce 100-ka changes in ice mass of noticeable amplitude and with about the right phase. If we now reset α_0 and α_3 to their previous values and add the insolation forcing to the inclination forcing, we obtain the fit to the SPECMAP $\delta^{18}O$ curve shown in Fig. 11c. By comparison with Fig. 5a, we find that the orbital inclination forcing led to a seemingly better simulation of the terminations. There is considerable power in the 100-ka band (Figs. 12a, b) as could have been anticipated from the the power spectra of the inclination parameter (Figs. 10c, d). Still, there is more power in the precessional and obliquity bands. If we invoke the ice-calving mechanism, we obtain complete deglaciations and a maximum in amplitude at the 100-ka period (Figs. 11d, 12c, d); but from Fig. 5b we see that the timing of the deglaciations is actually less accurate than without inclination forcing. Finally, we combine the inclination and insolation forcing, but

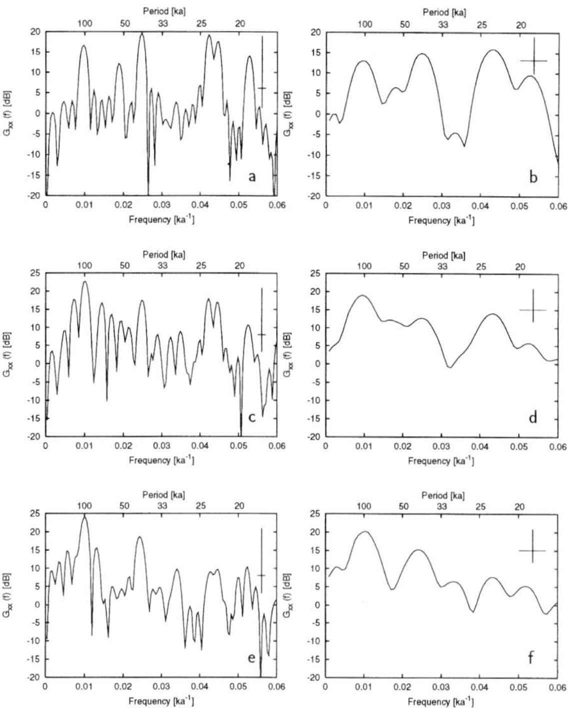

Fig. 12a–f Spectral analysis of the case 3 time-dependent solutions shown in Fig. 11 – with inclination forcing. **a** and **b** With $I' = I - I^{\star}$, including insolation forcing. **c** and **d** With $I' = I - I^{\star}$, including insolation forcing and ice-sheet calving. **e** and **f** With $I' = I - I^{\star}$, including insolation forcing and ice-sheet calving, eccentricity set to present-day value. The raw spectra are given on *the left-hand side* and the smoothed spectra are given on *the right-hand side*. Note that the power spectral amplitude $G_{\mathrm{xx}}(f)$ is plotted on a logarithmic scale and for different ranges

set the eccentricity to its present-day value $\epsilon_0 = 0.01672$, such that the precessional effect is no longer modulated. Figure 11e reveals that the fit to the SPECMAP stack is further degraded, although the 100-ka peak is now dominating the spectrum very clearly (Figs. 12e, f). This result does not change if the value of k_{R} is decreased in order to lower the level of the combined forcing.

6
The Climate Transitions in the Late Cenozoic

In the following we wish to illustrate the rich possibilities offered by the "full PCM". We neglect the elevation-temperature feedback ($k_{\mathrm{h}} = 0$), but include the influence of variations in atmospheric carbon dioxide concentration and mean ocean temperature ($k_{\mu} \neq 0,\ k_{\theta} \neq 0$). Furthermore, the time constant ϵ_2^{-1} is set at the value of 3 ka, which is probably more realistic . The resulting system of four time-dependent equations and three diagnostic equations has been studied extensively by Saltzman and Verbitsky (1992, 1993). When the ice-calving mechanism is neglected ($C = 0$), the PCM can generate a free oscillation of a period near 100 ka due to the instability introduced by a positive feedback in the global carbon cycle, with superimposed forced precessional and obliquity periods. When the ice-calving mechanism is permitted ($C \neq 0$), an intermittent, bistable solution is obtained. Such a solution is characterized by aperiodic transitions between episodes of near-40 ka-period oscillations (associated with the ice-calving instability) and near-100 ka-period oscillations (associated with the global carbon cycle instability).

We now assume that the tectonically-induced level of atmospheric CO_2 decreases linearly by about 100 ppm over the last 5 Ma (Saltzman and Verbitsky 1993). Thus, we take

$$\tilde{\mu} = \tilde{\mu}^* + \dot{\tilde{\mu}} t, \tag{7}$$

where $\tilde{\mu}^* = 253$ ppm and $\dot{\tilde{\mu}} = -20$ ppm/Ma, such that $\tilde{\mu} = 353$ ppm at $t = -5$ Ma. All other parameter values are also similar to those for case II of Saltzman and Verbitsky (1993) (see Table 1, case 4). In Fig. 13b we show the response of the "full PCM" to the combined insolation-carbon dioxide forcing ($k_I = 0$). As Saltzman and Verbitsky (1993) have pointed out, a remarkable aspect of this solution is that the imposition of a slow tectonic decrease of CO_2 transforms a previously chaotic solution into an ordered sequence of clearly defined regimes separated by sharp transitions.

With our choice of parameters we obtain a decent fit to the benthic $\delta^{18}O$ record from ODP site 806B (Fig. 13a) over the periods from 3 to 2 Ma BP and from about 1.2 Ma BP to the present. Between 2 and about 1.2 Ma BP the simulated ice mass is much larger than the one inferred from the $\delta^{18}O$ record. During this period the PCM simulates complete deglaciations which have no analogue in the climatic time series.

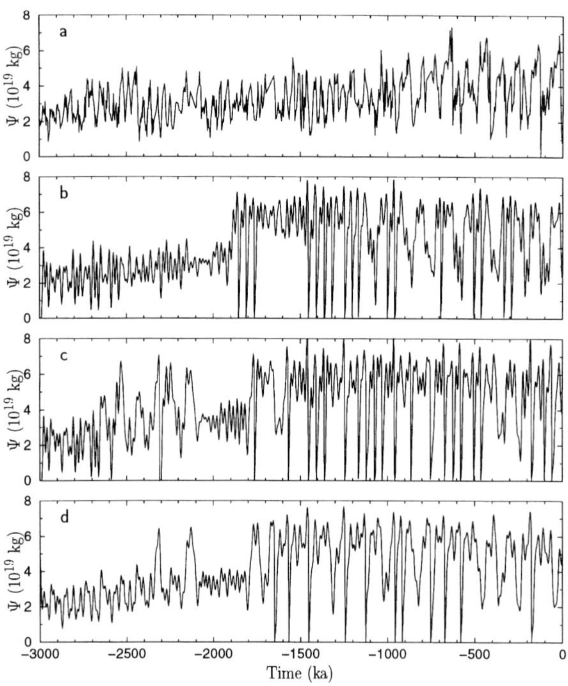

Fig. 13a–d The climate transitions in the late Cenozoic. **a** Benthic $\delta^{18}O$ record of ODP site 806B, scaled in terms of ice mass. **b** Case 4 time-dependent solution of the "full PCM" – global carbon cycle and heat storage in the ocean included, with insolation forcing and ice calving. **c** Same as **b**, but combined insolation and inclination forcing. **d** Same as **c**, but combined insolation and inclination forcing, eccentricity set to present-day value

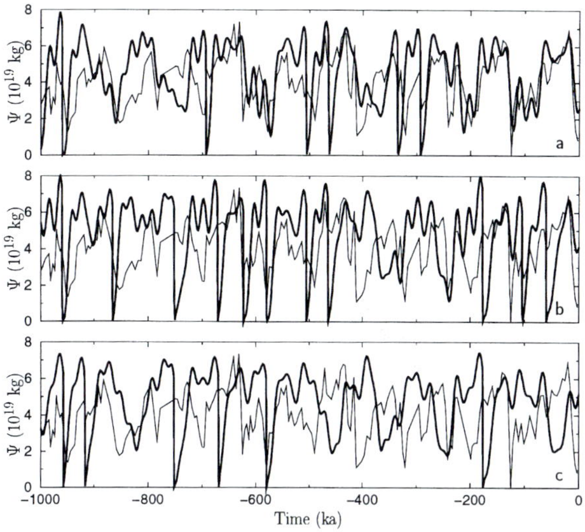

Fig. 14a–c Expanded views of the last 1 Ma of the case 4 time-dependent solutions of the "full PCM" shown in Fig. 13 – global carbon cycle and heat storage in the ocean included. **a** With insolation forcing and ice calving. **b** Combined insolation and inclination forcing, with ice calving. **c** Combined insolation and inclination forcing, with ice calving and eccentricity set to present-day value

If we add the inclination forcing to the combined insolation-carbon dioxide forcing ($k_I = -2\,^\circ C$), we obtain the solution for ice mass Ψ shown in Fig. 13c. Obviously, the inclination forcing can activate the global carbon cycle instability earlier than would otherwise be the case, well before 2 Ma BP. Furthermore, if the eccentricity is held fixed at its present-day value $\epsilon_0 = 0.01672$, the 100-ka glacial cycle starts as early as about 1.6 Ma B. P. and dominates most of the Pleistocene (Fig. 13d).

In Fig. 14 we give expanded views of the last 1 Ma of the solutions of the "full PCM" shown in Fig. 13. With only the combined insolation-carbon dioxide forcing, the agreement between the simulated and observed time series is satisfying (Fig. 14a). In particular, the timing of the terminations is good. With inclination added to the forcing this timing is less satisfactory (Fig. 14b). The simulated and observed time series even get out of phase if the eccentricity is held fixed (Fig. 14c). Again, this result does not change if the value of k_R is decreased relative to the value of k_I.

7 Discussion

The unconditional instability built into the Milankovitch template of Berger et al. (1994) leads to free oscillations which give so much inertia to the model that it can support the 100-ka climate cycle. The conditional instability contained in ice sheet-bedrock models such as the "reduced PCM" admits "quasi-free" oscillations which are a similar source of inertia. The existence of an instability of either type triggered during deglaciations seems to be required to account for the rapid retreats and the complete terminations recorded in deep-sea sedimentary cores, even in the case of inclination forcing that has considerable power in the 100-ka band. A notoriously difficult problem is the simulation of Termination V, which occurs at the base of isotope stage 11 at a time when eccentricity is very low (at about 400 ka BP; Imbrie and Imbrie 1980). The free and "quasi-free" oscillations underlying the solutions of the Milankovitch template (Paul and Berger 1997; Fig. 3) and the "reduced PCM" (Fig. 5b) can overcome this problem.

The first question that arises is what actually is the source of the large inertia inherent to the climate system? A good fit of the Milankovitch template to the target time series requires the "memory" of the climate system to cover a period of 57 ka (Berger et al. 1996; Paul and Berger 1997). The "reduced PCM" can only yield an oscillation of near-100 ka period with a long time constant for bedrock depression, $\epsilon_2^{-1} = 30$ ka. For a probably more realistic time constant of $\epsilon_2^{-1} = 3$ ka, the same model admits a free oscillation with a period closer to 40 ka than to 100 ka (Saltzman and Verbitsky 1992, 1993). The orbital inclination curve must be shifted by 33 ± 3 ka to give the best least-squares fit to the SPECMAP stack (Muller and MacDonald 1995). In this respect, we note that Eq. (6) does not allow for a mere shift of the orbital inclination curve. Rather it integrates the applied forcing over the elapsed time t with an exponential factor $\exp(-\alpha_3[t - t'])$. It is the damping constant α_3 that must be set to the large value of $1/100\ \text{ka}^{-1}$ in order to reduce the difference in phase between the PCM output and SPECMAP. Hence, in all these cases a time constant is assigned a value that is unrealistically large.

The answer to the above question may be that the large inertia inherent to the climate system results from a delicate balance of positive and negative feedbacks, and that either more spatial complexity or additional feedback mechanisms must be included to achieve such a balance. In contrast to the zero-dimensional "reduced PCM", the one-dimensional ice sheet-bedrock models of Pollard (1983) and Deblonde and Peltier (1991) yield realistic 100-ka climate cycles for a time constant of $\epsilon_2^{-1} = 5$ ka, if the ice sheet-calving mechanism or a related generalized meltwater parameterization is used. In this connection Pollard mentions that it is necessary to tune the crustal topography, such that calving is not initiated too easily – something that is possible in a one-dimensional model but cannot be done in a zero-dimensional model. In his model, calving affects only the outermost 50 to 150 km of the ice sheet, while in the "reduced PCM" the ice sheet becomes submerged completely below sea level and vanishes instantaneously (Saltzman and Verbitsky 1993).

Additional feedback mechanisms are provided by Eq. (7) and (8) of Saltzman and Verbitsky (1993) that govern the changes in atmospheric CO_2 concentration and mean ocean temperature in the "full PCM". The equation that represents the global carbon cycle contains five relatively unconstrained or "free" parameters. It is very difficult to judge whether the values assigend to these parameters are realistic or not. They are much more uncertain than those in the ice sheet-bedrock system, because at present it is impossible to calculate the fundamental fluxes leading to net changes of atmospheric CO_2 concentration on long time scales (Saltzman and Maasch 1991).

Another important feedback mechanism is basal sliding, which can occur in a one-dimensional ice sheet-bedrock model if thermodynamics is included. Oerlemans (1982) has shown that such a model can produce free oscillations with a period of about 100 ka, again for a time constant of $\epsilon_2^{-1} = 5$ ka.

The case 1 results with a long time constant for bedrock depression $\epsilon_2^{-1} = 30$ ka are in good agreement with similar experiments of Pollard (1983, Figs. 2b, 5) and Deblonde and Peltier (1991, Figs. 2a, c) on the one hand and the SPECMAP δ^{18}O record on the other hand.

The positive elevation-temperature feedback included in case 2 does not result in an improved simulation of the observed δ^{18}O record, while in all cases the ice-calving mechanism does. This means that it is the design of the negative feedbacks that determines how good the fit to the data is, one criterion for a good fit being how well the rapid deglaciations are simulated.

Case 3 shows that inclination forcing can help to achieve nearly complete terminations, although the ice-calving mechanism is still required to melt all ice. If the ice-calving mechanism is invoked, the timing of the terminations is less satisfactory with inclination forcing than without it.

We think that the use of raw spectra as advocated by Muller and MacDonald is problematic in comparing observed δ^{18}O data and simulated ice mass. This is because of the ever-present noise and the quasi-periodic nature of the insolation forcing as well as the inclination forcing. A high standard error in the amplitudes of the raw spectra does not allow a discussion on whether a peak near a period of about 100 ka is split or not. From the smoothed spectra of cases 1 and 3, we conclude that the widths of the peaks for the δ^{18}O data and the simulated ice mass are comparable, and that there is no excessive power in the 400-ka band, neither for the forced nor for the relaxation oscillator. However, we admit that a longer time series must be considered to properly assess a peak near a period of 400 ka.

In both parameterizations for the effect of changes in orbital inclination on temperature, we assume that times of high accretion are coincident with times of low temperature, at variance with Farley (1995), who found a 100-ka cycle of ^{3}He in sediment, but times of high accretion roughly coincident with the interglacials. Marcantonio et al. (1996) even dispute that the 100-ka cycle in IDP accretion is the cause of the observed helium-3 variability.

The second question that arises is what has actually caused the climate transitions in the late Cenozoic. On the one hand, Muller (1994) suggests that the sudden onset of the 100-ka cycle might have been caused by an increase in the amount of IDP or meteoroids at that time. On the other hand, the growth of large and unstable continental ice sheets is attributed to a slow decrease in the atmospheric CO_2 level, induced by tectonic processes (e.g., Saltzman and Maasch 1991; Saltzman and Verbitsky 1993; Raymo 1997). Both mechanisms are still speculative. The full PCM that we use in case 2 is built on the scond mechanism.

The transitions that are seen in Fig. 13b can be readily associated with the climate transitions in the late Pliocene and the mid-Pleistocene. The first occurs later than observed in the $\delta^{18}O$ data (at 2 Ma B. P. as compared to 2.6 to 2.4 Ma B. P.), the second earlier (at about 1.1 Ma B. P. – as compared to about 900 ka B. P. if the time domain is considered). Our set of parameters is designed to give a reasonable fit to the benthic $\delta^{18}O$ record from ODP site 806B. A different choice of parameters places these transitions closer to the observed dates (Saltzman and Verbitsky 1993). In any case, the PCM can qualitatively account for the abrupt climate transitions in the late Cenozoic as well as the major 100-ka climate cycles of the late Pleistocene. We note that the earlier climate transition is difficult to detect in the benthic $\delta^{18}O$ record shown in Fig. 13a. It indicates a more gradual rise in ice mass than the PCM predicts, and there is only a slight temporary increase in ice mass variance shortly after 2.5 Ma B. P.

We acknowledge that the conclusions that we draw from our results are semi-quantitative in nature since they are based on a comparison of simulated and observed time series by visual inspection. This method seems to be appropriate because the questions we have addressed are, in a sense, also semi-quantitative in nature. Thus, we are satisfied that there is no reason to abandon the traditional astronomical theory for the inclination-accretion hypothesis..

Finally, we wish to point out that the model of Saltzman and Verbitsky (1993) also yields the following result: if the anthropogenically induced level of atmospheric CO_2 were maintained over a long period of time, the climate system would be driven toward a stable state devoid of the natural ice age cycles (Saltzman and Maasch 1993). Berger and Loutre (1996) use a much more complex model to make a forecast for the next 130 ka, driven only by the astronomically induced changes in insolation and prescribed variations in atmospheric CO_2. In their experiments, the human burst of atmospheric CO_2 seems to be sufficient to lead to a melting of the Greenland ice sheet after 6 to 10 ka. They find that it takes at least 40 ka for the climate system to recover from what could happen over the next few centuries.

8 Conclusions

Our previous experiments with the Milankovitch template (Paul and Berger 1997), as well as our present experiments with the "reduced PCM" and the "full PCM", lead us to the following conclusions:

1 The 100-ka climate cycle may be explained in terms of either a conditional or an unconditional instability that admits free or "quasi-free" oscillations.
2 A simple parameterization of the elevation-temperature feedback does not result in an improved simulation of the observed $\delta^{18}O$ record.
3 Inclination forcing can help to achieve nearly complete terminations and increase the spectral power in the 100-ka band of the ice mass response. However, the ice-calving mechanism is still required to melt all ice and let the 100-ka peak dominate the power spectrum.
4 If the ice-calving mechanism is invoked, the actual timing of the terminations appears to be less satisfactory with inclination forcing than without it.

5 Orbital inclination may still be considered as an additional element of an astronomical theory of paleoclimates in an enlarged form, provided that evidence can be found for a true linking mechanism.
6 The spectral characteristics of ice sheet-bedrock models such as the PCM are similar to the $\delta^{18}O$ records: there is no significant splitting of the 100-ka peaks, their widths are comparable to the data and there is no excessive power in the 400-ka band.

Acknowledgements

We would like to thank Richard A. Muller for providing us with the orbital inclination data. A. P. would also like to thank Michael Schulz, who introduced him to the SPECTRUM program. He is very grateful to Andreas Manschke for helping him to process the data and preparing the figures presented here, and to Björn Grieger and Reiner Schlotte for carefully reading early versions of this manuscript. Finally, he would like to thank Klaus Herterich, who stimulated his interest in the astronomical theory of paleoclimates.

The thorough reviews by Christian Schäfer-Neth and Michael Schulz improved this manuscript very much, and we thank them for their effort.

This chapter is Contribution No. 122 of the Sonderforschungsbereich 261 at the University of Bremen.

References

Barron, E. J. (1985). Explanations of the Tertiary global cooling trend. *Palaeogeography, Palaeoclimatology, Palaeoecology 50*, 45–61.

Berger, A. (1988). Milankovitch theory and climate. *Reviews of Geophysics 26*, 624–657.

Berger, A., M. F. Loutre, and J. Laskar (1992). Stability of the astronomical frequencies over the Earth's history for paleoclimate studies. *Science 255*, 560–566.

Berger, A. L. (1978). Long term variations of daily insolation and Quaternary climatic changes. *Journal of the Atmospheric Sciences 35*(12), 2362–2367.

Berger, A. L. (1992). Astronomical theory of paleoclimates and the last glacial-interglacial cycle. *Quaternary Science Reviews 11*, 571–581.

Berger, A. L. and M.-F. Loutre (1991). Insolation values for the climate of the last 10 million years. *Quaternary Science Reviews 10*, 297–317.

Berger, A. L. and M.-F. Loutre (1996). Modelling the climate response to astronomical and CO_2 forcings. *C. R. Acad. Sci. Paris 323*, 1–16.

Berger, W. H. and G. Wefer (1992). Klimageschichte aus Tiefseesedimenten. Neues vom Ontong-Java-Plateau (Westpazifik). Naturwissenschaften 79, 541–550.

Berger, W. H., T. Bickert, M. K. Yasuda, and G. Wefer (1996). Reconstruction of atmospheric CO_2 from ice-core data and the deep-sea record of Ontong Java plateau: the Milankovitch chron. *Geologische Rundschau 85*, 466–495.

Berger, W. H., M. K. Yasuda, T. Bickert, and G. Wefer (1995). Brunhes-Matuyama boundary: 790 k.y. date consistent with ODP Leg 130 oxygen isotope record based on fit to Milankovitch template. *Geophysical Research Letters 22* (12), 1525–1528.

Berger, W. H., M. K. Yasuda, T. Bickert, G. Wefer, and T. Takayama (1994). Quaternary time scale for the Ontong Java Plateau: Milankovitch template for Ocean Drilling Program Site 806. *Geology 22*, 463–467.

Bickert, T., W. H. Berger, and G. Wefer (1997). The deep western equatorial pacific in plio-pleistocene times: Results from leg 130 (Ontong Java Plateau). *Paleoceanography*. Submitted.

Birchfield, G. E., J. Weertman, and A. T. Lunde (1981). A paleoclimate model of Northern Hemisphere ice sheets. *Quaternary Research 15*, 126–142.

Deblonde, G. and W. R. Peltier (1991). A one-dimensional model of continental ice volume fluctuations through the Pleistocene: Implications for the origin of the mid-Pleistocene climate transition. *Journal of Climate 4*, 318–344.

Farley, K. A. (1995). Cenozoic variations in the flux of interplanetary dust recorded by ^{3}He in a deep-sea sediment. *Nature 376*, 153–156.

Farley, K. A. and D. B. Patterson (1995). A 100-kyr periodicity in the flux of extraterrestrial ^{3}He to the sea floor. *Nature 378*, 600–602.

Foyle, B. and B. Haurwitz (1966). Noctilucent clouds. *Space Science Reviews 6*, 279–340.

Hays, J. D., J. Imbrie, and N. J. Shackleton (1976). Variations in the Earth's orbit: Pacemaker of the ice ages. *Science 194*, 1121–1132.

Imbrie, J., E. A. Boyle, S. C. Clemens, A. Duffy, W. R. Howard, G. Kukla, J. Kutzbach, D. G. Martinson, A. McIntyre, A. C. Mix, B. Molfino, J. J. Morley, L. C. Peterson, N. G. Pisias, W. L. Prell, M. E. Raymo, N. J. Shackleton, and J. R. Toggweiler (1992). On the structure and origin of major glaciation cycles. 1. Linear responses to Milankovitch forcing. *Paleoceanography 7* (6), 701–738.

Imbrie, J., J. Hays, D. Martinson, A. McIntyre, A. Mix, J. Morley, N. Pisias, W. Prell, and N. Shackleton (1984). The orbital theory of Pleistocene climate: support from a revised chronology of the marine δ^{18}O record. In A. Berger, J. Imbrie, J. Hays, G. Kukla, and B. Saltzman (Eds.), *Milankovitch and Climate. Part 1*, NATO Advanced Science Institutes Series C: Mathematical and Physical Sciences 126, pp. 269–305. Dordrecht, Netherlands: D. Reidel Publishing Company.

Imbrie, J. and J. Z. Imbrie (1980). Modeling the climatic response to orbital variations. *Science 202*, 943–953.

Laskar, J., F. Joutel, and F. Boudin (1993). Orbital, precessional, and insolation quantities for the Earth from −20 Myr to +10 Myr. *Astronomy and Astrophysics 270*, 522–533.

Maasch, K. A. (1988). Statistical detection of the mid-Pleistocene transition. *Climate Dynamics 2*, 133–143.

Marcantonio, F., R. F. Anderson, M. Stute, N. Kumar, P. Schlosser, and A. Mix (1996). Extraterrestrial ^{3}He as a tracer of marine sediment transport and accumulation. *Nature 383*, 705–707.

Milankovitch, V. (1995). *Milutin Milankovič, from his autobiography by his son, Vasko, and a preface by André Berger.* Katlenburg-Lindau: European Geophysical Society.

Mudelsee, M. and M. Schulz (1997). The Mid-Pleistocene climate transition: onset of 100 ka cycle lags ice volume build-up by 280 ka. *Earth and Planetary Science Letters 151*, 117–123.

Mudelsee, M. and K. Stattegger (1997). Exploring the structure of the mid-Pleistocene revolution with advanced methods of time-series analysis. *Geologische Rundschau 86*, 499–511.

Muller, R. A. (1994). Glacial cycles and orbital inclination. Report LBL-35665, Lawrence Berkeley Laboratory, University of California, Berkeley, California 94720, USA.

Muller, R. A. and G. J. MacDonald (1995). Glacial cycles and orbital inclination. *Nature 377*, 107–108.

Muller, R. A. and G. J. MacDonald (1997a). Glacial cycles and astronomical forcing. *Science 277*, 215–218.

Muller, R. A. and G. J. MacDonald (1997b). Simultaneous presence of orbital inclination and eccentricity in proxy climate records from Ocean Drilling Program Site 806. *Geology 25*, 3–6.

Oerlemans, J. (1982). Glacial cycles and ice-sheet modelling. *Climatic Change 4*, 353–374.

Paul, A. (1996). A seasonal energy-balance climate model for coupling to ice-sheet models. *Annals of Glaciology 23*, 174–180.

Paul, A. and W. H. Berger (1997). Modellierung der Eiszeiten: Klimazyklen und Klimaübergänge. *Geowissenschaften 15*, 20–27.

Pollard, D. (1982). A simple ice sheet model yields realistic 100 kyr glacial cycles. *Nature 296*, 334–338.

Pollard, D. (1983). A coupled climate-ice sheet model applied to the Quaternary ice ages. *Journal of Geophysical Research 88* (C12), 7705–7718.

Press, W. H., B. P. Flannery, S. A. Teukolsky, and W. T. Vetterling (1992). *Numerical Recipes in FORTRAN: The Art of Scientific Computing* (second ed.). Cambridge: Cambridge University Press.

Raymo, M. E. (1994). The initiation of northern hemisphere glaciation. *Annual Review of Earth and Planetary Sciences 22*, 353–383.

Raymo, M. E. (1997). The timing of major climate terminations. *Paleoceanography 12*, 577–585.

Ruddiman, W. F., M. E. Raymo, D. G. Martinson, B. M. Clement, and J. Backman (1989). Pleistocene evolution of Northern Hemisphere climate. *Paleoceanography 4*, 353–412.

Saltzman, B. and K. A. Maasch (1991). A first-order global model of late Cenozoic climatic change. *Climate Dynamics 5*, 201–210.

Saltzman, B. and K. A. Maasch, and M. Y. Verbitsky (1993). Possible effects of anthropogenically-increased CO_2 on the dynamics of climate: Implications for ice age cycles. *Geophysical Research Letters 20*, 1051–1054.

Saltzman, B. and M. Y. Verbitsky (1992). Asthenospheric ice-load effects in a global dynamical-system model of the Pleistocene climate. *Climate Dynamics 8*, 1–11.

Saltzman, B. and M. Y. Verbitsky (1993). Multiple instabilities and modes of glacial rhythmicity in the Plio-Pleistocene: a general theory of late Cenozoic climatic change. *Climate Dynamics 9*, 1–15.

Schulz, M. and K. Stattegger (1997). Spectral analysis of unevenly spaced paleoclimatic time series. *Computers and Geosciences*. In press.

Shackleton, N. J., A. Berger, and W. R. Peltier (1990). An alternative astronomical calibration of the lower Pleistocene timescale based on ODP site 677. *Transactions of the Royal Society of Edinburgh. Earth Sciences 81*, 251–261.

Sykes, M. (1988). *Astrophysical Journal L55*, 334.

Tauxe, L., T. Herbert, N. J. Shackleton, and Y. S. Kok (1996). Astronomical calibration of the Matuyama-Brunhes boundary: Consequences for magnetic remanence acquisition in marine carbonates and the Asian loess sequences. *Earth and Planetary Science Letters 140*, 133–146.

Thomas, G. E. (1991). *Reviews of Geophysics 29*, 553–575.

Tiedemann, R., M. Sarnthein, and N. J. Shackleton (1994). Astronomic timescale for the Pliocene Atlantic $\delta^{18}O$ and dust flux records of Ocean Drilling Program site 659. *Paleoceanography 9*, 619–638.

Milankovitch Cycles and Sequences: Two Different Stratigraphic Tools

W. Schwarzacher

1 Introduction

With the introduction of sequence stratigraphy, the word cycle has again become a term which is frequently used. Cycles have been classified in a very arbitrary way into orders of different length and several different orders can occur together in the stratigraphic record.

The study of repetition is important. Sedimentary series always contain a limited number of rock types and therefore random deposition will sooner or later lead to repetition. Repetitions however, can also be the outcome of a more orderly system. If one or more of the lithologies leads with preference to a definite successor, then repetition will occur regularly. Such cycles are generated by an oscillating system. There are therefore, two quite different groups of cycles caused either by oscillating systems or by random impulses. This difference is very important in explaining the origin of sediment sequences. We will discuss two types of stratigraphic cycles: sequences in their original definition as unconformity-bound units and Milankovitch cycles, which are genetically defined as being the result of orbital variations of the earth. Milankovitch cycles can form sequences, but this is not necessarily the case. Sequences, by implication, are formed by variations in the sea level, but we know very little about the causes of such variations. The astronomical mechanisms of insolation changes that cause the Milankovitch cycles, on the other hand, are relatively well known, but much less is known of how they will eventually produce stratigraphic cycles.

The differences between the two types of "cycles" have some very practical implications in stratigraphy. The most important advantage of Milankovitch cycles is that they represent constant or nearly constant time markers. Also, because they are the result of periodic or near-periodic changes in the sun planet system, they must have a worldwide effect, even if they are not recorded everywhere. The major disadvantage of Milankovitch cycles is the difficulty in recognising them. The most important advantage of sequences is their clear definition in stratigraphic terms which allows an accurate identification of sequence boundaries. Also useful is the relatively long time they represent, which makes interbasin correlation possible. With Milankovitch cycles, this has only been successful so far in the Pleistocene. Disadvantages are that we do not know a priori that sequences are synchronous worldwide, and it is also difficult to recognise sequence boundaries when they are removed from the margins of basins.

2
Sequences

Although sequences can be defined without any reference to actual processes, they are nevertheless always connected with relative sea-level fluctuations. The compilation of sequences from all over the world has led to the construction of a eustatic sea-level curve for the Mesozoic and Cenozoic (Haq et al. 1987). The curve provides data on the frequency and, to a lesser degree, also on the magnitude of sea level changes. The latter is estimated from the relative change of coastal onlap. Based on the magnitude of changes, three orders of cycles have been differentiated. The shortest fluctuation, which is called the third-order cycle, is grouped into supercycles or cycles of second order, and these form into even longer megacycles and megacycle sets, which together constitute the first order cycles. Third-order cycles are documented by definite sequence boundaries, but the existence of slower fluctuations relies entirely on the magnitude of sea level changes, which are very difficult to estimate quantitatively. The data in the chart do not show any fixed ratio between 3rd-and 2nd-order cycles and the number of short cycles within a 2nd-order cycle may vary from 2 to 9. The duration of the second-order cycles is also highly variable with a mean of 8 Ma. This high variability is reflected in the classification of cycles according to duration, which is given as 0.5 to 3 Ma for the third-order cycles, 3– 50 Ma for the second order cycles and 50+ Ma for the first order cycles.

Figure 1 shows the frequency distribution of the lengths of 3rd-order cycles from the Jurassic to the present. Considering the paucity of data, the graph could be interpreted as an exponential distribution which one would expect if the 3rd order boundaries were distributed at random. The data, however, are not sufficient to exclude other interpretations, such as highly damped self-oscillating systems which could be excited by random pulses. An attempt to obtain some information about the structure of the time sequence series was made by D. G. Smith (1994), who plotted the lengths of sequence intervals against absolute

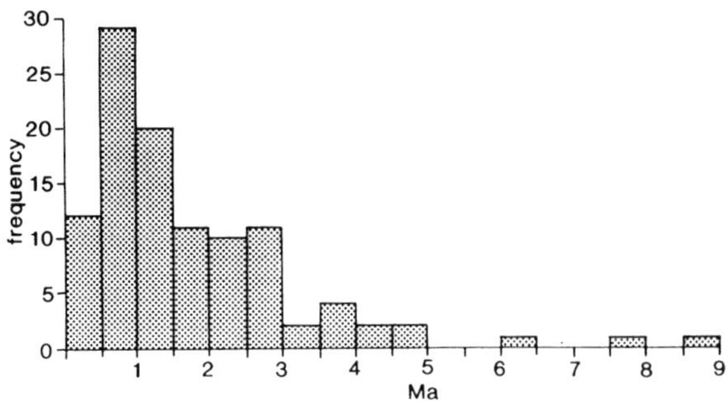

Fig. 1 Length of 3rd-order sequence cycles during the past 210 Ma.(Data from Haq et al. 1987).

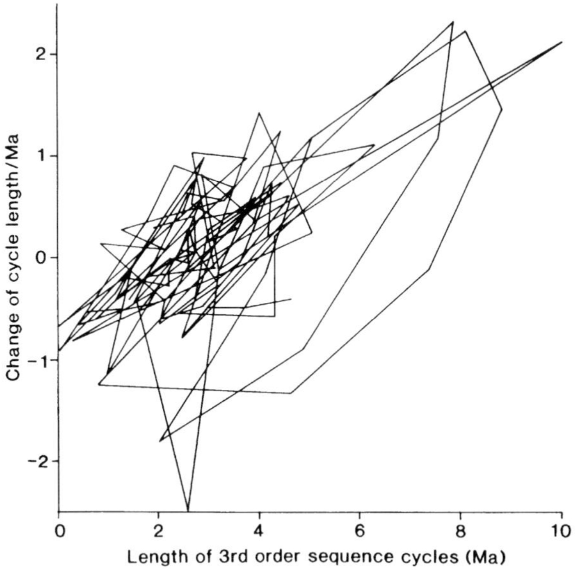

Fig. 2 Phase portrait of 3rd-order sequence cycles. The lengths of the cycles are plotted against the change in cycle length per 1 ma. This phase portrait should be compared with Fig. 3 and 4.

time. A phase diagram was then constructed by plotting each interval against its successor. A similar attempt is shown in Fig. 2., in which the time sequence series was obtained by interpolation at 0.5 Ma points using a cubic spline. The phase portrait in Fig. 2 was obtained by plotting the lengths of cycles against the rate with which the cycle length changed. The resulting phase path is quite irregular and one can hardly recognise any pattern of repetition. Smith found a certain amount of organisation in his phase portrait, but this may well be due to smoothing and possibly to the inclusion of the less well-documented Trias.

Even if it is not possible with the present data to decide if one is dealing with random events or a more organised system, it is interesting to speculate on how the approximate cycle length of 2 Ma could have originated. If the system was oscillating, then the length of this oscillation practically excludes any process which is exclusively linked to changes in the atmosphere or in the hydrosphere, since these change much more rapidly. Processes which provide such long time periods could be provided by the redistribution of materials on the earth's surface which could change the geometry of the ocean basins. Processes related to movements in the crust will be more effective however. It is interesting to speculate about the size of the systems which can generate such intervals. For example, if one assumes a progressive change which moves at 1 cm/year (which is a relatively slow spreading rate) the size of the cycle producing motor would be only 20 km. On the other hand, a larger system such as a 1000 km. wide convection cell, for example, would require much slower changes with rates at only 2 mm/year.

Shorter cycles have been called fourth- to sixth-order cycles or parasequences. Such fluctuations can have complex causes involving climate, tectonism and sedimentation. The interaction between these can form self-oscillating systems which can generate cycles. However, quasi-periodic cycles can only be generated by systems with the stability of the solar system and for this reason Milankovitch cycles are different.

3 Milankovitch Cycles

Milankovitch cycles are, by our definition, the result of changes in the earth's orbit. Such changes affect the amount of solar radiation received by the planet and this in turn, determines the climate to a large extent. The relationship between astronomical variables and climate is by no means simple, and various climatic models have attempted to explore it, but the details are largely unknown. The astronomical variations and the resulting insolation changes can be calculated with reasonable accuracy to at least 10 Ma BP (Berger and Loutre 1991).

The astronomical variables are a mixture of various harmonic functions (for example the different effects of the seven major planets) which result in a quasi-periodic process. Quasi-periodicity implies that the attractor representing the system must be multidimensional. A two dimensional phase portrait (see Fig.3) cannot represent the attractor completely but it clearly indicates the oscillating nature of the process. The periods of repetition in such a system change with time but they are predictable and therefore provide a time scale.

The astronomical signal is modified by the climate, by the complex way in which environments react to the climate and finally by the often very loose con-

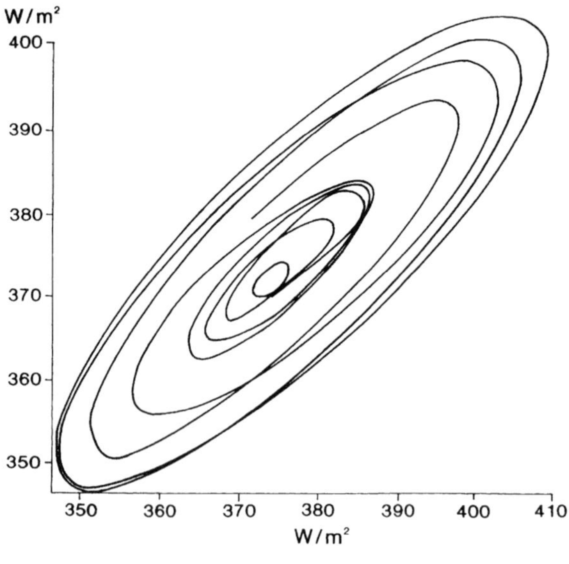

Fig. 3 Two-dimensional phase portrait of the astronomical signal. Mid-March insolation at 30 N is plotted against successive insolation values, which are shifted by 2 ka.

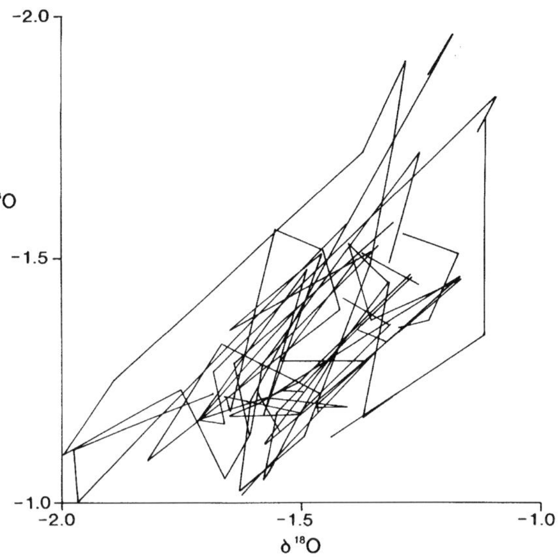

Fig. 4 Two-dimensional phase portrait of oxygen isotope values (core V28–239). The values on the y axis are shifted by 1 ka against the values on the x axis. It is still possible to recognise the elliptical phase path of the astronomical signal (Fig. 3).

nection between environment and sediment. Although climatic changes can control the sea level, Milankovitch cycles (as they are recorded in sediments) do not necessarily relate to sea-level fluctuations. Climate also has direct effects on sediment production and organic activity in a depositional environment.

If sedimentary properties reflect climatic conditions, they are called proxy indicators and a particularly useful indicator of this type is the isotope composition of the sediment. The study of Pleistocene cores has shown that the $\delta^{18}O$ anomalies clearly reflect the cold and warm interludes of this period. It is interesting to compare the phase portrait of such a record (core V 28-239 Shackleton and Opdyke 1976) with the astronomical data in Fig. 4. Although one can see similarities between the two, the $\delta^{18}O$ record contains many irregularities which must be part of the dynamic system which generated the sedimentary record. Nicolis (1987) made an interesting attempt to find the dimensionality of the attractor responsible for $\delta^{18}O$ variations in Pleistocene sediments. This analysis found a dimension of 3.1, which would suggest that the main features of the sediment-producing system can be understood in terms of deterministic mechanics and involving only a few variables. The fractal dimension 3.1 implies that the system could be chaotic. Mudelsee and Stattegger (1994) analysed a similar system and again found deterministic behaviour, based on a limited number of variables. Stochastic sequences which by definition are based on a large number of variables cannot account for the observed phase portraits.

The recognition of Milankovitch cycles must rely on three properties which are characteristic of such repetitions. These are the absolute time intervals of the Milankovitch periods, the frequency spectrum and the near periodic nature of the astronomical variations. Unfortunately, each of these properties will be changed by sedimentation and therefore each is difficult to recognise in the stratigraphic record.

4
The Milankovitch Spectrum

The astronomical system which is responsible for climate and sediment developments contains three main elements, eccentricity, obliquity and precession. Each of these consists of a complex frequency spectrum but as a first approximation, one can assume an average period of 21 ka for the precession, a period of 40 ka for the obliquity and periods of 100, 400 and 2000 ka for the eccentricity. Theoretically, insolation in low latitudes is largely determined by the precession signal, and in higher latitudes, the obliquity signal becomes more important. The eccentricity modulates both signals.

As already stressed, the relationship between theoretical insolation and climate is not linear but both the climatic and the stratigraphic records show similar periodicities to the astronomical record and preserve the frequency ratios of the various periods. In other words, the stratigraphic record, depending on its completeness, should have ratios between observed frequencies similar to those found in the Milankovitch spectrum. The ratios therefore can be used as criteria for recognising Milankovitch cycles.

The most commonly quoted ratio of Milankovitch cycles is the ratio of eccentricity to precession (which is approximately 1:5). This is indeed approximate and will fluctuate. For example, using insolation values for a latitude of 30° N calculated by Berger and Loutre (1991) for the interval of 2 Ma to 6 Ma before present, at 1-ka intervals, one finds 125 maxima. The amplitudes of the maxima again show maxima representing the eccentricity cycle. Figure 5 shows the frequency distribution of the number of shorter cycles between the maxima of the longer cycles. The data set has a clear mode at five short cycles but a large number of the larger eccentricity cycles are separated by only four and sometimes three and six short cycles. The variation is due to the quasi-periodic nature of the record.

Other possible ratios to be considered, are the ratio between the two strongest eccentricity periods (which is 1:4) as well as the ratios of obliquity to eccentricity

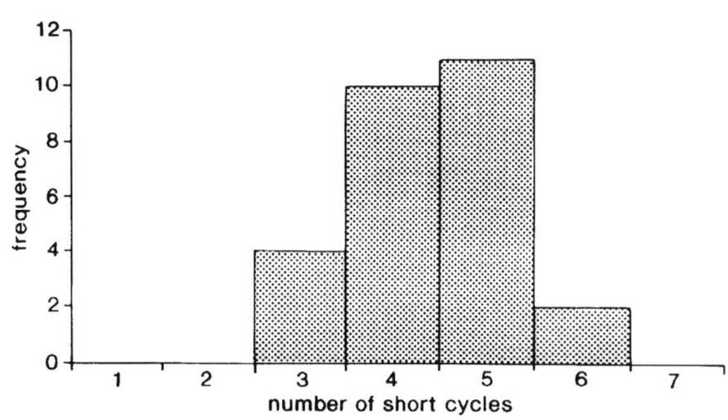

Fig. 5 Frequency distribution of the numbers of short cycles (precession) within longer cycles (eccentricity).

(which can be 1: 2 or 1:8). Since both precession and obliquity have changed during geological time, the ratios have also changed. According to calculations by Berger and Loutre (1994) the ratio of eccentricity to precession was approximately 1:4.8 in Upper Cretaceous time, 1:5.3 in Upper Carboniferous time and 1:5.7 during the Upper Ordovician.

5
Stratification Cycles, Bundles and Stacking Patterns

If detailed stratigraphic data are available, the best method for finding cyclicities and ratios between cycles is power spectral analysis. Very often, however, detailed and continuous quantitative data for such an analysis are not available and, for this reason, a study of the stratification pattern becomes important.

A particularly useful pattern of stratification is the grouping of identical beds into repeated cycles (Schwarzacher 1947). The bedding planes and beds within the groups are identical but the bedding plane which marks the boundary of the group is differently developed, very often as a master bedding plane (Schwarzacher 1958,1975). Geometrically, this pattern can be described by two types of boundary, let us say R and r and by sequences of the type R,r,r,r,R,r,r,r,R,.... The number of beds in each is constant or very nearly constant. This type of bedding is very well described by the term bundling of beds (Schwarzacher 1954) but for a precise description, the term stratification cycle should be used (Schwarzacher 1987). The term stacking pattern, which has been more recently used in the American literature, does not bring out the essential cyclic nature of this sedimentation pattern, but is useful in the description of other short cycles with definite sequences.

The reason why stratification cycles need special attention is that they are very difficult to explain by anything else than the effect of two near-periodic functions. To support this statement, we can use the following argument.

Assume that we have two types of events corresponding to the formation of boundaries R and r. There are four times as many events r as R, which means that the events could be arranged in a regular series: R,r,r,r,r,R,r,r,r,r,R, ... representing bundles of five. The probability (p)of obtaining R = 1/5 and the probability (q) of obtaining r = 4/5. The probability of obtaining purely by chance a single bundle R,r,r,r,r,R is equal to p^2 q . The probability of obtaining several such bundles by chance in succession, is astronomically small. The argument assumes complete statistical independence of every sedimentation process and this is unlikely. A more realistic model can be found by using the transition probabilities p(R,r) which can be written in matrix form:

$$\begin{array}{lllll}
p(R,R) & p(r,R) & .\,. & & . \\
p(r,R) & p(r,r) & p(r,r) & & . \\
p(r,R) & & & & . \\
. & & & & \\
p(r,R) & .\,.\,. & & & p(r,r),
\end{array}$$

and this will lead to a periodic chain, if it takes the form:

$$\begin{array}{llllll} 0 & p1 & 0 & 0\ldots & 0 & \\ 0 & 0 & p2 & 0 & . & . \\ . & & & & . & \\ . & & & & . & \\ 0 & & & & pj, & \end{array}$$

in which p1 to pj are either unity or square matrices with unity row sums (Schwarzacher 1975). In order to generate a period of j, the matrix must be at least of the dimension j * j. This also applies if the series is not strictly periodic but has a recurrence probability maximum of length j.

The probability matrix provides a descriptive model but it does not explain the mechanism which can produce this situation, in which identical states have different transition probabilities which are determined by their position in the chain. In this respect, stratification cycles are quite different from stacking patterns such as for example, the classical coal-measure cycle or the Bouma sequence. In any of these latter developments, the sediment sequences can be explained as being the result of a definite triggering event, such as a sea level change or a turbidity current. The triggering event can be either random or periodic, but unless a realistic mechanism is found to explain the constantly repeated triggering of bed formation at regular intervals (without involving periodic processes) stratification cycles are one of the best methods for recognising periodicity, and therefore for recognizing Milankovitch cycles. An obvious advantage of stratification cycles in demonstrating periodicity is that only the counting of beds between master bedding planes is involved and variations in bed thicknesses are of secondary importance

In a study specifically dealing with carbonate platforms, Drummond and Wilkinson (1993) suggest that stratification cycles can develop by the interaction of two periodic functions, one of which is a eustatic sea level change with constant frequency. The second cycle is generated by introducing a constant threshold which switches sedimentation on and off at intervals controlled by the available accommodation space, which in this unlikely model can never exceed a critical value (one metre) for any length of time. This model is an example of a mechanism which multiplies the frequency of the signal in the recorded sediment (see Sect. 6). Inasmuch as this mechanism leads to stratification cycles only if both the sea level change and the consequent interruptions of sedimentation are periodic, Milankovitch cyclicity is again the most likely explanation for this model.

A simple explanation for the formation of stratification cycles is shown diagrammatically in Fig. 6. The environmental variable is assumed to be represented by two (or more) periodic functions with different wavelengths and amplitudes but with a stable phase relationship. A bedding plane is formed whenever the signal reaches a critical value, whereby different values of the response thresholds can lead to the development of different stratification patterns. Obvi-

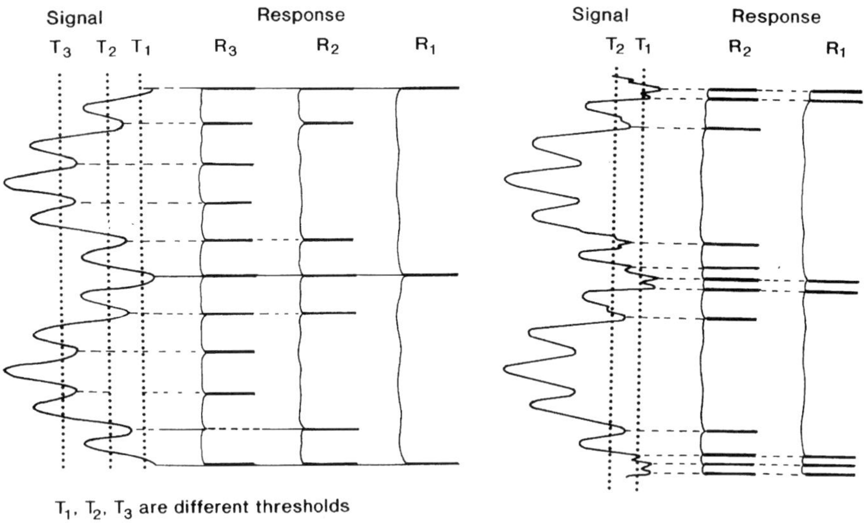

Fig. 6 A possible origin of stratification cycles. It is assumed that a signal must reach a certain threshold before a bedding plane is formed. On *the left hand side*, a modulated signal with three different threshold levels is shown. On *the right hand side*, a similar signal with noise is illustrated.

ously, either noisy signals or inaccurate recording (fluctuating response thresholds) can lead to stratification cycles in which the number of subdivisions fluctuates. Irregularities can lead to missing beds (missed beats) or they may insert extra beds. Some of this variation can be studied by tracing the same stratification cycle laterally. For example, in alpine Triassic limestones (Schwarzacher 1954), it was found that in north-facing exposures weathering was more severe and consistently more beds per bundle were recognised than in south-facing exposures. In a similar way, there were lateral variations in bed numbers in the Lower Carboniferous of Ireland (Schwarzacher 1975). Shaly developed facies always contained more beds per bundle than the less shaly limestone facies. Such observations clearly show that the number of beds in stratification cycles should be used with care when trying to establish the order of Milankovitch cycles and not every 100-ka eccentricity cycle will produce a bundle of five.

Lateral and vertical variations of cycle development are important indicators of environmental changes and are therefore of particular interest in the study of long-term sea level changes.

6 Cycles in the stratigraphic record

To obtain any clear understanding of the physical processes producing Milankovitch cycles, sequences, or any other type of cycle, it is necessary to have a time scale. In stratigraphy, this scale is largely provided by the thickness of the accumulated sediment. We know that over long periods of time and for particular environments, sedimentation can be regarded as being constant and thickness therefore is proportional to stratigraphic time. However, this relationship invariably breaks down and on a microscale, sedimentation is always discontinuous, and time (which is derived from thickness measurements) is unpredictable over short intervals. To establish the absolute duration of a cycle is therefore only possible when very good time control is available in the form of good radiometric data combined with good evidence of steady and continuous sedimentation.

It is well known that sedimentation rates are at least as sensitive to the environment as most other properties of the sediment. This means that any variation of the stratigraphic record (lithology, or biocontent) will also induce a variation in sedimentation rates and therefore distortions of the time scale. The relationship between environment, sediment material and sedimentation rates cannot be expected to be linear, and indeed it seems likely that the system contains stable, bistable and unstable elements which are subjected to random noise.

When the time development of cycles is studied, particular attention has to be paid to sedimentation gaps which may be due to sedimentation still stand or more commonly, to erosion. Figure 7 shows a simple example of a sinusoid signal and its stratigraphic record, which was formed under the assumption that the rate of sedimentation was directly proportional to the signal. When this signal is plotted against the cumulative sediment thickness, it forms a cycloid and if the signal exceeds a critical maximum sedimentation, then part of the record is missing. This is indicated by the loops in curve A of Fig. 7. The second example (curve B) assumes that only increasing rates of sedimentation resulted in sediment accumulation. This hypothetical case illustrates that a symmetrical signal can easily lead to an asymmetrical record. The relationship between environmental conditions and sedimentation, can not only change the shape of the signal, but it can also double or even multiply the frequency of the signal. A possible scenario for such a case is a biotope which is sensitive to extrema and which therefore could produce a double peak for each cycle of a sine wave. A reduction of frequency has already been illustrated by the example of the incomplete recording of stratification cycles.

As already mentioned, stochastic processes are important in producing stratigraphic sections. Random fluctuations are not only part of the deposition process but also of diagenesis and postdiagenetic preservation. They may be further increased by errors of observation. The magnitude of such disturbances can be estimated, to some extent, if data are available which represent identical time-spans in different localities (Schwarzacher 1975). The actual causes of random variations are usually too complex to be identified in detail. For the study of cycles, a knowledge of the random element is important because as before, it not only affects the recorded signal but also the thickness-related time scale. To use the terminology of radio communication, the stratigraphic signals are not only amplitude modulated by noise but also phase or frequency modulated.

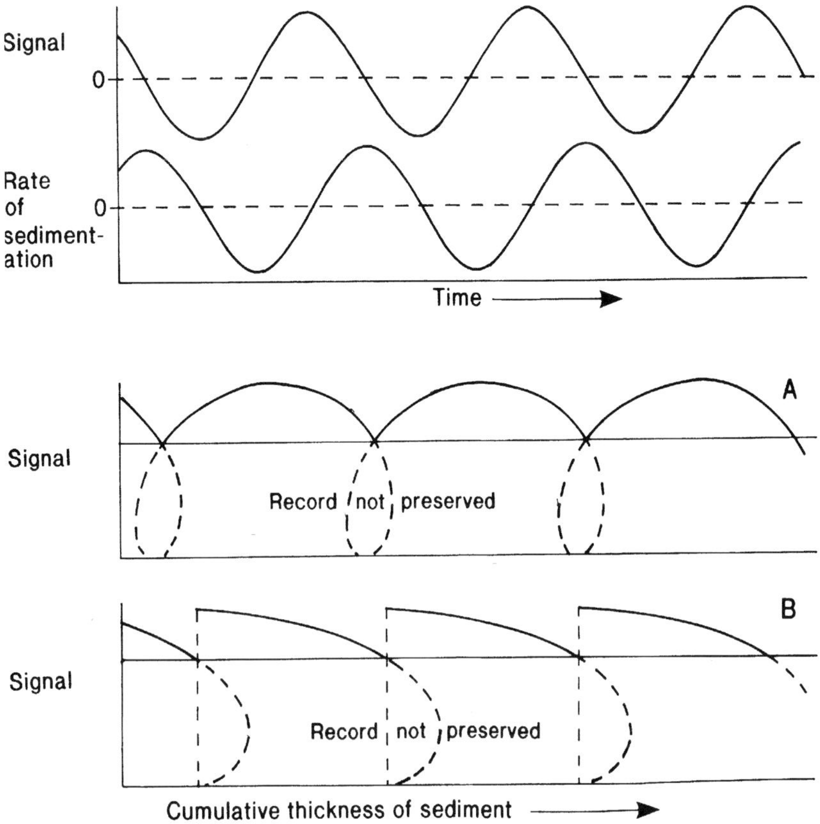

Fig. 7 The relationship between signal, sedimentation rate and accumulated sediment. Examples *A* and *B* illustrate how sedimentation gaps can develop.

Random fluctuations together with systematic changes of sedimentation rate and sedimentation gaps, can change the completeness of the stratigraphic section. By combining sequence stratigraphy with the analysis of Milankovitch cycles, it may be possible to estimate the completeness in quantitative terms.

7
Sequences and Milankovitch cycles

The magnitude of sea-level changes can be estimated from the stratigraphic profiles by determining coastal on lap and aggradation (Vail et al 1991). The sea-level changes are quite frequently several tens of metres. Glacio-eustatic sea-level changes are by comparison much smaller, being around 100 m during times of major glaciations and perhaps 0 to 20 m during normal times. The sea-level fluctuations on Triassic carbonate platforms have been estimated to be 10 m or less.

If a sedimentary sequence is predominantly determined by water depth, then the large fluctuations in sea level will clearly produce much better defined sedimentary cycles than the small fluctuations associated with Milankovitch cycles. One can argue, in a similar way, that any change of paleodepth will be less noticeable in deep water than in shallow water, although this is only partly true. Processes like sediment shedding, control of nutrients or transparency may originate in coastal areas and can have a considerable influence on deep water sedimentation. Nevertheless, the best environments for studying the interaction between sequences and Milankovitch cycles would appear to be shallow areas in which sequence boundaries are still developed but where sedimentation is continuous enough to record the Milankovitch cycles.

Of particular interest are studies of peritidal carbonate shelves and platforms. In these environments, sedimentation is almost exclusively determined by the available accommodation space, which is determined by the paleo water depth. If subsidence is constant, paleo water depth can be estimated by subtracting a constant trend from the cumulative sediment thickness curve (Schwarzacher 1975). Such Fischer plots have been used to reconstruct 3rd-order sea-level changes in cyclic Ordovician carbonates (Read and Goldhammer 1988) and similarly in Devonian carbonates (Elrick 1995). The time represented by the individual cycles in both studies is not definitely known, but they are believed to be of Milankovitch origin and therefore can be used for relative timing. The 3rd-order sea-level changes are found to be of the order of 20 m in amplitude but they are quite irregular and the intervals between successive high stands vary between 17 and 40 cycles. Errors can be introduced in this timing by incomplete sedimentation (missed beats) but the evidence strongly suggests that the 3rd-order fluctuations are nowhere near time-periodic.

A definite attempt to time sequences with Milankovitch cycles has been made by Strasser (1991). Cycles in a lagoonal to supratidal carbonate sequence of late Tithonian to early Berriasian age in the French Jura Mountains have been tentatively identified as 20-ka, 100-ka and 400-ka Milankovitch cycles. Furthermore, two sequence boundaries from the Haq et al (1987) chart could be recognised. The interval which, according to the global chart is 2.8 Ma, works out to be 3.6 Ma, according to the cyclostratigraphic timing. However, Strasser stresses the uncertainty not only in identifying the sequence boundaries, but also the uncertainty of having identified the Milankovitch cycles correctly. Additional sequence boundaries are suggested by Strasser and these are placed at intervals of four or at multiples of four 100-ka cycles. The status of these subsidiary sequences is not clear from his paper, as the 100-ka cycles are also referred to as sequences. Similar effects of sea-level fluctuations on the development of Milankovitch cycles have also been observed in the Swiss Jura Mountains (Pittet 1994).

An important new method of analysing stratigraphic sections is based on wavelet transforms which can provide frequency spectra of very short time intervals (Bolton et al. 1995; Prokoph and Barthelmes 1996).The method is ideally suited to locating changes in cyclicity which often coincide with sequence boundaries (B.Niebuhr 1996, pers.com).

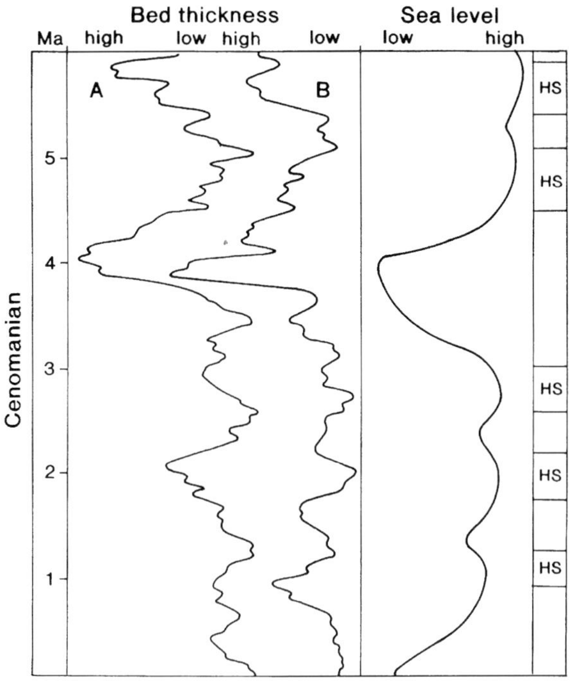

Fig. 8A, B Comparison of bed thickness measurements with a theoretical sea level curve. Bed thicknesses from the Cenomanian in the Gubbio district. **A** Contessa highway section **B** Mt. Petrano section. The time scale *on the left* is based on cycles which are believed to represent 100-ka eccentricity cycles. The positions of the sequence high stands *(HS)* are from Robaszynski et al. (1993).

Very little work has been done on comparing Milankovitch cycles with one- to third-order sequences in relatively deep water. An attempt to match sedimentation changes to an established sequence stratigraphy of the Cenomanian, is shown in Fig. 8 . The Cenomanian cyclostratigraphy of the Gubbio district in Italy has been reasonably well established (Schwarzacher and Fischer 1982, deBoer 1983, Schwarzacher 1994). Using the metre scale cycles (which are believed to represent 100 ka eccentricity cycles) as a scale, gives the length of time from the base of the Cenomanian to the top (Bonarelli anoxic event) as 5.8 to 6.2 Ma. In this environment, the sedimentation was not influenced by limited accommodation space and the nearly constant thicknesses of the metre cycles indicate constant sedimentation rates on this scale. Bedding of limestone shale couplets on a smaller scale of 10 to 20 cm is much more variable and this seems to indicate variable sedimentation conditions (Melnyk and Smith 1989). Although plots of bed thicknesses cannot yet be interpreted in terms of specific sedimentation conditions, the similarity between plots from different localities shows that they have real significance in interpreting sections. Plots from the Gubbio sections are compared with the sequence stratigraphy of the Cenomanian in central Tunisia (Robaszynski et al. 1993). Figure 8 shows the reconstructed sea level curve and the positions of the various high stands. The comparison with the thickness values makes it very likely that the strong regression in the upper half of the Cenomanian coincided with the development of more massive bedding in the Gubbio localities. Whether the three maxima which are particularly well developed in Fig. 8B, can be correlated with three high stands is doubtful. Detailed stratigraphic work and lateral tracing of cycles (Beaudoin 1992) is necessary before such comparisons can be conclusive.

8 Conclusions

The different origins of sequence and Milankovitch cycles strongly suggest that they should be treated as different stratigraphic units. Sequences representing first to third order cycles are believed to be due to tectonic processes which are most likely to involve the transport of mantle material. Dynamic systems of this type must be highly damped and therefore cannot sustain oscillations. The triggering of such sequences may well have been a random event. However, the record of sea level changes in the past is still very incomplete. This is largely due to the difficulty in estimating the magnitude of the eustatic component but also because accurate timing in shallow basins with multiple sedimentation gaps is difficult. Tracing sequence boundaries towards the centre of basins, together with the recording of Milankovitch cycles, can at least provide a more accurate time scale.

Shorter cycles may have had a variety of origins but near-periodic oscillations with periods of 2000 to 10 ka are almost certainly related to Milankovitch cycles. Although the astronomical origin of short cycles may be difficult to prove, this should always be attempted. Any cycle generated by an oscillating system is a potential time scale and is therefore of importance in the study of sequences. The reference to orders of cycles without giving evidence of how its alleged duration was determined and without examining the regularity of such "cycles", is not productive in quantitative stratigraphic studies. It may not yet be possible to prove the absolute synchroneity of global tectonic cycles, but if they are confused with Milankovitch cycles this will be even more difficult.

References

Beaudoin, B., 1992, The black cherts in the Scaglia bianca. Poster presentation, 29th. Int. Geol. Congress (Kyoto).

Berger, A. L., and Loutre, M. F., 1991, Insolation values for the climate of the last 10 million years. Quaternary Science Reviews v.10, p .297–317.

Berger, A. L., and Loutre, M. F., 1994, Astronomical forcing through geological time. In: de Boer, P. L., Smith, D. G., eds., Spec. Publ. Ass. Sediment. v.19, p. 15–24.

Bolton, E. W., Maasch, K. A., Lilly, J. M., 1995. A wavelet analysis of Plio-Pleistocene climate indicators: A new view of periodicity evolution. Geophys. Res. Lett., v.20, No.30, p. 2753–2756.

De Boer. P. L., 1983, Aspects of Middle Cretaceous pelagic sedimentation in S.Europe. Geol. Utraiectina 30.

Drummond, C. N., Wilkinson, B. H., 1993, Carbonate cycle stacking patterns and hierarchies of orbitally forced eustatic sealevel change. Jour. Sed. Pet., v. 63, p. 369–377.

Elrick, M., 1995, Cyclostratigraphy of Middle Devonian carbonates of the Eastern Great Basin. Jour. Sediment. Res., v. B65, p. 61–79.

Haq, B. U., Hardenbol, J., and Vail, P. R., 1987, Chronology of fluctuating sea levels since the Triassic. Science v. 235, p. 1156–1167.

Melnyk, D. H., and Smith, D. G., 1989, Outcrop to subsurface cycle correlation in the Milankovitch frequency band: Middle Cretaceous, central Italy. Terra Nova v.1, p. 432–436.

Mudelsee, M., and Stattegger, K., 1994, Plio/Pleistocene climate modeling based on oxygen isotope time series from deep-sea sediment cores: the Grassberger-Proccacia algorithm and chaotic climate systems. Math. Geol., v. 26, p. 799–815.

Nicolis, C., 1987, Climatic predictability and dynamical systems. In: Nicolis, C. and Nicolis, G., eds., Irreversible phenomena and dynamical systems analysis in geoscience. D. Reidel , Dordrecht, p. 321–345.

Pittet, B., 1994, Modele d'estimation de la susidence et des variations du niveau marin: Un exemple de l'Oxfordien du Jura suisse. Eclogae geol. Helv. v. 87/2, p. 513–543.

Prokoph, A., Barthelmes, F., 1996, Detection of non-stationarities in geological time series: wavelet transform of chaotic and cyclic sequences. Computers &Geosciences v.22, No 9

Read, J. F., Goldhammer, R. K., 1988, Use of Fischer plot to define third-order sea-level curves in Ordovician peritidal carbonates, Appalachians. Geology. v. 16 p. 895–899.

Robaszynski, F., Hardenbol, J., Caron, M., Amedro, F., Dupuis, C. , Gonzales Donoso, J. M., Linares, D., and Gartner, S., 1993, Sequence stratigraphy in a distal environment: The Cenomanian of the Kalaat Senan Region (central Tunisia). Bull. Centre Rech. Explor.-Prod. Elf - Aquitaine, v. 17, p. 395–493.

Schwarzacher, W., 1947. Uber die sedimentare Rhytmik des Dachsteinkalkes von Lofer. Verh. Geol. Budesanstalt (Austria) 1947, p 175–188.

Schwarzacher, W., 1954, Die Grossrhytmik des Dachsteinkalkes von Lofer. Tschermaks mineral. petrogr. Mitt., v.4, p,44–54.

Schwarzacher, W., 1958, The stratification of the Great Scar Limestone in the Settle district of Yorkshire. Liverpool and Manchester Geol. Jour., v. 72, p. 124–142.

Schwarzacher, W., 1975, Sedimentation Models and Quantitative Stratigraphy. Elsevier, Amsterdam, 382 pp.

Schwarzacher, W., 1987, The analysis and interpretation of stratification cycles. Paleoceanography v.2, p. 79–94.

Schwarzacher, W., 1994, Cyclostratigraphy of the Cenomanian in the Gubbio district, Italy : a field study. In : de Boer, P. L., Smith, D. G., eds., Spec. Publ. Ass. Sediment. v.19, p. 87-97.

Schwarzacher, W., and Fischer, A. G., 1982, Limestone-shale bedding and perturbations in the earth's orbit. In: Einsele,G. Seilacher, A., eds, Cyclic and Event Stratification. Springer, Berlin, p. 72–95.

Shackleton, N. J., and Opdyke, N. D., 1976, Oxygen isotope and magnetic stratigraphy of Pacific core V28-239 late Pliocene to latest Pleistocene. Geol. Soc. America, Mem. 145, p. 449–464.

Smith, D. G., 1994, Cyclicity or Chaos? Orbital forcing versus non-linear dynamics. In: de Boer, P. L., Smith, D. G., eds., Spec. Publ. Ass. Sediment. v.19, p. 531–544.

Strasser, A., 1991, Lagoonal-peritidal sequences in carbonate environments: autocyclic and allocyclic processes. In: Einsele, G., Ricken, G., Seilacher, A., eds. Cycles and events in Stratigraphy. Springer , Berlin Heidelberg, p. 709–721.

Vail, P. R., Audemard, F., Bowman, S.. M., Eisner, P, N., and Perez-Cruz, C. 1991, The stratigraphic signatures of tectonics, eustasy and sedimentology - an overview. In: Einsele, G., Ricken, G., Seilacher, A., eds. Cycles and events in Stratigraphy. Springer, Berlin Heidelberg, p. 617–659.

Scaling and the Palaeogeographical Distribution of Stratigraphic Events

F.P. Agterberg and F.M. Gradstein

1 Introduction

The RASC computer program for ranking and scaling of biostratigraphic events (Agterberg 1990; Gradstein et al. 1985) can be used to construct a composite zonation (standard) for the sedimentary record of a basin. This zonation of nonrecurrent stratigraphic events is primarily based on averaged first and last occurrences of fossil taxa in the stratigraphic record but other types of events (log-markers, seismic events, sequence boundaries) can be incorporated. The basic requirement for an event to be included in the final zonation is that it can be uniquely identified in at least one stratigraphic section or well. The zonation is initially constructed using the observed superpositional relationships for coexisting pairs of taxa that are relatively abundant. Events that can be determined with certainty in the sections are weighted more strongly. Selected rare events that occur in one or a few sections only within the basin but are of global significance, can be inserted into the final solution by invoking the RASC unique (rare) events option.

Recent examples of application of the RASC method include Gradstein and Bäckström's (1996) study on Cainozoic biostratigraphy and palaeobathymetry of the northern North Sea and Haltenbanken using microfossils in 26 wells, and Gradstein and Agterberg's (in press) paper dealing with the underlying Cretaceous in the same basin. The second paper introduces event variance analysis which is based on deviations between observed and predicted occurrences. The purpose of the present paper is to discuss new methods of scaling and correlation developed after 1994, and to show how logistic regression can be used to construct palaeogeographic patterns of events.

Intervals between successive events in the standard can be graphed in a dendrogram. For example, Fig. 1 shows the Albian-Cenomanian scaled optimum sequence, offshore mid-Norway, as derived from 87 dinoflagellate, foraminiferal and a log event in 29 wells (cf. Gradstein and Agterberg, in press). Clusters in the dendrogram represent stratigraphic events that are approximately coeval on the average. Gaps between clusters can indicate unconformities accompanied by stratigraphic hiatuses. The biggest breaks in Fig. 1, occurring at events 92 and 52, coincide with latest Albian and mid Cenomanian highstands, respectively, well-known from NW European sequence stratigraphy.

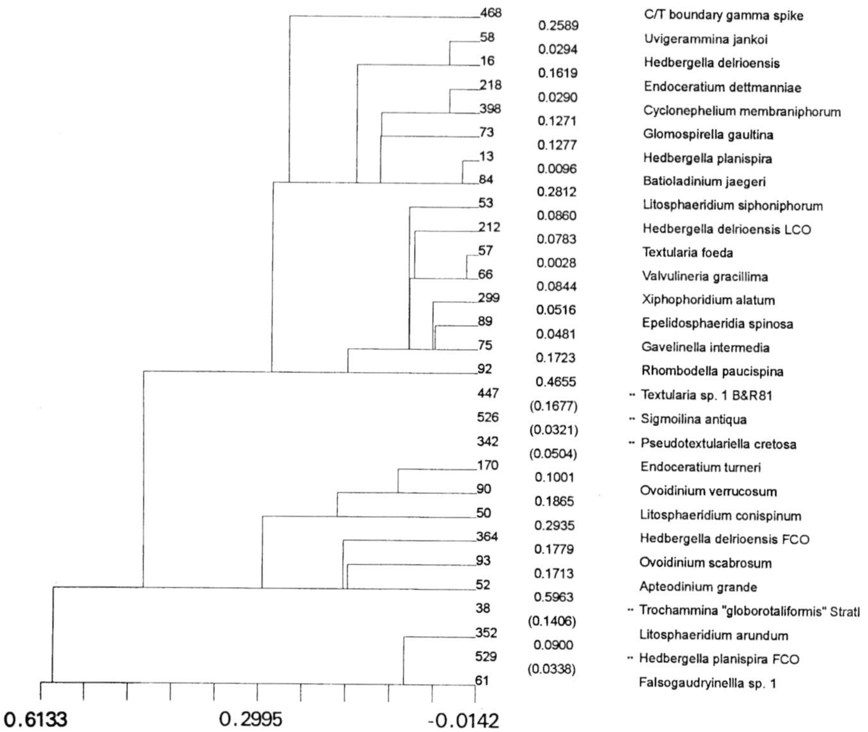

Fig. 1 Albian-Cenomanian Scaled Optimum Sequence, offshore mid-Norway. In this dendrogram, the interevent distances are plotted on the scale *to the left*. Unique (rare) events are preceded by *double asterisks*. Part of RASC solution with k_c=6 and m_c=3. (cf. Gradstein and Agterberg, in press)

Interevent distances are computed from the relative frequencies of locally observed superpositional relations (event above/below another event, or pair of coeval events) for the relatively abundant taxa. In Fig.1 these intervals are subdivided into subintervals if they contain unique (rare) events. The resulting scaled optimum sequence is a composite standard which can be used for zonation and correlation between sections.

RASC has many points of similarity with graphic correlation (Mann and Lane, eds. 1995), a method for constructing a composite standard and stratigraphic correlation of wells introduced by Shaw (1964). The main difference in approach is that RASC is based on averaging fossil events from many sections, a procedure initiated by Hay (1972), whereas graphic correlation works toward the composite standard by starting with two sections and adding other sections successively. Also, Shaw's method of graphic correlation uses estimates of the true first and last occurrences of fossil taxa, instead of average event positions.

2 Correlation and Standard-Error Calculation (CASC)

In CASC, which is a companion program of RASC, all individual sections are compared to the standard. This is accomplished by fitting a curve (line of correlation) in an XY plot where X represents the composite standard value (sum of interevent distances measured from origin set at first event) and Y the relative stratigraphic order value of an event in the section. Deviations between the observed position of an event and its average position in the standard (its probable location on the line of correlation) can be random or indicative of palaeoecological trends. Events with relatively small standard deviations and widespread geographic distribution are useful as stratigraphic time markers. Events with large standard deviations are either poor markers because of significant biostratigraphic uncertainty, or indicate time-transgressive behavior in which the event becomes systematically younger or older according to a geographic trend.

The probable locations of events can be used to correlate sections with one another. Good markers yield good correlations with few if any cross-overs of lines connecting either their observed or their estimated positions in the wells. This type of uncertainty can be expressed by means of error bars. The standard deviation of the probable value of an event on the line of correlation can be estimated and projected along the depth scale (in meters). Probable values and observed values have different standard deviations. Both can be multiplied by 2.0 to obtain approximate 95% confidence intervals.

For each section an YZ plot is obtained where Y, as before, represents relative stratigraphic order of an event and Z its depth (in meters). In general, there is relatively little scatter in this plot and an interpolation spline curve (line of observation) can be used for the relationship between Y and Z. Contrary to the line of correlation which normally is fitted to scattered points falling within a relatively broad zone, and is often approximately straight or only moderately curved, the line of observation can be constructed through all observed points. It tends to be wiggly and reflects local facies and sedimentation rates. An error bar initially estimated along the Y-axis can be transformed into a generally asymmetrical error bar along the depth axis of a section. The resulting error bars have different widths, not only for different events in the same section, but also for the same event in different sections. They provide 95% confidence intervals for the observed and the estimated average locations of the events in the sections.

Figure 2 provides an example of CASC using three events in four wells from the data set used to obtain Fig. 1. In the top part, the observed positions of two events (205: last occurrence of *Heterosphaeridium difficile*, and 212: last common occurrence of *Hedbergella delrioensis*) are connected by lines, and accompanied by their estimated 95% confidence intervals which have been constructed around the probable positions for these events. In Well No. 3, the observed position of event 205 is slightly above the upper limit of its 95% confidence interval. This does not necessarily mean that event 205 is anomalous in this well,

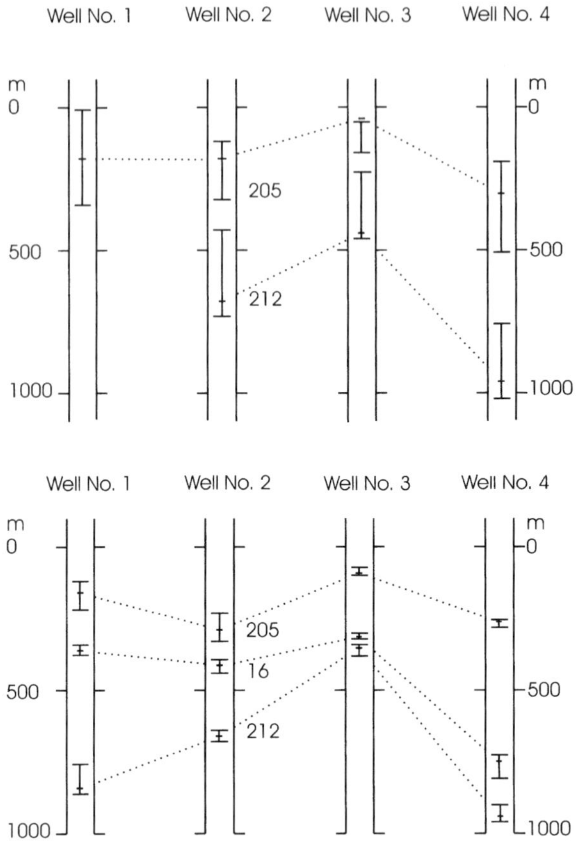

Fig. 2 Biostratigraphic correlation with error bars (95% confidence limits) obtained by CASC for three events in four wells (based on Albian-Cenomanian Scaled Optimum Sequence, Fig. 1). *Top diagram* Observed data: last occurrence of *Heterosphaeridium difficile* (*205*) and last common occurrence of *Hedbergella delrioensis* (*212*) with *error bars* around probable positions. *Bottom diagram* Probable positions for same two events and last occurrence of *Hedbergella delrioensis* (*16*) with error bars. The *error bars* for probable positions are shorter

because, under the assumption of normal frequency distributions for all events, approximately 5% of plotted locations would fall outside their 95% confidence limits. However, if a point is relatively far away from its error bar (e.g., 4 or more standard deviations from its probable position), an anomalous situation (e.g., reworking) would be indicated.

Error bars (95% confidence limits) for the probable locations of three events are shown in the bottom part of Fig. 2 (event 16: last occurrence of *Hedbergella delrioensis* was added). These error bars tend to be narrower than those for the observed positions.

3 The Frequency Distribution of Events

RASC and CASC operate in the direction of the axis of time. All observed superpositional relations between fossil events are projected onto a single axis for relative time. The events are scaled along this time axis by performing calculations on the relative frequencies of the superpositional relations. (An event occurring above another event in a section is scored as 1; below as 0; and coeval as 0.5.) Suppose that two events co-occur in r sections, and that one of the two events has total score of s with respect to the other one, then the relative frequency of this superpositional relation is $p = s/r$. In scaling each p-value is converted into a z-value using the standard normal distribution (probit transformation).

In practice, the distribution of fossil taxa across a region can be highly irregular resulting in relatively small frequencies of individual events observed in sections or wells, and even smaller frequencies for pairs of events co-occurring in the same section. In general, the frequency distribution of microfossil taxa observed in sections is positively skewed: most taxa occur in one or a few sections only, and relatively few taxa occur in many sections. In RASC, two threshold parameters are set for scaling: (1) k_c for minimum number of sections in which an event should occur, and (2) $m_c<k_c$ for minimum number of sections in which a pair of events should occur before it is included for statistical analysis. This is because z-values based on p-values for small samples may not be useful because their standard deviations are too large.

On the other hand, because of the very small frequencies of co-occurrences of events, scaling may not be possible for a given data set unless m_c is set equal to a small value. For this reason, it is desirable to use a scaling method which is as robust as possible, even when m_c is as small as 1 or 2. In RASC this is accomplished by weighting the frequencies according to sample size and by using further threshold parameters automatically set in the computer program.

For n events, the maximum number of superpositional relations that can be used to estimate the distance between two successive events (say i and j) is $n^* = n-1$. Each usable pair of events consisting of i or j paired with a third event (say k) results in a value $x_{ij} = z_{ik}-z_{jk}$ which is one of the maximally n^* x_{ij}-values (indirect distance estimates) averaged to estimate the interevent distance between i and j along the RASC scale. Weights assigned to the x_{ij}-values during the averaging depend on the frequencies of the superpositional relations between all sets of three events (i, j and k) used. An x_{ij}-value is only estimated when the frequency of at least one of the pairs of events (i and k, or j and k) is not less than m_c. The case $i=j-1$ with $k=0$ is special and results in a direct estimate of the x_{ij}-value when i and j co-occur at least m_c times. If $n^*<10/m_c$, the interevent distance is set equal to zero.

In practice, n^* is much smaller than its maximum value ($=n-1$) for the following two reasons: (1) the total number of pairs of events involving i or j is reduced by one for each value x_{ij} that cannot be computed, because one or both superpositional relations with k are missing; (2) if i and j occur stratigraphically above (or below) k in all sections where the two events co-occur with k, the resulting two frequencies for superpositional relations are both equal to 1 (or 0), and again x_{ij} cannot be computed.

4 Palaeogeographic Distribution Patterns of Taxa

Even when k_c is set relatively high, the probability that two taxa co-occur may remain very small. One possible explanation of this is that many taxa existed within restricted geographic domains occupying parts of the entire study area. If the two domains of two taxa had little overlap, very few sections would contain both taxa. Because of the small frequencies of taxa and pairs of taxa in wells (and smaller frequencies of superpositional relations between events), it is, at present, not feasible in most practical applications to establish reliable two-dimensional patterns that could be incorporated in RASC for further analysis. However, it is useful to study such patterns in their own right, because this provides guidelines for selecting wells for analysis and the setting of threshold parameters.

Geographic distribution patterns of fossil events can be studied separately to establish their traceability across the study region. As indicated by the positive skewness of frequency distribution for number of different taxa per section, the main difficulty to be resolved in establishing traceability is to eliminate the effect of relatively low probability of observing an event in a well for most taxa. Not detecting a taxon in the samples for a given well does not mean that the taxon did not exist at the location where the well was drilled. The probability to be estimated can be regarded as the product of the probability of being within the domain where the taxon existed and the probability of detecting it. The logistic model can be used for estimating probability of occurrence of a fossil event at any sampling point including the locations of wells within the study area.

Bonham-Carter et al. (1986) applied correspondence analysis to foraminiferal data from 36 offshore wells on the Labrador Shelf, Grand Banks, and Scotian Shelf for biostratigraphic correlation and for investigating systematic trends in distribution related to palaeogeography. Best results were obtained by applying this technique to assemblage zones for separate time-slices defined on clusters in a RASC zonation. Geographic trends in faunal distribution, differing according to latitude, were shown to exist. Their reordered data matrices (taxa versus scores along first axis of correspondence analysis) clearly indicate that there are three types of taxa: those present in wells randomly distributed across the entire study area, and those definitely missing in wells either in the southern or the northern part of the study area (Bonham-Carter et al. 1986).

Suppose that the geographic locations of the wells are coded as x (for easting or longitude) and y (northing or latitude), and that the presence/absence data for a taxon are coded as 1 (present) or 0 (absent). Then the following logistic equation can be fitted for each taxon: $Pr(\text{taxon is present}) = \exp\{f(x,y)\}/[1+\exp\{f(x,y)\}]$ where Pr denotes probability and $f(x,y)$ represents a function of the geographical co-ordinates. The two functions applied in this paper are (I) $f(x,y)=a+bx+cy$, and (II) $f(x,y)=a+by$. In these expressions a, b and c are the coefficients of the logistic model which can be estimated by the maximum likelihood method. According to Model II, the probability changes as a function of latitude only. Model I allows for change in the east-west direction as well. Use of the logistic model for this type of trend surface analysis (using Model I or II) has two advantages: (1) the probability being estimated cannot exceed 1, nor can it be negative; and (2) the trend surface (two-dimensional pattern of the probability of occurrence) either consists of a single plateau with approximately constant probability of occurrence of a taxon in the entire study area, or there are two plateaus separated by a relatively steeply dipping transition zone. In the second case, one of the plateaus has zero probabilities for non-occurrence of the taxon, whereas the second one has probabilities that increase in the direction perpendicular to the transition zone.

Figures 3 and 4 show results obtained by applying the preceding logistic model to several taxa from the Middle-Late Eocene time-slice of Bonham-Carter et al. (1986, see their Figs. 1 and 9 for names of wells and other additional information). In order to facilitate comparison between presences and absences for different taxa, all probabilities have been classified according to the same scheme (three classes: 0.00–0.05, 0.05–0.20, and 0.20–1.00), and are shown at the well sites only. The first taxon (148: *Globigerinatheka index*) is geographically restricted to the southernmost part of the study region. It is a southerly planktonic species influenced by a proto-Gulf Stream. Within its area of occurrence it was relatively abundant as it was observed in 5 of the 9 wells on the Scotian Shelf and southern Grand Banks. According to the logistic model, its transition zone lies just south of the large cluster of wells on the Grand Banks. The second example is also for a planktonic species (79: *Globigerina tripartita*) with transition zone north of the Grand Banks. Its probability of detection on the Grand Banks and Scotian Shelf is relatively small as it was observed in 5 of 22 wells only.

Also in Fig. 3, *Turborotalia pomeroli* (no. 33) is an example of a ubiquitous benthonic taxon. It was observed in as many as 11 of the 36 wells, and according to the logistic model it existed across the entire study area with approximately constant probability of detection. During Middle-Late Eocene, a large percentage of taxa were restricted northerly benthonics, reflecting the fossiliferous, thick terrigenous mudstone sequence observed in northern wells. *Plectofrondicularia* Sp. 1 (no. 41) is a benthonic form that probably did not exist on the Scotian Shelf and southern Grand Banks. Two other examples of northern benthonics are *Eponides* Sp. 5 (no. 74) and *Anomalina* Sp. 1 (no. 173) both shown in Fig. 4. Contrary to the four patterns of Fig. 3, which are based on Model I, these first two patterns in Fig. 4 are for Model II.

In general, Models I and II provide nearly identical estimated probabilities at the well sites. However, when the total number of wells in which a taxon was observed is relatively small, Model II may give better results, because it has two coefficients only. On the other hand, when the number of wells with a taxon is relatively large, Model I may be better. These statements can be checked by visual comparison of the patterns resulting from Models I and II, and by statistical analysis. The pattern labelled 74A in Fig. 4 was obtained by Model I. The small estimated probability (+ sign) at one of the well sites where it is known to occur, and the minor extension of the plateau with relatively large probabilities of occurrence to the northern Grand Banks indicate that its transition zone would

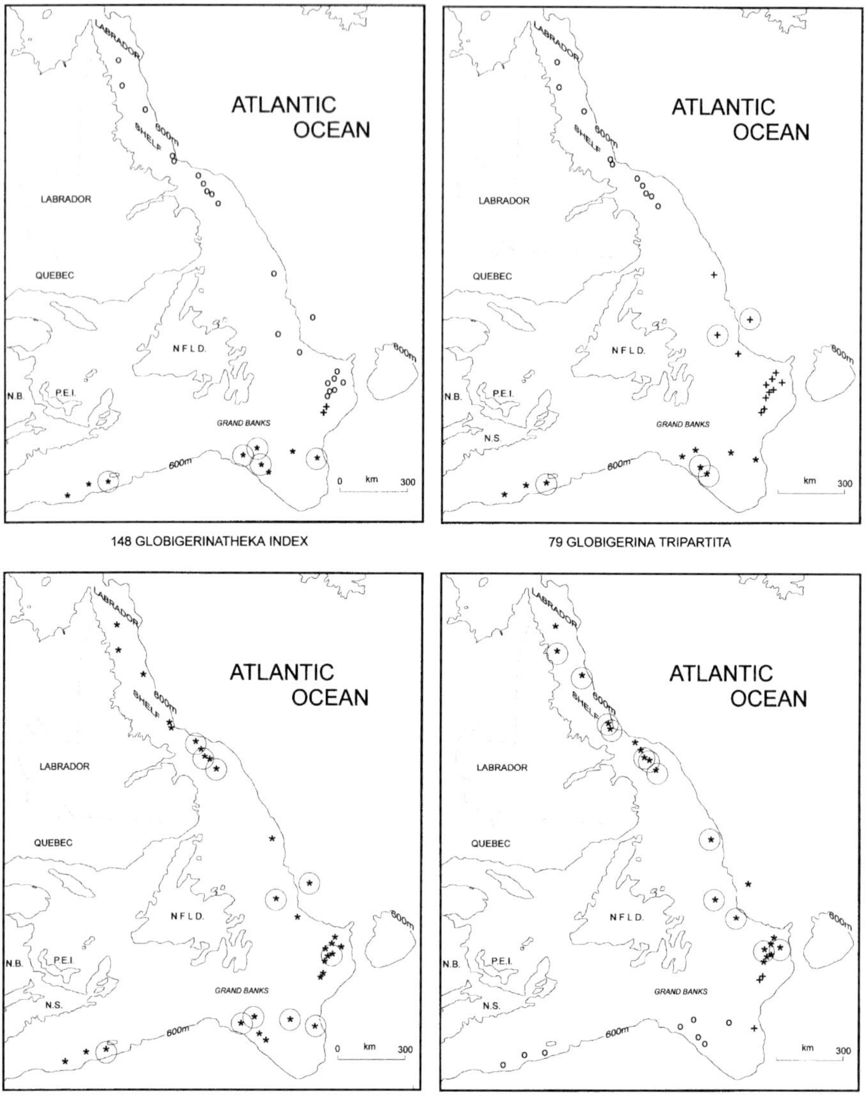

Fig. 3 Palaeogeographical distribution patterns obtained by logistic regression (Model 1) of presence/absence data for two planktonic Foraminifera (148 *Globigerinatheka index*, and 79 *Globigerina tripartita*) and two benthonics (33 *Turborotalia pomeroli* and 41 *Plectofrondicularia* Sp. 1) in data set of 36 wells from Bonham-Carter et al. (1986). Presence of taxa in wells is marked by *large circles*. *Symbols* represent probabilities classified as o (0.00–0.05), + (0.05–0.20) or * (0.20–1.00)

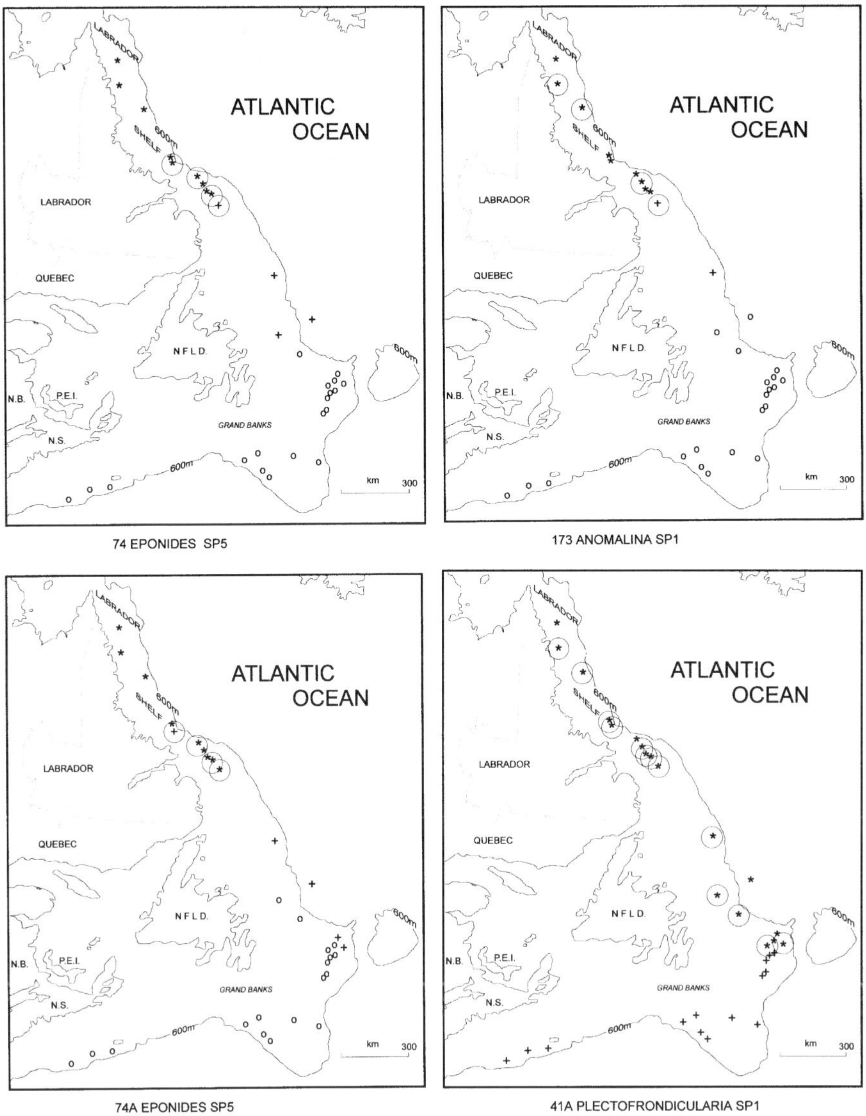

Fig. 4 Logistic patterns (Model 2, *upper part*) for two benthonics (74 *Eponides* Sp. 5 and 173 *Anomalina* Sp. 1). The *two patterns in the lower part* represent alternative solutions marked by an A obtained by other logistic models. The trends in these other two patterns are less likely to be realistic (see text for further explanations)

trend northnorthwest. This trend is not supported by the data. Pattern 41A obtained by Model II also seems to extend too far southward.

The statistical significance of coefficients in a logistic model can be tested by means of an ordinary chi-squared test for goodness of fit, or by means of the deviance statistic. By adding a single coefficient to the model (*c* when expanding Model II to Model I), both test statistics are asymptotically distributed as chi-squared with a single degree of freedom. For *Eponides* Sp. 5 (no. 74), the two test statistics are 1.63 and 0.94, respectively. Neither exceeds the 95% and 99% confidence limits which are 3.84 and 6.63, respectively. Consequently, pattern 74 is more acceptable than 74A. On the other hand, for *Plectofrondicularia* Sp. 1 (no. 41) the test statistics are 5.72 and 1.48. Although this result is not conclusive, the fact that one of two test statistics exceeds the 95% confidence limit suggests that pattern 41 is slightly better than 41A. Thus the results of the statistical significance test agree with those obtained by visual inspection of the patterns.

5 Conclusions

Recent developments of the RASC method for ranking and scaling of biostratigraphic events (Gradstein and Agterberg, in press) include variance analysis of deviations between observed fossil events and best-fitting lines of correlation in the stratigraphic sections used. Standard deviations events can be calculated and used to identify fossil events that are good time markers. This information can also be used as input for the CASC computer program for correlation and standard-error calculation. For a new version of the RASC and CASC computer programs, see Agterberg and Gradstein (1997).

Palaeogeographic distribution patterns of taxa used to define biostratigraphic events can be reconstructed by means of logistic regression analysis. The results shown in this paper were derived using the program LOGDIA (Agterberg 1989) but the authors are preparing a new computer program (TRACE) for traceability of fossil events.

References

Agterberg, F.P., 1989, LOGDIA – Fortran 77 program for logistic regression with diagnostics: *Computers & Geosciences*, vol. 15, no. 4, pp. 599–614.

Agterberg, F.P., 1990, *Automated Stratigraphic Correlation*: Elsevier, Amsterdam,424 p.

Agterberg, F.P., and Gradstein, F.M., 1997, RASC and CASC – Biostratigraphic zonation and Correlation Software (Computer programs and user guide), Ottawa, Canada.

Bonham-Carter, G.F., Gradstein, F.M., and D'Iorio, 1986, Distribution of Cenozoic Foraminifera from the northwestern Atlantic Margin analyzed by correspondence analysis: *Computers & Geosciences*, vol. 12, no. 4B, pp. 621–635.

Gradstein, F.M., and Agterberg, F.P., in press, Uncertainty in stratigraphic correlation: Proceedings, *"High Resolution Sequence Stratigraphy"*, Norwegian Petroleum Soc. Conference, Stavanger, November 1995, Elsevier, Amsterdam.

Gradstein, Felix, and Bäckström, Sven, 1996, Cainozoic biostratigraphy and palaeobathymetry, northern North Sea and Haltenbanken: *Norsk Geologisk Tidsskrift*, vol. 76, pp. 3–32.

Gradstein, F.M., Agterberg, F.P., Brower, J.C., and Schwarzacher, W., 1985, *Quantitative Stratigraphy*: UNESCO, Paris, and Reidel, Dordrecht, 598 p.
Hay, W.W., 1972, Probabilistic stratigraphy: *Eclogae Geol. Helv.*, vol. 65, pp. 255–266.
Mann, K., and Lane, R., Editors, 1995, *Graphic Correlation*, SEPM (Soc. for Sedimentary Geology), Special Publication no. 53, Tulsa, Oklahoma, 263 p.
Shaw, A.B., 1964, *Time in Stratigraphy*: McGraw-Hill, New York, 365 p.

Quantitative Stratigraphy of Sedimentary Sequences: A Case Study of the Middle Ordovician Event

L. Ainsaar, T. Martma, T. Meidla, M. Rubel and N. Sidaravičienė

1 Introduction

The irreversibility of evolution is one of the most reliable tools to date geological events, which makes the palaeontological record valuable in reconstruction of the history of sedimentary basins. The resolution of palaeontological time scales can be improved by applying methods of quantitative stratigraphy, due to more rigorous and/or more complete approach by using the distribution of taxa as a reliable tool for correlation, i.e. a time scale (Shaw 1964; Guex 1987; Agterberg 1990). Moreover, the ordinal scales of taxa which are independent of the stratigraphical framework can be used to describe the dynamics in the diversity of taxa and, to some extent, the evolution of the corresponding groups. Such an opportunity was noted by Paul (1985), but mostly ignored in studies on radiation and extinction of taxa, in spite of the actuality of the problem (Sepkoski 1995a; Holland 1989).

In this chapter, an algorithm for the construction of the palaeontological time scale (Rubel and Pak 1984) is used to process distributional data of ostracodes from several Ordovician core sections in the East Baltic. The changes in origin and extinction rates of ostracodes during the Ordovician period are discussed, based on the approach of quantitative stratigraphy. The age of a remarkable middle Ordovician biotic event in Baltoscandia is compared with the age of a geochemical event – deviation in carbon isotopic composition of seawater – using a regional ostracode succession constructed by a deterministic algorithm as a time scale. The relationship of the faunal changes with the oceanographic event is demonstrated and the contemporaneity of geochemical signals in different sections is proved.

2 The Baltoscandian Palaeobasin: Geological Setting

In the Ordovician, the Baltoscandian Palaeobasin represented an epicontinental sea on the Baltica palaeocontinent situated in Southern Hemisphere. With the northward drift of Baltica, the palaeolatitude of the central part of the basin changed from approximately 50° to 20° during the Ordovician period (Torsvik

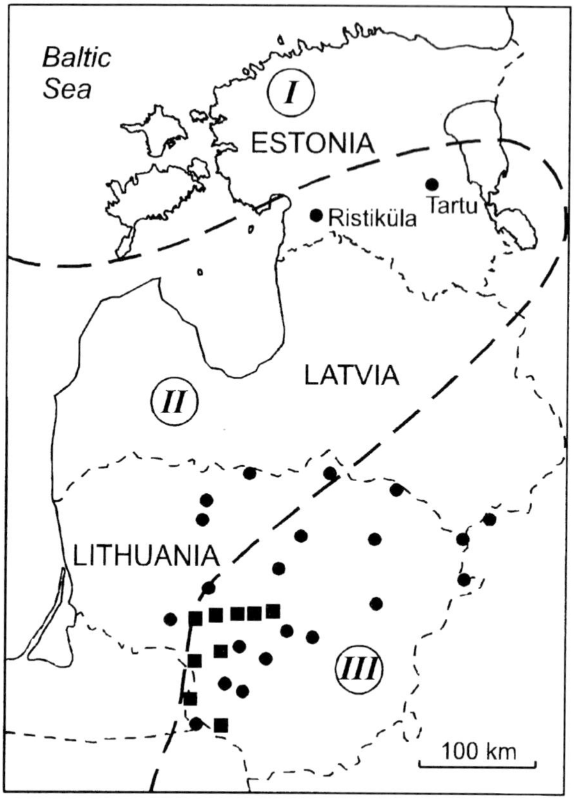

Fig. 1 Approximate boundaries of the post-Tremadocian Ordovician confacies belts according to Jaanusson (1976) in the East Baltic area (I North Estonian Confacies Belt; II Central Baltoscandian Confacies Belt; III Lithuanian Confacies Belt) and location of the boreholes (filled dots single boreholes; filled squares borehole groups; for details see Sidaravičienė 1992)

et al. 1996). The Ordovician carbonates of the Baltoscandian shelf sea (epicontinental sea) were deposited over a large area in Scandinavia and the East Baltic. According to fossil and lithofacies, the large-scale zonation of the palaeobasin comprises major confacies belts (Männil 1966; Jaanusson 1976; Fig. 1). The post-Tremadocian Ordovician of the East Baltic area (see Fig. 2) represents a nearly complete carbonate sequence (Männil and Meidla 1994; Hints et al. 1995) with a total thickness between 70 and 240 m. Stratigraphical gaps occur in the sections of the marginal belts (North Estonian and Lithuanian confacies belts), whereas deposition in the depressions (e.g. in Latvia and in western Lithuania, assigned to the Central Baltoscandian Confacies Belt) was almost continuous. The marginal, shallower belts are characterized by limestones formed mainly in open shelf or upper to middle ramp settings, usually below a fair-weather wave base. The Lithuanian Confacies Belt includes the Middle Lithuanian "trough", occupying most of eastern Lithuania and being characterized, during most of the Ordovician, by higher deposition rates of argillaceous-carbonate material. The Central Baltoscandian Confacies Belt is mainly characterized by argillaceous-carbonate sedimentation taking place under conditions of lower ramp or deeper.

BRITISH SERIES	*BALTIC SERIES, STAGES*	
	HARJU	PORKUNI
ASHGILL		PIRGU
		VORMSI
		NABALA
	VIRU	RAKVERE
		OANDU
CARADOC		KEILA
		HALJALA
		KUKRUSE
LLANDEILO		UHAKU
		LASNAMÄGI
LLANVIRN		ASERI
	OELAND	KUNDA
		VOLKHOV
ARENIG		BILLINGEN
		HUNNEBERG
TREMADOC		VARANGU
		PAKERORT

Fig. 2 Ordovician stratigraphy of the East Baltic region. Correlation with British series according to Männil and Meidla (1994).

The correlation along the confacies belts is based on an extensive biostratigraphical data set (for reviews, see Kaljo and Nestor 1990; Männil and Meidla 1994) and is well elaborated in the East Baltic. However, the correlation along the facies gradient remains problematic in several particular stratigraphic levels, including the upper Viruan.

3 The Oandu Event

A pivotal moment in the development of the Baltoscandian Palaeobasin was recognized by numerous palaeontologists in late Viruan. A remarkable extinction event at the end of the Keilan age and a subsequent short-term (about 1 ma) episode of extremely low microfaunal diversity is documented in North Estonia (Hints et al. 1989; Kaljo et al. 1995, 1996). It is expressed as an interval containing remarkably few taxa, followed by the appearance of a new, presumably, immigrated fauna (Männil 1966; Jaanusson 1976). Such a turnover was named after the regional stage the Oandu crisis (Kaljo et al. 1995, 1996).

This biotic event coincided approximately with a regressive episode (Kõrts et al. 1991; Ainsaar et al. 1996) and changes in carbonate sedimentation pattern of the region (Männil 1966; Hints et al. 1989; Nestor 1990). These events, associated with suggested changes in climate (Jaanusson 1973; Webby 1984; Hints et al. 1989), led to a remarkable facies differentiation in the late Viruan-Harjuan (Männil 1966; Nestor 1990). According to Kaljo et al. (1996), the late Keilan extinction and the Oanduan crisis of acritarch and chitinozoan communities in northern Estonia may be related to a global drop in diversity of different invertebrate groups, described from the late Caradoc by Sepkoski (1995b).

Recent studies by Ainsaar et al. (1995) on stable isotopic composition of carbonate rocks in South Estonian core sections reported a significant positive shift in carbon isotopic composition (expressed as $\delta^{13}C$) in the upper part of the Keila Stage. The isotopic shift, which is supposed to reflect a change in ocean water chemistry, is stratigraphically close to faunal changes in the studied sections (Fig. 3). For the interpretation of such a coincidence, the events need to be dated to reveal their order or contemporaneity. The existing descriptions of faunal dynamics, based both on the regional stages (Hints et al. 1989) or detailed logs of selected single section (Kaljo et al. 1995, 1996) are too rough or too local, respectively, for reliable dating of the events. In this case, the necessary reliability for dating the short-time isotopic shift has been achieved by the use of ostracode data, organized by means of a quantitative stratigraphical method (Rubel and Pak 1989).

4 Ostracode Data

Ostracodes have been chosen for the construction of the regional scale for the East Baltic area. The choice was mainly based on the suitability of microfauna for characterizing the subsurface area and on the large data set available. The distribution data of 255 ostracode species from the 48 Ordovician core sections of Lithuania (see Fig. 1) range from the Volkhov to the Porkuni Stage (Sidaravičienė 1992, 1996). The construction of the ostracode scale was based on a data set, which includes local ranges of species collected in the Lithuanian and Central Baltoscandian confacies belts of Lithuania (see Fig. 1). The scale is applied for dating of particular levels in the Keila, Oandu and Rakvere stages in two South Estonian sections, Ristiküla (Ainsaar et al. 1996) and Tartu. Ostracodes from these sections are included into the construction of the scale. The same intervals of the sections were analyzed for stable isotope composition. Ristiküla and Tartu boreholes are situated near the transition between the North Estonian and Central Baltoscandian confacies belts (see Fig. 1). They represent a nearly complete succession of the studied interval, on the whole corresponding to major regressions (Ainsaar et al. 1996).

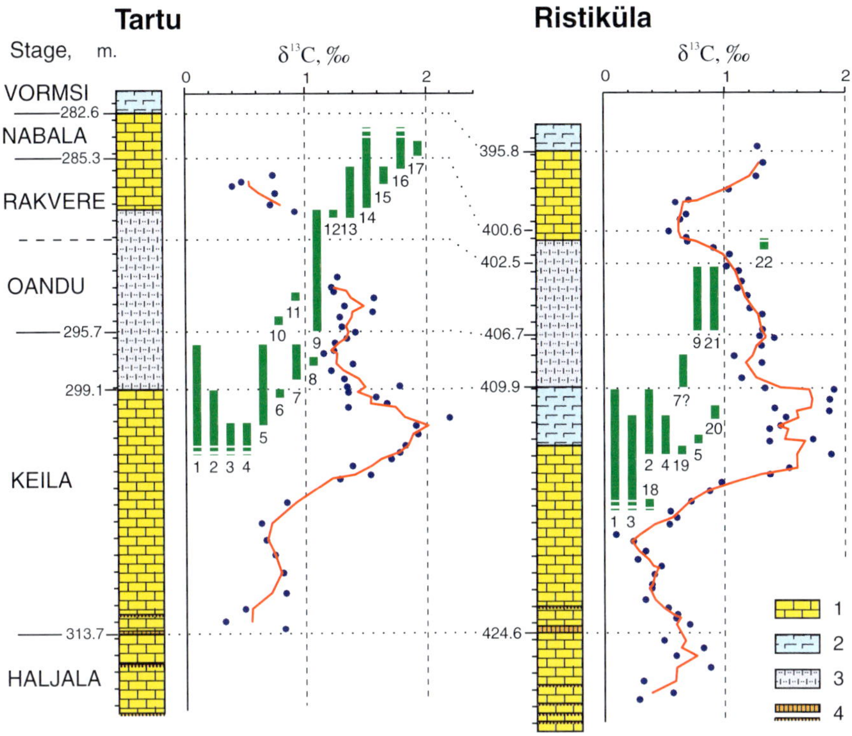

Fig. 3 Carbon isotopic composition of carbonates and distribution of selected ostracode species in the Tartu and Ristiküla core sections. The isotopic curves (heavy red lines) are constructed by three-point running averages of the d13C values (blue dots). Dotted lines Biostratigraphic correlations. Ostracodes: 1 *Pedomphalella egregia*; 2 *Sigmoopsis rostrata*; 3 *Trianguloschmidtella triangulata*; 4 *Hesperidella esthonica*; 5 *Consonopsis consona consona*; 6 *Pelecybolbina graesgardensis*; 7 *Tetrada (Tetrada) pseudoiewica*; 8 *Polyceratella aluverensis*; 9 *Klimphores minimus*; 10 *Pelecybolbina illativis*; 11 *Consonopsis zastrowensis*; 12 *Estoniosylthere longata*; 13 *Bairdiocypris indeterminatus*; 14 *Sigmobolbina camarota*; 15 *Daleiella admiranda*; 16 *Microcheilinella lubrica*; 17 *Gotlandina caudica*; 18 *Piretia erinacea*; 19 *Bichilina prima*; 20 *Scrobisylthis reticulatus*; 21 *Pelecybolbina sp. ind.*; 22 *Pelecybolbina pelecyoides*. Lithology: 1 limestone, 2 marl; 3 dolomitized quartzose siltstone; 4 K-bentonite

5 Isotopic Data

Due to the tectonic stability of the region, the Palaeozoic rocks covering the southern flank of the Baltic Shield have not experienced deep burial and strong tectonic stress, (e.g. Munnecke and Samtleben 1996). This generally good state of preservation, a prerequisite of stable isotope study, may be locally perturbed by secondary (re)crystallization, e.g. dolomitization or the occurrence of crack zones may occur in certain levels.

The stable carbon isotope composition of the carbonate rocks of the Haljala to Nabala stages in South Estonia has been examined altogether in 113 samples from the Ristiküla and Tartu core sections. The samples are taken from a 34 m interval in the Ristiküla core and a 28 m interval in the Tartu core section (Fig. 3). Bulk-rock samples were crushed and material for isotope analysis was selected, avoiding obvious veins or burrows. The dolomitization rate of the carbonate has been estimated with XRD method, whereas dolomite formed mostly 1–5% (occasionally up to 45%) of the carbonate in limestones and marls, and 20–90% of carbonate in the intercalations of dolomitized quartzose siltstones. Recrystallization in meteoric or burial environments is much more likely to preserve the carbon isotopic signatures than the oxygen values, and carbon values may be preserved even in bulk dolomites (Marshall 1992). The $\delta^{13}C$ value in the studied samples has low positive correlation with dolomite/calcite ratio (r=0.45), referring to a lack or minimum depletion effect of dolomitization. In our interpretation, we consider the $\delta^{13}C$ values to reflect the original composition of ancient seawater.

The carbonate samples were powdered to the grain size of <10 mm, reacted with 100% phosphoric acid at 100 °C for 15 min, and CO_2 analyzed using the Finnigan-Mat Delta E mass spectrometer. Carbonate $\delta^{13}C$ was detected. In Fig. 3, the $^{13}C/^{12}C$ ratio is given as per mil deviation from the PDB standard; the reproducibility of the results is higher than 0.1‰.

The curves of $\delta^{13}C$ in Ristiküla and Tartu sections are remarkably similar (Fig. 3). In both sections a positive shift (~2‰) occurs in the upper part of the Keila Stage. Upsection of the maximum, the $\delta^{13}C$ values gradually decrease upward, reaching the initial values in the Oandu and Rakvere stages. The dating of the excursions in the studied sections, being based on the traditional estimates of the ranges of taxa in stage level resolution, could not provide a necessary precision level in the boundary interval of the Keila and Oandu stages.

6
Construction of the Scale

The local ranges of ostracode species in each section are given according to their observed occurrences. Beside the sampling effects, such ranges depend on several factors: organic evolution, environmental changes and sedimentological phenomena. The first and last occurrences of species in any pair of sections are practically never exactly in the same order and none of the individual range charts can be taken as a reliable time scale. Obtaining a time scale means constructing an uncontradictory, composite succession from all observed local ranges. For this purpose, the total ranges of species and their mutual position in relation to a universal time axis must be estimated. This can be achieved by different algorithms in quantitative stratigraphy and results in creation of ordinal or even regular time scales for the correlation and dating. The same scale can be used also for tracing radiation or extinction events. This approach has the advantage of not being influenced by subjective stratigraphic correlations.

The deterministic algorithm by Rubel and Pak (1984) and the corresponding computer programme DISTR (Pak 1989) have been used to estimate the total ranges of the studied ostracode species and to arrange their appearances/disappearances into an ordinal time scale. This algorithm extends the local ranges of each species, according to their observed coexistence with other species, which in turn are arranged by their persistent mutual positions in all the studied sections. It should be emphasized that not all taxa studied can be arranged into a single succession. This is due to lacking and contradictory relations between the taxa in real data sets. This results, in practice, in the presence of multiple possible time scales for a single data set.

In the latest modification of the algorithm, preferences given to certain formal properties of the taxa (see Rubel and Sarv 1996) allow the formation of the possible successions to be controlled. A succession can be constructed preferably of taxa which occur in a greater number of sections. As a rule, such a succession will contain comparatively fewer taxa, which have often relatively long ranges. The scale formed by such long-ranging species is less detailed than the scales created under different preset conditions. If preference is given to taxa having less contradictory and uncertain relations to the others, the succession usually contains more taxa with relatively short ranges, being arranged into more numerous unitary associations. In this case, the average and particular frequencies of the taxa included into the scale are lower, which restricts the possibilities to locate the corresponding datum planes in sections.

7 Ostracode Successions

Aspects of evolution of ostracodes in the Ordovician of the East Baltic have been discussed in several papers (Sarv 1968; Hints et al. 1989; Meidla and Sarv 1990; Sidaravičienė 1992). In the development of this group, the Oandu crisis is the most remarkable event in the post-Tremadoc Ordovician, resulting in the formation of a uniform late Ordovician ostracode fauna (Meidla 1996).

In the present chapter, the results obtained with the DISTR programme under different strategies, are discussed. The statistics on the results is given in Table 1. The scale (succession) with the greatest number of taxa and datum planes includes only 38% of all taxa studied, while 54% of taxa have been included at least into one of these successions. However, all three successions describe the temporal alteration of origins and extinctions of taxa in the same way – all have a remarkable change in their middle part. It is expressed, first of all, by a reflection of the radiation of taxa, marking the beginning of the second period in the development of the biotic history of the Baltoscandian Palaeobasin and coinciding with the Oanduan migration (Jaanusson 1976) or Oandu crisis (Kaljo et al. 1995 1996).

A principal chart of total ranges of ostracodes constructed by the algorithm is shown in Fig. 4. The dynamics of appearances and disappearances of species, in the three successions constructed, can be summarized in a diagram in Fig. 5.

Table 1 Processings of distributional data of 255 ostracode species from 50 sections according to three different strategies. The strategy 100 prefers the species occurring in the highest number of sections (= frequency), strategy 1 gives no preferences based on frequency. Ranges are determined by number of successive unitary associations (rows of range charts)

Processing	1	2	3
Strategy	100	10	1
Scaled species	84	95	97
Datum planes	35	40	44
Mean frequency	14.202	9.705	8.06
Mean range	5.679	4.232	4.43

Generally the appearances/disappearances of ostracode species are evenly distributed through the Ordovician sequence, except the Oandu crisis time. No notable extinction peak can be recorded close to this level. The extinctions are slightly prevailing in the upper part of the scale, but not concentrated to a particular level. The Oanduan radiation itself, following the late Keilan extinction event, may be compared with the widespread radiation in the earliest Silurian, which is well described in the Baltic ostracode data set but remains outside of our study (for a review see Meidla 1996).

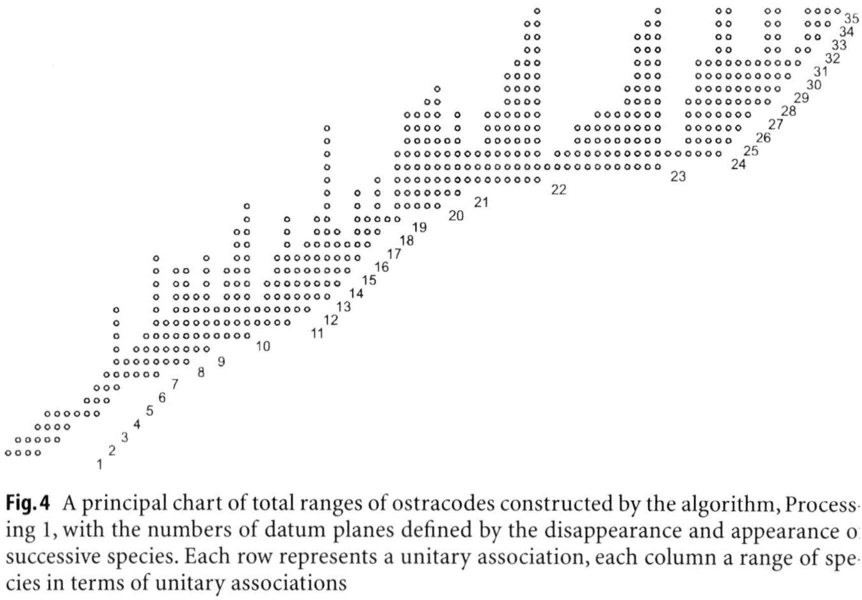

Fig. 4 A principal chart of total ranges of ostracodes constructed by the algorithm, Processing 1, with the numbers of datum planes defined by the disappearance and appearance of successive species. Each row represents a unitary association, each column a range of species in terms of unitary associations

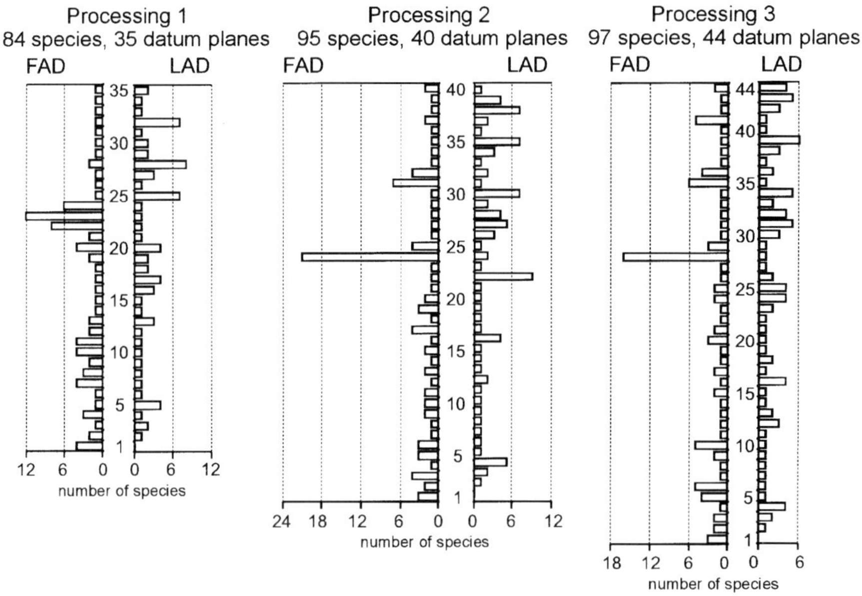

Fig. 5 Dynamics of appearances (FAD) and disappearances (LAD) for each datum plane Processings 1, 2 and 3

8
Dating of the Events and Discussion

The ostracode successions presented here are, to some extent, comparable to traditional biozonations (e.g. Sidaravičienė 1996) and tied to the chronostratigraphic standard, as demonstrated by Rubel and Sarv (1996). A time scale constructed on the basis of the faunal data consists of a succession of unitary associations, each of them being characterized by a unique set of coexisting species. The succession of associations can be used for defining datum planes, which can be recognized in the sections and used for dating purposes.

According to the scale of most widespread taxa, containing 84 species, the isotopic excursions in the Tartu section coincide with the datum planes No 17 to 20 (Fig. 6). The same datum planes in the Ristiküla section correspond to the maximum values of the isotopic curve there. Therefore, the isotopic maximum in Tartu section can be considered as contemporaneous with that in the Ristiküla core, although the latter is extended into a wider interval. In both cores, the massive radiation follows the isotopic excursion and can be dated by datum planes occurring higher in the section.

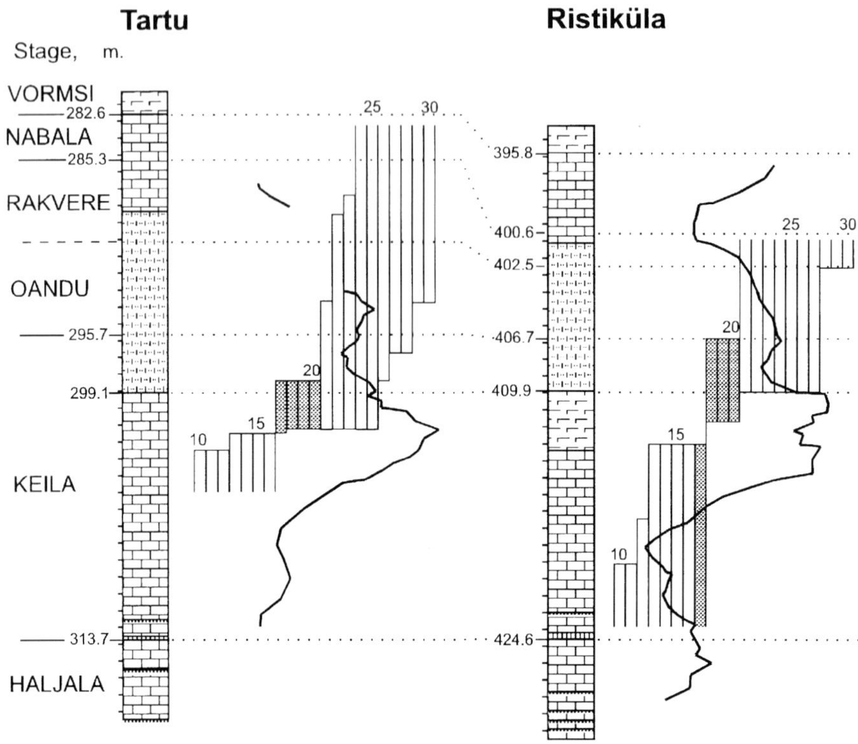

Fig. 6 Dating of the isotope curves by means of datum planes defined by the succession of 84 ostracode species (see Fig. 4). The isotopic $\delta^{13}C$ maximum in the Tartu section can be dated by datum planes Nos. 17 to 20 (shaded). The same planes correspond to the maximum values of the isotopic curve in the Ristiküla section

A comparable isotopic event has been recorded also outside the Ordovician Palaeobasin of Baltoscandia, from Caradocian carbonates of the North American mid-continent (Hatch et al. 1987; Jacobson et al. 1995). In the latter area, a positive excursion in carbon isotope composition of approximately the same magnitude has been recorded at about the same stratigraphical level. In this context, a large-scale, perhaps global event related to the Oandu biotic crisis can be suggested, as it was closely related to the changes in ocean water chemistry.

9 Conclusions

1. The outline of the ostracode species dynamics during the Ordovician in the Baltoscandian Palaeobasin is reflected in the structure of an ordinal time scale, constructed by means of a deterministic algorithm, independent of local stratigraphical framework.

2. The isotopie excursions in both studied Estonian sections are contemporaneous, coinciding with an episode of increased extinction rate of ostracodes.
3. The ostracode radiation event in the late Viruan was preceded by a positive shift in carbonate stable carbon isotope ratio ($\delta^{13}C$).

Acknowledgements

This study was supported by the Estonian Science Foundation (grant No. 951). The authors wish to thank F. Gradstein for the critical reading of the manuscript and useful suggestions.

References

Agterberg, F. P., 1990. Automated Stratigraphic Correlation. Developments in Palaeontology and Stratigraphy, 13. Elsevier, Amsterdam, 424 p.

Ainsaar, L., Kirsimäe, K. and Martma, T., 1995. Sedimentological and chemical changes associated with the Oandu Event (Caradoc, Ordovician) in Estonia. In: Kohonen, T. and Lindberg, B. (eds). The 22nd Nordic Geological Winter Meeting, 8–11 January 1996, Turku, Finland. Abstracts. Turku, p. 4.

Ainsaar, L., Kirsimäe, K. and Meidla, T., 1996. Regression in Caradoc: Evidences from southwestern Estonia (Ristiküla core). In: Stouge, S. (ed.). WOGOGOB-94 Symposium. Working Group on Ordovician Geology of Baltoscandia, Bornholm – 94. Geological Survey of Denmark and Greenland, Report 98, p. 5–12.

Guex, J., 1987. Correlations biochronologiques et associations unitaires. Presses Polytechniques Romandes, Lausanne, 244 p.

Hatch, J. R., Jacobson, S. R., Witzke, B. J., Risatti, J. B., Anders, D. E., Watney, W. L., Newell, K. D. and Vuletich, A. K., 1987. Possible late Middle Ordovician carbon isotope excursion: Evidence from Ordovician oils and hydrocarbon sourcerocks, Mid-Continent and East-Central United States. American Association of Petroleum Geologists Bulletin, 71, p. 1342–1354.

Hints, L., Meidla, T., Nõlvak, J. and Sarv, L., 1989. Some specific features of the Late Ordovician evolution in the Baltic basin. Proceedings of the Academy of Sciences of the Estonian SSR. Geology, 38, p. 83–87.

Hints, L., Meidla, T. and Nõlvak, J., 1995. Ordovician sequences of the East European Platform. Geologija, 1994, 17, p. 58–63.

Holland, C. H., 1989. Synchronology, taxonomy and reality. In: Chaloner, W. G. and Hallam, A. (eds). Evolution and Extinction. Cambridge University Press, p. 263–277.

Jaanusson, V., 1973. Aspects of carbonate sedimentation in the Ordovician of Baltoscandia. Lethaia, 6, p. 11–34.

Jaanusson, V., 1976. Faunal dynamics in the Middle Ordovician (Viruan) of Balto-Scandia. In: Basset, M. G. (ed.). The Ordovician System: Proceedings of a Palaeontological Association Symposium, Birmingham, September 1974. University of Wales Press, Cardiff, p. 301–326.

Jacobson, S. R., Finney, S. C., Hatch, J. R. and Ludvigson, G. A., 1995. *Gloeocapsomorpha prisca*-driven organic carbon isotope excursion, late Middle Ordovician (Rocklandian), North American mid-continent: new data from Nevada and Iowa. In: Cooper, J. D., Droser, M. L., and Finney, S. C. (eds). Ordovician Odyssey: Short Papers for the Seventh International Symposium on the Ordovician System, Las Vegas, Nevada, USA, June 1995. The Pacific Section Society for Sedimentary Geology, Fullerton, p. 305–308.

Kaljo, D. and Nestor, H. (eds), 1990. Field Meeting, Estonia 1990. An Excursion Guidebook. Tallinn, 209 p.

Kaljo, D., Nõlvak, J. and Uutela, A., 1995. Ordovician microfossil diversity patterns in the Rapla section, northern Estonia. In: Cooper, J. D., Droser, M. L., and Finney, S. C. (eds). Ordovician Odyssey: Short Papers for the Seventh International Symposium on the Ordovician System, Las Vegas, Nevada, USA, June 1995. The Pacific Section Society for Sedimentary Geology, Fullerton, p. 415–418.

Kaljo, D., Nõlvak, J. and Uutela, A., 1996. More about Ordovician microfossil diversity patterns in the Rapla section, northern Estonia. Proceedings of the Estonian Academy of Sciences. Geology, 45, p. 131–148.

Kõrts, A., Männil, R., Põlma, L. and Einasto, R., 1991. Epochs and depositional environments of kukersite-type (algal) organic matter accumulation in the Ordovician and Silurian of Estonia. In: Kaljo, D., Modzalevskaya, T. and Bogdanova, T. (eds). Major Biological Events in Earth History. Transactions of the XXXII Session of All-Union Palaeontological Society. Tallinn, p. 87–95 (in Russian).

Männil, R., 1966. Evolution of the Baltic Basin during the Ordovician. Valgus, Tallinn, 200 p. (in Russian with English summary).

Männil, R. and Meidla, T., 1994. The Ordovician System of the East European Platform (Estonia, Latvia, Lithuania, Byelorussia, parts of Russia, the Ukraine and Moldova). In: Webby, B. D., Ross, R. J. and Yong Y. Zhen (eds). The Ordovician System of the East European Platform and Tuva (Southeastern Russia): Correlation Charts and Explanatory Notes. International Union of Geological Sciences Publication, 28, part A, 52 p.

Marshall, J. D., 1992. Climatic and oceanographic isotopic signals from the carbonate rock record and their preservation. Geological Magazine, 129, p. 143–160.

Meidla, T., 1996. Late Ordovician Ostracodes of Estonia. Fossilia Baltica, 2, 222 p.

Meidla , T. and Sarv, L., 1990. Ostracodes. In: Kaljo, D. and Nestor, H. (eds). Field Meeting, Estonia 1990. An Excursion Guidebook. Tallinn, p. 68–71.

Munnecke, A. and Samtleben, C., 1996. The formation of micritic limestones and the development of limestone-marl alternations in the Silurian of Gotland, Sweden. Facies, 34, p. 159–176.

Nestor, H., 1990. Basin development and facies models. In: Kaljo, D. and Nestor, H. (eds). Field Meeting, Estonia 1990. An Excursion Guidebook. Tallinn, p. 33–36.

Pak, D. N., 1989. Construction of paleontologic timescale by means of linear criterion. In: Oleynikov, A. and Rubel, M. (eds). Quantitative Stratigraphy – Retrospective Evaluation and Future Development. Institute of Geology, Academy of Sciences of the Estonian SSR, Tallinn, p. 61-71 (in Russian with English summary).

Paul, C. R. C., 1985. The adequacy of the fossil record reconsidered. Special Papers in Palaeontology, 33, p. 7–15.

Rubel, M. and Pak, D. N., 1984. Theory of stratigraphic correlation by means of ordinal scales. Computers and Geosciences, 10, p. 97–105.

Rubel, M. and Sarv, L., 1996. Reconstruction and use of the succession of East Baltic Silurian ostracodes. Proceedings of the Estonian Academy of Sciences. Geology, 45, p. 177–188.

Sarv, L., 1968. Ostracodes Craspedobolbinidae, Beyrichiidae and Primitopsidae of Silurian of Estonia. Valgus, Tallinn, 103 p. (in Russian with English summary).

Sepkoski, J. J., Jr. 1995a. Patterns of Phanerozoic extinction: a perspective from global data bases. In: Walliser, O. H. (ed.). Global Events and Event Stratigraphy. Springer, Berlin, p. 35–51.

Sepkoski, J. J., Jr., 1995b. The Ordovician radiations: Diversification and extinction shown by global genus-level taxonomic data. In: Cooper, J. D., Droser, M. L., and Finney, S. C. (eds). Ordovician Odyssey: Short Papers for the Seventh International Symposium on the Ordovician System, Las Vegas, Nevada, USA, June 1995. The Pacific Section Society for Sedimentary Geology, Fullerton, p. 393–396.

Shaw, 1964. Time in Stratigraphy. McGraw-Hill Book Co., New York, 365 p.

Sidaravičienė, N., 1992. Ordovician Ostracodes of Lithuania. LitNIGRI, Vilnius, 309 p. (in Russian).

Sidaravičienė, N., 1996. Lietuvos Ordoviko Ostrakodai, Biostratigrafija. Lietuvos Geologijos Institutas, Vilnius, 82 p.

Torsvik, T. H., Smethurst, M. A., Meert, J. G., Van der Voo, R., McKerrow, W. S., Brasier, M. D., Sturt, B., A. and Walderhaug, H. J., 1996. Continental break-up and collision in the Neoproterozoic and Palaeozoic - A tale of Baltica and Laurentia. Earth-Science Reviews, 40, p. 229–258.

Webby, B. D., 1984. Ordovician reefs and climate: a review. In: Bruton, D. L. (ed.). Aspects of the Ordovician System. Palaeontological Contributions from the University of Oslo, 295, p. 89–100.

Topic C:
Space-Time Models of Basin Fill

Use of Sedimentary Simulations for Dating Sequence Boundaries and Measuring the Size of Eustatic Sea Level Changes: an Example from the Neogene of the Bahamas

C. G. St. C. Kendall and A. Sen

1 Introduction

Two major problems which confront sequence stratigraphers interpreting seismic lines are (1) the ages of sequence boundaries (Miall 1990); and (2) the size of eustatic sea-level changes (Burton et al. 1988). Sedimentary simulation can be used in conjunction with sequence stratigraphy to interpret seismic cross sections to solve these problems. To this end, a seismic cross section that records the Neogene carbonate fill of the West Andros Channel of the Bahamian Platform was interpreted, and a sedimentary simulation based on empirical modeling reproduced this interpretation. This match between the interpretation and the seismic has enabled the age dating of sequence boundaries and determined that the size of the eustatic sea-level excursions for this time period matches the sea-level events on the Haq et al. (1987) chart.

At the heart of this study is the recognition of eustatic events, evidenced by the presence of synchronous sedimentary sequences and the unconformities that bound them (Vail et al. 1978). These eustatic signals produce changes in the accommodation for sedimentary fill and have a worldwide extent. Their chronostatigraphic correlation is dependent upon reliable time markers spaced sufficiently close in time to bracket the sediment packages formed in response to changes in sea level. The amplitude of these eustatic events presents an enigma, since these cannot be determined independent of models for tectonic behavior and sedimentation. The result is that, while sea-level charts can be created, the amplitudes of given events on these charts are dependent on assumptions concerning the rates of subsidence and sediment accumulation. Unfortunately, there are no direct methods available to measure the amplitudes of sea-level variations. This is because there is no datum available to measure from, since the earth surface constantly moves in response to (1) sediment compaction, (2) isostatic response to sediment loads, and (3) thermal tectonic movement (Burton et al. 1988). Thus, the relative sea level position is dependent on tectonic behavior and eustatic position, so the size of either of these two variables can be measured only by assuming a model for the other's behavior. Methods which attempt to indirectly measure sea level have to assume models of tectonic behavior. Such methods include tide gauges, strandline position (which additionally assume a continental relief and/or tectonic behavior), paleobathymetry, seismic sequence onlap, stacked subsidence curves, and the matching of sequence geometries with graphical simulations (Burton et al. 1988). Despite the fact that sea-level amplitudes cannot be measured independently, stratigraphic predictions based on sea-level curves or tectonic models of behavior are reproducible and verifiable away from areas of interest. This is because the onlapping or downlapping of the sediments involved is dependent on the rates of sedimentation, the tectonic movement and sea-level position. Hence, if rates of subsidence and of carbonate accumulation are constant for several cycles of eustatic sea level, then the frequency and amplitude of the onlapping sequence geometries will be the product of the frequency and amplitude of the changes in eustatic sea level. The hypothesis presented here is that if the sequence geometries of a sedimentary simulation generated with the Haq et al. (1987) chart match the interpreted sequence on the seismic, then the ages of these latter sequences can be assumed to match those of the simulation. The shallow water carbonate platforms of Bahamas provides a perfect opportunity to test the hypothesis.

2 The Assumptions

To build reasonable simulations of stratigraphic interpretations that capture sea-level events, geologic sections are required for which it can be assumed that there was a low rate of uniform subsidence and a high rate of constant sediment accumulation for several sea-level cycles of the Haq et al. (1987) curve. Examples of the response to this kind of stratigraphic signal can be seen on the seismic sections of the Lower Cretaceous of offshore South Africa (Fig. 1) and the onshore seismic of the National Petroleum Reserve of Alaska (Fig. 8 in Bird et al. 1992). It can also be recognized on the seismic record for the Neogene of the Bahamas platform (Fig. 2) and the Neogene section displayed in the cliffs of the island of Mallorca, offshore eastern Spain (Fig. 3). In all these locations, the rates of subsidence appear to be low while the sedimentation rates are high. The result is that the changing onlapping position of the sequence geometries of these areas are assumed to be produced by sea-level changes.

With the absence of absolute sea-level positions for a paleoshore, it was assumed that should the rate of sedimentation be sufficiently high, then any shoreward accommodation was filled to sea level. When this happens the shelf margin can then be used as a proxy of the sea-level position. This assumption is not unreasonable for most carbonate shelves, but when it is applied to clastics confirmatory paleobathymetric markers are needed. In the case of carbonates the reasonableness of this assumption can be seen expressed in the seismic sections of the Neogene of the Bahamas (Fig. 2) and the cliffs of Neogene sediments exposed in the cliffs of the island of Mallorca (Fig. 3). The sedimentary sections developed in both these locations are the product of carbonate accumulation ,rates which were high enough to fill the accommodation space up to sea level during

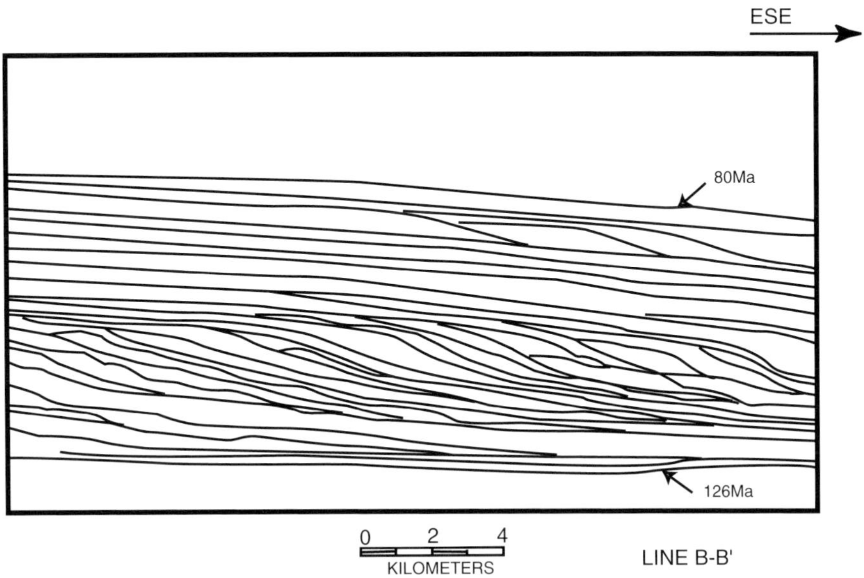

Fig. 1 Part of interpreted seismic section *(line B-B')* of Lower Cretaceous from Southern Platmos subbasin, off-shore South Africa showing prograding clinoform sediment wedges. (After Brown et. al. 1995)

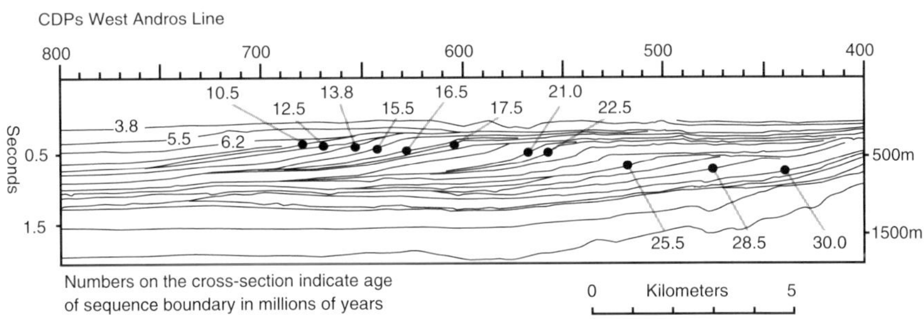

Fig. 2 Interpreted seismic section from the Bahamas

each sea-level cycle. In both cases, the Neogene section is expressed by prograding clinoforms. In the Bahamas, this effect can be seen on both the western side of the bank and in an interior sea, the Straits of Andros (Eberli and Ginsburg 1987, 1989).

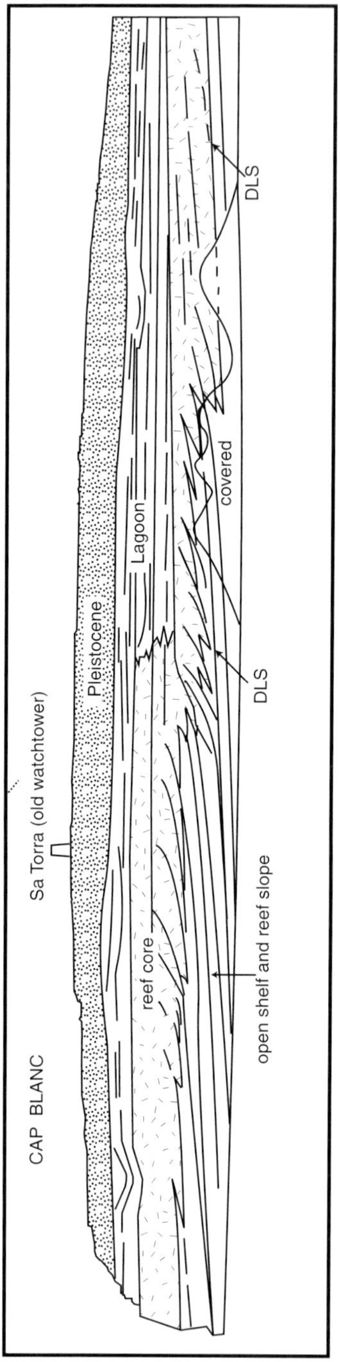

Fig. 3 A schematic map of the Miocene reef complex and Cap Blanc, Spain (CS condensed section; DLS downlap surface). (After Pomar 1993)

3
Sequence Stratigraphy and Simulation of the Andros Channel

3.1
Introduction to the Data Set

The sequence stratigraphy of the Andros Channel involves an interpretation of carbonate platform development, based on seismic data and limited well control from northwestern Great Bahamas Bank (Fig. 4). This Bahamian seismic data set consists of a cross-bank profile (Fig. 2) that was part of an approximately 700-km grid of unmigrated, multichannel seismic profiles. The top 1.7 s (two-way travel time) were used for the study. This seismic profile is tied to the Great Isaac well at the northwestern edge of the Great Bahamas Bank. The time/depth conversion from the Great Isaac well indicates that the entire Cenozoic history is recorded in these upper sections (Tator and Hatfield 1975; Schlager et al. 1988). The cuttings from this well give some of the initial information on the lithology and age of the reflectors (Schlager et al. 1988; Eberli and Ginsburg 1987, 1989). More recently, two continuously cored wells positioned on the cross-bank profile record the Late Miocene-Recent progradation of the margin (Eberli et al. 1994).

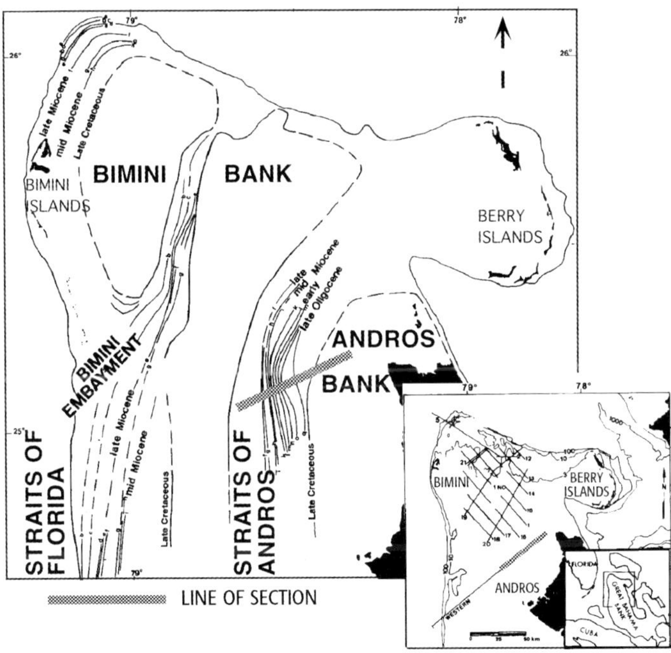

Fig. 4 Map of Great Bahamas Bank showing location of seismic section modeled in this chapter

Seismic profiles of the northwestern Great Bahamas Bank have been interpreted to document the lateral growth potential of isolated platforms that were welded together by margin progradation to form larger banks (Eberli and Ginsburg 1987, 1989). The mechanism responsible for an evolution from aggradation to progradation was thought to be sediment overproduction on the platform (Hine et al. 1981, Willer et al. 1990). Excess sediment was transported offbank and caused a decrease in accommodation space on the marginal slope. Progradation occurred in pulses that are interpreted to result from third-order sea-level fluctuations (Eberli and Ginsburg 1987, 1989). Limited well control has prevented confirmation of these hypotheses, but sedimentary simulation can be used to demonstrate the proposed theory for platform evolution is reasonable, and matches the sea-level events on the Haq et al. (1987) charts (Eberli et al. 1994).

3.2 Seismic Stratigraphic Interpretation

Eberli and Ginsburg (1987) had previously correlated sequence boundaries on the seismic section across the Andros Channel to the Haq et al. (1987) chart. In the present study, the line was re-interpreted. Initially second-order unconformities were identified on the basis of their very extensive erosional character and their correlation to the second-order sea-level events of the Haq et al. 1987 chart (Fig. 5). Third-order unconformities were then identified enclosing their equivalent seismic sequences. The ages of these latter sequences were bracketed with the ages determined from the second-order type 1 unconformities, and correlated with the third-order, type 1, unconformities of the Haq et al. (1987) chart (Fig. 5). As can be seen from the interpretation of the seismic, the major events on the Haq curve have produced distinct stratigraphic signals (Fig. 3). For instance, following the sea-level fall at 30 Ma in early Oligocene times, a major unconformity was created. This separates the Upper Neogene carbonate accumulation from the rest of the Tertiary and the Cretaceous. Similarly, a major fall in sea level occurred at 10.5 Ma and this also has a distinct seismic expression as a major unconformity. It is apparent that the sea level fell below the shelf margin with the resulting unconformity bracketing the series of third-order sea-level events between 10.5 and 30.0 Ma. While examining the geometric position of these latter sequences with respect to the shelf margin, and counting them, it can be seen that more unconformities can be identified than there are sea-level events on the Haq et al. (1987) chart. It is suspected that some of these interpreted sequences may be a result of separating the products of both low and high stand cycles of sea level as sequences. Despite these additional sequences, there appears to be a general match of the onlapping relationship of the different sequences on the seismic (Fig. 2) and the extent of basinward retreat of shorelines, with the amplitude of the sea-level events (Fig. 5). Having made the sequence stratigraphic interpretation, a simulation was developed.

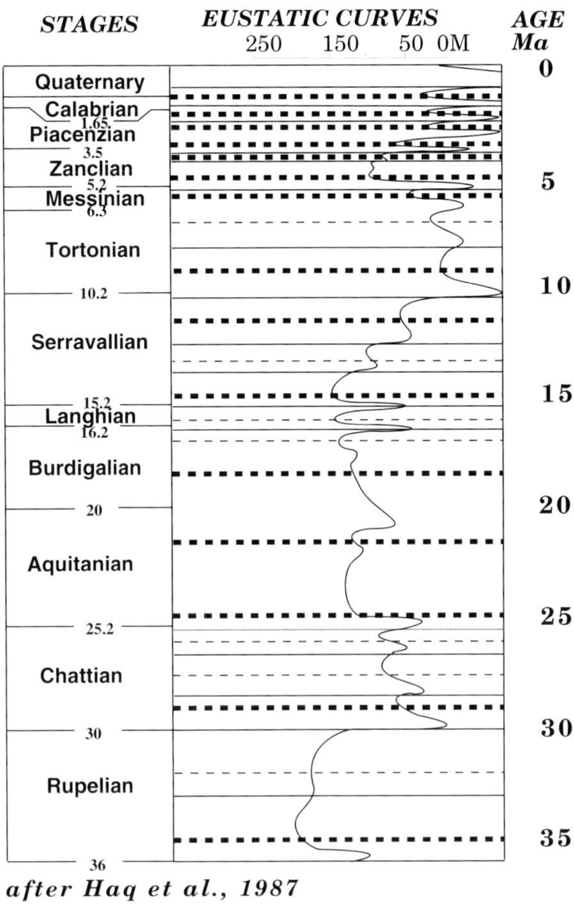

Fig. 5 Chart of Neogene eustatic events from 36 Ma to present. (After Haq et. al. 1987)

3.3 Simulation Input Parameters

Time Interval. The last 30 Ma of platform evolution of the Bahamas/Straits of Andros was modeled using time steps of 250 ka duration. The time taken for the simulation to execute was directly effected by these time steps. A duration of 250 ka produced 120 time steps which were considered essential for the proper resolution of the simulation. (Table 1a)

Initial Basin Surface. The initial depositional surface of the Straits of Andros used in the simulation was derived from the seismic line. This was the sequence boundary which had the interpolated age of 30 Ma (Fig. 2). The time/depth conversion of this surface and the other sequence boundaries above it were deter-

Table 1(a-f) Different inputs parameters used during simulation of Andros Channel (see text for details)

a) **Time interval**

Time steps in simulation	
Duration	Number
250	12

b) **Initial basin surface**

Distance (km)	Depth (m)
0	-370
18	-370
26	-71
28	-21

c) **Tectonic movement**

Subsidence rates (m/Ka) at specified locations as a function of time			
Time (Ma)	Location across the basin in km		
	1.0000	17.0000	30.0000
-30.0	-0.0090	-0.0090	-0.0090
-18.0	-0.0090	-0.0090	-0.0090
-10.0	-0.0090	-0.0090	-0.0090
-06.5	-0.0345	-0.0345	-0.0310
-06.0	-0.0580	-0.0580	-0.0485
-04.0	-0.0567	-0.0567	-0.0470
00.0	-0.0385	-0.0385	-0.0350

d) **Benthic carbonate depostion (BCD) rates**

BCD rates (m/Ka) at different depths as a function of times					
Depth (m)	Time in million years				
	-30.00	-17.00	-14.00	-03.00	00.00
-300	0.00	0.00	0.00	0.00	0.00
-200	0.07	0.07	0.06	0.04	0.07
-50	0.07	0.07	0.06	0.04	0.07
-15	0.35	0.35	0.06	0.04	0.35
-00	0.40	0.40	0.06	0.04	0.40

e) **Pelagic carbonate deposition (PCD) rates**

Variation in PCD rates (m/Ka) across the basin as a function of time								
Time (Ma)	-30.00	-24.50	-10.50	-10.00	-4.50	-3.50	-2.50	0.00
Rate	0.0280	0.0217	0.0100	0.008	0.008	0.008	0.080	0.080

f) **Carbonate Parameters**

Talus/turbidite depositional angle	20°
Percent talus	30%
Percent turbidite	70%
Talus penetration distance	1 m
Turbidite penetration distance	10 m

mined from the velocity profile of the Great Isaac well (Eberli and Ginsburg 1987, 1989). Only the east side of the Straits of Andros basin was considered for the simulation. This basin was initially formed as an asymmetric trough approximately 30 km wide with a depth of approximately 370 m (Table 1b). As the program calculates the area of sediments per column across the basin, the basin was divided into 60 columns of 0.5 km width.

Eustatic Sea Level. The Haq et al. (1987) sea-level curve (Fig. 5) was used for modeling the Bahamas/Straits of Andros dataset. The size of the sea-level excursion directly controlled the high-frequency geometries of the sequences.

Tectonic Movement. The subsidence history of the Bahamian/Straits of Andros dataset was derived directly from the seismic section and was initially modeled by the simulation with no sediment fill. The thickness of the sediment measured on the seismic crossing the platform was used as a first approximation to determine the changes in the rate of subsidence. It can be seen that these rates changed several times through the 30-Ma period, and at different locations across the basin, reflecting low frequency changes in accommodation space at different time intervals as well as different locations. A constant rate of 0.009 m/ka was used for 30 to 10 Ma and faster rates has to be used thereafter in order to ensure that there would be enough space available to accommodate the carbonates as they were generated (Table 1c).

Carbonate Deposition. Benthic carbonate production rates mimic the response to photosynthesis of carbonate-producing organisms. Most of the sediment was produced by organisms that were dependent upon light, so production rates decrease rapidly with increasing water depth (Schlager 1981). The simulation modeled accumulation, not production. This accumulation was modeled as from a combination of benthic and pelagic sources. In the simulation, the resulting geometries were very sensitive to small changes in the accumulation rate with depth and time. Note that the accumulation rate fell rapidly in water deeper than 50 m. (Table 1d)

Pelagic deposition is a critical source of carbonate sediment and comes directly from the water column. In the simulation this pelagic rain is defined as accumulation which varied as a function of time (Table 1e). This rate was used to modulate the progradational fill of the basin by the benthic carbonate, enabling the filling the downslope basin so the progradation could occur. It is believed that in nature this pelagic material included the mud-sized aragonite needles that were transported offbank on the Great Bahamas Bank, probably during highstands of sea level on the banktop (Wilber et al. 1990) and in offbank positions during lowstands.

Carbonate Parameters. As in nature, the simulation limits the growth of carbonate buildup crests to sea level. Excess carbonate accumulation, which would cause the buildups to rise above sea level, was transported off the buildup and deposited as talus and turbidite. The simulation algorithm assumes that all the carbonate talus of the margin came from the "reef" crest. The carbonate parameters were such that the angle of repose was 20°, and the distance of transport of the apron was 1 km for the apron of sediments and 10 km for turbidites (Table 1f). These values were set at the beginning of each simulation run. The percentage of the talus that was transported downslope off the carbonate platform into the basin was an input parameter of 20% (Table 1). By determining the respective amounts of backreef and talus deposition, windward versus leeward effects were modeled. The angle of repose of 20° was important for deposition of these mass gravity flows. Carbonates have a high angle of repose, which decreases with cohesiveness (Kenter and Schlager 1989).

Experiments with the Simulation. Several experiments were performed in order to fine tune the simulation. Once the shape of the initial basin surface was determined, the elevation of that surface had to be positioned in such a way that at the end of the simulation run there was an exact fit of the sedimentary fill of each of the sequences to the accommodation. Using Haq et al. (1987) sea-level curve as input, this match was dependent upon both subsidence rates and rates of carbonate accumulation. Attempts were made to keep the subsidence constant through several sea-level cycles. Ultimately a value of 0.009 m/1000 years was selected and this value was kept constant from 30–10 Ma. Thereafter, the subsidence rate had to be increased, especially after 6 Ma, to enable the exact fill of sediments for that time interval. Additionally erosional surfaces were created at the interpreted type 1 sequence boundaries of 10.5 and 3.5 Ma. These erosional events are obvious on the seismic and the 3.5 Ma boundary has a particularly irregular erosional surface.

Once subsidence was determined, the rate of benthic carbonate accumulation and that of pelagic deposition were established individually and coordinated with each other. Then two sets of experiments were performed. In one, the benthic carbonate rates were varied with a constant rate of pelagic accumulation, and in another, with slightly different constant values of benthic carbonate rates, the rate of pelagic accumulation was changed. In each case the resulting simulation was matched step by step with the seismic interpretation.

Though the rates of accumulation for pelagic and benthic carbonate rates were varied considerably, the experiments showed that once a combination of the two sources of carbonate produced the best match between the simulation and seismic, it was very difficult, if not impossible to further change the rates to produce better matching simulation outputs. This combination of benthic and pelagic rates produced the correct dimensions for the progradation and aggradation of the sediment wedges without oversteepening the prograding clinoforms. The timing and sequences of progradation was controlled by the size and timing of the sea-level excursions. Attempts were also made to develop similar geometries, with different sea-level curves and different values of same input parameters. While it was possible to produce similar geometries it was, however, not possible to match the output to the seismic interpretation.

3.4
Results from the Bahamas Simulation Execution

When the simulation is examined in detail, it is possible to see that each one of the sequences produced, matches those on the seismic (Fig. 6). These are in order, sequences from 30 to 28.5 Ma 28.5 to 25.5 Ma; from 25.5 to 22.5 Ma; from 22.5 to 21 Ma; from 21 to 17.5 Ma; from 17.5 to 16.5 Ma; from 16.5 to 15.5 Ma;

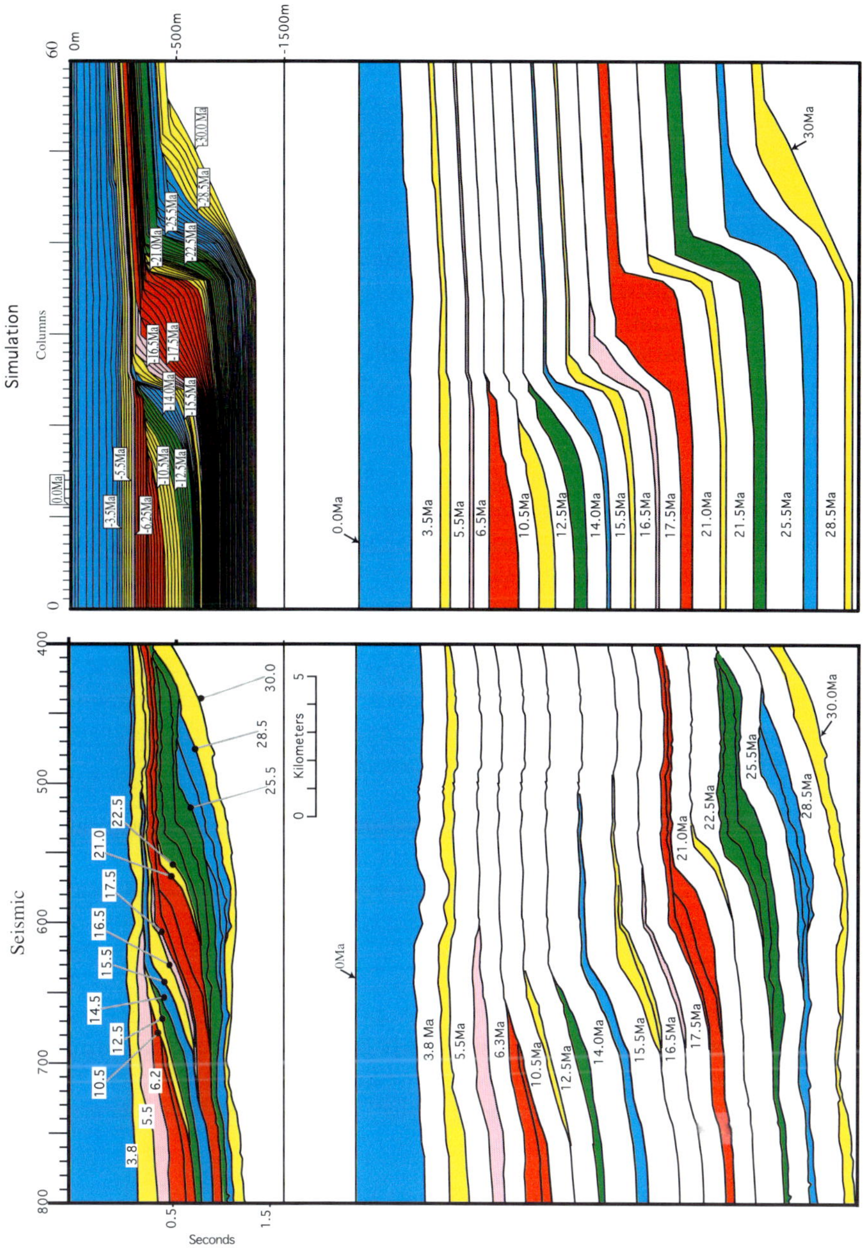

Fig. 6 Comparison of the simulation with the seismic for individual sequences. Each *date* indicates the age of the boundary above and below it (except for 30 and 0 Ma). Time steps shown in the main simulation are removed later for clarity. See text for details

from 15.5 to 13.8 Ma; from 13.8 to 12.5 Ma; from 12.5 to 10.5 Ma; from 10.5 to 6.2 Ma; from 6.2 to 5.5 Ma; from 5.5 to 3.8 Ma; and from 3.8 to the present day. It can be seen on the seismic that the geometries associated with each sea level event match the simulation with remarkable resolution (Fig. 6).

The illustrations in Fig. 6 track the geometry and timing of the evolving sedimentary fill of the Neogene sedimentary fill of the West Andros channel.

30-28.5 Ma. The geometry of this sequence was the product of the low subsidence rates of 0.09 m/ka and high rates of shallow water benthic carbonate accumulation. The initial rise in sea level on the Haq et.al. (1987) chart at this time caused aggradation on and onlap across the shelf margin which was followed by progradation during a sea-level fall. At this time the accommodation was reduced, and the shallower water depths caused faster accumulation rates.

28.5-25.5 Ma. The low subsidence rate, as well as the carbonate accumulation rate, were maintained. Aggradation and progradation matched the sequence associated with the time interval 30-28.5 Ma on the seismic interpretation. The accommodation produced by this sea-level event was reduced, since the sea level was unable to rise far across the shelf margin.

25.5-22.5 Ma. A larger rise in the sea level at the beginning of this interval caused carbonate aggradation. The subsidence and carbonate accumulation rates (both benthic and pelagic) matched those of the previous interval. The extensive carbonate onlap of the shelf was driven by a sea-level rise.

22.5-21 Ma. The same rates for subsidence and carbonate accumulation rates were maintained. The sea level shows a slight fall at the end of this period, thus producing a little progradation of the carbonate margin. The ramps produced are steeper than those we have on the seismic, but the amount of progradation was the same.

21-17.5 Ma. The same rates for subsidence and benthic carbonate depositional rates were maintained. The sea level slowly rose and both progradation and aggradation took place. A wide lowstand wedge was produced, because the sea level did not rise sufficiently to onlap the shelf and enable sediment to accumulate here.

17.5-16.5 Ma. The initial half of this sequence was marked by a sharp increase in sea level followed by a sharp decrease in the later half. As a result, the simulation shows an aggradation and onlap of the shelf followed by progradation. The same rates for subsidence was maintained. The deep water benthic carbonate rate remained the same, but the shallow water rate started to decline. The basin was too deep for much progradation to take place, and this also produced a steeper ramp.

16.5- 15.5 Ma. The beginning of this time period was marked by a drop in sea level which is followed by a marked rise and the beginning of another fall. The same rates of subsidence were maintained, and the shallow water benthic as well as pelagic showed a decline in accumulation rate.

15.5-14 (13.8) Ma. The sea level rose initially but remained constant in the later part of this period, aggradation and progradation continued with minor onlap of the shelf margin. The same rates for subsidence were maintained, while the rate of accumulation of shallow water benthic as well as pelagic carbonate continued to steadily decline.

14 (13.8)-12.5 Ma. Sea-level falls caused the basin margin to prograde. The same rates for subsidence and carbonate depositional rates were maintained. No onlap of the shelf margin occurred.

12.5-10.5 Ma. The sea level remained constant before falling drastically at the end of this sequence. The basin margin showed progradation. The same rates for subsidence and benthic carbonate accumulation rates were maintained. The pelagic rate dropped to 0.008 m/ka. The 10.5 Ma surface is marked by an erosional surface.

10.5-6.5 Ma. The sea level continued to fall below the shelf margin and to maintain the match between the seismic and the simulation, the subsidence rate was increased to –0.03 m/ka at the end of this time period. The same rate of carbonate accumulation was maintained to fill the offshore basin and maintain the correct amount of progradation of the carbonate margin.

6.5-5.5 Ma. The sea level fell a little during this interval and extended the ramp-like shallow progradation. The higher rate of subsidence and similar low rate of carbonate accumulation was maintained to match the previous interval, the simulation was more difficult to match to the seismic because rates of carbonate accumulation were now difficult to tune to produce the correct geometric response.

5.5-4(3.8) Ma. There was a small, gradual rise in sea level. The same rate of subsidence was maintained and the rate of carbonate accumulation remained the same as in the previous sequence interval. A highly eroded surface was prescribed for the end of this period. This produced a good match in the thickness for the onlap and aggradational geometries seen on the seismic.

4(3.8)-0 Ma. The thickness or the final basin fill of the shelf was obtained by keeping the rate of subsidence constant while increasing the rates of pelagic and benthic carbonate accumulation. The pelagic accumulation rate was increased to 0.08 m/ka while the shallow water benthic rates returned to the initial high values.

4 Discussion and Conclusions

The laterally stacked sequences of a seismic section are the product of sea-level changes whose signal can be identified by making a sequence stratigraphic interpretations (Vail et al. 1988). As Eberli et al. (1994) showed earlier a sedimentary simulation can successfully reproduce the geometries seen on a seismic line. The seismic data documents the fill of the Straits of Andros and the sediments are expressed as a series of onlapping and downlapping wedges with various angles of slope. The simulation output produces a match with the seismic interpretation. The simulation helped to quantify individual factors controlling aggradation and progradation in the Bahamas. It showed that different basin depths effected the timing and extent of progradation. It also showed that there was a close balance between aggradation and progradation, and that small changes in the rate of relative sea-level movement and/or carbonate accumulation rates caused immediate switches from aggradation to progradation of the margin. Progradation took place after the space on the upper slope had been reduced and the sediment transported offbank could fill the remaining space and extend the platform margin farther basinward. In particular, progradation was triggered by sea-level drops that shifted sediment production and accumulation to the margin slope. Carbonate production rates similar to modern rates were required to produce the sediment necessary for progradation, which suggests that carbonate production has been consistently high since the early Tertiary. At the same time, repeated exposure and erosion have reduced the overall preservation and decreased the overall accumulation rate.

Thus, progradation occurred in pulses which are recorded on the seismic lines and are confirmed by the simulation as a succession of prograding and sigmoidal sequences, with each sigmoid believed to have been formed as the result of a single cycle of sea-level fall and rise (Eberli and Ginsburg 1989). Each prograding sequence was up to 500 m thick and probably consisted of an offlapping complex of reefal carbonates covered by calcareous sand. Eberli and Ginsburg (1989) thought that during the transgressive stages, marginal reefs were established and then buried during the subsequent highstand, when abundant sediment was produced on the flooded bank and transported to and off the leeside of the bank. Their interpretation was based on findings from the leeside of the modern bank where early Holocene reefs are covered by offbank transported sand (Hine et al. 1981). The two 1990 core borings on the western margin of the Great Bahamas Bank have confirmed this interpretation. This justified the use of benthic accumulation as well as a background pelagic rain to produce the geometries during simulation. Interestingly the simulation suggested progradation of the bank margin continued at sea-level lows when the platform was exposed.

For a carbonate shelf setting with a low rate of subsidence and a high rate of sedimentation, a very clear stratigraphic signal is produced by a particular sea-level curve. This requires that the rate of carbonate sedimentation was such that the accommodation was filled to sea level so that the sediment surface on the shelf can be taken as a proxy of sea level.

For such a case it can be concluded that when the rates of subsidence and carbonate accumulation were constant during several sea-level cycles, a match in the frequency and amplitude of the onlapping geometries of seismic and simulation suggest that the frequency and amplitude of eustatic events of the Haq et al. (1987) input curve is close to reality. This example from Bahamas appears to prove that this is true for the Neogene.

The encouraging results of this simulation of a carbonate platform suggest that simulations should be performed for other areas to see if cross sections can be reproduced. Indeed, having established the dimensions of the events on the Haq sea-level chart, we plan to make a simulation study of the Neogene clastic fill of the Baltimore Canyon. So far, the resulting simulation matched the seismic interpretation of Greenlee et al. (1992), and confirmed the ages of sequence boundaries and so determined the size of local tectonic events.

The Haq eustatic chart might also be used as an input to simulate interpreted cross sections of mixed carbonate/clastic fills. We are considering simulating the early Cretaceous of the Neuquen Basin of Argentina. In this case, with all the control variables undetermined, a complex mix of varying rates of sedimentation, and tectonics are expected to produce a match between the simulation and this cross-section.

Acknowledgments

We would like to acknowledge John Reistroffer and Parvita Siregar for their initial interpretations of the seismic cross section. Dan Gill is acknowledged for his critical reviews of the manuscript. We sincerely thank Chris Crescini and Kurt Johnston for their thoughtful criticism and discussion, Charlton Purvis and Jay Gardner for system maintenance and My Tran for helping in preparation of the text.

References

Bird, Kenneth J. and Molenaar, M. C., 1992, The North Slope Foreland Basin, Alaska: in R. W. Macqueen and D. A. Leckie eds., Foreland Basins and Fold Belts, American Association of Petroleum Geologists, Memoir 55, p.3–363–393.

Brown, L. F., Benson, J. M., Brink, G. J., Doherty, S., Jollands, A., Jungslager, E., H. A., Keenan, J. H. G., Muntingh, A., and Van Wyk, N. J, .S., 1995, American Association of Petroleum Geologists Studies in Geology 41, 184 pages.

Burton, R., Kendall, C. G. St. C., and Lerche, I., 1987, Out of our depth: on the impossibility of fathoming eustasy from the stratigraphic record: Earth Science Reviews, v. 24, p.237–277.

Eberli, G. P. and Ginsburg, R. N., 1987, Segmentation and coalescence of platforms, Tertiary, NW Great Bahama Bank: Geology, v. 15, p. 75–79.

Eberli, G. P. and Ginsburg, R. N., 1989, Cenozoic progradation of NW Great Bahama Bank – A record of lateral platform growth and sea level fluctuations: SEPM Special Publication 44, p. 339–355.

Eberli, Gregor P., Whittle, G. L., Kendall, C. G. St. C., Cannon, R. L., and Moore, P. D., 1994, Testing a seismic interpretation of the great Bahama bank with a computer simulation: American Association of Petroleum Geologists Bulletin, v. 78, p. 981–1004.
Greenlee, S. M., Devlin, W. J., Miller, K. G., G. S. Mountain, and P. B. Flemings, 1992, Integrated sequence stratigraphy of Neogene deposits, New Jersey continental shelf and slope: comparison with the Exxon model,: GSA Bull., v. 104, p. 1403–1411.
Haq, B. U., Hardenbol, J., and Vail, P. R., 1987, Chronology of fluctuating sea levels since the Triassic (250 million years ago to present): Science, v. 235, p. 1156–1167.
Hine, A.C., Wilber, R.J., Bane, J. M., Neumann, A. C., and Lorenson, K. R., 1981, Offbank transport of carbonate sands along open, leeward bank margins: northern Bahamas: Marine Geology, v. 42, p. 327–348.
Kenter, J.A.M., And Schlager, W.,1989, Comparison of shear strength in calcareous and siliciclastic marine sediments: Marine Geology, v.88, p. 145–152.
Miall, A. D., 1990, Principles of sedimentary basin analysis: Springer-Verlag, New York, NY, 668 pages.
Pomar, L., 1993, High-resolution sequence stratigraphy in prograding Miocene carbonates: Application to seismic interpretation: in R. G. Loucks and J. F. R. Say, eds., Carbonate Sequence Stratigraphy, American Association of Petroleum Geologists, Memoir 57, p.389–407.
Schlager, W.,1981, The paradox of drowned reefs and carbonate platforms: Geol. Soc. Amer. Bulletin, v. 92, p. 197–211.
Schlager, W., Bourgeois, F., Mackenzie, G. and Smit, J., 1988, Boreholes at Great Isaac and Site 626 and the history of the Florida Straits. In Austin, J.A.,Jr., Schlager, W., et al., Proc. ODP, Sci. Results, 101, College Station, TX. (Ocean Drilling Program) p. 425–437.
Tator, B.A., And Hatfield, L.E., 1975, Bahamas present complex geology: Oil and Gas Journal, v. 73, p. 172–176.
Vail, P. R., 1988, Seismic stratigraphy interpretation procedure. In: "Atlas of Seismic Stratigraphy", Bally A.W. (ed.), Am. Assoc. Petrol. Geol. Studies in Geology, v. 27, p. 1–10.
Vail, P. R., Mitchum Jr., R. M., Todd, R. G., Widmier, J. M., S. Thompson III, J.B. Sangree, J. B., Bubb, J. N., and Hatlelid, W. G., 1977, Seismic stratigraphy and global changes of sea level. in C.E. Payton ed., Seismic Stratigraphy – Applications to hydrocarbon exploration. American Association Petroleum Geologists Mem. 26; p. 49–212.
Wilber, R. J., Milliman, J. D., and Halley, R. B., 1990, Accumulation of Holocene banktop sediment on the western margin of Great Bahama Bank: Modern progradation of a carbonate megabank: Geology, v. 18, p. 970 – 975.

High-Resolution Forward Stratigraphic Modeling of Ca2-Carbonate Platforms and Off-Platform Highs (Upper Permian, Northern Germany)

K. Leyrer, C. Strohmenger, K. Rockenbauch and T. Bechstaedt

1 Introduction

The Zechstein 2 Carbonate, also called Ca2 or Stassfurt-Carbonate, is Germany's most prolific carbonate gas play. An important target for exploration in Germany for more than half a century, the Ca2 has been penetrated by several hundreds of wells.

Throughout the Southern Zechstein Basin the Ca2 is sandwiched between the anhydrites of the first Zechstein cycle (Werra Anhydrite, or A1) and the anhydrites of the second Zechstein cycle (Basal Anhydrite, or A2). The depth of the Ca2 reservoir varies between 2000 and 4500 m in the western parts of Germany and 1500 and 3000 m in eastern Germany. The Ca2 depositional patterns were strongly controlled by the preexisting morphology, which mainly evolved during the formation of the A1 (Strohmenger et al. 1993c, 1996b). The thickness of the Ca2 ranges from 20 to 80 m in a platform position and reaches a maximum thickness of more than 250 m in a slope position several hundreds of meters basinward of the A1-platform margin (Fig. 4). In basinal environments, the Ca2 thickness decreases rapidly and generally does not exceed 10 m.

Detailed investigations on the basement of the Ca2 in Northwest Germany show that the Zechstein Conglomerate (Z1C), which either overlies the Early Permian Rotliegendes or Late Carboniferous strata (Fig. 1), reaches a maximum thickness of only about 0.5 m and therefore is not able to create a distinct morphology above the more or less plain topography which characterizes the landscape in Northwest Germany after Rotliegend time. The following intervals, which are called Mutterflöz Carbonate (T1Ca) and Copper Shale (T1), together reach a maximum thickness of several meters and are also without any influence on the overlying strata. The Zechstein Limestone (Ca1) is the first formation in Late Permian time which reaches thicknesses of some tens of meters. The controlling factors of these unequal sediment distributions mostly remain unclear, but the chemical composition of the water enabled several organisms like algae and brachiopods to build up reef-like communities, e.g. in the Thuringian (Peryt 1986) and Emsland areas. However, the morphological highs of the Ca1 are not reflected by the thickness patterns of the Werra Anhydrite (A1), which overlies the Ca1.

The A1 represents the first evaporite member in the succession of Upper Permian sediments in northern Central Europe and, due to its ability to build up maximum thicknesses of about 300 m, it was able to develop a distinct morphology characterized by a transition from platform to slope and basin depositional environments (Fig. 4). This transition is characterized by decreasing sediment thickness from platform (250–300 m) to basin (about 50 m) and by changing sedimentary structures. Anhydrites in platform position often show salina- (swallow-tail anhydrite), and sabkha-type (chicken-wire anhydrite) textures. Although the A1 anhydrite presumably was deposited mainly within water depths of a few centimeters to a few tens of meters, detailed investigations show that even minor sea-level fluctuations can be recognized in these thick anhydrite successions. The deeper-marine anhydrites of slope and basin position are mainly dark-colored and show typical millimeter-laminations (Linien-type anhydrite). In platform-position, the overlying Ca2 shows a variety of different shallow-water facies types (Fig. 5), but with increasing water depth in upper parts of the slope the sediment becomes more uniform. The deeper parts of the slope and also the basin sediments of the Ca2 are composed of millimeter-thick, rhythmically interbedded carbonates with high organic content, which results in fine laminations similar to the underlying deeper-water anhydrites of the A1. The Ca2-sealing anhydrites (A2, or Basal Anhydrite) of the second Zechstein cycle reach a maximum thickness of about 60 m on the Ca2-platform. Due to progradation, the thickness of the A2 sediments reaches its maximum of about 100 m several hundred meters basinward of the Ca2 platform edge and decreases in a basinward position to less than 5 m.

Sediments younger than the A2 do not influence this simulation study, and are not described in detail. An overview of the Upper Permian lithostratigraphy is shown in Fig. 1.

Although our understanding of the Ca2 has improved in recent years, many important issues like the quantification of sedimentation-controlling parameters and their relative impact on sedimentation during Ca2 time are not completely understood. Additional problems arise when comparing the Ca2 sediments at different locations along the southern margin of the Southern Permian Basin. Comparisons of Ca2 core data from Northwest German exploration areas (e.g. South Oldenburg) with data from the East German Ca2-wells in the area of Southeast Brandenburg show significant differences in sedimentary history. These differences affect the topographic settings (Fig. 4, 6, 7), the (vertical) facies patterns, and the thicknesses of different facies members. Because these differences determine the spatial distribution of potential hydrocarbon reservoir facies, they have to be taken into account to optimize prediction of hydrocarbon reservoirs. High-resolution forward stratigraphic modeling is one of the most suitable procedures for understanding these differences, and their impact on sedimentation (Leyrer et al. 1995, 1996).

This chapter shows the results of 2-D forward stratigraphic modeling of the Ca2 in two different gas-producing areas (South Oldenburg area, Northwest Germany, and Southeast Brandenburg area, East Germany).

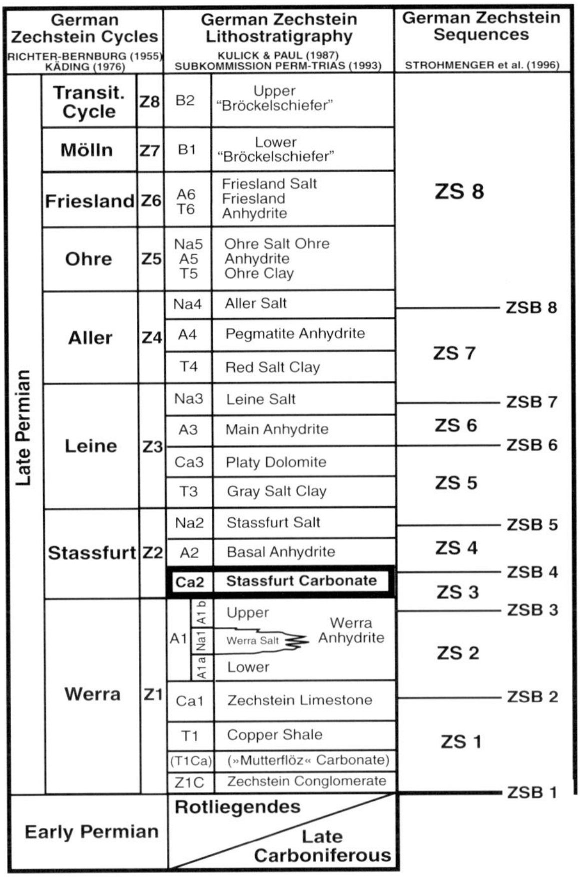

Fig. 1 Lithostratigraphic subdivision and depositional sequences of Upper Permian in Central Europe. Partly simplified after Richter-Bernburg 1955; Käding 1976; Kulick and Paul 1987; Subkommission Perm-Trias 1993; Strohmenger et al. 1996

The focus of the study is to test the recent sequence stratigraphic interpretation of the Ca2 (Strohmenger et al. 1996a, Fig. 4), to quantify factors governing sedimentation and their impact, and, as a consequence, to improve the prediction of reservoir facies distribution.

2 Basic Data Set

The recognition and adequate understanding of depositional sequences are basic preconditions for designing optimal models for predicting sequences, facies patterns and reservoir distribution (Wanless 1991). The Ca2 meets these conditions.

The basic data for the high-resolution forward stratigraphic modeling of the Ca2 come from two areas, the northwest German South Oldenburg area and the east German southeast Brandenburg area (Fig. 2). The Ca2 in both areas has been an important target for exploration for decades; therefore, a lot of modeling-relevant data exist. These data, including lithologic and facies interpretations and also a sequence stratigraphic interpretation of the Ca2, provide the basis for the performance of a high-resolution forward stratigraphic modeling. One part of these data can serve as inputs while other parts act as a control data set for the simulation results. The calibration of several sedimentation-governing parameters is possible and helps to simulate Ca2 reservoir facies distribution at other localities characterized by lower-quality or even missing data. Apart from well data, published data also were used to simulate the Ca2 sedimentary history. This integrated approach makes it possible to avoid unrealistic simulation results and creates a synthesis of all previous results.

Data from almost 200 wells were used, including detailed studies on lithology, pore type and texture classification, sedimentary and diagenetic structures and the classification of particles (e.g. Mausfeld (1987), Mausfeld and Zankl (1987), Huttel (1989), Mausfeld and Huttel (1991), Below (1992), Steinhoff (1993), Strohmenger et al. (1993a, b, 1996a, b), Strohmenger and Strauss (1996). In addition, detailed analyses of core material and the integration of well data, both of which led to the current depositional and sequence stratigraphic model of the Ca2 (Strohmenger et al. 1996a; Fig. 4), were used to test the model, and, after confirmation, as inputs for simulation.

2.1 A1/Pre-Ca2 Topographic Setting

Compared to siliciclastic sediments, carbonates are influenced more readily by topographic setting. Furthermore, topography and sedimentation interact because the deposition of sediments may change topographic settings considerably.

In the South Oldenburg area, Ca2-sedimentation is characterized by a platform-slope-basin configuration and the topographic difference between platform top and basin floor is about 300 m (Fig. 4). The underlying A1 platform extends laterally for at least a few thousand square kilometers and is rimmed by a slope with inclination angles of up to 10°. It consists of anhydrites (A1, or Werra Anhydrite), which have been karstified along the platform and the platform margin during sea-level lowstand (Zechstein sequence boundary 3, ZSB3) at the beginning of Zechstein sequence 3 (ZS3). The resulting surficial segmentation of the platform led to intensive lateral gradations of Ca2 facies representing sediments of the early highstand systems tract.

The pre-Ca2 topographic setting in the southeast Brandenburg area generally corresponds to that of south Oldenburg. However, important differences exist in the morphologic differentiation of the platform, because the southeast Brandenburg area is characterized by a highly dissected platform setting (Fig. 6). In addition, a morphological dissection of the Ca2-slope in southeast Brandenburg led to the existence of isolated highs (detached platforms), called off-platform highs (Strohmenger et al. 1993c, 1996b). Due to differences in the thickness of the Werra-Anhydrite (A1), these off-platform highs can be subdivided in high-relief types (Z1/A1-thickness about 300 m) and low-relief types (Z1/A1 thickness about 200 m, Strohmenger et al. 1993c, 1996b). As core investigations and simulation results show, the low-relief off-platform highs were flooded during Ca2 transgressive systems tract (TST), and subsequently drowned during Ca2 maximum flooding (condensed section representing Ca2 highstand systems tract deposits). In contrast, the high-relief off-platform highs and the platform in the Southeast Brandenburg area, like the Northwest German platform, were flooded only during Ca2 maximum flooding, and record sedimentation throughout Ca2 highstand systems tract (HST). This allows a comparison of sedimentary processes and developments of both areas.

Core studies show that Ca2 facies development is driven by the thickness distribution of the underlying A1. The Ca2 platform facies is restricted to the A1 platform area and simulation shows that a progradation of Ca2 facies (Ca2 highstand systems tract) first happened after 400 ka of Ca2 time (Ca2 maxium flooding surface, Figs. 15, 23). This point in Ca2 time marks when the slope sediment accumulation rate exceeds that of the platform. One of the main reasons for this relatively long timespan with nearly no progradation (Ca2 transgressive systems tract) seems to be the slope of the A1 platform, which ranges from 2° to probably more than 10°. As simulation results show, these slopes may have prevented autochthonous deposition of Ca2 transgressive systems tract deposits along the upper A1-slope, but resulted in the redeposition of these sediments at deeper slope positions. Additionally, the water chemistry in this restricted Southern Zechstein Basin was characterized by a high content of sulfate ions, at least at the end of A1time and the beginning of Ca2 time. Therefore, carbonate production rates may have been low during early Ca2 time. Furthermore, the water body likely was stratified at the transition time between A1 and Ca2. This would allow different mineralogies and production rates depending on the depth of the water column. Decreased sediment production in basinal position (starved basin) may also lead to decreasing progradation (Harris,1989).

2.2 Ca2 depositional model (Figs. 4, 7)

Detailed examination of vertical platform facies successions shows, that these sediments are generally comprised of two major shallowing upward cycles (Ca2-cycle II and III, Fig. 4) and four subcycles or parasequences (PS 4–7). The higher-order cycles represent the Ca2 highstand systems tract and are characterized by a transition from high-energy (representing coated-grain facies or ooid facies) at the base of the subcycle to low-energy algal-laminated facies and tidal flat facies at the uppermost parts of the subcycle. These four subcycles are thought to be the result of higher-order sea-level fluctuations and therefore are called parasequences (Strohmenger et al. 1996a; Strohmenger and Strauss 1996). The preservation of four parasequences on the Ca2-platform gives valuable information about eustatic sea-level fluctuations, their amplitudes, and impacts on sedimentation. Because these parasequences are short-term phenomena, they are not driven by "sudden" tectonic movements, and only the normal background subsidence has to be taken into account when estimating their real amplitude for simulation.

The Zechstein 2 Carbonate is characterized by five main facies: platform, upper slope, middle slope, lower slope and basin (Fig. 4, 7). In addition, these facies have been divided into 26 subfacies considering water-energy-related parameters like component and bedding types (Mausfeld (1987), Mausfeld and Zankl (1987), Huttel (1989), Mausfeld and Huttel (1991), Below (1992), Strohmenger et al. (1993a, b, 1996a, b), Strohmenger and Strauss (1996). Due to their sensitivity to environmental changes, Ca2 platform carbonates include 14 different subfacies, which belong to supratidal, intertidal and shallow subtidal paleogeographic positions (Strohmenger et al. 1996a; Strohmenger and Strauss 1996).

The Ca2-slope, which extends from the platform-margin down to the toe-of-slope, can be subdivided into upper slope (one subfacies), middle slope (four subfacies) and lower slope (five subfacies). In a basinward direction the toe-of-slope is followed by the basinal facies with two subfacies (Strohmenger et al. 1996a; Strohmenger and Strauss 1996).

Although the program PHIL 5.1 cannot simulate the 26 Ca2-subfacies types, the algorithms used to calculate depositional facies successions apparently include the main factors controlling Ca2 sedimentation patterns. Mainly considering energy levels and inclination angles of the sedimentation area in its calculations, PHIL 5.1 assigns simulated carbonate sediments to seven depositional environments and can distinguish up to nine carbonate lithology types. A very good match with Ca2 core information, particularly on the platform, can be reached using a combination of lithology and depositional facies simulation results.

In slope sediments, the facies interpretation of the Ca2 core material is mainly based on the observed bedding types, lamination patterns and turbidite occurrence. These facies patterns simply reflect the relative paleotopographic position of sediments on the Ca2 slope. Although PHIL 5.1 cannot analyze and evaluate bedding types within its simulation, the program's algorithms use the inclination of the depositional surface, and, incorporating the principles of user-defined cellular automata (Marco Polo Software 1994), calculate the amount of accumulated sediment. A sediment package which exceeds the stability conditions of the calculated cell will be removed and deposited in a cell positioned somewhat downward at the slope. Therefore, PHIL 5.1 can recognize the conditions necessary for turbidite occurrence, and simulates their existence, frequency, and thickness. Consequently, even without the ability to simulate bedding types, a correlation of core observations and simulation in Ca2 slope and basinal environments is possible by analyzing turbidite occurrences with regard to the slope position of sediments.

3
Input Data

To obtain simulation results which adequately match the geological data, the first source of input data should be the basic data set. Because this data set is incomplete, the missing input data have to be estimated.

Missing data mostly affect important sedimentation-governing parameters like sea-level amplitudes and/or their frequencies, spatial and temporal subsidence rates, and sediment production rates/sediment supply. For modeling purposes, most of these data require a detailed quantification. However, even the extensive Ca2 data set provides only a few reliable data and a lot of important parameters had to be estimated during numerous modeling iterations. In the present paper only the most important input parameters, which are time, subsidence rates, sea-level fluctuations and sediment production rates, will be discussed.

3.1 Time Setting

To assign the amount of time, that geological formations need for their build-up is generally very difficult, because sedimentation is governed by many different variables (Schlager 1993). Estimation of these processes and especially their impact on the resulting sediments is further complicated by frequent changes in sediment mineralogy as is the case in the Upper Permian of northern Central Europe. The Upper Permian in this area is characterized by an alternation between evaporite and carbonate formations, which means a change in water chemistry and production rates. Customary methods of time estimation, like biostratigraphy, do not apply in most of the Upper Permian strata because time periods are too short and fossils are extremely rare. A petrographic method of estimating time in (deep-water) evaporites is counting varves (Richter-Bernburg 1985). Although varves seem to be induced by yearly processes and therefore are able to record very short-term phenomena, they are dependent on steady-state conditions and consequently restricted to both deeper-water environments ("Linienanhydrit") and time periods without changes in the sedimentary system. However, a typical feature of evaporites is the differential precipitation rate, so thickness patterns give no hints on sedimentation periods (Füchtbauer 1968). As a result of his recently performed integrated time analysis Menning (1995) suggests a duration of about 7 Ma. for the seven common Zechstein cycles (Z1–Z7). According to stratigraphic simulations, this assumption seems to be reasonable. The Stassfurt cycle Z2 (Stassfurt-Carbonate, or Ca2 + Basal Anhydrite, or A2 + Stassfurt Salt, or Na2) was estimated by Menning (1995) to last about 1.5 Ma. Following our simulation results, this seems to be somewhat too short. Taking into account the duration of the observed sea-level fluctuations and estimating carbonate production rates within a reasonable range, the Ca2 itself needs about 1.5 Ma. Assuming shorter time periods would result in higher frequencies of sea-level fluctuations and higher carbonate production rates, both of which are unrealistic in such a restricted sedimentary basin.

3.2 Subsidence Rates

The North European Zechstein Basins are genetically classified as resulting from rifting. They formed along intracontinental convergence belts (Sengör 1995).

The subsidence of both the Northern and Southern Permian Basin was mainly a consequence of the decay of Stephanian-Autunian induced thermal anomalies and began during the Saxonian (Ziegler 1990). Subsidence in main parts of the basins resulted from thermal contraction of the lithosphere and the loading of the basins by water and sediments (Ziegler 1988). The occurrence of minor syn-Zechstein faulting is restricted to a few areas along the southern margin of the Southern Permian Basin. Ca2 sedimentation in the investigated areas indicates no synsedimentary faulting and therefore, as a first approach, tectonic subsidence during Ca2 time was assumed to be linear.

Tectonic subsidence rates within PHIL 5.1 are specified at hingepoints along the model profile (Marco Polo Software 1994) and are linearly interpolated between these points. The hingepoints may be user-defined and PHIL 5.1 allows at any time to vary both the temporal and spatial distribution of subsidence rates. In addition, PHIL's algorithms allow specification of the lithosphere's flexural loading (e.g. flexural parameter and mantle density) and compaction (defined by initial porosity and compaction rate) for up to 24 different lithologies.

A common way of calculating subsidence rates is by backstripping overlying geological formations. As many case studies show, this generally works well in younger sediments, but with increasing sediment age the calculation error grows rapidly. Backstripping of the Ca2 sediments in northern Central Europe faces many problems and therefore the results of such calculations should be used only as a general guide. Recently performed backstripping of Ca2 sediments in parts of Northwest Germany showed subsidence rates of about 1.5 cm/ka up to 3 cm/ka for sediments in Ca2 platform position and 7–11 cm/ka for those in basin positions. This fits well to earlier calculations of Füchtbauer (1964), who suggested a subsidence rate of 2 cm/ka for Ca2 platform sediments in Northwest Germany. Simulations performed using these subsidence rates show reasonable results. Major and sudden changes in subsidence are not indicated. Varying subsidence rates during numerous simulation runs showed the sensitivity of the Ca2 sediments to subsidence rate changes and made a quantification of subsidence possible. Simulated linear subsidence rates of 1 cm/ka to 2.5 cm/ka in the platform realm, 4 cm/ka to 8 cm/ka in slope position and 10cm/ky to 12 cm/ky in basinal areas are within a reasonable range and result in the best fit between model results and core data. In addition to the amount of subsidence for each area the difference in rate between platform and slope are important. In eastern Germany, difference in subsidence rates is one of the factors ruling the development of the topographic setting, since it is fundamentally important for the development of off-platform highs. In comparison to the northwest German Ca2 platform, these islands, which are thought to have established themselves on older (pre-Zechstein) horst-positions (Van der Baan 1990; Strohmenger et al. 1993a, 1996b) or on volcanic highs of Carboniferous/Lower Permian age, show increased subsidence rates in slope and basinal areas. This increase is fundamentally important for the islands to maintain their isolated character and results in a characteristic facies distribution in the early Ca2 time.

3.3 Sea Level

Although the intensive Zechstein research of the past 10 years has shed new light on topics like relative shifts of coastal onlap and Zechstein sequence stratigraphy in general (Tucker 1991; Strohmenger et al.1993a, b; 1996a, b) the mechanisms responsible for sea-level fluctuations are not clear and a quantification of sea-level change amplitudes and frequencies was not possible so far.

However, the high-resolution forward stratigraphic modeling of the Ca2 provides the possibility to fix amplitudes and frequencies of sea-level change, which affect Ca2 sedimentation at least throughout the southern margin of the Southern Permian Basin, since the recently established sequence stratigraphic framework (Strohmenger et al. 1996a) and the resulting relative sea level interpretation can first be tested by comparing model results and "reality" at different locations. If the model results resemble the actual data, the values used for sea-level frequencies and amplitudes provide one reasonable set of input values.

The simulation program PHIL 5.1 either uses digitized eustatic sea-level curves as input or constructs a curve by using input sea-level cycle periods, their amplitudes, phases and character (sawtooth or sinusoidal).

The Permian period is generally assumed to be a time of sea-level fall, since the formation of the supercontinent Pangea was nearly finished (first-order lowering). This class of sea-level order, which seems to be caused by accretion and splitting of supercontinents (Vail et al. 1991), is far beyond the time scale of high resolution stratigraphic modeling and therefore a quantification of duration and amplitudes was not possible in this study.

A sinusoidal cycle with a period of 20 Ma and a magnitude of 100 m (e.g. Revelle 1990) was chosen for the second-order eustatic sea-level curve (Table 1). In accordance with the general sea-level fall in Permian times, this cycle had to be shifted in its phase about 160°, which results in a slight sea-level fall of second-order magnitude during the simulated Ca2 time (Strohmenger et al. 1996a).

Third-order sequence cycles are fundamental units in sequence stratigraphy and are generally characterized by a duration between 1 and 10 Ma (Plint et al. 1992). Vail et al. (1991) suggest a duration of 0.5–5.0 Ma. In the reconstruction of the Ca2 sea level this cycle plays a major role, since it governs the main sequence stratigraphic units. The cycle duration of 1.1 my (Table 2) was chosen according to the observed Ca2 facies and sequence stratigraphic patterns which

Table 1 Duration of sea-level orders

Order	PLINT et al. (1992) Duration	Reconstr. Ca2 sea level Duration
2	10–100 Ma	20 Ma
3	1–10 Ma	1.1 Ma
4	200–500 ka	335 ka
5	10–200 ka	167 ka

Table 2 Specifications of the constructed Ca2 sea-level

Cycle (order)	Period (ka)	Range (m)	Phase Shift (deg)	Asymm. cycle	Percentage of first segment
2	20000.0	100.000	160.000	No	
3	1100.00	50.0000	–232.00	Yes	15.00
4	335.000	20.0000	-200.00	Yes	80.00
5	167.000	8.00000	0.00	Yes	80.00

suggest the development of somewhat more than one complete third-order sequence. This cycle duration allows the formation of a transgressive systems tract, a highstand systems tract, and also a lowstand systems tract overlying an intra-Ca2 sequence boundary. Because our investigations on Ca2-sediments suggest an asymmetric cycle development, the cycle was reconstructed as a weak "sawtooth", which means that it contains some smooth parts derived from a sinusoidal cycle. However, the reconstructed third-order cycle is characterized by a phase shift of -232° and a first segment percentage of 15%, which leads to a long-term sea-level rise of about 900 ky during which the (late lowstand-?), transgressive- and highstand systems tracts of the Zechstein sequence ZS3 occurred (Figs. 10, 11). The following striking drop in sea level results in reworking of parts of the prograded Ca2-highstand systems tract deposits. The subsequent sea-level rise caused a more or less *in-situ* resedimentation of the previously reworked sediments. Reaching the Ca2-platform top again, the slow, but steady, sea-level rise causes the precipitation of salina- and sabkha-type anhydrites (A2-salina and -sabkha). As numerous simulations show, this saw-toothed eustatic cycle of third-order magnitude and duration is of decisive importance for the observed sedimentation patterns of the Ca2. It is highly probable that its obvious sawtooth-shape was to a large extent caused by the restricted water exchange, which characterized the Southern Permian Basin and consequently led to a delayed reaction to sea-level changes of the open-marine Tethys.

Fourth-order cycles have a duration of 200–500 ka (Plint et al. 1992) and are thought to be induced mainly by climatic changes. These changes are caused by cyclic perturbations in orbital parameters of the Earth, and a main parameter seems to be the changing volume of ice caps due to orbital control factors (Imbrie and Imbrie 1979). Fourth-order sea-level changes belong to the Milankovitch cycles (Milankovitch 1941) and according to the Vail nomenclature, the sedimentary result of these cycles and possible higher-order cycles should be called parasequences (Van Wagoner et al. 1990). Because the imprint of fourth-order cycles on Ca2-sedimentation patterns clearly can be subdivided from those of fifth-order cycles, the sequences caused by changes of fourth-order magnitude in this chapter are called cycles (Ca2cycle I, II, III and Ca2/A2-cycle I, II) and correspond to the Ca2-cycles described by Strohmenger et al. (1993c, 1996a, b) and Strohmenger and Strauss (1996). Fourth-order cycles, also called high-frequency sequences (Kerans and Fitchen 1995), are estimated to last about 335 ka each, and are characterized by an amplitude of about 20 m with a phase shift of -200° (Table 2). These features and their superimposition on the third-order sawtooth cycle lead to a subdivision of the third-order cycle. As simulations show, the time between the transgressive systems tract (TST) of Zechstein sequence ZS3 and the early highstand systems tract (HST) of Zechstein sequence ZS4 is characterized by more than four fourth-order cycles (Figs. 10, 11). Three of the Ca2 cycles occur during the slow, but steady sea-level rise caused by the third-order cycle during Zechstein sequence ZS3 (late lowstand?), transgressive and highstand systems tract (Ca2 cycle I, II and III according to Strohmenger et al. 1993a, b, c, 1996a, b). One cycle is situated in the lowstand systems tract (LST) of Zechstein sequence ZS4 (Ca2/A2 cycle I) and an additional one (Ca2/A2 cycle II), which was not modeled completely, occurs during the late lowstand and the following early transgressive systems tract during A2 time of Zechstein sequence ZS4. Although the fourth-order cycles, in general, follow the sea-level trend given by the third-order cycle, they have their own considerable impact on Ca2 sedimentation, as they are responsible for the buildup of major cycles (e.g. the two major shallowing-upward cycles (Ca2 cycle II and III) preserved on the Ca2 platform (Fig.5). Furthermore, each of the fourth-order cycles in Ca2 platform position reflects a rapid initial transgression and a long stage of slow sea level fall. That is why, in contrast to the reconstruction of the third-order sequence, the fourth-order cycles are characterized by an "80% sawtooth", which means a short and rapid transgression followed by a long stage of slow sea-level fall.

Fifth- and higher-order cycles are thought to last between 10 and 200 ka (Plint et al. 1992) and belong, like fourth-order cycles, to the Milankovitch frequencies. Apart from the already described fourth-order cycles, fifth-order cycles play a major role in Ca2 sedimentation. They form small-scale shallowing-upward cycles, called parasequences (Strohmenger et al. 1996a; Strohmenger and Strauss 1996), and are an important parameter, especially in platform sedimentation during Ca2 time. Although the reconstruction of the sea level suggests the occurrence of at least nine fifth-order cycles during the time span between the early TST of Zechstein sequence ZS3 and the early TST of Zechstein sequence ZS4, only those four which shape the Ca2-highstand systems tract are obviously preserved (PS 4, 5, 6, and 7). Those three fifth-order parasequences (PS 1, 2, and 3) which occur during the Ca2 transgressive systems tract of Zechstein sequence ZS3 mainly affect Ca2 sedimentation on the preexisting A1 slope. However, as simulations show, the shallow-water components of these sediments were reworked immediately after their deposition and probably resedimented as turbidites along the Ca2 slope. The last 500 ka of Ca2 time (LST and early TST of Zechstein sequence ZS4) are characterized by the occurrence of another two complete fifth-order parasequences (PS1 and PS2 within Ca2/A2 cycle I) and parts of a third one (PS3 within Ca2/A2 cycle II). Parasequences PS1 and PS2 of Ca2/A2 cycle I developed during Ca2 low-stand systems tract of Zechstein sequence ZS4 (Ca2 lowstand wedge, LSW). With the onset of Ca2/A2 cycle IIa the Ca2/A2 platform of Zechstein sequence ZS4 was finally flooded (A2 transgressive systems tract). All fifth-order sequences are reconstructed by having a frequency of 167 ka and an amplitude of about 8 m (Table 2). According to observed Ca2 facies patterns, they had to be chosen as sawtooth-cycles with their peak at 80% of first segmentation. In contrast to the fourth-order cycles no phase shift was used. However, the *in-situ* sedimentary preservation of fifth-order parasequences clearly depends on their position on the sea-level curve, which is mainly determined by third- and fourth-order rises and falls. In slope areas, fifth-order cycles (typically short-term sea-level rises) often result in increased slump and turbidite rates (e.g. Ca2 cycle Ia, Ib) due to their unstable position and unfavourable water energy regimes. Furthermore, simulations show that the effects of fifth-order (and higher-order) sea-level fluctuations often are strengthened by fluctuations of fourth- and lower-order cycles, which has to be taken into account when analyzing Ca2 facies patterns.

Very likely, the described fifth-order parasequences are superimposed additionally by higher-order cycles. These higher-orders were not considered in the simulation, since no correlatable periodic sedimentary evidence can be found in the studied Ca2-cores. The reconstructed Ca2 sea-level, which consists of four combined and superimposed cycles (Table 2), shows a strikingly good fit to the Ca2 data set.

3.4 Sediment Production Rates

Within PHIL 5.1, carbonate production is mainly a function of water depth and is defined as an exponential distribution curve. For production rates, PHIL 5.1 assumes three main sedimentation realms: shelf, shelf margin and the periplatform area. Each of the three has its own distribution curve, whose shape is governed by the rate and depth of maximum sedimentation, and the width of the production function. As shown in Fig. 3, the assumed carbonate production rates used for the simulation of the Ca2 differ considerably in their spatial distribution. Simulation shows that sediment production during Ca2 time in Northwest Germany generally exceeds that of eastern Germany by about three to four times and is estimated at about 70 cm/ka in shelf areas, 60 cm/ka in shelf margin areas and 20 cm/ka in periplatform settings. These data are only valid during Zechstein sequence ZS3 transgressive and highstand systems tracts because production rates decrease during the last 500 ka of Ca2-sedimentation (LST and TST of Zechstein sequence ZS4) in Northwest Germany. Production rates during the LST of Zechstein sequence ZS4 are estimated to be 50 cm/ka in shelf settings, 40 cm/ka in shelf margin depositional environments and about 10 cm/ka in periplatform areas. Due to the fact that the actual eastern German Ca2 sediments are generally thinner, the production rates are assumed to be considerably lower than those of Northwest Germany. Furthermore, the simulated sedimentary successions of late Ca2 age (LST and TST of Zechstein sequence ZS4) in eastern Germany indicate no substantial change in Ca2 carbonate production rates. The production rates in the simulated eastern German setting are estimated to be about 20 cm/ka in shelf environments, 15 cm/ka in the shelf margin area, and less than 5 cm/ka for the periplatform realm (Fig. 3). Compared to the northwest German setting, the main reasons for these lower production rates seem to be unfavourable energy conditions (generally higher water energy, which leads to reworking and resedimentation of unconsolidated sediments) and the rate and spatial distribution of tectonic subsidence.

Evaporites occur commonly in Upper Permian sediments and a reasonable simulation of Ca2 sediments cannot be done without taking into account the existence of anhydrites below and above Ca2 strata. Within PHIL 5.1 the sedimentation rate of evaporites is a function of water depth, which is determined from the user-defined maximum evaporation rate, the duration of the simulated time step and the water column thickness (Marco Polo Software 1994). In the program, the deposition of evaporites is restricted to areas deeper than the upper tidal range. The evaporite options implemented in PHIL 5.1 are not tenable to simulate pure evaporite formations like the Werra Anhydrite (A1). Furthermore, the A1 lacks data necessary for reasonable modeling. However, simulations of the Ca2 show that the change from Ca2 carbonates to evaporites of the overlying Basal Anhydrite (A2) and mixed carbonate-evaporite successions, mostly of short-term durations, can be modeled. Using estimated evaporite production rates between 15 cm/ka and 120 cm/ka during the simulated time span (early TST of Zechstein sequence ZS3 to early HST of Zechstein sequence ZS4), simulation results do not show any sedimentation of evaporites until the TST of Zechstein sequence ZS4 (Figs. 19, 27). This fits very well to Ca2 core observations in both of the studied areas. Varying the model's input parameters suggests that the amplitude and frequency of the relative sea-level fluctuations is an important factor in the production and sedimentation of evaporites, especially during the LST and TST of Zechstein sequence ZS4. Once deposited, even thin evaporite layers often change inclination angles of mainly shallow-water areas due to their rapid sedimentation.

In addition, to quantify the above-mentioned parameters, high-resolution stratigraphic modeling requires a fixing of adequate time increments for carbonate production calculations (carbonate production time step). Because setting a large time step will not record short-term influences and their effects on carbonate production, and is not able to explain the development of the geometry of a carbonate body sufficiently, a time step of 5 ka was chosen for calculation of carbonate production during Ca2 time in both areas.

Sedimentation is commonly ruled by more parameters than those mentioned above and PHIL 5.1 contains many additional options to define and quantify sedimentary processes adequately. Variables like lithospheric flexure loading, gradients of depositional profiles, gravity flow and erosion rates and evaporation rates were also quantified in the study, but it would go far beyond the scope of this chapter to demonstrate their estimation and impact on Ca2 sedimentation.

4 Basin Modeling

The main goals of high-resolution forward stratigraphic modeling of the Ca2 with PHIL 5.1 were to:
1. explain the observed facies distribution,
2. quantify the sedimentation-governing parameters and, as a consequence,
3. improve the prediction of reservoir facies distribution.

Although the modeled areas are quite different in terms of topographic setting, they show striking similarities in their sequence stratigraphic evolution. Both areas were modeled in equal time steps of 10 ka resulting in 150 time slices. In order to give a sufficient overview, the simulated development of the Ca2 sedimentary succession is described at each parasequence boundary. The simulation results are partly shown in Figs. 12–27. As already mentioned, the basement (Werra Anhydrite, or A1) topography was provided as input (Figs. 12, 20) because inadequate modeling options exist for simulating the development of pure evaporite formations.

4.1 Parasequences 1–3 (TST of Zechstein sequence ZS3)

Parasequences 1–3 comprise the TST of Zechstein sequence ZS3, and, according to simulation results, are established during a time span of approximately 400 ka. Due to the fact that sea level remains in slope position during this part of Ca2 time, sedimentation occurs only on the slope. As a consequence, no aggradational and only slight progradational patterns are developed during the TST.

Parasequence 1 (PS1; Figs. 13, 21)
Core investigations and simulation results suggest a substantial drop in sea-level in both the south Oldenburg area and the southeast Brandenburg area at the end of A1 time (A1-HST of Zechstein sequence ZS2). During most of the LST of Zechstein sequence ZS3 sea level stayed in positions near the middle and upper parts of the A1 slope and was characterized by only small-scale fluctuations.

The modeling of the Ca2 starts within the last few 10000 years of parasequence 1 (PS1, or lower part of Ca2 cycle Ia). Simulation shows a slow sea-level rise occurring during PS1, which is interpreted as belonging to the early TST of Zechstein sequence ZS3. Due to the steep inclination of the A1 slope, the sediments generated in this stage are deposited in a very unstable position and are subjected to erosional processes mainly governed by water energy. Because this energy regime has its greatest influence in the upper parts of the water body, the sediments affected are mainly high-energy shallow-subtidal carbonates. In Northwest Germany this leads to a nearly complete *in-situ* reworking and resedimentation in basinward positions of the sediments generated during the late parts of the first parasequence. In contrast, most of the carbonate sediments of

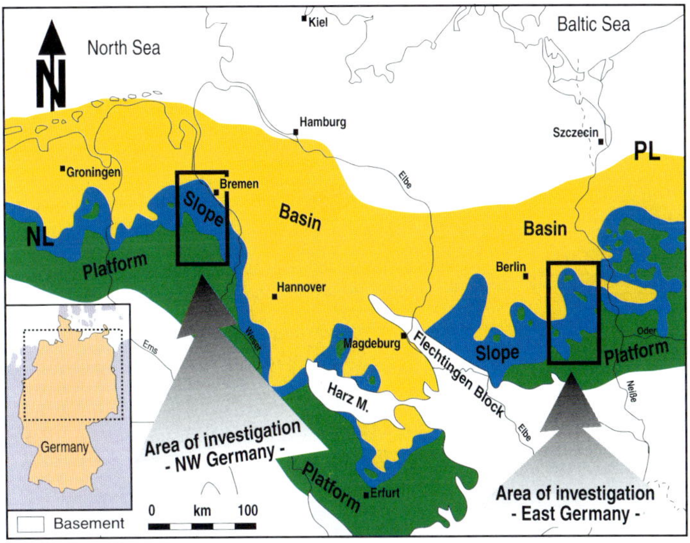

Fig. 2 Ca2 deposits along the southern margin of the Southern Permian Basin, (NL the Netherlands; PL Poland. (After Strohmenger et al. 1996)

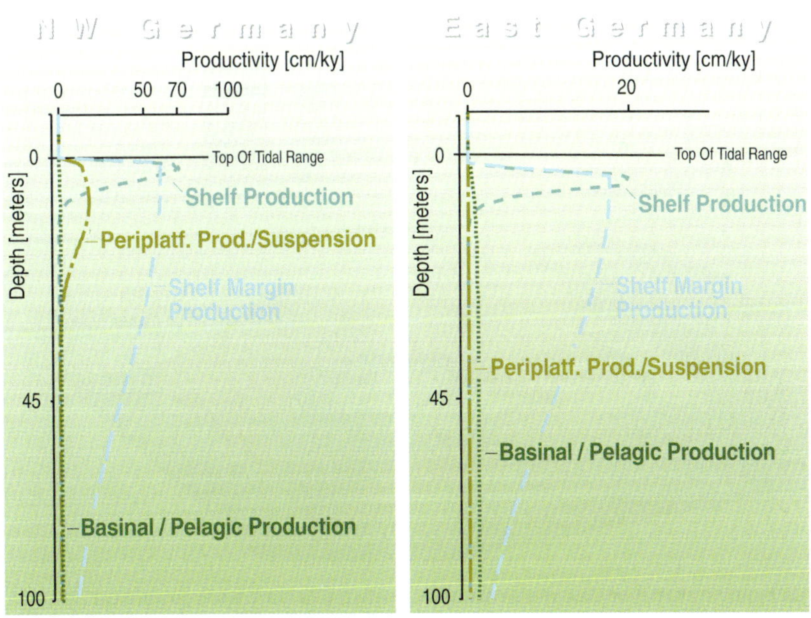

Fig. 3 Ca2-carbonate production rates in NW Germany (left) and E Germany (right)

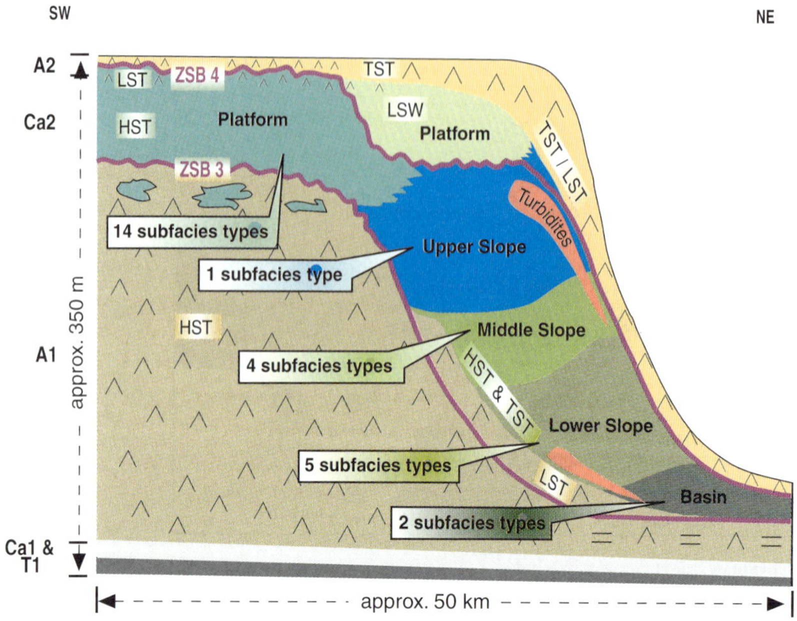

Fig. 4 Schematic sketch of Ca2 facies succession, NW Germany. (After Strohmenger et al. 1996 a)

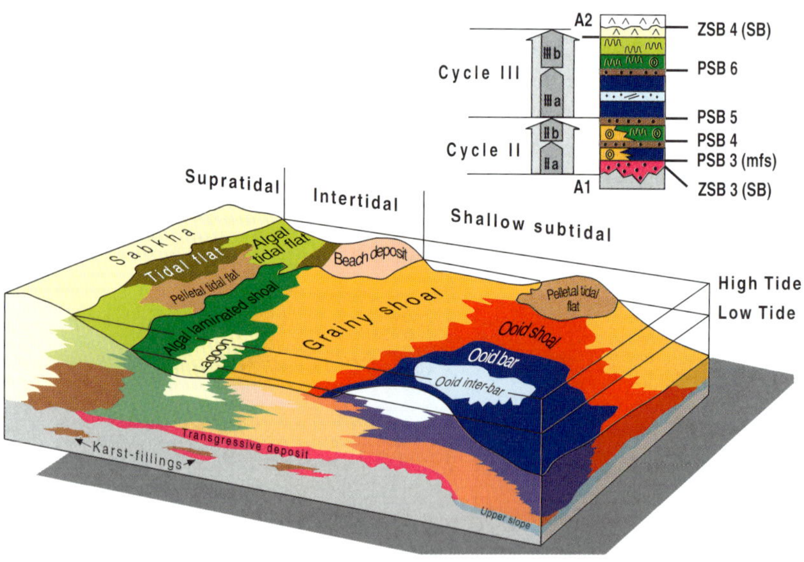

Fig. 5 Schematic sketch of Ca2 platform subfacies distribution. Idealized vertical Ca2 subfacies succession shows four shallowing-upward cycles (Ca2 cycles IIa, IIb, IIIa and IIIb). (After Strohmenger et al. 1996a)

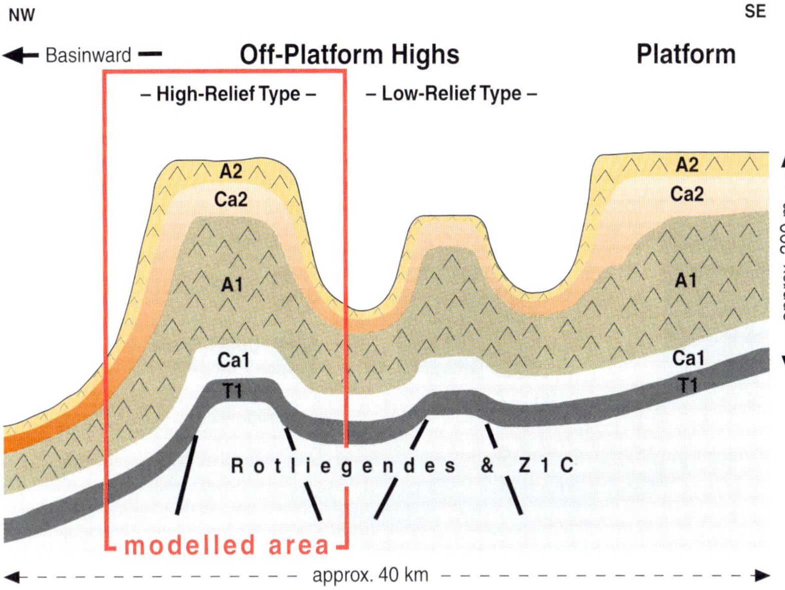

Fig. 6 Schematic sketch of Ca2 deposital environments in eastern Germany

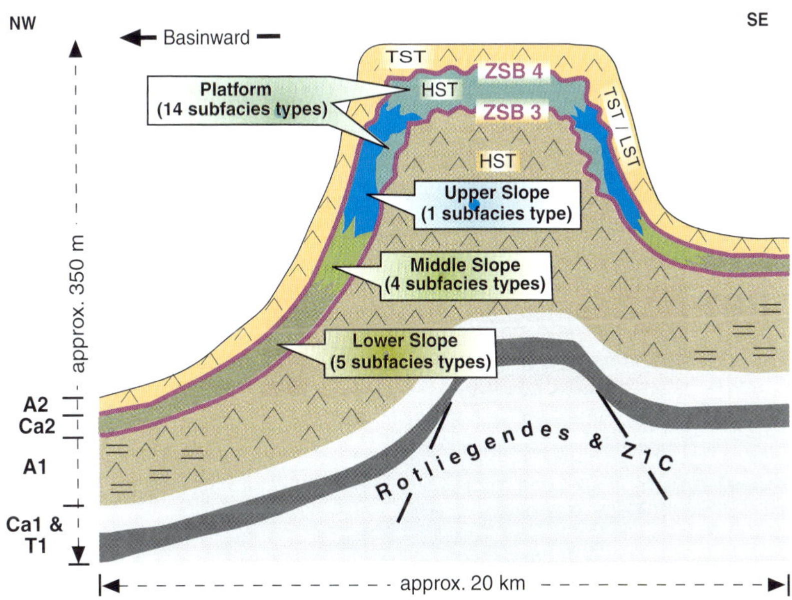

Fig. 7 Schematic sketch of Ca2 high-relief off-platform high facies succession, eastern Germany

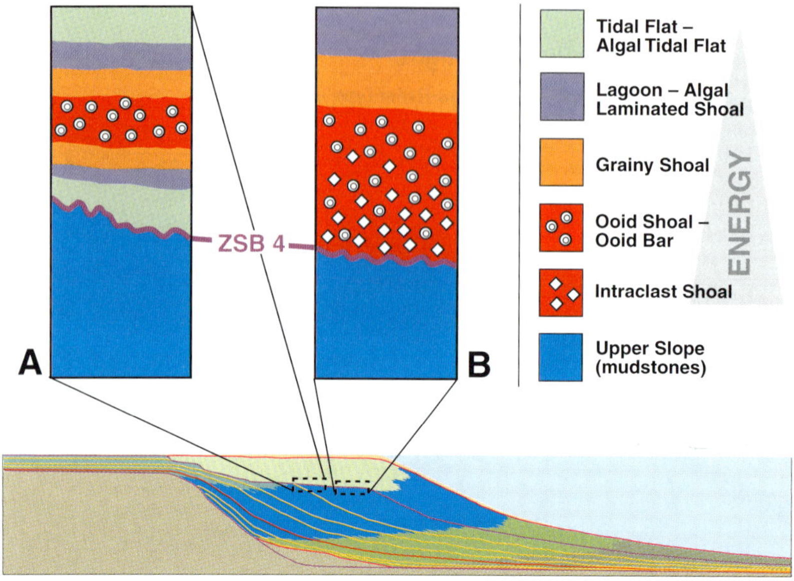

Fig. 8 Facies succession in upper Ca2 sediments in slope position (HST of ZS3 to TST of ZS4), NW Germany

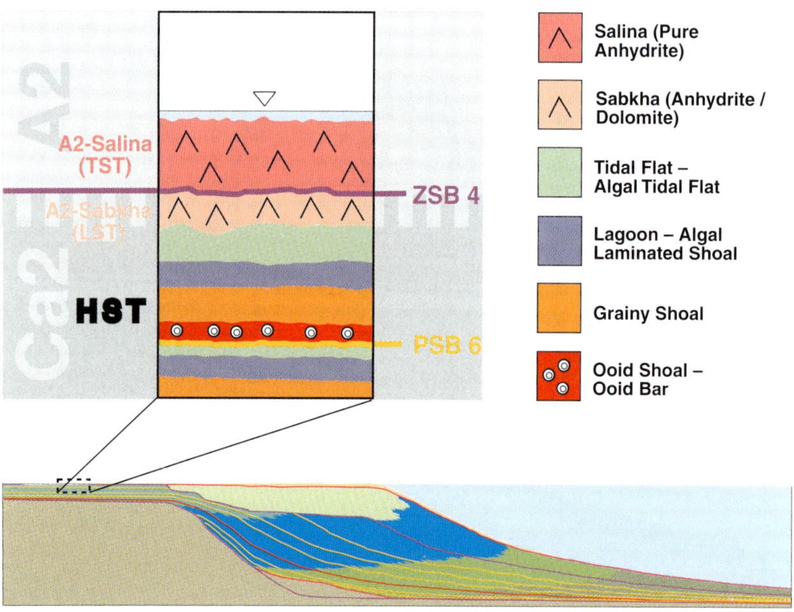

Fig. 9 Facies succession in upper Ca2 sediments in platform position (HST of ZS3 to TST of ZS4), NW Germany

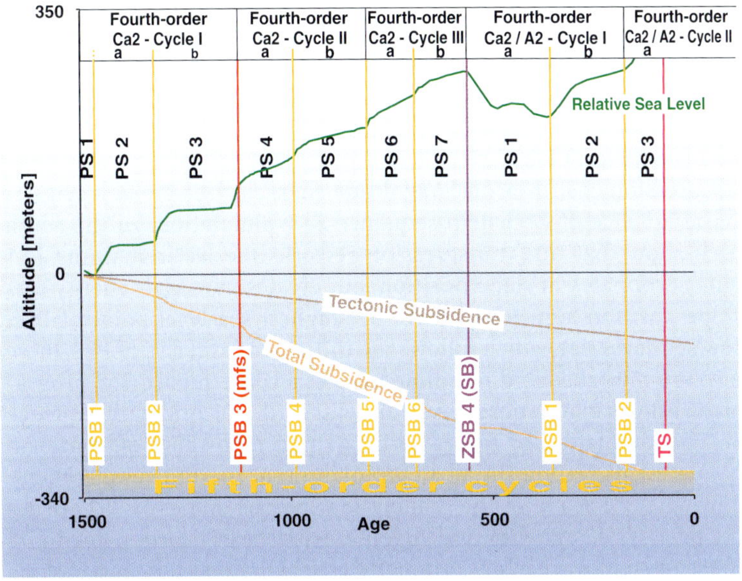

Fig. 10 Reconstructed Ca2 sea-level curve, NW Germany

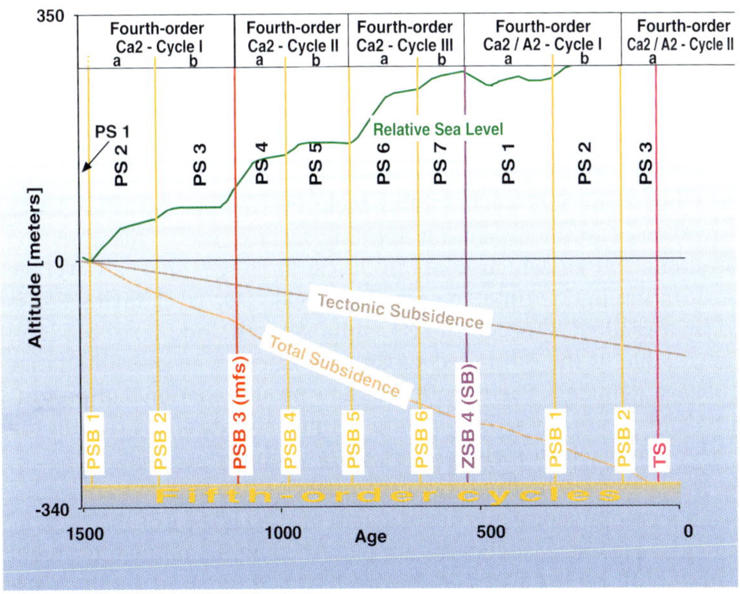

Fig. 11 Reconstructed Ca2 sea-level curve, East Germany

PS1 are not reworked along the slope of the platform and high-relief off-platform highs in East Germany. The main reasons for this are higher subsidence rates in slope positions and differences in water energy regimes (reefal buildup facies, moderate water energy), resulting in greater segmentation of the slope surface.

Parasequence 2 (PS2; Figs. 14, 22)
An initial transgression of fifth-order magnitude characterizes the first stage of the second parasequence of Zechstein sequence ZS3. Simulations of the south Oldenburg area show that this transgression reworks the last remaining PS1 shallow-marine carbonates in slope position, but creates new high-energy carbonates in slope position somewhat above the highest sea-level point of PS1. Due to the slow sea-level rise, which follows the initial sea-level rise of PS2, these carbonates also are reworked and resedimented. As a consequence, the progradation in Northwest Germany caused by sediments of the second parasequence is very small. The simulation of the eastern German high-relief off-platform high (and the platform) setting shows a continuation of the PS1 development. Somewhat above the PS1 sea level, the shallow-marine carbonates of PS2 are deposited (upper part of Ca2 cycle Ia), and, together with the PS1 carbonates, build a stacked pattern of shallow-marine carbonates (reefal buildups) in upper-slope position. The high tectonic subsidence rates, which mainly influence sedimentation in slope realms of the high-relief off-platform highs, lead to a rapid "downward movement" of those shallow-marine carbonates. Consequently, these carbonates are not exposed to high erosional energy for a long time and are not reworked. The water energy in general seems to be lower in easterb Germany, and therefore the early consolidated shallow-marine carbonates resist erosional processes of the subtidal environment. Because erosion mainly affects unconsolidated sediments in upper and middle parts of the slope, the rate of progradation is similiar to that in Northwest Germany.

Parasequence 3 (PS3; Figs. 15, 23)
Parasequence 3 builds up Ca2 cycle I (Ca2-cycle Ib) and its top (PSB3) was picked, for convenience, at the maximum flooding surface (mfs). This gives PS3 a duration of about 200 ka, which is about 30–50 ka longer than an average parasequence duration in Ca2 time. In eastern Germany, sea level during PS3 is characterized by a gradual, short-term rise and a smooth, long-term sea-level fall, which results, due to the estimated subsidence rates, in an overall relative sea-level fall of several meters. Facies patterns show no aggradational or progradational trends and produced sediments are completely reworked during the following transgression, which includes the mfs of Zechstein sequence ZS3. Simultaneously, the preserved high-energy carbonates of PS1 and PS2 are sealed (drowned) by upper-slope sediments deposited in the final stages of PS3. Reaching the mfs, the first carbonate sediments are deposited on top of the high-relief off-platform high built by the Werra-Anhydrite (A1). Simulation results show that these sediments belong completely to intertidal and shallow-subtidal facies

types. In platform position, typical facies successions in cores of both study areas show transitions from karstified A1 anhydrites to high-energy carbonates, indicating transgressive deposits with carbonate and anhydrite clasts overlain by shallow-subtidal or intertidal facies.

In contrast to the eastern German setting, the northwest German sea level shows a rapid and more significant rise at the beginning of PS3. The shallow-water carbonates of the preceding parasequence are reworked during this initial transgression. Due to moderate subsidence rates, the following sea-level fall is of small magnitude and initiates the reworking of main parts of the PS3 shallow-water carbonates. At the end of PS3 sea level rises again and, with flooding of the karstified top of the A1 platform, the mfs of Zechstein sequence ZS3 is established.

4.2
Parasequences 4–7 (PS4–PS7; Figs. 16, 17, 24, 25)

According to simulation results, parasequences 4–7 are generated during a time period of about 600 ka. They represent the highstand systems tract (HST) of Zechstein sequence ZS3 and their platform sediments form the main reservoir rocks within the Ca2. The impact of two fourth-order cycles causes two main shallowing-upward trends (Ca2 cycle II and Ca2 cycle III). The first two parasequences preserved in platform position are PS4 (Ca2 cycle IIa) and PS5 (Ca2 cycle IIb). The simulation of the East German high-relief off-platform highs shows that the first notable aggradation on top of the affected off-platform highs occurs during PS4, after the establishment of the mfs. The regressive phase of sea level during PS4 is characterized by a smooth shape and quite short duration, which leads to the generation of only thin intertidal successions on top of PS4. Although the fifth parasequence (PS5) during Ca2 high-stand systems tract begins with a short-term and smooth transgressive stage, the following sea level fall has a considerable impact on sedimentation and results in an erosional surface at the end of PS5. The main reason for this drop in sea level is the coincidence of regression phases of both fourth and fifth order. Starting with PS6 (Ca2 cycle IIIa) the fourth and fifth orders coincide in their transgressive stages, leading to a long-term sea level rise during which the briefly exposed platform sediments are flooded again. Due to this sea level rise, PS6 consists mostly of thick, grain-dominated shallow-subtidal sediments generated during a catchup phase of sedimentation. Although accumulation accelerates in this situation and results in aggradation of several meters up to several tens of meters, the progradation rate remains low, mainly because of unfavorable current conditions. This process continues for the remainder of Zechstein sequence ZS3-time (upper part of PS6 and the entire PS7). Therefore, in most areas on top of the high-relief off-platform highs, sedimentation no longer outpaces sea level during the highstand systems tract of Zechstein sequence ZS3. The occurrence of a type-1 sequence boundary (Zechstein sequence boundary ZSB4) on top of PS7 marks the end of the highstand systems tract of Zechstein sequence ZS3 and introduces Zechstein sequence ZS4 with a major drop in sea level (Figs.10, 11).

In the northwest German platform setting, the general sedimentation patterns are similiar to those of eastern Germany, given that the sequence stratigraphic framework and the reconstructed sea level have the same characteristics. However, due to lower water energy, lower subsidence rates and higher sediment production rates, depositional results vary to some extent. Progradation and aggradation rates are much higher in NW Germany than in eastern Germany and sedimentation nearly always keeps pace with sea-level during the HST of Zechstein sequence ZS3. Consequently, the four shallowing-upward trends (PS4 to PS7) caused by fifth-order sea-level fluctuations during the highstand systems tract of Zechstein sequence ZS3 are clearly visible in Ca2 platform sediments of Northwest Germany, and both vertical and lateral facies successions show intensive interfingering (Fig. 5). The sea level lowstand at parasequence boundary PSB5 affects Ca2 platform sediments for only a short time. The subsequent rise of sea level at the initial stage of PS6 is not as impressive as in eastern Germany. Simulation results in both study areas indicate that the end of Zechstein sequence ZS3 is reached after about 900 ka of sedimentation.

4.3
Parasequences Ca2/A2-1 and Ca2/A2-2 (Figs. 18, 26)

The first cycle of Zechstein sequence ZS4 (Ca2/A2-cycle I) contains the two parasequences Ca2/A2-1 and -2 and builds up the main parts of the LST of Zechstein sequence ZS4. Being superimposed on the regressive part of the simulated third-order cycle, both parasequences are generally characterized by a major and rapid drop in sea level. Only the intercalated marine flooding surface, which marks the start of parasequence Ca2/A2-2, indicates a short-term sea level rise. However, due to its position on the Ca2 slope and the following slow transgression phase, this surface is not preserved in the sedimentary record.

The East German high-relief off-platform highs are affected by this major drop in relative sea level (about 30–50 m) during the LST of Zechstein sequence ZS4, resulting in erosional processes mainly in the realm of the Ca2 platform. In addition, minor erosion and reworking takes place in upper slope sediments. The products of these processes are resedimented at the base of slope and build up, together with in-situ carbonates, a lowstand wedge several meters thick.

In contrast to this, the drop in relative sea level at the initial stage of Zechstein sequence ZS4 has a more dramatic impact in the northwest German setting. Due to lower subsidence rates, sea level drops about 50–70 m in this area. Besides this, the sea-level rise at the beginning of parasequence Ca2/A2-2 leads to intensive reworking and nearly in-situ resedimentation of the highstand systems tract of Zechstein sequence ZS4. Being superimposed on a third-order sea level rise, the transgression continues throughout the second parasequence, resulting in continued reworking (and in-situ resedimentation) of most of the prograded Ca2-platform sediments. At the base of the LSW, this leads to a distinct facies contact of upper slope mudstones and high-energy shoal and bar sediments containing reworked intraclasts (Fig. 8B). This erosive contact at the base of the LSW also can be recognized in cores, and is even more obvious in a more platformward position because facies change from upper slope mudstones to intertidal sediments (tidal flat or algal tidal flat; Fig. 8A).

As suggested by Strohmenger et al. (1996a), the slow sea-level rise during the late LST also causes a gradual but important diagenetic change in previously unaffected landward sediments on the Ca2 platform (Fig. 9). Due to an inferred change in water chemistry during the late LST, the uppermost layers of the Ca2-highstand carbonates are nearly completely altered to anhydrites of sabkha type (alterations of anhydrite and dolomite). This suggests that the sabkha-type anhydrites and the upper parts of the Ca2-LSW carbonates are synchronously generated, and their formation mainly depends on their paleotopographic position.

4.4 Parasequence Ca2/A2-3 (Figs. 19, 27)

The rise in sea level continues through the third parasequence of Zechstein sequence ZS4. This parasequence belongs to the second Ca2/A2-cycle and generates the uppermost part of the Ca2 lowstand wedge, which is bound by the Ca2 transgressive surface (TS) of Zechstein sequence ZS4. This marine-flooding surface marks the end of Ca2 time and is overlain by the first sedimentary anhydrites (salina-type anhydrites) of the A2 (Basal Anhydrite, Fig. 9). The simulation ends in the late stages of parasequence Ca2/A2-3 and shows only the first 70 ka of A2 sedimentation.

In both Northwest Germany and eastern Germany, the transgressive surface TS of Zechstein sequence ZS4 (Figs. 18, 26) marks the change from carbonate to evaporitic sedimentation (salina-type anhydrites; Fig. 9). A main reason for this change in mineralogy is thought to be a change in water chemistry. Given the reconstructed sea level, the formation of main parts of the A2 and the entire overlying Na2 salt takes place during the regressive stage of a third-order sea level cycle superimposed on the regressive trend of the second-order cycle. Consequently, this leads to a general lack of freshwater input into the Southern Permian Basin and favors the deposition of evaporites. In addition, a very slow, but long-term, sea level rise might also support evaporitic sedimentation. This is reflected in the simulation, because the initial formation of anhydrites at the beginning of A2 time relates to a very slow sea-level rise during the first stages of Ca2/A2 cycle IIa.

In Northwest Germany, the basinward parts of the Ca2 platform affected by LST sedimentation experienced no diagenetic changes during the late LST, and therefore show a transition from mostly intertidal Ca2 carbonates to anhydrites of salina type (pure anhydrites). In contrast, those parts of the Ca2 platform affected by diagenetic changes during the late LST show a transition from Ca2 sabkha-type anhydrites to salina-type anhydrites of early A2 (Fig. 9). The eastern German high-relief off-platform highs, which are generally affected by higher subsidence rates, do not show excessive alterations of Ca2 strata during the sea-level rise at the end of the LST of Zechstein sequence ZS4. The transition in these areas commonly takes place from mostly intertidal Ca2 carbonates to pure anhydrites, indicating a salina environment.

5
Results and Conclusions

The high-resolution forward stratigraphic modeling of the Ca2 platform in the northwest German south Oldenburg area and the Ca2 high-relief off-platform highs in the eastern German southeast Brandenburg area gives insights on factors ruling sedimentation during Ca2 time.

As shown by the results of detailed core description and supported by the simulation results, the general facies distribution and, more importantly, the reservoir facies distribution within the Ca2 are mainly driven by the topography of the underlying anhydrite (Werra-Anhydrite, or A1). In addition, the modeling shows that differential subsidence rates (Sect. 3.2) contribute to the shaping of the different depositional systems during Ca2 time. The assumed carbonate production curves (Sect. 3.4; Fig. 3) show substantial differences between Northwest Germany and eastern Germany during Ca2 deposition. northwest German Ca2 carbonate production rates are assumed to be considerably higher than those in eastern Germany, and show a substantial decrease at the end of the HST of Zechstein sequence ZS3 (after about 900 ka of Ca2 sedimentation). Since this decrease cannot be recognized in the investigated eastern German area, it is thought to be linked to the different paleogeographic positions of each study area. In the case of the eastern German southeast Brandenburg area, the subsidence rates also contribute to the preservation of a highly dissected morphology characterized by the existence of a platform and two types of off-platform highs. Interacting with sea level, different subsidence rates in this region also influence the deposition and preservation of primary reservoir sediments at the upper slope area during the first two parasequences (PS1 and PS2, Sect. 4.1).

The drop in sea level which marks the end of Zechstein sequence ZS3 is slower and of minor impact in areas directly connected to the Tethys, which acts like an open marine system. Even minor sea-level rises add a certain amount of normal marine water to the restricted water body near the connection site and the impact of evaporation on the water chemistry is weakened considerably in these places. However, the impacts of such processes depend on the direction and power of water flow, and generally vanish with increasing distance from the open marine system. This idea is supported by comparing Ca2 sediments in the two study areas since the southeast Brandenburg area is situated considerably nearer to the South Polish opening to the Tethys than the northwest German South Oldenburg area. The higher water currents in this southeastern-most part of the Southern Permian Basin contribute to the buildup of isolated carbonate bodies (off-platform highs) and widely prevent the accumulation of sediments in slope and basin areas of these islands.

The late LST, TST and HST of Zechstein sequence ZS3 are built up of seven shallowing upward cycles (parasequences), which mark minor flooding events and are characterized by clearly asymmetrically shaped (sawtooth) cycles with their maximum after about the first quarter of their time (Table 2). This results in a rapid sea-level rise followed by a long time of slow regression during each parasequence. In sediment production terms, this means a catchup phase (production of shallow subtidal sediments) after a lag time due to the rapid rise at the beginning of each parasequence. About three quarters of the duration of each parasequence is dominated by regression and is represented mostly by various intertidal deposits. Carbonate production rates during this stage are assumed to decrease progressively due to the lack of normal marine water input and high evaporation rates.

Modeling has also shown that vertical differences in water chemistry play a major role in Ca2 sedimentation patterns in the Southern Permian Basin. Modeling results indicate a significant stratification of the water body, which results in the sedimentation/precipitation of different lithologies during the same time. One example of this is the precipitation of anhydrites in some deeper water areas during parts of Ca2 time when carbonates were forming in shallow-water areas.

Core studies suggest a sea level curve with at least four different orders of fluctuations (second to fifth) with asymmetric shapes for the third, fourth and fifth-order fluctuations. Due to the lengy durations of the first and second-order sea level fluctuations there is not much evidence for the shape of these curves during the simulated time period. However, a second-order sea level highstand is thought to have caused the sedimentation of the Copper Shale (T1; Strohmenger et al. 1996a). As a consequence, the second-order sea level is characterized by a regressive development during Ca2 time. The asymmetric shape of the third, fourth and fifth-order sea level curves is thought to be caused mainly by the isolated position of the Southern Permian Basin, which was connected to the open marine system Tethys in the area of southeast Poland. Both the investigated sediment successions and the results of the stratigraphic modeling suggest a widely restricted communication between both marine systems, possibly due to barriers, which have influenced in- and outflow rates considerably. With increasing distance from the open-marine system, the imprints of these flooding events show an increasing delay in time and a decreasing potential of preservation of at least sixth- and higher-order cycles, which has to be taken into account in simulation procedures. The asymmetric shape of the reconstructed sea level curves are additionally emphasized by restricted outflow rates, which led to slow and long-lasting sea level falls.

Although the simulations of the Ca2 sedimentary history cannot prove cause-and-effect relationships, they give new insights into the highly complex Ca2 sedimentation system. The sensitivity to variations in input data offers possibilities to test different depositional models and the core-analysis-based Ca2 sequence stratigraphic framework. In addition, it allows estimation of the relative influences of sediment-controlling parameters during Ca2 time in different paleotopographic settings. The quantification of most sedimentation-relevant factors leads to an improved understanding of the complex sedimentation history of the Ca2 and enables a considerably improved prediction of reservoir facies distribution. Finally, it reduces exploration risks, even in largely uninvestigated areas with few available data.

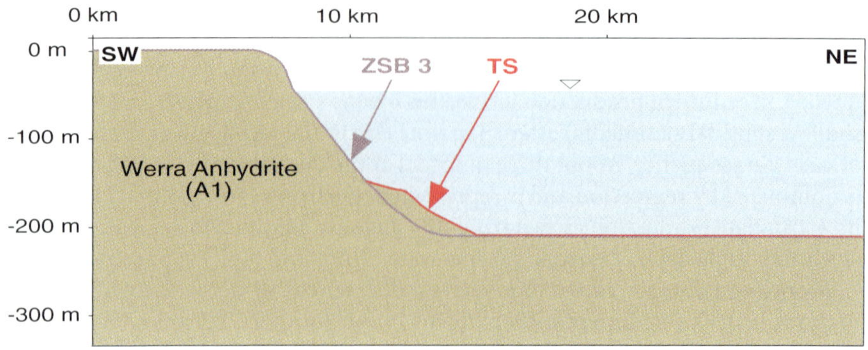

Fig. 12 Basement of Ca2, NW Germany

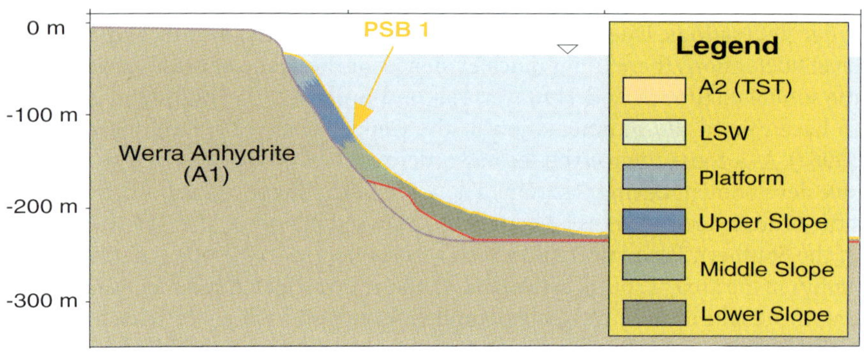

Fig. 13 PS 1, NW Germany

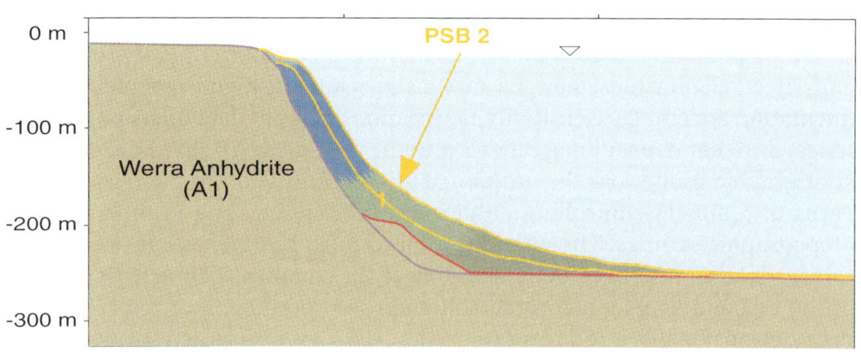

Fig. 14 PS 2, NW Germany

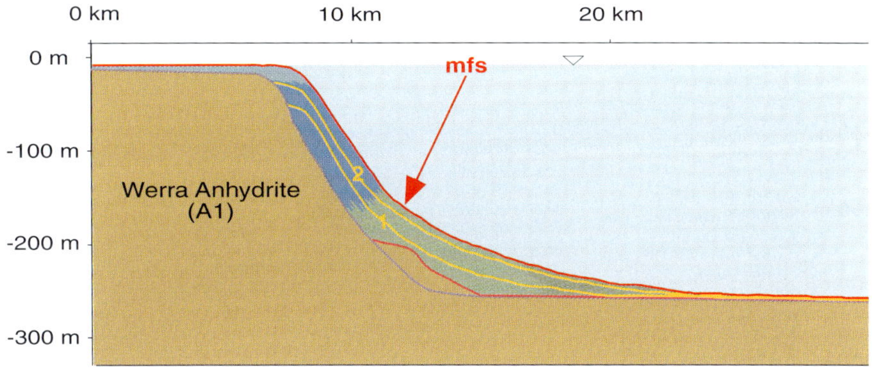

Fig. 15 PS 3 (mfs), NW Germany

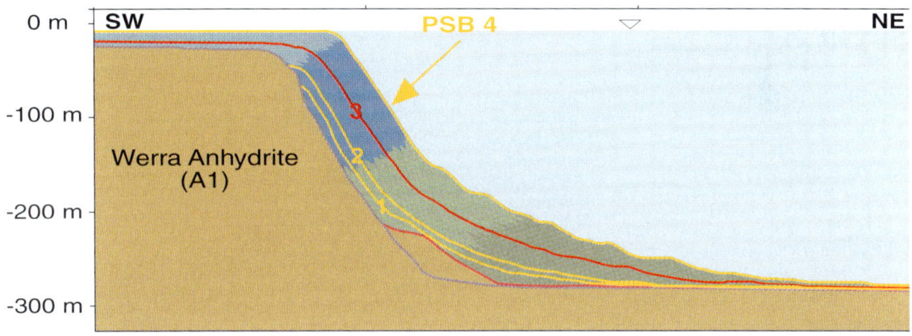

Fig. 16 PS 4, NW Germany

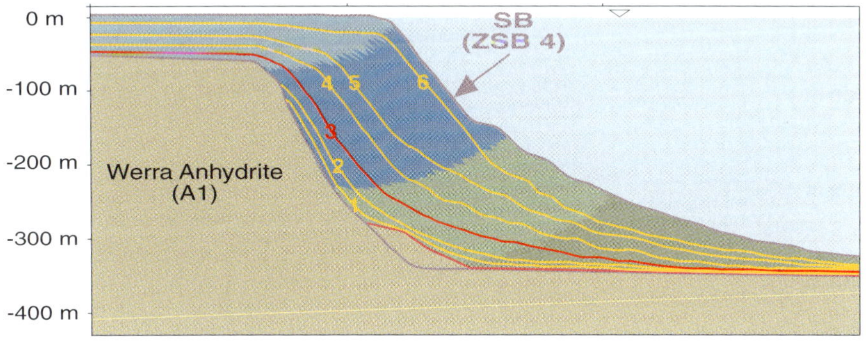

Fig. 17 PS 7 (ZSB 4), NW Germany

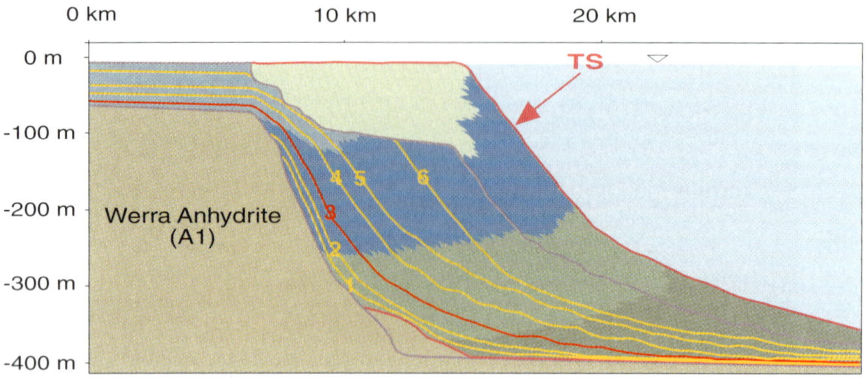

Fig. 18 Top Ca2, NW Germany

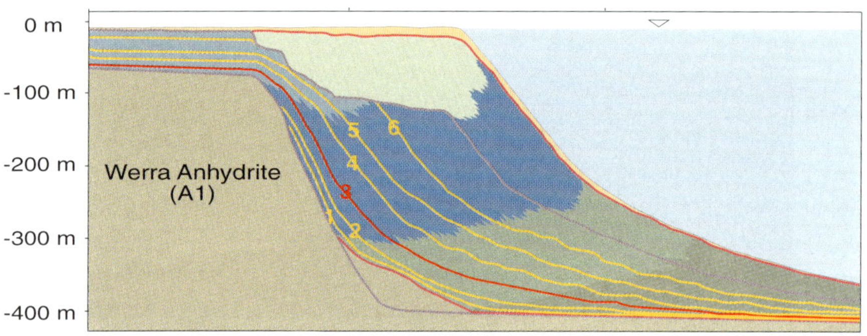

Fig. 19 Ca2 LSW and A2 LST/eTST, NW Germany

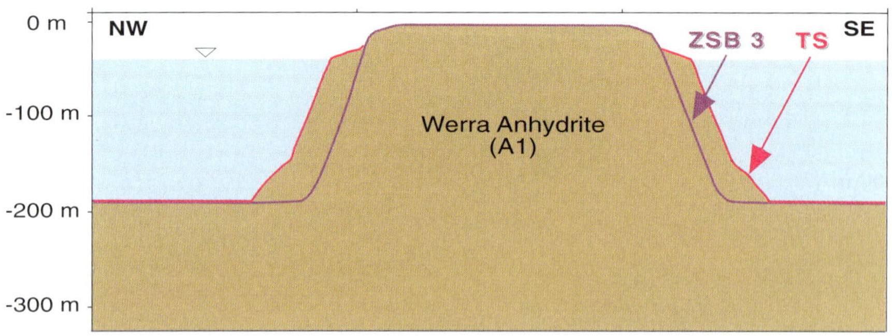

Fig. 20 Basement of Ca2, East Germany. See Fig. 13. For legend

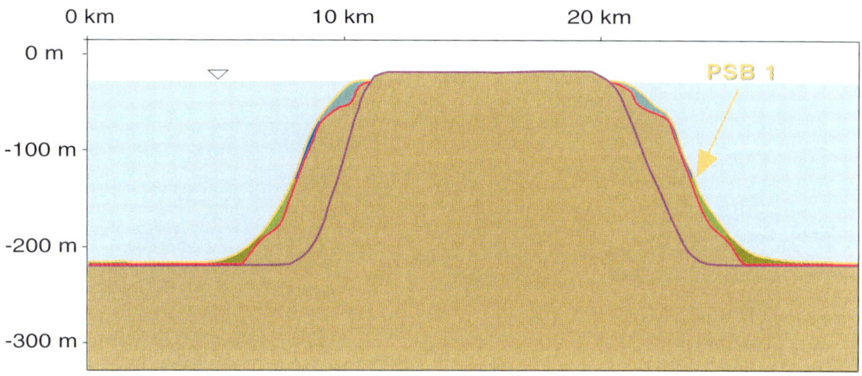

Fig. 21 PS 1, East Germany

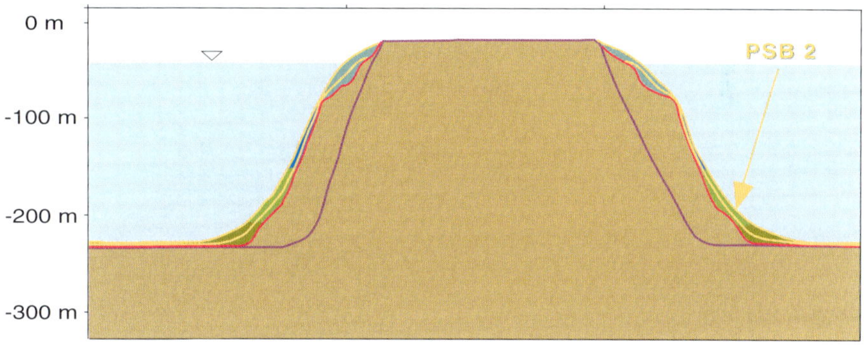

Fig. 22 PS 2, East Germany

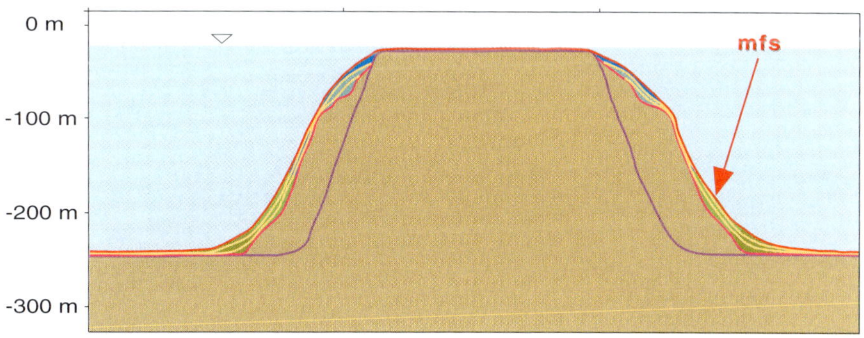

Fig. 23 PS 3 (mfs), East Germany

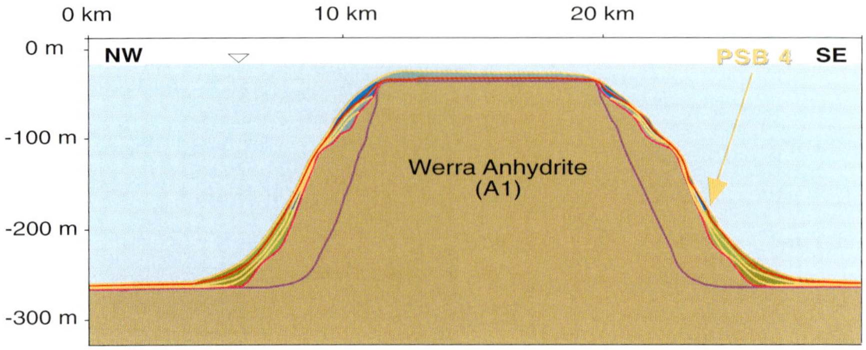

Fig. 24 PS 4, East Germany

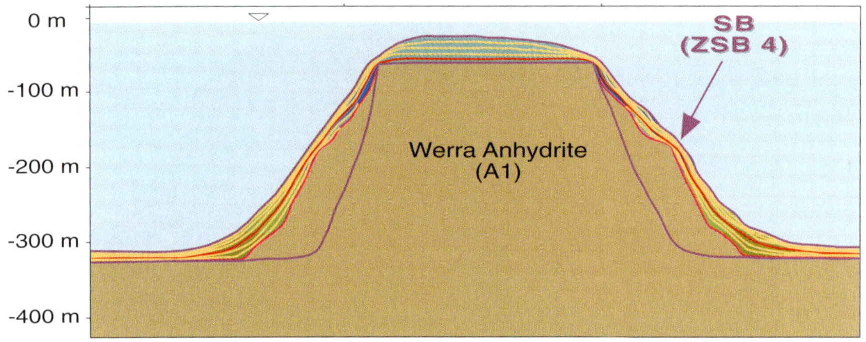

Fig. 25 PS 7 (ZSB 4), East Germany

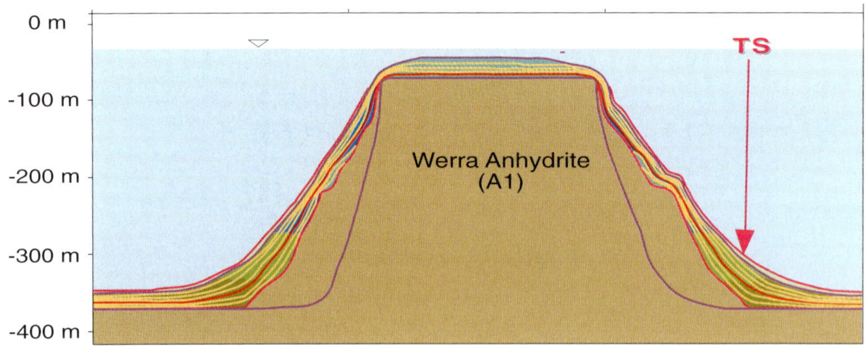

Fig. 26 Top Ca2, East Germany

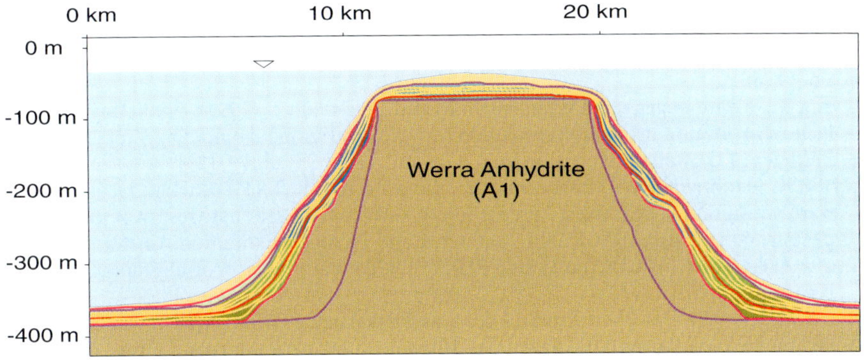

Fig. 27 Ca2 LSW and A2 LST/eTST, East Germany

Acknowledgements

PHIL is a trademark of PetroDynamics Inc. (formerly MarcoPolo Software).
BEB Erdgas and Erdöl GmbH, Hannover / Germany is thanked for financial and data support.
Erdoel-Erdgas Gommern GmbH, Gommern / Germany is thanked for providing data and the permission for publication.
For valuable discussions the authors thank Thomas Aigner, Scott A. Bowman, Werner Dauben, Chris Kendall, Werner Leu, Eberhard Plein, Ingo Steinhoff and Walter L. Watney.
We also thank Karen M. Love for her review of this paper.
We gratefully acknowledge the management of BEB Erdgas and Erdoel GmbH, as well as the W.E.G. publication committee, for permission to publish this paper.

References

Below, A. (1992): Fazies- und geochemische Diagenesestudie im Zechstein 2-Karbonat (Ca2) Nordwestdeutschlands.- Rep. Geol. Pal. Inst. Univ. Kiel, 60, 147 p.

Füchtbauer, H. (1964): Fazies, Porosität und Gasinhalt der Karbonatgesteine des norddeutschen Zechsteins.- Zt. dt. geol. Ges., 114, 484–531.

Füchtbauer, H. (1968). Carbonate sedimentation and subsidence in the Zechstein Basin (northern Germany).- in Müller, G. and Friedmann, G.M.(Eds.): Recent Developments in Carbonate Sedimentology in Central Europe.- Springer, Berlin, 196 - 204.

Harris, M.T. (1989): Modeling carbonate-progradation geometry and sediment-accumulation rates - a comparison of "MARGIN" results with field data.- in: Franseen, E.K. and Watney, W.L (Eds.): Sedimentary modeling: Computer simulation of depositional sequences.- Kansas Geol. Survey, Subsurface Geology Series, 12, Lawrence, Kansas, 11–13.

Huttel, P. (1989): Das Stassfurt-Karbonat (Ca2) in Süd-Oldenburg - Fazies und Diagenese eines Sediments am Nordrand der Hunteschwelle.- Göttinger Arb. Geol. Paläontol., 39, 94 p.

Imbrie, J. and Imbrie, K.P. (1979): Ice ages: Solving the mystery.- McMillan Press, New York, 224 p.

Kerans, C. and Fitchen, W.M. (1995): Sequence hierarchy and facies architecture of a carbonate-ramp system: San Andres Formation of Algerita Escarpment and Western Guadaloupe Mountains, West Texas and New Mexico.- Bureau of Economic Geology, Report 235, Austin/TX, 86 p.

Leyrer, K., Strohmenger, C., Rockenbauch, K. and Bechstädt, T. (1995): Predicting reservoir facies distribution using high resolution forward stratigraphic modeling (Upper Permian Zechstein 2 Carbonate, North Germany) (abstr.).- AAPG Bull., 79, 1231.

Leyrer, K., Strohmenger, C., Rockenbauch, K. and Bechstädt, T. (1996): High resolution forward stratigraphic modelling - prediction of Zechstein 2 Carbonate reservoir quality, Upper Permian (abstr.).- EAGE 58th Conference and Technical Exhibition, Amsterdam, L 043.

MARCO POLO SOFTWARE, Inc. (1994): PHIL™ 1.5 Manual - Tools for stratigraphic analysis.- Houston, TX, U.S.A., 110 p.

Mausfeld, S. (1987): Der Plattformrand des Stassfurtkarbonats (Ca2) südlich von Oldenburg: Sedimentologie, Fazies und Diagenese.- Unpubl. Ph.D Thesis, Marburg, 176 p.

Mausfeld, S. and Huttel, P. (1991): Facies development on a prograding ramp: The Stassfurt Carbonate Formation (Ca2) of South Oldenburg (NW Germany).- Zbl. Geol. Paläont., Teil I/4, 1057–1072.

Mausfeld, S. and Zankl, H. (1991): Sedimentology and facies development of the Stassfurt Main Dolomite in some wells of the South Oldenburg region (Weser-Ems area, NW Germany).- in: Peryt, T.M. (Ed.): The Zechstein Facies in Europe.- Lect. Notes Earth Sci., 10, 123–141.

Menning, M. (1995): A numerical timescale for the Permian and Triassic periods: An integrated time analysis.- in: Scholle, P.A.; Peryt, T.M. and Ulmer-Scholle, D.S. (Eds.): The Permian of Northern Pangea. Vol.1.- Springer, Berlin, 77–97.

Milankovitch, M. (1941): Kanon der Erdbestrahlung und seine Anwendung auf das Eiszeitenproblem.- Acad. R. Serbe Spec. Ed., 133, 633 p.

Peryt, T.M. (1986): Zechstein Stromaria (= Archaeolithoporella)-cement reefs in Thuringia.- N. Jb. Geol. Paläont. Mh., 1986, 307–316.

Plint, A.G., Eyles, N.G., Eyles, C.H. and Walker, R.G. (1992): Control of sea-level change.- in: Walker, R.G. and James, N.P. (Eds.): Facies Models. Response to sea level change.- Geol. Assoc. of Canada, Newfoundland, Canada, 15–25.

Revelle, R.R. (1990): Sea-Level Change.- Studies in Geophysics, Washington, D.C., National Academy Press, 246 p.

Richter-Bernburg, G. (1985): Zechstein-Anhydrite: Fazies und Diagenese.- Geologisches Jahrbuch, A 85, 3–82.

Sengör, A.M.C. (1995): Sedimentation and tectonics of fossil rifts.- in: Busby, C.J. and Ingersoll, R.V. (Eds.): Tectonics of Sedimentary Basins.- Blackwell Science, 53–117.

Schlager, W. (1993): Accommodation and supply - a dual control on stratigraphy.- Sed. Geol., 86, 111–136.

Steinhoff, I. (1993): Komponentenanalyse und ihre Beziehung zur Porositätsentwicklung im Zechstein 2 Karbonat (Oberperm Nordwestdeutschlands).- Unpubl. Dipl.-Thesis, Hamburg, 146 p.

Strohmenger, C., Jäger, G., Mitchell, J.C., Love, K.M., Antonini, M., Gast, R., Van de Pol, E., Rockenbauch, K. and Steffan, M. (1993a): An integrated approach to Zechstein Ca2 Carbonate reservoir facies prediction in the South Oldenburg area, Upper Permian, Northwest Germany (abstr.).- AAPG Bull., 77, 1668.

Strohmenger, C., Love, K.M., Mitchell, J.C. and Rockenbauch, K. (1993b): Sedimentology and Diagenesis of the Zechstein Ca2 Carbonate, Late Permian, Northwest Germany (abs.).- AAPG Bull., 77, 1668.

Strohmenger, C., Voigt, E. and Zimdars, J. (1993c): Einfluß von Eustasie und Paläorelief auf die sedimentologische und diagenetische Entwicklung der Zechstein 2 Karbonate (Ober-Perm, Nordost-Deutschland).- Erdöl, Erdgas, Kohle, 109, 445–450.

Strohmenger, C., Antonini, M., Jäger, G., Rockenbauch, K. and Strauss, C. (1996a): Zechstein 2 Carbonate reservoir facies distribution in relation to Zechstein sequence stratigraphy (Upper Permian, Northwest Germany): An integrated approach.- Bull. Centres Rech. Explor.-Prod. Elf-Aquitaine, 20, 1–35.

Strohmenger, C., Voigt, E. and Zimdars, J. (1996b): Sequence stratigraphy and cyclic development of Basal Zechstein carbonate-evaporite deposits with emphasis on Zechstein 2 off-platform carbonates (Upper Permian, Northeast Germany).- Sed. Geol., 102, 33–54

Strohmenger, C. and Strauss, C. (1996): Sedimentology and palynofacies of the Zechstein 2 Carbonate (Upper Permian, Northwest Germany): Implications for sequence stratigraphic subdivision.- Sed. Geol., 102, 55–77.

Tucker, M. E. (1991): Sequence stratigraphy of carbonate-evaporite basins: Models and application to the Upper Permian (Zechstein) of northeast England and adjoining North Sea.- J. Geol. Soc. London, 148, 1019 - 1036.

Vail, P.R., Audemard, F., Bowman, S.A., Eisner, P.N. and Perez-Cruz, C. (1991): The stratigraphic signatures of tectonics, eustasy and sedimentology.- in: Einsele, G., Ricken, W. and Seilacher, A. (Eds.): Cycles and events in stratigraphy.- Springer, Berlin, 617–659.

Van Der Baan, D.(1990): Zechstein reservoirs in The Netherlands.- in: Brooks, J. (ed.): Classic petroleum provinces.- Spec. Publ. Geol. Soc. London, 50, 379–398.

Van Wagoner, J.C., Mitchum, R.M., Campion, K.M. and Rahmanian, V.D. (1990): Siliciclastic sequence stratigraphy in well logs, core and outcrops: Concepts for high-resolution correlation of time and facies.- AAPG Methods in Explor. Ser., 7, 55 p.

Wanless, H.R. (1991): Observational foundation for sequence modeling.- in: Franseen, E.K.; Watney, W.L.; Kendall, C.G.St.C. and Ross, W. (Eds.): Sedimentary modeling: Computer simulations and methods for improved parameter definition.- Kansas Geol. Survey Bull., 233, 43–62.

Ziegler, M.A. (1989): North German Zechstein facies patterns in relation to their substrate.- Geol. Rdsch., 78, 105–127.

Ziegler, P.A. (1990): Tectonic and palaeogeographic development of the North Sea rift system.- in: Blundell, D.J. and Gibbs, A.D. (Eds.): Tectonic evolution of the North Sea rifts.- Clarendon Press, Oxford, 1–36.

Validation of Sediment Accumulation Regions in Kansas, USA

W. L. Watney, J. Kruger, J. C. Davis, J. Harff, R. A. Olea, and G. C. Bohling

1 Introduction

Recognition of recurrent structural deformation and sedimentation in cratonic shelves is not new and these processes affect these areas worldwide (Merriam and Förster, submitted b). "Plains-type folds" have been described in the literature since the turn of the century. Structural reactivation continues to be documented in the northern Midcontinent, USA and refined in timing, spatial distribution, and causal relationships (Gerhard et al. 1982; Baars and Watney 1991; Berendson 1997; Watney et al. 1997; Merriam and Förster; submitted b). This chapter further documents km-scale reactivation features.

Regionalized classification based on thicknesses of genetic stratigraphic units was used to differentiate sediment accumulation regions (SARs) on the Kansas shelf, bordering the Arkoma and Anadarko basins (Fig. 1a,b); Watney et al. 1995; 1997). Regionalized classification is a statistical procedure previously described by Harff and Davis (1990), that divides an area into regions that are as homogeneous as possible while as distinct as possible from other regions. Each of the 15 different km-sized regions is defined by a similar thickness of six genetic stratigraphic intervals, including four successive Late Pennsylvanian (late Carboniferous) high-frequency, genetic stratigraphic units (GSUs) (Fig. 2) that are part of the Nuyaka Creek genetic set (GS) (Fig. 2). Two other genetic units included in the regionalized classification are divisions of the Muncie Creek genetic set that overlies the Nuyaka Creek GS. The divisions consist of a lower carbonate-dominated interval and an upper clastic-dominated section. These divisions reflect significant changes in internal lithofacies of GSUs included in the Munice Creek GS.

The GSUs range in thickness from 10 to 150 ft (3 to 50 m). They are bounded by thin flooding units, commonly overlain by a thin, radioactive condensed section (black shale). The succeeding lithofacies depends on shelf location and vary from clastics to carbonates indicating an upward shallowing trend culminating in subaerial exposure and often paleosol development. Subaerial exposure affects most of the shelf except along its lowest reaches (Watney et al. 1989). GSUs are also associated with pronounced shifts of lithofacies across the shelf and result in non-Waltherian facies successions. Spatial variations in the lithofacies

Fig. 1 a Location map of Kansas in US, b Regionalized map showing sediment accumulation regions (SARs) accompanied by *bars* showing average thicknesses of stratigraphic units and genetic sets used in the classification

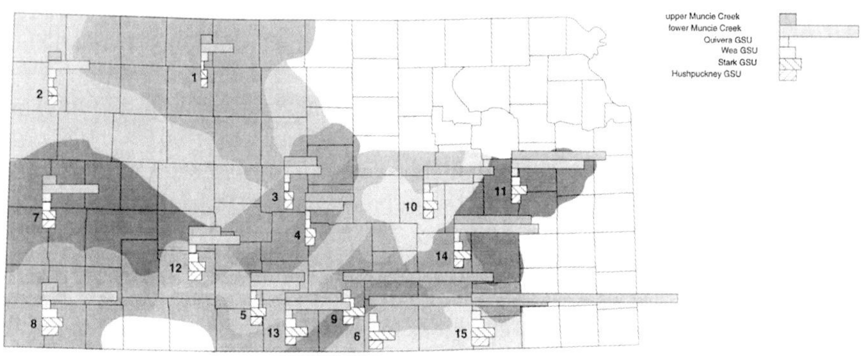

have been demonstrated across the shelf and are believed to reflect (1) changes in shelf elevation and what sea-level history is "seen" during marine innundation at those locations, (2) changes in rates of subsidence and sediment preservation potential, and (3) sediment supply (Watney et al. 1989).

The GSU corresponds closely to cyclothems traditionally believed to be controlled by glacio-eustacy (Heckel 1977; Watney 1980; Boardman and Heckel 1989). The condensed units are believed to reflect rapid and usually deep, regional innundation of the shelf resulting in loss of carbonate environments and slow accumulation of organic rich shales (Watney et al. 1995). The radioactive condensed sections are the most persistent interval on well logs and thus provide practical markers and are used to distinguish GSUs. Subaerial exposure often resides on subtidal deposits indicative of forced regression. These "coherent" packages of strata supply some of the thickness data for the analysis. GSUs, originally distinguished from core and outcrop studies, were correlated among 3096 wireline wells, covering 116000 km^2.

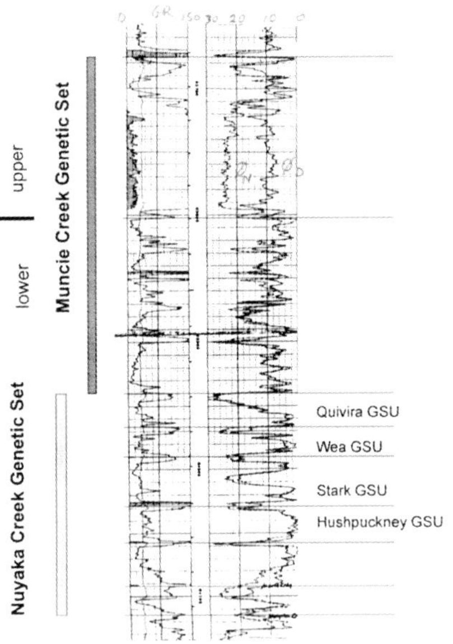

Fig. 2 Well log showing genetic stratigraphic units (GSUs) and genetic sets (GSs) used in the analysis

A genetic set consists of five to seven genetically related GSUs ranging in thickness from 100 to 300 ft (30 to 90 m). Genetically related GSUs forming the sets have been previously established for the Upper Pennsylvanian through (1) examination of stratal stacking patterns on the shelf margin in southern Kansas, bordering the Anadarko and Arkoma basins and (2) analyzing profiles of Th:U ratios that serve as redox (oxidation/reduction) profiles. The Th:U ratio serves as a proxy for the level of innundation of the shelf, e.g., subtidal deposits that were more reducing (more organic rich) and less oxidizing paleosols are indicative of greater innundation. The results of these Th:U profile analyses and the stratal stacking drew the same conclusions (Watney et al. 1995).

GSUs and GSs serve as alternatives for closely associated, temporally distinct, unconformity-bounded depositional sequences and sequence sets. Subaerial exposure surfaces can be easily identified in cores and surface exposures within these rocks and have been recognized to be widely correlatable. The subaerial exposure surfaces are usually within a meter of the flooding surface and condensed sections (Watney et al. 1995). However, the exposure surfaces cannot be reliably identified from well logs outside of lithology tools such as neutron-density, and photoelectric curves used in combination with spectral gamma ray (Bohling et al. 1996). Wireline logs used in the study are primarily gamma ray and neutron-density porosity. Thus, it is necessary in the regional subsurface investigation to rely on recognition of flooding units and condensed sections as boundaries of GSUs rather than depositional sequences. Previous mapping and analysis of these genetic stratigraphic packages resolved meaningful patterns and trends in sedimentation and associated shelf configuration because of their close approach to temporal distinction and genetic coherency (Watney et al.

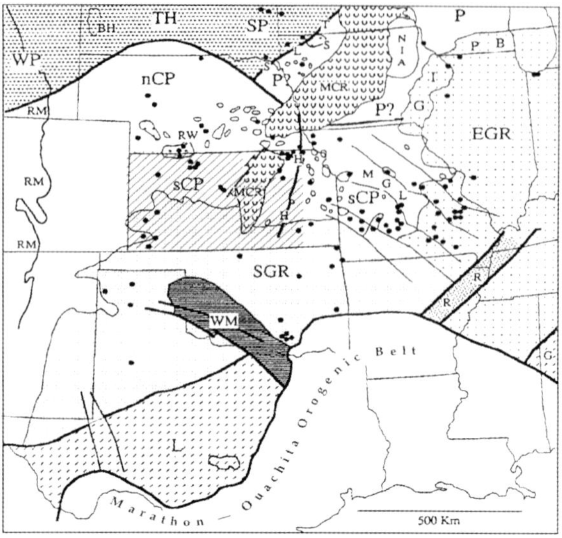

Fig. 3 Major Precambrian geologic features of the western Midcontinent (after Van Schmus et. al. 1993). Study area is shown by *cross-hatched pattern*. Also indicated is approximate location of Paleozoic orogenic belt

1995). However, it is difficult to integrate isopachous data and discern significant, persistent patterns among the maps.

Regionalized classification provided a way to establish spatial patterns in thickness changes in a manner similar to manually overlaying transparencies of isopachous maps and looking for coincidences (Watney et al. 1997). The methodology of regionalized classification applied to genetic units allows us to model large amounts of empirical stratigraphic data (Watney et al. 1997). However, the shelf elements (called sediment accumulation regions or SARs) defined by regionalized classification must be tested and validated. In particular, changes in internal stratigraphy and lithofacies need to be compared to the SARs. This is one of the objectives of this chapter; another is to substantiate correlations of SARs with the Precambrian basement using potential fields mapping.

Regionalization is based on the initial assumption that the spatial distributions of the data, in this instance the spatial variation of thicknesses, are discontinuous. This is in sharp contrast to contouring, where continuity is presumed. Because the approach is nontraditional, there is a need to examine the boundaries of the SARs to determine if boundary changes are not merely artifacts of the statistical procedure, in other words, to confirm that sharp boundaries do represent geological changes.

The SARs are referred to as sediment accumulation realms (Watney et al. 1997), because they are based strictly on thickness. Watney et al. (1997) showed that the SARs are closely correlated to three major Precambrian basement crustal provinces as classified by Van Schmus and others (1993) – Central Plains orogen, southern granite-rhyolite, and Midcontinent Rift (Fig. 3). Comparison of basement-related lineaments derived from gravity and magnetics with regionalized classification will help substantiate the apparent correlations with basement heterogeneity.

The challenge in genetic and sequence stratigraphy is to detail three dimensional changes to develop models of sedimentation for use in prediction. Models that can be applied to petroleum geology in a maturely developed province such as Kansas necessarily need to be three-dimensional and consider large volumes of data in order to be most useful and successful. Regionalized classification is an inverse procedure that extracts patterns that can be used to develop a tectono-stratigraphic model tailored to the local setting.

2 Methodology

The validation and testing of SARs and their boundaries is addressed using two approaches. First, a test is made of the significance of SARs and their boundaries in controlling the internal stratigraphy and lithologies associated with the stratigraphic interval used in the classification. This test is accomplished through the use of a regional, 300-km long color gamma-ray cross- section with each well supported with analyses of accompanying neutron-density porosity logs. This section crosses eight SARs.

Gamma ray logs were digitized from 25 wells with the interval of interest, ranging in thickness from 100 to 350 m . The natural gamma radiation is recorded by a potassium iodide crystal in the wireline tool that responds to radiation contributed by potassium, uranium, and thorium. The magnitude of the gamma radiation is associated with major lithologies – shale, sandstone, and carbonate rock.

Color is assigned to the gamma-ray intensities and presented as a cross-section to facilitate recognizing lithologic changes. The PfEFFER software is used to read digital LAS (log ascii standard) files and to construct the color cross-sections. Well logs are digitized by scanning the logs into TIFF files which were importing into a software package, Neuralog, used to digitize the log image on a screen. Digitized log ascii standard files were created that were read into the Excel-based log analysis software to generate the cross-sections.

A second approach in substantiating SARs is through their comparison with gravity and magnetic data. The digitized boundaries of the SARs were overlain on maps of the (1) residual Bouguer gravity map of Kansas with second-order regional trend removed and vertical sun-shading (Xia et al. 1995; Fig. 4) and (2) aeromagnetic map of Kansas, reduced to pole, and to a horizontal plane and with vertical sun-shading (Yarger 1983; Fig. 5). These potential fields data were imported into a workstation for image analysis and overlain with other data.

Regionalized classification is conducted as two fundamental phases, typification and regionalization. Typification is accomplished using cluster analysis, independent of the locations of the wells. Regionalization is conducted by estimating the probabilities that a specific well location should be assigned to each possible class, represented by an individual cluster. The sum of these probabilities will be 1, so the well location is certain to have some assignment. Maps are then made on the probability of membership of every location into each class. Every point on the map is assigned to the class for which the probability of member-

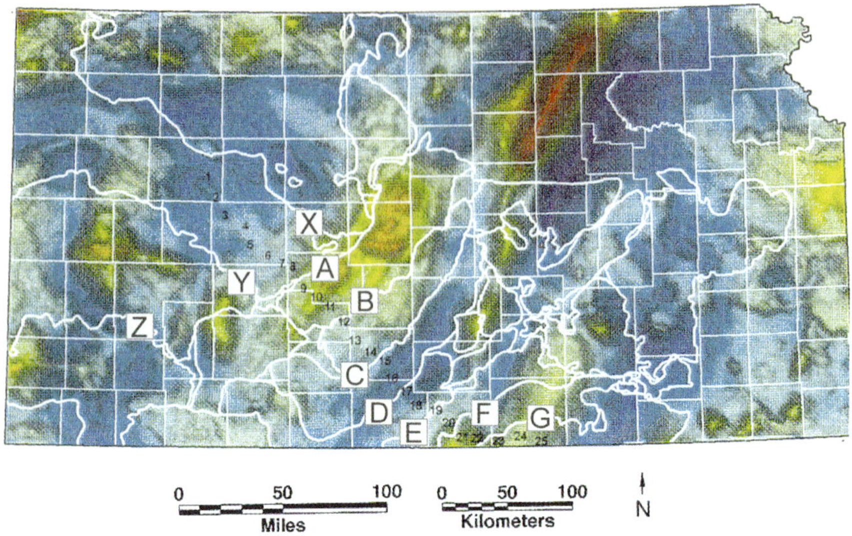

Fig. 4 Residual Bouguer gravity map of Kansas (2nd order polynomial surface removed) with overlays of SAR boundaries (*white*). Vertical sun-angle shading provides pseudo-relief on the gravity map. *Darker shading* indicates steeper gravity gradients. *Red* Highest gravity values; dark blue lowest gravity values. Grid for map from Xia et al. (1995; KGS Map M-41E). Differences in the gravity values are caused by density changes below the surface and reflect changes in rock type and/or structure. Some of the changes are related to vertical offset of the Precambrian basement surface and density changes within the Phanerozoic section, but most reflect different rock types within the basement. Some of the different rock bodies which cause the anomalies reach the basement surface, but some (particularly those with longer spatial wavelengths) do not. Higher gravity gradients indicated by *darker shading* represent the probable location of basement faults, shear zones or other lithologic boundaries between the rock bodies causing the different gravity anomalies, and/or vertical offsets within the basement. In many areas, SAR boundaries show an excellent correlation with the gravity anomaly map and in particular the higher gravity gradients. This suggests that Precambrian basement structures have been reactivated during the Phanerozoic and influence the location of the SAR boundaries. Wells 1 through 25 comprising cross section shown in Fig. 6 are labeled on this map. SAR boundaries are labeled *A* through *G* and *X*, *Y*, and *Z* in-vicinity of cross-section index

ship is the greatest. A map of class membership should form geographic regions containing only wells belonging to one class if the regionalization is valid. An invalid regionalization will produce regions containing mixtures of well classes.

Regionalized classification begins with a single region encompassing all of the data. In this study, regions were subdivided until a balance was achieved between the numbers of divisions and their geologic significance. The process of increasingly subdividing the study area produces regions that represent diminishing contrasts in thicknesses between each other. The procedure is subjective and subject to the bias of the classifier. An optimum number of regions could have been obtained statistically by maximizing the differences between the regions compared to variation within a region, but this has not yet been attempted.

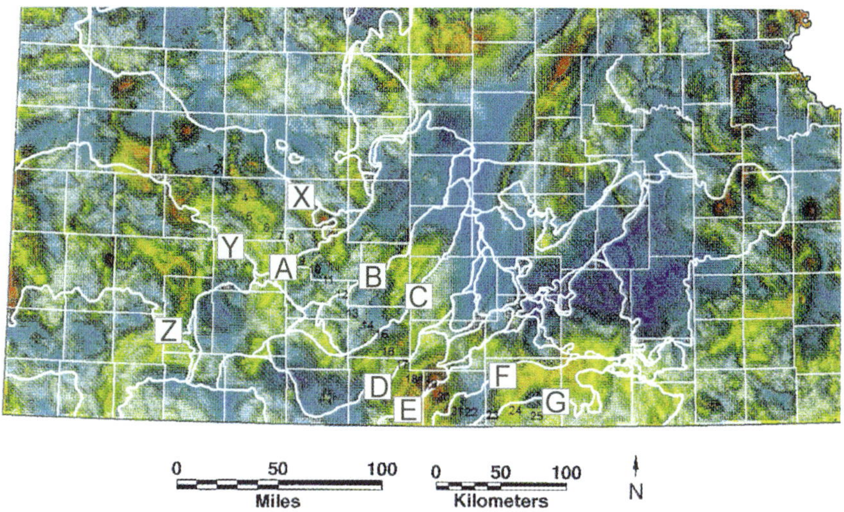

Fig. 5 Aeromagnetic map of Kansas reduced to the pole and reduced to a horizontal plane, with overlays of SAR boundaries. Vertical sun-angle shading provides pseudo-relief on the magnetic map. *Darker shading* indicates steeper magnetic gradients. *Red* Highest magnetic values; *dark blue* lowest magnetic values. Grid for map from Xia et al. (1995; KGS Map M–41D). Differences in the magnetic values are caused by magnetic susceptibility changes (variations in magnetic mineral content such as magnetite) below the surface and reflect changes in rock type and/or structure. Some of the changes are related to vertical offset of the Precambrian basement surface, but most reflect different crystalline rock units within the basement, particularly in the upper part of the basement. Since gravity and magnetics measure different physical properties there is not always a good correlation between the two. However, many anomaly boundaries and gradient trends are the same in both the magnetic and gravity maps. As in the gravity map, steeper gradients in the magnetic map correlate with faults, shear zones, or other lithologic boundaries within the upper part of the Precambrian basement. As in the gravity map, SAR boundaries show an excellent correlation with magnetic anomalies, particularly the higher magnetic gradients, suggesting further support for basement reactivation. The displays were made using ER Mapper software from Earth Resource Mapping. Wells *1 through 25* comprising cross section shown in Fig. 6 are labeled on this map. SAR boundaries are labeled *A* through *G* and *X*, *Y*, and *Z* in vicinity of cross-section index

In an extreme case one could have as many subdivisions as there are samples. Fifteen were chosen here as larger divisions did not produce spatially significant new regions, i.e., the major regions stayed consistent in size and shape. Until other statistical measures are developed to assist in optimization of the number of regions, further testing of the regions must be done by geologic verification.

3 Results

3.1 Regionalization

Areas of similar thicknesses identified from the regionalized classification are rhombic and arcuate shaped, 10s of km in size and referred to as sediment accumulation regions (SARs) (Fig. 1b). Each region is an area where thicknesses are more consistent among the six stratigraphic intervals included in the analysis. Each region is separated by narrow bands of changes in thickness that are not classified into any region.

3.2 Color Gamma-Ray Cross-Section

A regional color cross-section was used to examine internal stratigraphic changes within and across SARs. The color cross-sections depict changes in natural gamma radiation. Ten color levels are used to partition the stratigraphic interval such that basic lithologic changes can been seen. Lithology assignments are assisted with the use of additional well log data (neutron-density porosity) for each well. Most yellow and pink intervals are limestone and blue and purple are clastics. Condensed sections with high gamma ray are colored dark blue gray to white. Correlation lines connect correlatable condensed sections in GSUs with a few exceptions – the Toronto, Haskell, and Stanton Limestones.

Major limestones associated with GSUs are labeled in this section. The genetic sets GSs shown include the Nuyaka Creek overlain by the Muncie Creek, delimited by prominent radioactive condensed sections. GSUs comprising the GSs are delimited for the most part by additional thin radioactive condensed sections. The GSUs used in the regionalized classification are named in the Nuyaka Creek GS as Hushpuckney, Stark, Wea, and Quivira. These names refer to the condensed sections that form the lower boundary of each GSU. The Munice Creek GS is delimited by the Muncie Creek and Heebner condensed sections.

The northwest-to-southeast cross section (Fig. 6) is discussed in the context of successive SAR boundaries which cross the section at positions starting from the shelfward northwestern side.

The SAR boundary labeled A, farthest northwest, is associated with an abrupt southeastward thickening of the shale (blue) in the Douglas Group and a limestone in the lower Munice Creek GS, the Farley Limestone GSU. The Iola and Dewey Limestones are thicker locally west of boundary A. SAR boundary B is associated with initial southward thinning and eventual downlapping of the Oread Limestone immediately southeast of boundary B. The shale-dominated Douglas Group between the Oread and Stanton Limestones thickens abruptly southeast of boundary B. Gamma radiation intensity is elevated in the Douglas Group immediately before it thickens to the southeast at boundary B, suggesting a cor-

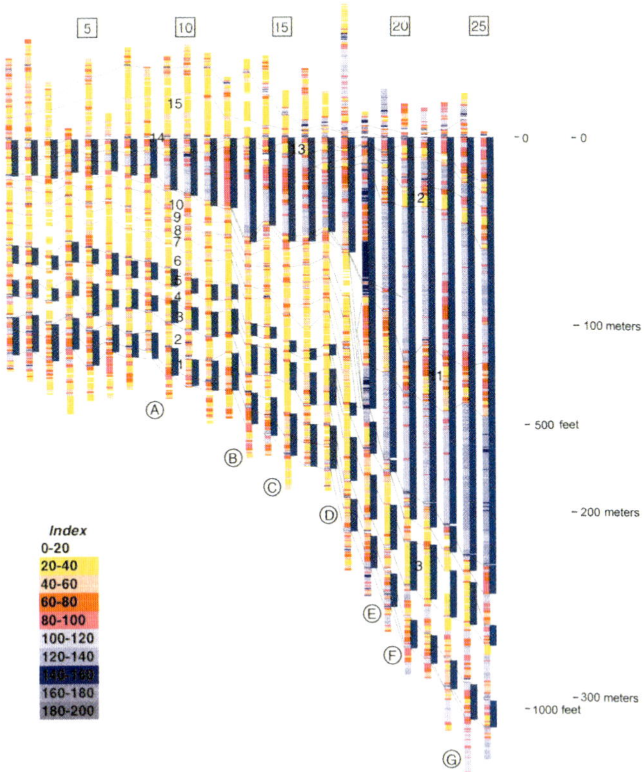

Fig. 6 Northwest-southeast stratigraphic cross-section. Datum is base Heebner Shale (unit 14; top Muncie Creek Genetic Set). The section length is 200 miles (320 km). Vertical scale shown in feet and meters along *left margin*. Vertical exaggeration is approximately 1100 times. Twenty-five wells are used indexed at top of cross section in *numbers surrounded by a box*. *Numbers* correspond to well numbers in Fig. 4 and 5. Natural gamma-ray data are presented in **color**. Ten-level color scale located in *lower left*. *Warmer and lighter colors* correspond to lower gamma ray. *Yellow and pink intervals* are limestone and blue and purple are clastics. Condensed sections with high gamma radiation are colored *dark blue-gray to white*. *Correlation lines* connect correlatable condensed sections and flooding units in GSUs. Lithostratigraphic units (formation level) which are part of GSUs are numerically coded within the cross section (*1* Hertha limestone; *2* Swope limestone; *3* Dennis limestone; *4* Cherryvale Formation; *5* Dewey limestone; *6* Iola limestone; *7* Argentine limestone; *8* Farley limestone; *9* Platsburg limestone; *10* Stanton limestone; *11* Stalnaker sandstone; *12* Douglas limestone; *13* Toronto limestone; *14* Heebner shale; *15* Oread limestone). Munice Creek Shale is thin elevated gamma ray shale above Dewey limestone (*5*). Nuyaka Creek Shale is thin shale following lowermost correlation line below Hertha limestone (*1*). Intervals bounding genetic sets are identified as follows: Nuyaka Creek genetic set (Nuyaka Creek Shale to Muncie Creek Shale) and Muncie Creek genetic set (from Muncie Creek Shale to Heebner Shale). Previous work has established correlation of wireline logs to numerous cores and outcrops. Biostratigraphic correlations, where available, have confirmed physical correlations. Cross-section was generated using PfEFFER software as a single Excel spreadsheet. Visual Basic module to generate cross section was written by Saibal Bhattacharya

relation with change. The Farley Limestone also thickens abruptly southeastward at boundary A. Dennis Limestone is thick and the accompanying density-neutron logs indicate thick, high porosity immediately to the east of boundary B.

SAR boundary C borders the northwestern side of the location where the Oread Limestone thins and apparently begins to downlap unto its underlying radioactive condensed section. Farther southeast the Oread Limestone is entirely pinched out to where the overlying radioactive condensed section (Queen Hill) rests on a condensed section which normally lies below the Oread Limestone. The Oread Limestone is thick and porous immediately northwest of the C boundary. The location of this downlap is also near the position where the Douglas Limestone in the Douglas Group first appears. The Hertha Limestone locally thickens and is lower in gamma ray radiation at boundary C.

SAR boundary D is the sight of probably the most dramatic changes in thickness. The limestone unit overlying the Oread Limestone (Deer Creek Limestone) thins and apparently downlaps at this position. The Douglas Limestone first appears at this position and continues to thicken southeastward. The lower portion of the Muncie Creek genetic set comprised of carbonate units thickens nearly 40% across this boundary. The Stanton Limestone at the top of the lower portion of the Muncie Creek genetic set thickens 150% while the Iola Limestone situated at the base of this set thins southward across the D boundary. Below the Muncie Creek genetic set, the Nuyaka Creek genetic set remains fairly constant in thickness. However, the Cherryvale Formation, residing in this lower set, thickens 20% on the southeast side of boundary D.

SAR boundary E is associated with another abrupt thickening in the Muncie Creek GS, but this time all the limestones that had thickened dramatically at boundary D have now downlapped unto the Muncie Creek condensed section immediately northwest of boundary E. The Upper Muncie Creek genetic set that is clastic dominated thickens nearly 200% southeastward replacing the section that is equivalent to the carbonates in the lower Muncie Creek GS, which have appears to have downlapped. A thick sandstone called the Stalnaker develops at and southeast of boundary E at a stratigraphic position adjacent to the midportion of the carbonate bank before it downlapped. The Douglas Limestone in the Upper Muncie Creek GS is also characterized by major southeastward thickening and lowering of the gamma radiation indicative of cleaner (less shaly) carbonate. Other internal changes in the Douglas interval include increased gamma radiation shelfward (northwest) of boundary E, closely resembling the change seen near boundary B. Modest thickening occurs in the Nuyaka Creek GS across boundary E. The Dennis Limestone in the Nuyaka Creek GS thickens and becomes much lower in gamma radiation to the east of boundary E. The Swope Limestone also thickens and transforms from a carbonate-dominated interval on the shelf (northwest) to become a mixed clastic and carbonate interval southeast of boundary E. This results in an overall change in the Nuyaka Creek GS from mixed and evenly distributed carbonate and clastics to the northwest to clastic-dominated in the lower half and carbonate-dominated in the upper portion southeast of boundary E.

SAR boundary F is associated with abrupt southeastward thickening of the shale between the Toronto and Douglas Limestone units. The upper Muncie Creek GS abruptly thickens to the southeast, thickening basinward in steps that began so noticeably at boundary D.

SAR boundary G is associated with abrupt thickening of the shale between the Heebner Shale, a condensed section, the datum of the cross section, and the succeeding condensed section, the Queen Hill Shale. The Nuyaka Creek GS continues to thicken southeastward, but more significantly, the lithofacies in this interval dramatically changes near boundary G. Southeast of boundary G, the carbonate units that remain in the Nuyaka Creek GS at this low position on the shelf pass abruptly southeastward to clastics. Unlike the apparent downlapping of carbonates in the lower Muncie Creek genetic set, each individual GSUs appears to be correlatable across the boundary linking carbonate and clastic facies of each. The actual internal temporal relationships are not known, but it appears from outcrop analogs that the clastics came later at the end of each GSU as reciprocal sedimentation during low-stand conditions (Watney et al. 1989). The Dennis Limestone is very clean on the shelfward side of boundary G before its abrupt change to sandstone with its attendant higher gamma radiation.

4 Discussion of the Northwest to Southeast Cross Section

4.1 Nuyaka Creek genetic Set

The Upper Pennsylvanian section undergoes significant lithofacies and thickness changes along this cross-section ranging from thinner carbonate-dominated GSUs on the northwestern, the higher shelf side, to thicker clastic-dominated GSUs along the basinward side in the southeast. The Nuyaka Creek GS is more uniform through the entire section compared to the Muncie Creek GS, yet with notable changes in lithofacies and, locally, thickness within particular GSUs in conjunction with region boundaries. Mixed clastics and carbonates are common in the Nuyaka Creek GS along the northwestern reaches of the cross-section with the lower and upper sections containing the most clastics. In contrast, the central and lower shelf region in the southeast is carbonate-dominated. Only the upper portion of the Nuyaka Creek GS is dominated by carbonates in the lower reaches of the shelf. The lithofacies is entirely comprised of clastics on the lowest reach of the shelf, probably representing a boundary between shelf and basin.

Lower shelf carbonates in the Stark GSU are unusually thick compared to landward sites and extended out over the thicker lower clastics, suggesting lateral accretion later in the Nuyaka Creek genetic set. This could be attributed at least in part to a fall in relative sea level or an extended time of the seafloor in the photic zone at that elevation of the shelf during the Stark GSU.

The overall lithofacies and stratigraphic pattern in the Nuyaka Creek GS suggests transgression–regression. Transgressive clastics dominate the base of the genetic unit shelfwide, with larger amounts of clastics supplied early and late in the cycle along the landward (northwestward) end. This also suggests a proximal source of clastics at this location. The pattern suggests a probable relative rise and fall in sea level. Previous studies demonstrate that these landward clastics are comprised of mixed red and green silty deposits with considerable evidence for episodic subaerial exposure (Watney 1980; Watney et al. 1997). In contrast, the lower clastic section is thicker on the lower shelf and basin margin, dominating the lithofacies in this region up through the Hushpuckney GSU. The thicker lower clastics on the lower shelf are possibly the result of sustained deeper water that precluded development of carbonate sediments.

4.2
Munice Creek Genetic Set

In general, the Muncie Creek GS exhibits much more variation in thickness and lithofacies than the Nuyaka Creek GS in the area of the cross-section. Thicknesses increase markedly from shelf to toward the basin. The base of the Muncie Creek GS is a prominent condensed section of the Muncie Creek Shale. Carbonate facies of the lower Munice Creek GS are displaced landward some 100 km from the carbonate bank margin of the underlying Nuyaka Creek GS to near SAR boundary D (Fig. 5). The backstepping is also associated with landward displacement of more carbonate-dominated lithofacies compared to the Nuyaka Creek GS. The landward mixed clastic-carbonate lithofacies appears to have been removed from the area of the cross section. The carbonate margin is very abrupt occurring in less than 2 km with an estimated slope of 3°. The carbonate GSUs thicken at the margin before downlapping into the basinward clastics. The exact horizon of downlap of each GSU is conjectural until more biostratigraphic data are obtained.

Clastic proportions increase toward the top of the lower carbonate-dominated Muncie Creek GS. The lower carbonate interval is abruptly overstepped by clastics that extend considerably landward, thinning and lapping out slightly landward from the SAR A boundary. This lapout is 200 km landward from the carbonate bank margin in the Nuyaka Creek. Backstepping and inferred relative sea-level rise was interrupted by fluctuations in sea level resulting in accumulation of sandstone in front of the lower Muncie Creek GS carbonate bank (Winchell 1957). This has been effectively described as low-stand deposition, in part, the result of sediment bypassing through incised valleys cut into the shale succession on other parts of this shelf farther to the east (Archer et al. 1994). Also, the Douglas Limestone appears to represent sea-level fluctuation, lapping out onto the mid-shelf position as sediment accommodation is apparently lost.

The geometry of the Douglas Limestone and the lower reaches of the Toronto Limestone parallel each other and could also be described as offlap, suggesting that they represent lateral accretion events (GSU scale?) within the upper

Muncie Creek GS. This would reflect infilling of the lower shelf and possibly represent a relative fall in sea level. Again, the sites of significant stratigraphic changes closely correspond to SAR boundaries.

The Muncie Creek GS represents another period of transgressive (backstepping) and regression (lateral accretion). The first part of the interval is dominated by carbonate deposition that led to development of a prominent carbonate bank margin. The succeeding clastics stepped over the carbonates and also filled in the basin in front of the carbonate shelf margin, suggesting flooding and backstepping followed by low-stand infilling of the basin. The geometries of the Douglas and Toronto Limestones suggest offlapping and lateral accretion of the clastic-dominated interval, suggesting probable low-stand conditions late in this genetic set prior to the next prominent marine flooding.

The Muncie Creek genetic set is brought to a close by another prominent condensed section, the Heebner Shale. The Heebner GS is initiated by a backstepping event and apparent drowning of the lower shelf. The Oread Limestone in the Heebner GSU downlaps in the midportion of the cross section immediately southeast of boundary C. The carbonate lithofacies overlying the Queen Hill GSU step slightly basinward to boundary F before it, too, downlaps. Clastics are abundant in the lower shelf, suggesting that the lower shelf remained drowned for extended times during accumulation of the Heebner GS.

It is evident from the above discussion that the SARs and SAR boundaries are important features associated with significant changes in the corresponding stratigraphy of the GSUs and GSs used to obtain the thickness data. Lithofacies change abruptly at SAR boundaries including downlapping of carbonate bank margins into later clastic filled basin areas, mixed clastic–carbonate lithofacies to clean carbonate, and onlapping of limestone. Relative changes in sea level were used to explain many of the relationships. However, sites of stratal changes closely follow structural elements, SARs, that probably reflect changes in elevation of the depositional surface and changes in subsidence rate. SARs apparently also affect the distribution of sediment by influencing the pathways for clastic sediment transport and patterns of lateral accretion of carbonate lithofacies. Eustacy is probably an important process in the formation of these deposits, but local changes in shelf configuration apparently exert important controls themselves on the relative sea-level history and strongly control the nature of observed sedimentation.

5 Gravity and Magnetic Maps

Although regional Precambrian lithology and structure have been interpreted from the gravity and magnetic data of Kansas (Yarger 1983) the studies did not address the issue of Phanerozoic basement reactivation and the development of sediment accumulation regions. The presence of the SARs and their impact on lithofacies distribution reveals a complex three-dimensional view of apparent basement reactivation. Previous work (Watney et al. 1997) has established that the SARs correspond to Precambrian terranes classified by lithology, geochronology, and structure. Further resolution can be provided by examination of the magnetics and gravity information in this area and comparing trends and patterns to basement geology and SARs. Two maps were selected to examine potential fields and SAR relationships. The first map is the residual Bouguer gravity map of Kansas with second-order regional trend removed and vertical sun-shading (Fig. 3). This map is based on a previously published contour map from data points every 1 x 1 km in eastern Kansas and every 1 x 2 km in western Kansas (Xia et al. 1995). The other map is the residual aeromagnetic map of Kansas, reduced to pole and with vertical sun–shading (Fig. 4). This map is based on a 1 x 1-km grid constructed for a previously published contour map from E-W flight lines flown 2 km apart (Xia et.al. 1995). The first part of the discussion is referenced to SAR boundaries A through G encountered along the northwest to southeast cross-section, beginning on the west side with SAR boundary A. The second part of the gravity and magnetic map discussion includes three additional SAR boundaries from western Kansas, X, Y, and Z.

Boundary A is located at the edge of the northeast-northwest Midcontinent Rift deformation and roughly associated with a northeast-southwest gravity high. Sun–shading suggests northeast–southwest structural trends that are subparallel to boundary A. The trends may represent Precambrian faults and mafic dikes. The magnetics map shows a similar correlation of the trends suggesting that they are shallow Precambrian features. Both maps suggest that boundary A is part of a major structural and compositional change in the basement.

Boundary B has similar gravity and magnetic trends as boundary A. The northern part of the boundary roughly corresponds to the eastern margin of a large gravity high mentioned previously. The trend of the boundary is also subparallel to magnetic and gravity lineaments probably associated with faulting and intrusive activity related to Midcontinent rifting.

Boundary C, like boundaries A and B, follows the same northeast-southwest orientations of the Midcontinent rifting. Boundary C is oblique to the eastern margin of the gravity high, but appears to follow a strong northeast-southwest magnetic lineament toward the middle of the boundary. The northern portion of boundary C, although oblique to the eastern margin of the gravity high, appears to follow a more subtle lineation that cross-cuts a flanking gravity low. Although the previous boundaries A, B, and C are slightly oblique to the major magnetic and gravity lineaments between them, NE-SW-trending segments of this boundary appear to correlate roughly with the trend of individual lineaments. These northeast-southwest-trending segments seem to be offset from one another along northwest-southeast-trending magnetic and gravity lineaments. The NW-SE-trending offsets may correlate with faults or shear zones associated with the Central Plains orogen, or extension related the Southern Granite–Rhyolite Province.

SAR boundary D follows the northeast-southwest-trending gravity and magnetic lineaments slightly better than A, B, or C. The curvature in the northeast portion of the boundary appears to follow similar curvature of the axial gravity high interpreted as gabbros and/or basalts of the Midcontinent Rift System. Regions bounded by B, C, and D boundaries also roughly correspond to the Peace Creek Tectonic Zone recognized by Berensen (1997), which he attributed to reactivation of the Midcontinent Rift System.

In southern Kansas, the southwestern portion of boundary E also seems to follow northeast–southwest trends associated with the Midcontinent Rift and appears to cut across a saddle between two gravity highs along the rift trend, as axial gravity high north of the boundary trends to a more northern direction. The northern extent of boundary E, however, curves to an easternly, then a southeasterly, trend as it approaches the northern margins of a gravity high, which is probably associated with deeper mafic intrusive activity along the Midcontinent Rift System. The curve of this boundary also appears to correlate with the northern margin of the Southern Granite-Rhyolite Province, which is interpreted from previous work on magetics (Yarger 1983) and age dating of well cuttings (Van Schmus 1992; Van Schmus and Bickford eds. 1993; Van Schmus, Bickford, and Turek, in press).

Boundary F is much like boundary E. The southwestern parts of boundary F follow the northeast-southwest Midcontinent Rift structures and curves to the east along the northern portions of this boundary. The northern portions of this boundary also appear to follow the northern margins of the large magnetic high probably associated with the Southern Granite-Rhyolite Province.

Boundary G appears to cut obliquely across the Midcontinent Rift System as the boundary trends in a east-northeast, west-southwest orientation. In addition, this boundary is somewhat irregular towards the east. However, the northwest-southeast trend of the northern margin of this boundary roughly correlates with northwest-southeast structural lineaments as well as gravity and magnetic lineaments. The easternmost part of this boundary turns abruptly to a northeast–southwest trend, and may be related to structures associated with the easternmost flank of the Midcontinent Rift System.

A second set of SAR boundaries are described from western Kansas, boundaries X, Y, and Z. The second set contrasts with the first set as they trend perpendicular to them. SAR boundary X closely follows arcuate-shaped gravity anomalies probably created by structural deformation and granitic intrusions associated with the Central Plains Orogen. This boundary also closely corresponds to a trend of isolated magnetic highs, interpreted as epizonal granitic intrusives (Yarger 1983) that occur near the northern margin of the Southern Granite-Rhyolite Province and may be close to the same age (Van Schmus et al. 1993). Previous workers have suggested that the northern margins of the Southern Granite-Rhyolite Province may simply be an erosional boundary. Close correspondence between the northern margin of the Southern Granite-Rhyolite Province, both gravity and magnetic trends and accordant sites of Paleozoic structural reactivation, i.e., SAR boundaries, suggests that this boundary may in fact be partly structurally controlled.

Boundary Y is another arcuate trend. It closely follows the arcuate trend of gravity and magnetic anomalies to the north between itself and boundary X. Although the region associated with this boundary terminates in the southeast, the trend of boundary Y continues to the southeast, forming the northern and southern margins of several other regions. This continuing trend follows arcuate gravity and magnetic anomalies. The eastern half of boundary Y also appears to correspond to the northern margins of several large gravity highs that are probably caused by mafic intrusions. This boundary also appears to correspond closely to the southern margin of several magnetic highs.

SAR Boundary Z resides in the southwest corner of Kansas in the heart of the Hugoton Embayment, the site of significant current oil and gas development. The northeastern part of boundary Z appears to follow northwest-southeast-trending gravity and magnetic lineaments, probably associated with both Central Plains Orogen and Southern Granite-Rhyolite Province structures. The eastern margin of this boundary turns abruptly to a southerly trend at the eastern margin of a ring-like magnetic anomaly which is possibly associated with a caldera formed during evolution of the Southern Granite-Rhyolite Province. If this truly is a caldera, its faulted margins, associated with the ring-like magnetic anomaly, may have controlled the location of the eastern part of boundary Z.

The northern part of boundary Z trends roughly east-west and may be roughly associated with the northern margin of several east-west-trending gravity and magnetic anomalies. This northern part of boundary Z turns abruptly southward along the western margin of a prominent gravity high, probably created by gabbroic intrusives.

The regional structure of the Precambrian basement appears to correlate with the arcuate shape of the northern margin of the Hugoton Embayment and all three of the SAR boundaries. X, Y, and Z, as well as the arcuate trends of the gravity and magnetic anomalies, Precambrian basement configuration, and Paleozoic structures including SAR boundaries. This suggests that Precambrian structures were responsible for both the regional shape of the Hugoton Embayment and the trend of the regionalization boundaries.

The SAR boundaries vary in their orientation, and with the types of Precambrian features with which they correlate. The stratigraphic expression also varies spatially, depending on what spatial location and genetic set of which it is a part. The causes of the SAR development are clearly related to basement heterogeneity and likely represent interactions with far-field stresses that influence which basement element is reactivated. However, the correspondence of the SAR with the gravity and magnetic lineaments is not always one-to-one, probably because of the often time-oblique nature of far-field stress orientation relative to the orientation of basement structures.

6 Discussion

Phanerozoic sedimentary rocks form a thin veneer on the Precambrian basement in Kansas and, based on results of this study, show that their accumulation is strongly influenced by differential subsidence associated with heterogeneities in the basement. A geostatistical integration of thicknesses of six temporally distinct genetic stratigraphic units resolves km-sized regions containing similar sediment accumulation (sediment accumulation regions, SARs). Rhombic and arcuate shapes of these SARs correspond closely to gravity and magnetic trends and lineaments and known structure in the Precambrian basement.

A gamma-ray cross-section reveals that SAR boundaries affect all scales of stratigraphic units including GS, GSU, and smaller-scale stratigraphic units and lithofacies. In particular, SAR boundaries control the location of prominent carbonate margins consisting of single and stacked GSUs. Recurrence of carbonate margins at the same location in different GSs suggests consistency in the site of sediment starvation and drowning of the shelf. Moreover, the linear trends of these boundaries and their correspondence with basement structure strongly links them to sites of basement reactivation. Sediment accumulation regions appear to be validated from this investigation.

GSs are defined as coherent packages of related GSUs. They exhibit abrupt transgression and more gradual regression, characterized by backstepping of carbonate-dominated strata and lateral accretion of the same. The duration of a GS is estimated to be approximately 1.5 Ma (Watney et al. 1995) and is of duration similar to episodic foreland basin subsidence resulting from thrusting and downloading along collisional margins (Quinlan and Beaumont 1984). However, stratal patterns in the greater Pennsylvanian indicate innundation of the Midcontinent US and other shelves and basins worldwide, corresponding to the Absaroka cratonic sequence and inferred eustatic rise associated with it (Haq et al. 1987). The GS, also classified as a third-order cycle, exhibits an apparent tectonic signal here, but may also be the result of a superimposed eustatic influence. Experimentation with a two-dimensional stratigraphic simulation modeling has reproduced the observed succession and carbonate margin. The backstepping and drowning of the lower shelf were modeled using a rapid (5 to 10 m/ka) 15 m rise in relative sea level coupled with high frequency glacio-eustatic sea level oscillations of 110 m during each GSU (Watney et al. 1995). Results of modeling imply that tectonics or eustacy could explain the observed two-dimensional geometries. However, when characterizing the shelf in three dimensions as was done in this current investigation, reactivation of basement enters in strongly as a control. Episodic structural fragmentation of the shelf is suggested here to be as important as eustacy in developing genetic and sequence stratigraphic models. It may be necessary to draw on greater stratigraphic intervals and utilize larger areas to establish 3-D control in order to develop empirical models that can better discern cause and effect and, in turn, improve our attempts at developing predictive stratigraphic models.

Utilization of a series of genetic units, representing coherent, temporally distinct depositional packages, reveals an evolution in the shelf configuration, each unit analogous to a strobe, freezing the motion in that time interval. In contrast, aggregate stratigraphic intervals that are not time-distinct will blur or remove the effect of the changing shelf.

Future studies are needed to examine the nature and timing of structural movement and sedimentary response during contrasting periods of tectonism, e.g., pre and post foreland basin development. Also, studies are needed to further document the levels of coherency of stratigraphy, lithofacies, and diagenesis in the SARs and refine predictive models for resource exploration.

7 Conclusions

1. High-frequency, carbonate-dominated genetic stratigraphic units (GSUs) and genetic sets (GSs) are marine to nonmarine cycles bounded by condensed sections. Elevated natural gamma radiation of condensed sections facilitates mapping in the subsurface.
2. Regional mapping of thickness and lithofacies in GSUs and GSs provide detailed temporal (as proxies for associated depositional sequences and sequence sets) and three-dimensional spatial variations of the Kansas shelf configuration bordering the Anadarko and Arkoma basins.
3. Regionalized classification of GSU and GS thickness data is used to create geometric models of basin/shelf subdivisions recognized here as sediment accumulation regions (SARs), km-scale rhombic and arcuate-shaped crustal blocks.
4. Color gamma-ray log cross-sections illustrate changes in stratigraphy and lithofacies corresponding closely to SARs.
5. SARs correspond closely with magnetic and gravity anomalies and lineaments as well as Precambrian geology as determined from well control. This suggests that the SAR boundaries are strongly influenced by Precambrian structure, particularly reactivated basement faults.
6. Locus of major stratigraphic and lithofacies change is strongly dependent on inferred differential, episodic subsidence. Subsidence appears to be a major influence on the GSUs and GSs, in spite of apparent high magnitude/high frequency eustacy.
7. SAR, GSU and GS development may have significant implications in the exploration for stratigraphic traps (abrupt change in shelf configuration and stratigraphy/lithofacies) and evaluation of fluid migration routes (fracture systems).

Acknowledgements

ERMapper from Earth Resource Mapping was used for gravity and magnetic image analysis. Gravity and magnetic images were generated by Joe Kruger while a visiting scientist at the Kansas Geological Survey. Lea Ann Davidson is thanked for her assistance in manuscript preparation. Appreciation is extended to PfEFFER software group for support in developing color cross section module, particularly Saibal Bhattacharya. Thanks is also given to Julie Grauer, Ken Stalder, and Dana Adkins-Heljeson for help and suggestions in experimenting with color cross section generation.

References

Archer, A.W., Lanier, W.P., and Feldman, H.R., 1994, Stratigraphy and depositional history within incised-paleovalley and related facies, Douglas Group (Missourian/Virgilian; Upper Carboniferous) of Kansas, U.S.A., in Boyd, R., Zaitlin, B.A., and Dalrymple (eds.), SEPM Special Publication 51, Incised Valley Systems: Origin and Sedimentary Sequences, p. 176–190.

Baars, D.L., and Watney, W.L., 1991, Paleotectonic control of reservoir facies: Kansas Geological Survey Bulletin 233, p. 253–262.

Berendsen, P., 1997, Tectonic evolution of the Midcontinent Rift System in Kansas: in, Ojakangas, R.W., Dickas, A.B., and Green, J.C., eds., Middle Proterozoic to Cambrian Rifting, Central North America, Geological Society of America, Special Paper 312, Boulder, CO, p. 235–241.

Boardman, D.R., II, and Heckel, P.H., 1989, Glacial-eustatic sea- level curve for early Late Pennsylvanian sequence in north- central Texas and biostratigraphic correlation with curve for midcontinent North America; Geology, v. 17, p. 802-805.

Bohling, G.C., Doveton, J.H., and Watney, W.L., 1996, Systematic identification of sequence stratigraphic units from wireline logs: in, Pacht, J.A., Sheriff, R.E., and Perkins, B.F., Stratigraphic Analysis Utilizing Advanced Geophysical, Wireline, and Borehole Technology for Petroleum Exploration and Production: Seventeenth Annual Research Conference, Gulf Coast Section, Society of Economic Paleontologists and Mineralogists Foundation, Houston, Tx., p. 29–37.

Gerhard, L.C., Anderson, S.B., LeFever, S.B., Carlson, C.G., 1982, Geologic development, origin, and energy mineral resources of Williston Basin, North Dakota: American Association of Petroleum Geologists, Bulletin, v. 66, p. 989–1020.

Haq, B.U., Hardenbol, J., and Vail, P.R., 1987, Chronology of fluctuating sea levels since the Triassic: Science, v. 235, p. 1156–1167.

Harff, J. and Davis, J.C., 1990, Regionalization in geology by multivariate classification: Mathematical Geology, v. 22, p. 925–936.

Heckel, P.H., 1977, Origin of phosphatic black-shale facies in Pennsylvanian cyclothems of midcontinent North America: American Association of Petroleum Geologists, Bulletin, v. 61, p. 1,045–1,068.

Merriam, D.F., and Förster, 1994, Precanbrian basement control on "Plains-type folds" (compactional features) in the Midcontinent region, USA: Proceedings of the 11th Conference on Basement Tectonics, p. 1–18.

Merriam, D.F., and Förster, A., 1997, Origin, development, and distribution of "Plains-Type Folds" (compactional features) in the Cherokee Basin of the American Midcontinent: American Association of Petroleum Geologists, Transactions of the 1997 AAPG Mid-Continent Section Meeting, p. 52–56.

Quinlan, G.M. and Beaumont, C., 1984, Appalachian thrusting, lithospheric flexure, and the Paleozoic stratigraphy of the eastern interior of North America: Canadian Journal of Earth Sciences, v. 21, p. 973–996.

Van Schmus, W.R., 1992, Tectonic setting of the Midcontinent Rift System, in Geodynamics of rifting, Volume II, Case history studies on rifts: North and South America, Ziegler, P.A., ed., Tectonophysics, v. 213, p. 1–15.

Van Schmus, W. R., (co-editor), Bickford, M.E. (co-editor), et al., 1993, Chapter 4: Transcontinental Proterzoic Provinces, in, Reed, J.C., Jr., et al., editors, Precambrian: Conterminous U.S. Boulder, Colorado, Geological Society of America, The Geology of North America, v. C-2, p. 171–334.

Van Schmus, W.R., Bickford, M.E., and Turek, A., in press, Proterozoic geology of the east-central midcontinent basement, in van der Pluijm, B.A., and Catacosinos, P., eds., Basement and basins of eastern North America, Geological Society of America, Special Paper 308.

Watney, W.L., 1980, Cyclic sedimentation of the Lansing and Kansas City Groups (Missourian) in northwestern Kansas and southwestern Nebraska - A guide for petroleum exploration: Kansas Geological Survey, Bulletin 220, 72 p.

Watney, W.L., French, J.A., Doveton, J.H., Youle, J.C., and Guy, W.J., 1995, Cycle hierarchy and genetic stratigraphy of Middle and Upper Pennsylvanianstrata in the upper Mid-Continent, in Hyne, N., ed., Sequence Stratigraphy in the Mid-Continent, Tulsa Geological Society, Special Publication #3, p. 141–192.

Watney, W. L., Davis, J.C., Olea, R. A., Harff, J., and Bohling, G.C., 1997, Modeling of sediment accommodation realms by regionalized classification: Geowissenschaften, v. 15, p. 28–33.

Winchell, R. L, 1957, Relationship of the Lansing Group and the Tonganoxie (ìStalnakerî) sandstone in south-central Kansas: Kansas Geological Survey, Bulletin 127, p. 123–152.

Yarger, H. L., 1983, Regional interpretation of Kansas aeromagnetic data: Kansas Geological Survey Geophysics Series 1, 35 p.

Xia, J., Miller, R., and Steeples, D., 1995, Aeromagnetic map of Kansas, reduced to a horizontal plane and reduced to the pole, scale 1:1,000,000: Kansas Geological Survey, Map Series M-41.

Xia, J., Miller, R., Steeples, D., and Adkins-Heljeson, D., 1995, Residual Bouguuer gravity map of Kansas, the second-order regional trend removed, scale 1:1,000,000: Kansas Geological Survey, Map Series M-41.

Methods Used in a Mass-Balance Study of Holocene Sediment Accumulation on the Southern North Sea Coast of Germany

C. Hoselmann and H. Streif

1 Introduction

A quantitative analysis of the accumulation of coastal Holocene deposits is carried out in the form of a mass-balance study for the lowland area between the Ems and Weser Rivers on the North Sea coast of Germany. The area investigated comprises various geomorphological elements and depositional environments such as the East Frisian barrier islands, sheltered and open tidal flats, bay flats of Dollart and Jadebusen, estuaries of the Ems and Weser Rivers, and extensive coastal marshlands (Fig. 1). From this point of view, this area can serve as a model for general mass-transport and accumulation of sediment connected with transgressions over coastal lowlands.

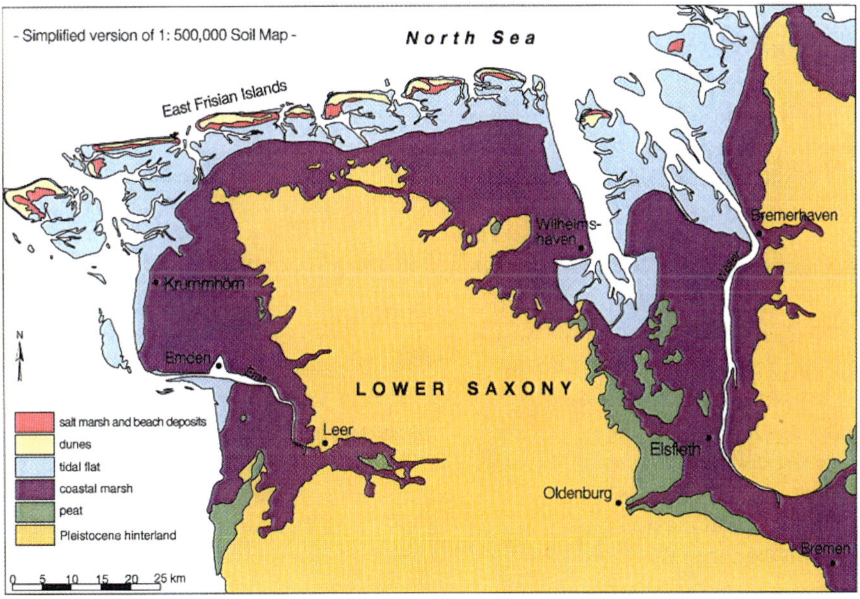

Fig. 1 Sketch map of the investigated area in Lower Saxony

The study is a contribution to a Priority Program of the Deutsche Forschungsgemeinschaft with the title Changes of the Geo-Biosphere in the Course of the last 15 000 Years. Under this aspect, the mass-balance study can also be regarded as a contribution to the Core Project Land Ocean Interaction in the Coastal Zone of the International Geosphere Biosphere Program.

2
Evolution of the Coastal Holocene

A relative sea-level rise of 100 to 130 m affected the southern part of the North Sea basin and adjacent lowland areas in the course of the last 18 000 years of the Weichselian late glacial and Holocene periods. Radiocarbon-dated transgression levels, however, exist only for the past 8600 years and permit a detailed reconstruction of the sea-level rise from about –45 m to its present-day elevation (Behre et al. 1984; Streif 1990; Menke 1996). The evolution of the coastal landscape took place in the final stage of the Holocene transgression under the influence of a sea-level rise of about 25 m during the last 7500 years.

Prior to 8000 B.P., a hilly Pleistocene landscape covered with forest and local peat bogs existed in the present day coastal lowland area. At about 7500 B.P., the first incursions of brackish water reached the area now occupied by the East Frisian Islands and were followed by the inundation of the entire coastal region. In the course of this event, which took only 5500 years, tidal erosion affected mainly the seaward part of the inundated area, locally forming deeply incised gully systems. In extensive areas, however, the original relief of the pre-existing landscape is preserved underneath a wedge-shaped body of Holocene coastal deposits. The transgression and the accompanying sediment accumulation were almost completed around 2000 B.P. Since then, repeated storm surges affected various locations along the coast and formed tidal bays which expanded far into the coastal marshes. Most of these bays silted up under natural conditions or were systematically dyked during medieval and modern times.

The general geological features of the wedge-shaped body of coastal Holocene deposits are depicted in a schematic cross section from the North Sea to the Pleistocene hinterland (Fig. 2). At its seaward margin, in the region of the East Frisian barrier islands, the Holocene is made up of 25 to 35 m thick marine and beach sands. They are overlain by locally up to 25-m-thick dune sands derived from adjacent beach areas. Thinner Holocene sequences consisting of sandy, silty, and clayey tidal-flat to brackish deposits occur beneath the present-day tidal-flat area and coastal marshland. The fine-grained allochthonous clastic material stems from a marine or fluvial source and accumulated in the coastal zone in the course of the Holocene transgression. Additionally, peat layers, which were formed in coastal fens, occur at the base of the sequence of tidal clastics or are intercalated in it. These autochthonous organic deposits become thicker towards the landward margin of the coastal lowland area, and in many places continuous peat sequences abut onto the Pleistocene hinterland.

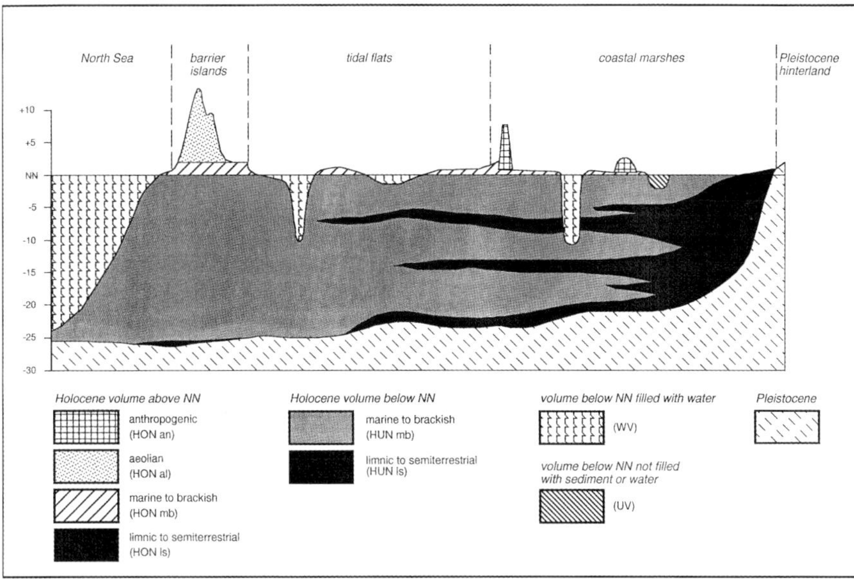

Fig. 2 Schematic cross-section through the coastal Holocene depicting the deposits and units considered in the mass-balance study

3
Fundamental Aspects and Basic Data

The mass balance study aims at a quantification of the volume of the wedge of Holocene coastal deposits. In addition, information on the spatial extension and thickness of the sediment and peat layers are used to calculate the total volumes of allochthonous clastic and autochthonous organic units that make up the coastal Holocene. Comparison of these figures with the clastic suspension loads of the Ems, Weser and Elbe Rivers during the past 7500 years permits quantification of mass transport from marine and riverine sources into the coastal zone.

The basic data used in this study consist of records from boreholes sunk in the coastal area. This information is stored in the database of the Geological Survey of Lower Saxony using the symbols in the geological symbol key (Preuss et al. 1991). Standardisation of these highly variable borehole records was achieved by using the lithological classification system established by Barckhausen et al. (1977). This system makes it possible to subdivide the recorded sections into lithological sequences and then to classify the entire section into 1 of 3 main profile types and 12 subordinate profile types.

The computer programs DASP (Kühne 1983, Homann and Huy 1985) and GIROS (Preuss 1988, 1992) were used to plot this information, as a basis for manual construction of structure contour maps. The lower boundary of the Holocene wedge was reconstructed and displayed on 1:25 000 structure contour maps. Maps on the same scale showing the distribution of the clastic sediments and peat layers present a picture of the internal makeup of the coastal Holocene. The surface relief of the accumulation wedge in the subaquatic, eulitoral and terrestrial environments was derived from hydrographic soundings and topographic levelling. This provided a good basis for determination of the volumes of the different units of the coastal Holocene.

4 Geometry of the Accumulation Wedge

For the mass-balance study, it was necessary to subdivide the sediment wedge into a series of units, which are depicted in a schematic cross-section from the North Sea to the Pleistocene hinterland (Fig. 2).

4.1 Units Used in the Calculation

Separate calculations were carried out to determine the volumes of Holocene sediments above and below German zero datum NN (Normalnull), which is nearly identical with mean sea level in this region. The sediment volumes above and below NN (HON and HUN, respectively) are both mainly made up of marine to brackish deposits (HON mb and HUN mb), but also contain lacustrine to semiterrestrial deposits (HON ls and HUN ls). The aeolian dune deposits (HON al) on the East Frisian Islands, as well as anthropogenic structures (HON an) such as dwelling mounds, dykes and dumps of dredged material, are limited to positions above NN.

The volume of water bodies (WV) in the foreshore area, tidal gullies, estuaries, and freshwater lakes is of considerable importance for the mass-balance study. Less important is the volume of morphological depressions (UV) in the coastal marshes in which the land surface may lie 2 m below NN and which are not filled with sediments or water. For the calculation of the total volume of the Holocene deposits, the values of WV and UV have to be subtracted from the sum of the other units (HON+HUN).

4.2 The Surface of the Accumulation Wedge

In the terrestrial area (East Frisian Islands, salt marshes, and dyked coastal marshland), the surface of the accumulation wedge is identical with the land surface (Fig. 2) and can be derived from the topography or digital elevation models. The ordnance survey of Lower Saxony currently offers two types of data sets (Washausen 1992). The Digital Elevation Model DEM 50 is based on a 50 × 50-m grid and is available for the whole of Lower Saxony. Each grid node has one elevation above NN with an accuracy of a few meters and sometimes > |10| m. Thus the DEM 50 is unsuitable for studies in coastal lowland areas.

The DEM 5 is based on a 12.5 × 12.5-m grid and contains much higher-quality data; the elevation values have an accuracy < |0.5| m. A 3 D plot representing a 4 km^2 area of marshland at Krummhörn on the Ems estuary made from this kind of digital data is shown in Fig. 3. The figure demonstrates the high vertical resolution of this model by clearly depicting a morphological depression and an artificial dwelling mound in the flat coastal marshland. Consequently, the DEM 5 is an excellent data base for calculation of the HON and UV volumes.

As far as possible, DEM 5 data were used in this research project for reconstructing the surface morphology of the coastal marshes and East Frisian barrier islands. However, DEM 5 covers only 10% of the study area; thus, the remaining land surface had to be reconstructed on the basis of other topographical data. Well locations were one source which contributed a number of scattered elevation values. The remaining areas were covered by manually digitizing contour lines and point elevations on topographical map sheets. At least ten points per km^2 were digitized from 1:25 000 maps to reconstruct the relief of the flat coastal marshlands. The morphology of the dune landscape on the East Frisian Islands required a much denser data set, which was obtained by digitizing approximately 250 points per km^2 from 1:5000 topographical map sheets.

The surface relief of subaquatic areas (foreshore zone of the barrier islands, tidal gullies, estuaries, and lakes) is the same as the sediment-water interface (Fig. 2). There is a similar boundary in the intertidal zone. The surface of the sediment wedge in both these areas was reconstructed on the basis of the digital coastal model (DIGEKÜ; Schleider and Buziek 1991). Supplementary digital data sets were obtained from the Dutch-German research project WADE (Wadden Sea morphological development due to an accelerated relative sea-level rise; Niemeyer et al. 1995). The remaining gaps were filled by digitizing contour lines on 1:25 000 coastal map sheets, which were published by the board of trustees for research in coastal engineering. Thus, at least 10 elevation values per km^2 are available for reconstructions of the surface relief in subaquatic areas and in the intertidal zone.

4.3 The Base of the Accumulation Wedge

The base of the accumulation wedge (Fig. 2), which corresponds to the interface between Holocene and Pleistocene deposits, was reconstructed on the basis of borehole data. Over much of the coastal zone, this boundary is equivalent to the surface of the original Pleistocene landscape. In other areas, especially in the seaward zone of the coastal lowlands, the original landscape underwent drastic modification by tidal erosion during the Holocene transgression. The basal boundary of the accumulation wedge can be reconstructed in two ways. One is by manual construction of structure contour maps, the other is with the aid of computer programs and based directly on the depth of the Pleistocene-Holocene boundary from borehole data.

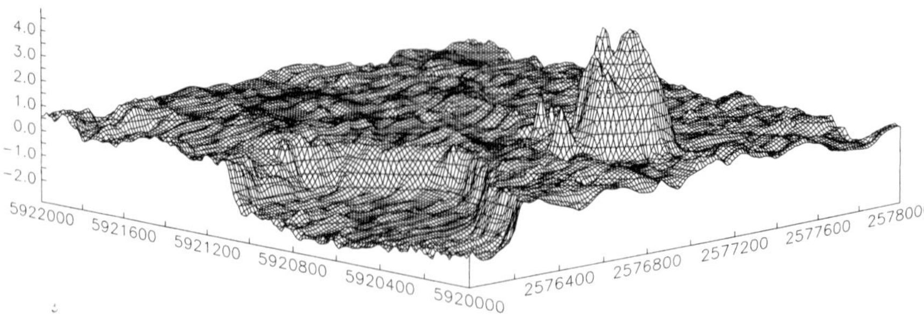

Fig. 3 Three-dimensional graphical plot of the Digital Elevation Model (DEM 5) of a 4 km^2 area in the Krummhörn region on the lower Ems, map sheet 2508/30 (Groß Midlum). This example demonstrates the high resolution of the DEM 5 and shows the morphological depression of the Freepsumer Meer (max. depth -2.47 m NN), which is not infilled with sediments or water; the anthropogenic dwelling mound of Groß Midlum (maximum elevation +4.87 m NN) is visible in the flat landscape of coastal marshland

5 Comparison of Methods for Determining the Volume of Holocene Deposits below NN

Different methods were employed to determine the volume of Holocene deposits below NN. A short description of the methods and a comparison of the results obtained is given below.

1. Calculations on the basis of manually constructed structure contour maps:
- The miner's method of calculating reserves:
  a) Planimetry of the structure contour map.
  b) Calculation of the volumes of successive 1-m-thick layers.

- Computer methods:
  a) Digitizing of structure contour maps with ARC/INFO.
  b) Transfer of digitized isolines to the Interactive Surface Modeling (ISM) program.
  c) Volume calculation of the HUN volumes with ISM.

2. Calculations on the basis of borehole data with the aid of various computer programs:
- Interactive Surface Modeling (ISM), version 7.1, on VAX with VMS 6.2.
- MacGRIDZO, version 3.62, on Apple Power Macintosh with MacOS 7.5.
- Surfer for Windows version 6 on IBM-compatible PC with Windows 3.1.

The conventional way to carry out a mass-balance study of the Holocene is to construct structure contour maps of the base of the accumulation wedge manually. Experience has shown that manual construction, which is admittedly subjective, provides us with a picture that comes close to the natural situation. In these constructions, for instance, isolated sites at which the base of the Holocene was encountered in an unusually deep position in a borehole may be interpreted as belonging to a narrow gully. By attributing an isolated point to an extensive gully system, however, an individual depth value receives a new and wider spatial significance. Digitizing of the contour lines produced a large number of points arranged along these lines, comprising a very high data density. For the purpose of the study, these scattered data had to be transformed into a regular grid.

Compared with manual construction, computer programs make exclusive use of borehole data regardless of geological or morphogenetic aspects. The scattered depth values obtained from borehole records are transposed directly onto a regular grid, which is most suitable for volume determinations.

Important for converting these two sets of digital data into regular grids (Fig. 4) are grid spacing, number of boreholes per grid node, and the selected gridding algorithm (e.g. kriging, inverse distance, weighted least squares, triangulation, and minimum tension gridding).

For comparison of these two procedures and the results obtained, a section was chosen from map sheet 2716 (Elsfleth) on the lower reaches of the Weser River. This area (105.6 km^2) is characterised by a variable relief of the base of the Holocene, which displays a depth range of between −0.85 and −16.3 m NN. In this area a total of 535 boreholes, an average of 5.1 boreholes per km^2, penetrated the coastal Holocene sequence down to its base.

The results obtained using the above methods were compared with results of a method involving planimetry, which is normally used for calculating mining reserves. A PLANIX 5000 precision planimeter made by the Tamaya company was used to traverse along the structure contour lines. This was done three times, and then the mean was used to calculate the volumes of sediment in the depth intervals between 0 and −1 m NN, −1 and −2 m NN, −2 and −3 m NN down

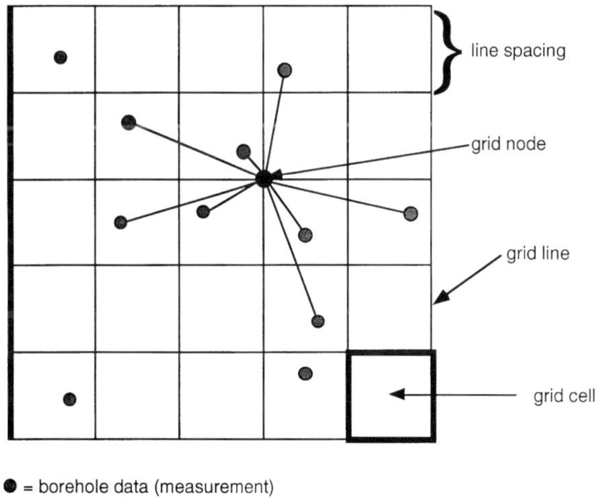

Fig. 4 Schematic graph of parameters, which have to be considered in calculations of grid nodes

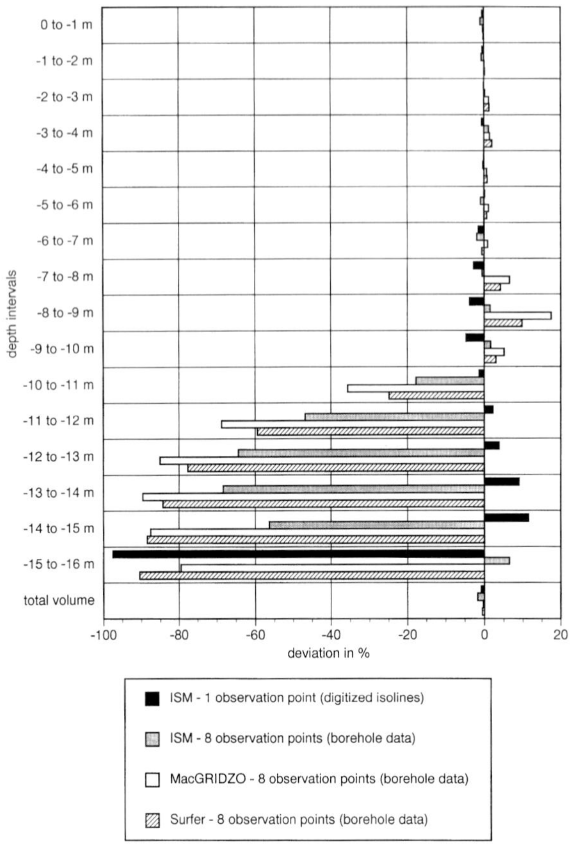

Fig. 5 Comparison of the results of volume calculations carried out with different computer programs for an area of 105.6 km^2 on map sheet 2716 (Elsfleth). The graph shows the percentage divergence from results obtained by evaluation of a structure contour map using a planimeter

to −15 and −16 m NN. In this way, a total volume of 805 873 740 m^3 of Holocene deposits below NN (HUN) was obtained for the test area.

The computer-aided transformation of the scattered data from digitized isolines and borehole records into a square grid in both cases produced a total of 301 grid lines in the X and Y directions with 90 601 grid nodes. Since the data set derived from the digitized isolines had a very high data density (115 786 points), only the nearest point to each grid node was used. In contrast, borehole data produced only 535 points; the depth values obtained from the eight closest boreholes to each grid node was used (Fig. 4).

The grids obtained with different computer programs were used to determine the volumes of the deposits in the depth intervals between 0 and −16 m NN, as well as for the total volume of Holocene deposits below NN (HUN). Fig. 5 shows the percentage to which these calculated volumes diverge from corresponding values obtained with the planimeter technique. It is clearly visible that the larger differences occur only in the depth range below −10 m NN, which was penetrat-

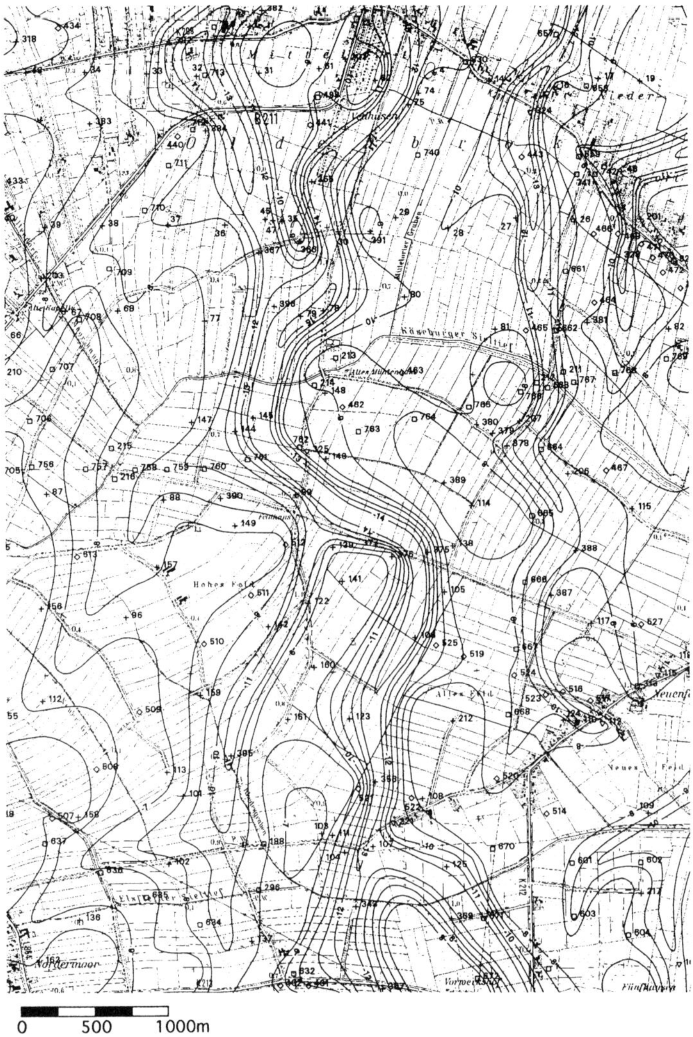

Fig. 6 Detail of the manually constructed structure contour map of the base of the Holocene accumulation wedge. Topographic base: 1:25 000 map sheet 2716 (Elsfleth). (With permission of the Niedersächsisches Landesverwaltungsamt – Landesvermessung – dated 29 Nov. 1996 Az.: B4 - 2005/97)

ed by a small number of boreholes. Manual construction of the structure contour lines (Fig. 6) infers an extensive gully system. In contrast, computer gridding directly based on borehole data is not affected by morphologic interpretation and therefore produces a series of isolated, deep depressions ("bull's-eye" artifacts). Although mathematically correct, these elements have little similarity with the natural land forms known from the study area or adjacent areas. It can also be seen in the diagram that large discrepancies in the results obtained by the

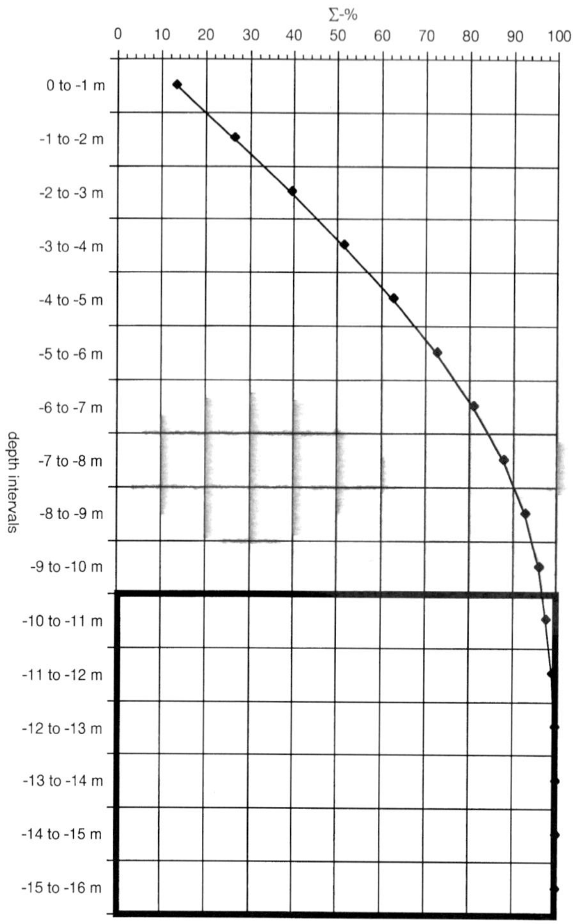

Fig. 7 Summing up of the volumes calculated for the 1-m depth intervals between 0 and -16 m NN using the miner's method for calculating reserves, sheet 2716 (Elsfleth)

different programs occur only in depth intervals between -10 and -16 m NN. However, it has to be pointed out that the deposits in this depth interval make up only 1.94% of the total Holocene volume below NN (Fig. 7) .

With respect to the total volume, the results obtained with different computer programs show no significant discrepancies (Fig. 5). This is remarkable, because all programs applied use different gridding algorithms. ISM uses a method called minimum tension gridding (Interactive Surface Modeling User's Guide 1990). The MacGRIDZO gridding algorithm uses the weighted least squares method (MacGRIDZO User Manual 1994). In this test, Surfer for Windows uses a kriging gridding algorithm (SURFER for Windows User's Guide 1995).

Additional test series were carried out with MacGRIDZO to find out to what extent certain parameters influence the results of volume calculations. The

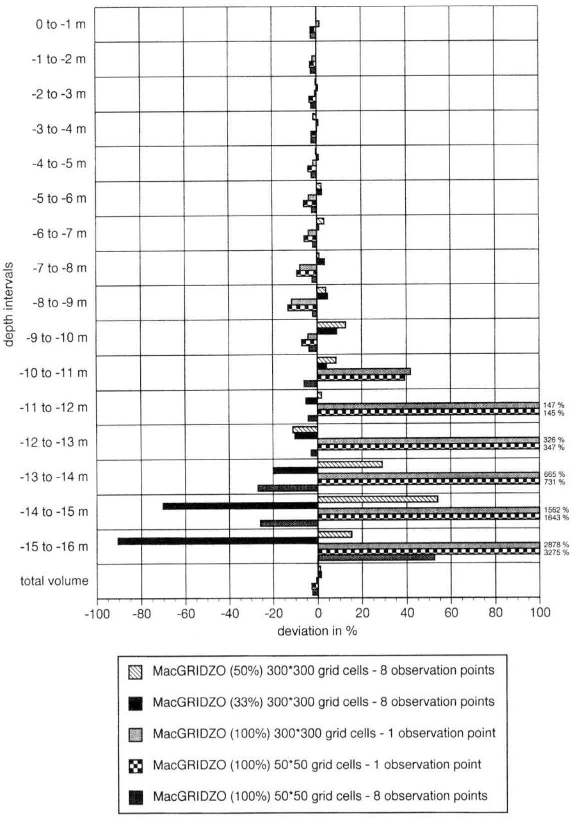

Fig. 8 Comparison of results obtained from volume calculations with the MacGRIDZO computer program on sheet 2716 (Elsfleth). The differences refer firstly to volume calculations using different numbers of boreholes (100, 50 and 33% of the total number of boreholes). Secondly they are affected by calculations using one or eight depth values per grid node, and 301 grid lines in the X and Y directions

variations of the parameters included: (1) number of boreholes (100, 50 and 33% of the total number of boreholes), (2) boreholes per grid node, and (3) line spacing. Again the differences in the depth intervals between 0 to -10 m NN are insignificant (Fig. 8). Below -10 m NN large positive deviations were obtained. This is especially true for calculations that use the depth value of one single borehole per grid node. Probably this has to be attributed to the fact that in this case the calculation of a single grid node is dominated by data obtained from a small number of boreholes that intersected the base of the Holocene below -10 m NN. In the case of gridding with borehole data using 8 holes per grid node, a single borehole with the base Holocene below -10 m NN has only a minor influence on depth obtained at the grid node. Further, it should be pointed out that calculations of the total volume below NN with 100, 50 and 33% of the total number of boreholes show deviations of less than 3%. Therefore it can be concluded that, with the methods applied, a relatively small data base is sufficient for reliable determinations of volumes of the coastal Holocene.

6 Summary and Conclusions

The mass-balance study funded by the Deutsche Forschungsgemeinschaft deals with a quantification of Holocene sediment accumulation in the coastal zone between the estuaries of the Ems and Weser Rivers (Lower Saxony). This region is highly suitable for this kind of research because it is built up of various coastal sedimentary environments comprising barrier islands with sheltered and open tidal flats, bay flats (Dollart, Jadebusen), and estuaries (Ems and Weser) probably involving different accumulation processes. The Holocene accumulation wedge in this region was formed within the past 7500 years under the influence of a sea-level rise of 25 m. Because of the great variety in sedimentary environments investigated, the study can serve as a model for general mass-transport and sediment accumulation connected with transgressions over coastal lowlands. Making use of a great many published radiocarbon dates, the mass-balance study also offers the opportunity to work out the interaction between changing rates of sea-level rise and corresponding accumulation rates of coastal deposits. It is also hoped that quantitative studies of this kind will be useful for predicting the future development of coastal lowland areas under the influence of changing rates of sea-level rise.

As far as available, digital data from various coastal authorities in Lower Saxony were used to reconstruct the surface of the Holocene sediment wedge and to determine the volume of a series of units. Hydrographic soundings were used to calculate the volumes of water bodies below NN. In the tidal-flat area similar data were used to reconstruct the surface morphology and determine the volumes of water below NN and sediment above NN. Topographic surveying data were used as a basis for determining the volume of morphological depressions in the marshland that lie below NN and are filled neither by water nor sediment. The volume of coastal deposits above NN was calculated using data from the same source. Any gaps in the above data sets were filled by systematic digitizing of elevation points and contour lines on 1:5000 and 1:25 000 topographic maps. The digital data sets compiled in this manner contained at least 10 elevation points per km^2 in the coastal marshlands, tidal flats, and subaquatic zone. A much denser coverage was used in the case of the sand dunes on the East Frisian Islands, where over 250 elevation points per km^2 were digitized.

The sediment volume below NN was calculated on the basis of boreholes, which were sunk in the coastal area and which penetrated the entire sequence of Holocene deposits. As a test of accuracy, (Figs. 5 and 8) volume calculations were carried out with various computer programs (ISM, MacGRIDZO, and Surfer) and gridding algorithms (minimum tension gridding, weighted least squares, and kriging). Additional tests dealt with the different evaluation methods based on manual construction of structure contour maps, planimetry and digitizing of these isolines, and calculation of volumes directly from borehole data. Further studies dealt with the influence of certain parameters (line spacing, number of evaluated depth values) on calculations of a single grid node. The results of these tests carried out on map sheet 2716 (Elsfleth) demonstrate that large deviations occur only in the volumes of depth intervals below -10 m NN. This effect can be explained as follows. Only a small number of boreholes intersected the base of the Holocene in an unusually deep position, i.e. below about -10 m NN; these were interpreted by the author of the manually constructed structure contour maps as belonging to a system of deep, narrow gullies. Any such interpretation is disregarded by computerized grid calculations that are directly based on borehole data and which in consequence produce a series of isolated, deep, and relatively wide structural depressions ("bull's-eye" artifacts). However, the volume of sediment in the depth interval -10 to -16 m NN is relatively unimportant as it comprises less than 2% of the total Holocene volume below NN.

In order to use a standardised procedure and to monitor the results obtained, it was decided to calculate the Holocene volume below NN on the basis of both data sources, the manually constructed structural contour maps and directly from borehole data. ISM proved to be a suitable computer program, because it works on a VAX under the VMS operating system and allows the user to calculate the grids and volumes in batch jobs. This is necessary because of the huge data files. Furthermore, ISM is the only one of the programs used that allows blanking out of certain areas (e.g. Pleistocene hinterland). In the course of gridding, this function replaces the node values by zero. A grid spacing of 40 m proved to be appropriate for all computations. Because of the huge number of measurements in the case of the digitized contour lines, only one depth value was considered for each grid node. In contrast, grid calculations based only on borehole data made use of eight depth values for each grid node.

At present the volume of the Holocene sediment wedge has been determined for over 50% of the research area (total area 4700 km^2) in the coastal lowlands of Lower Saxony. By the end of 1997, the calculations and interpretation of the results will have been completed

Acknowledgements

The authors would like to thank the Deutsche Forschungsgemeinschaft for their financial support within the scope of the Priority Program Changes in the Geo-Biosphere in the Course of the last 15 000 Years. We would also like to thank the Niedersächsisches Landesverwaltungsamt - Abteilung Landesvermessung -, the Wasser- und Schiffahrtsdirektion Nordwest in Aurich, the Bundesanstalt für Wasserbau in Karlsruhe, and the Forschungsstelle Küste Norderney des Niedersächsischen Landesamtes für Ökologie for making available the digital elevation models for parts of the coastal marshland, tidal flats, and subaquatic areas.

References

Barckhausen, J., Preuss, H. & Streif, H. (1977): Ein lithologisches Ordnungsprinzip für das Küstenholozän und seine Darstellung in Form von Profiltypen.- Geol. Jb., **A 44**: 45-77, 7 Fig., 3 Tab.; Hannover.

Behre, K.-E., Dörjes, J. & Irion, G. (1984): Ein datierter Sedimentkern der südlichen Nordsee.- Probl. Küstenforsch. südl. Nordseegebiet, **15**: 135-148, 6 Fig., 2 Tab.; Hildesheim.

Homann, H.-H. & Huy, D. (1985): DASP-Beschreibung für Anwender (VAX-Version).- 283 pp., 17 Fig.; Hannover (unpublished report NLfB).

Interactive Surface Modeling (1990): User's Guide - Release 7.1.- Dynamic Graphics, Inc., 604 pp.; Alameda.

Kühne, K. (1983): DASP - Ein System zur Verwaltung und Auswertung geowissenschaftlicher Daten.- Geol. Jb. **A 70**: 41-59, 9 Fig.; Hannover.

MacGridzo (1994): User Manual - Version 3.62.- RockWare, Inc., 273 pp.; Golden.

Menke, B. (1996): Palynologische Untersuchungen des Vibrokerns "Gauss 1987/5" aus der südlichen Nordsee.- Geol. Jb., **A 146**: 177-182, 1 Fig.; Hannover.

Niemeyer, H.D., Goldenbogen R., Schroeder, E. & Kunz, H. (1995): Untersuchungen zur Morphodynamik des Wattenmeers im Forschungsvorhaben WADE.- Die Küste, **57**: 65-94, 23 Fig., 2 Tab.; Heide i. Holst.

Preuss, H. (1988): Map Construction using Advanced Raster Techniques.- Geol. Jb., **A 104:** 187-195, 7 Fig.; Hannover.

Preuss, H. (1992): Mapping using Integrated Raster and Vector Data.- Geol. Jb., **A 122:** 167-176, 6 Fig.; Hannover.

Preuss, H., Vinken, R. & Voss, H.-H. (1991): Symbolschlüssel Geologie.- 328 pp., 1 Fig., 21 Tab.; Hannover.

Schleider, W. & Buziek, G. (1991): Digitales Geländemodell Küste (DIGEKÜ) - Grundlagen und Aufgabe.- Die Küste, **52:** 139-144, 2 Fig.; Heide in Holstein.

Streif, H. (1990): Das ostfriesische Küstengebiet.- Slg. Geol. Führer, **57**, 2 Aufl.: 376 pp., 48 Fig., 10 Tab., 1 encl.; Berlin - Stuttgart (Borntraeger).

Surfer for Windows (1995): User's Guide - Version 6.- Golden Software, Inc., 528 pp.; Golden.

Washausen, M. (1992): Digitale Geländemodelle (DGM) in Niedersachsen - Qualität, Aufbau, Nutzung.- Nachrichten Niedersächs. Vermessungs- u. Katasterverwaltung, **42.** Jg., 3: 177-203, 13 Fig.; Hannover.

Sedimentological Control Parameters and Compaction – a Complex Model of Porosity Distribution in Holocene Muds of the Arkona Basin (Western Baltic)

I. Gustavs

1
Introduction

Detailed knowledge of the lateral and vertical distribution of liquid and solid sediment phases is a prerequisite for the development of mass-balance models. The Institute of Baltic Sea Research Warnemünde (IOW) started a thorough investigation of porosity and related physical and sedimentological properties of Holocene muds of the western and southern Baltic Sea with the objective of:

- Mathematically describing the vertical and lateral distribution of porosity.
- Generating a complex model to portray porosity as a function of compaction and sedimentological parameters.
- Extending the framework of information about compactional and related parameters by experimental means of oedometer testing.

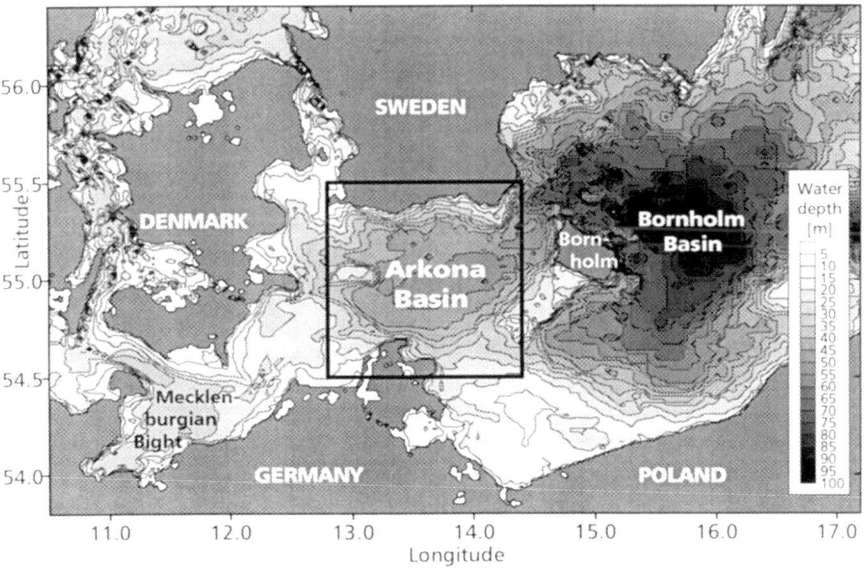

Fig. 1 Area of investigation. (Bathymetric data from Seifert and Kayser 1995)

- Theoretically calculating the effects of compaction on sediment thickness and fluid motion within an idealized mud sequence.

This chapter is a partial account of such study, concentrating on the successive development of a complex porosity model for the uppermost half meter of Holocene muds in the Arkona Basin (Fig. 1), and an application of this model and a scenario of compaction effects to an idealized mud sequence.

2 Factors that Influence Porosity

One can divide factors controlling porosity into two main classes. The first class groups all facies components, retaining those properties of the solid compounds that influence the water-binding capacity of the sediment. For instance, porosity is directly proportional to grain-size sorting and content of organic matter, biogenic silica and clay minerals, while it shows an indirect proportionality to mean grain diameter.

The second group consists of diagenetic factors. Ignoring chemical effects playing a minor part in the stage of early diagenesis, this class consists of physical factors that are acting in connection with increasing burial and through time.

3 Compaction and its Classical Modeling

Physical diagenesis or compaction results in a continuous porosity decrease with increasing burial. This process, being mainly a function of effective pressure and time, causes pore water movements and a reduction of sediment thickness. The most simple and most frequently used mathematical description of compaction is an empirical, in most cases exponential function in a porosity-depth diagram, called compaction curve (Athy 1930; Weller 1959; Roll 1974; Sclater and Christie 1980; Baldwin and Butler 1985; Huang and Gradstein 1990). This method is still the most common way of mathematically considering compaction effects for mass balances (e.g. Wold 1992).

Two parameters characterize the compaction curve – the initial porosity at the sediment surface, and the rate of porosity reduction with depth (compaction constant). Usually, these two parameters are constant and specific for each sediment type. This assumption implies a homogeneous composition of each sediment type that is not fulfilled in the case of Baltic muds.

A second fact to be considered is the interrelation between depth interval and compaction curve parameters. The so-called compaction constant is not strongly constant, but depends on the depth interval of measured porosities (near-surface, some meters, or deep burial). The wide disparity between the compaction constants for kilometer intervals estimated by Sclater and Christie (1980) of about 0.005 km^{-1}, and values of about 0.05 m^{-1} (=50 km^{-1}) for the near-surface meter interval estimated from our own measurements for Baltic muds illustrates this behaviour.

4 Data

The research is a continuation of a similar survey done by Endler (1990), who measured physical and sedimentological properties on mud samples from the southern Arkona Basin. To obtain more generalized models, the database has

Table 1 Characterization of data sets

Reference	No. of locations	Sampling depth (cm)	Sample interval (cm)	No. of measurements	
				Physical	Sedimentological
Gustavs	35	25 … 60	2 … 3	583	326
Endler	29	20 … 60	3 … 4	470	470
Archive	302	3	3	–	302

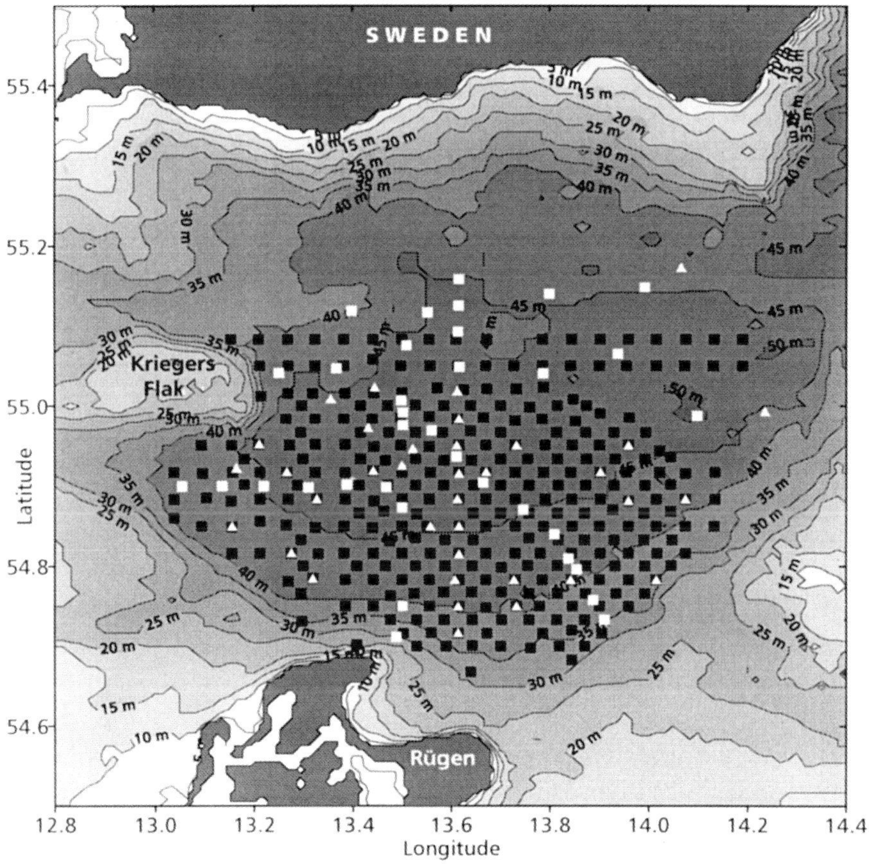

Fig. 2 Sampling locations in the Arkona Basin. *White squares* New measurements; *white triangles* Endler (1990); *black squares* archive data; *isolines* water depth

been extended with more measurements. Endler's, as well as the new measurements, are mainly taken from the uppermost half meter of sediment. In the statistical analysis, we avoided mixing of the two data sets, because of relevant procedural differences (for instance, pipette method versus laser particle size analyzer).

A third set of mud data was taken from geochemical archive data of the IOW. These data describe sedimentological parameters of topsoil samples of the southern Arkona Basin. Because they were measured using the same methodology as Endler (1990), one can use these data as proxy parameters for the application of the porosity model derived from Endler's data.

The measured physical properties of interest in our study are water content and bulk density which were used to estimate porosity. The sedimentological variables of interest are the content of organic carbon, and the grain size distribution. A short characterization of the these two data sets is given in Table 1. Sampling locations are shown in Fig. 2.

5 Development of a Mud-Specific Model

The objective of developing a mud-specific model is twofold. First, the model should result in a better understanding of porosity distribution and porosity changes by separating and quantifying several diagenetic and facial influences. Secondly, the model should facilitate the reconstruction and prediction of porosity values from proxy parameters.

The ideal model should meet three basic demands:

- be as accurate as possible,
- retain as many parameters as necessary,
- use commonly available parameters only.

As already mentioned, there is a problem of inconsistency among the different measurement methods of the "old" data (Endler, archive) and "new" data. This inconsistency is particularly important in grain-size analysis that strongly depends on the estimation method. The statistical analysis would lose much of its accuracy by mixing data sets. For successive model improvement, we employed Endler's data because later we will apply the model using his data base to the geochemical data base.

5.1 Application of the Empirical Model

This step of the modeling was aimed to deliver information about the accuracy of the classical way of compaction modeling for near-surface muds. Does the accuracy of the model justify the use of such a function in mass balances?

The simplest model is

$$\Phi = \Phi_0 e^{-Bz}, \tag{1}$$

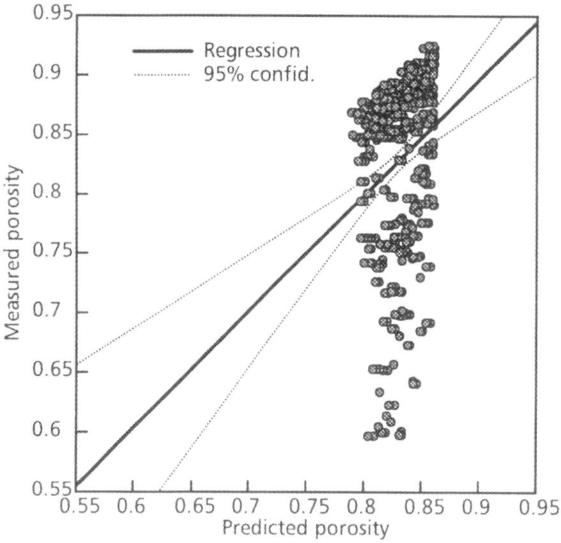

Fig. 3 Empirical porosity-depth, predicted versus observed values

where Φ is porosity (fraction 0-1), $\Phi_0 = 0.863$ is initial porosity at z=0 cm, B= 1.525×10^{-3} is the depth-dependent compaction constant, and z is depth in centimeters. Figure 3 shows a diagram of measured versus modeled values. As this graphic and the quality criteria of $R = 0.25$ and $R^2 = 0.06$ clearly indicate, there is only a weak correlation between measurements and model values. The model explains only 6% of the measured porosity variations. One reason for this poor quality is that bulk densities vary between 1.1 g/cm^{-3} in the center and 1.6 g/cm^{-3} at the margin of the basin. Therefore, depth is not an equivalent for the compaction-causing part of porosity variations.

5.2 Application of the Physical Model

To better explain the compaction factor of porosity, we tried a pressure-based model of the form

$$\Phi = \Phi_0 e^{-bp} , \quad (2)$$

where $\Phi_0=0.899$ is initial porosity, $b=1.528\times10^{-4}$ is the pressure-based compaction constant, and p is the effective pressure (Pa). The quality criteria of $R= 0.68$ and $R^2=0.45$ as well as the diagram of measured versus modeled values in Fig. 4 indicate the significant improvement of model accuracy. Yet the unexplained part of the natural variations is more than 50%.

The reason for this behavior lies in a strong non-diagenetic part of porosity variations, a significant part of the facial component that is caused by sedimen-

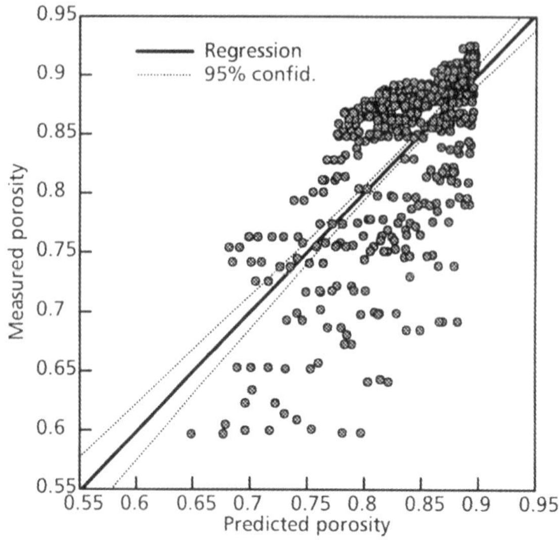

Fig. 4 Physical porosity-pressure model, predicted versus observed values

tological variations in a wide range (organic matter 1...8%, mean grain diameter 10...70 µm) within this type of sediment.

5.3 Development of a Physical-Sedimentological Model

Use of sedimentological attributes is essential to properly model porosity. Because of the strong influence of facies factors on porosity, especially in near-surface muds without significant compaction effects, sedimentological composition is the most important factor controlling porosity. Despite that this has been common knowledge for a long time, there is a lack of information about the exact influence of these components. Thus, the following step of modeling should help answering the questions: is porosity mainly controlled by grain-size distribution or by content of organic matter? At what depth does compaction become the most important factor? The next step towards a better matching of measured and modeled values takes the form

$$\Phi = \left(f_0 + f_c c_{org} + f_m x_m + f_s \sigma\right) e^{-bp}, \tag{3}$$

where c_{org} is the content of organic carbon (weight percent, %), x_m is the mean grain diameter (µm), σ is the standard deviation of grain size distribution (µm), and f_0=0.868, f_c=1.21×10^{-2}, f_m=-5.99×10^{-3}, f_s=4.53×10^{-3}, and b=7.2×10^{-5}. Now R=0.97 and R^2=0.94. From the quality criteria as well as from the diagram in Fig. 5 it is evident that by integrating sedimentological parameters into the pressure-based compaction model one obtains a powerful, high-resolution model that explains 94% of the natural porosity variation.

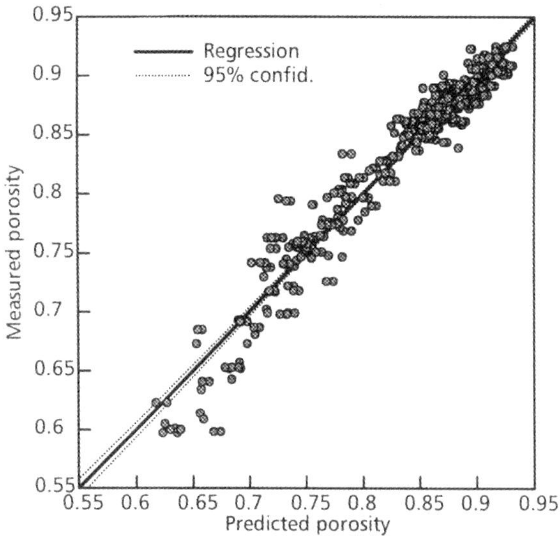

Fig. 5 Physical-sedimentological porosity model, predicted versus observed values

One can calculate the influence on porosity of each factor by multiplying the estimated factors with the range of the different parameters. These values show that porosity at about equal parts is governed by grain size distribution, content of organic matter and effective pressure. In other words, sedimentological composition clearly dominates the porosity distribution near the sediment surface, while effective pressure at a depth of about 2 m and deeper becomes the most important factor.

Table 2 Coefficients of the physical-sedimentological porosity model

Data (reference)	f_0	f_c	f_m	f_s	b
Endler	0.868	1.21×10^{-2}	-5.99×10^{-3}	4.53×10^{-3}	7.2×10^{-5}
Gustavs	0.789	2.46×10^{-2}	-6.27×10^{-3}	4.43×10^{-3}	4.1×10^{-5}

Table 3 Influence of components on porosity

Component	Min...max, range (Endler data) Min...max, range (this work)	Change of Φ (parts of 1.0)
C_{org}	0...10(%), 10 0...8(%), 8	0.10 0.20
Mean	10...50(μm), 40 8...68(μm), 60	0.25 0.35
σ	10...30(μm), 30 5...45(μm), 40	0.10 0.10
p_{eff}	0...2000(Pa), 2000 0...2000(Pa), 2000	0.15 0.10

6
Application of the Physical-Sedimentological Model

The set of measured porosities at the sediment surface in the Arkona Basin includes 64 locations (white squares in Fig. 2). To achieve a higher resolution in lateral direction before creating a regional model, we increase the number of data points by porosity modeling. One can add 302 measurements by considering the IOW data base (gray squares in Fig. 2). This resulted in a significant densification in the data coverage, especially in the southern part of the study area.

In the next step, after calculating initial porosities, the coincidence of measured and modeled data was assessed by a graphical method. For this purpose, the numerous modeled porosities were interpolated and plotted in form of an isoline map, over which the measured porosities were projected as numbers.

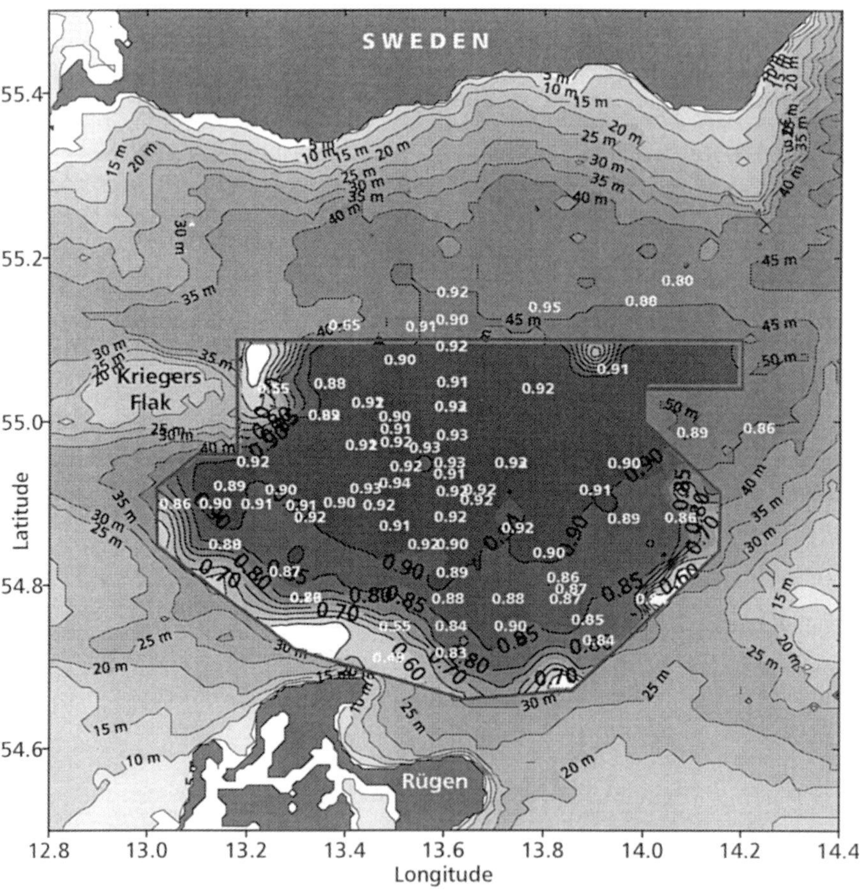

Fig. 6 Modeled and measured porosities. *Isolines with black numbers* Modeled porosities from archive proxy data; *white numbers* measured porosities; *background* water depth

Figure 6 shows the resulting map. This illustration clearly indicates that in most cases the measured data are exactly equal to the respective areas in the isoline map. This encouraging result justified establishing a combined data set, enclosing the modeled data as well as the porosity measurements.

7 Regional Modeling of Initial Porosity by Means of Trend Surface Analysis

Systematic decrease of porosity from high values in the center to low values at the margin of the basin (see Fig. 6) led to the hypothesis of a second-order trend in the data. We ran trend-surface analysis to confirm this hypothesis.

7.1 Trend-Surface Analysis

In the case of this study, the trend surface takes the form of an elliptic paraboloid given by

$$\Phi_0(x, y) = -10301.4 - 2.7x + 375.5y + 2.7xy - 0.4x^2 - 3.4y^2, \quad (4)$$

where x and y are the geographical longitude and latitude. The ellipsoid, shown in Fig. 7, is elongated in east-west direction, with the center situated in the middle of the Arkona Basin. The axes run nearly parallel to the longitude and latitude.

7.2 Analysis of Residuals

According to the F-test, the trend surface is a significant descriptor of the data. Nevertheless, the R^2 of 0.5 indicates important differences between modeled and measured data. Are these differences distributed randomly or do they show a systematic behaviour? In order to answer this question, we analyzed the distribution of the residuals:

- low differences between model and measurement (gray),
- measurement is at least 6% lower than model (white),
- measurement is at least 6% higher than model (black).

The distribution of the residual is given in Fig. 7. Most of the locations, covering especially the center of the basin, show differences less than 6%. Higher differences appear mainly at the margins. The residuals alternate between positive values in the deeper parts (SE, SW, NE) and negative values near-shore and around submarine hills (S, W, E), precisely tracing the even center and the morphologically differentiated margin of the basin. They clearly show the expected dependence of initial porosity from morphological features of the basin.

Thus, the surface modeling the variation in porosity forms a convex ellipsoidal surface with a slight continuous transition from higher to intermediate values in the central part, but with a steeper and undulating course at the margin of the Arkona Basin.

8 Effects of Compaction on Sediment Thickness and Pore Water Movement

As an extension of numerical procedures (Gustavs 1992), we created a coupled steady-state model of sedimentation, compaction, and pore-water movement to study the effects of compaction on growing sedimentary sequences. The model is based on the relation between effective pressure and porosity and assumes a homogenous accumulation of mud at a constant sedimentation rate during time.

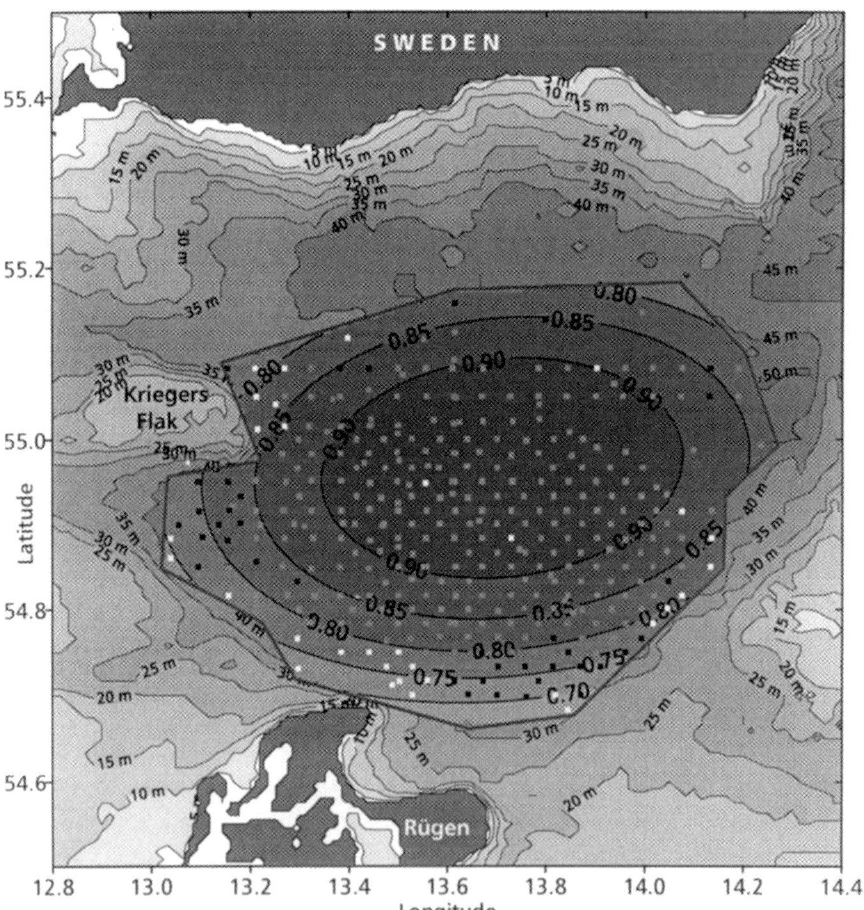

Fig. 7 Trend surface (isolines) and residuals of initial porosity Φ_0(symbols, see text)

The input parameters were set as follows: initial porosity $\Phi_0 = 0.9$,pressure-based compaction constant $b=2.5\times10^{-5}$ Pa^{-1}, sedimentation rate $SR=1$ mm/a. The stepwise calculation was performed at 10-year intervals with sedimentary layers marked every 1000 years. The duration of the sedimentation was limited to 8000 years. Figure 8 shows the results of the simulation with the solid basement and the sediment surface as reference horizon, respectively.

8.1 Effects on Sediment Thickness

A duration of 8000 years of sedimentation in combination with a sedimentation rate of 1 mm/a without compaction effects would result in a sediment thickness of exactly 8 m. As illustrated by Fig. 8, the thickness of the entire sediment sequence is reduced to about 4m at the end of the time interval. In this case, compaction reduces the whole mud layer to about half its uncompacted thickness.

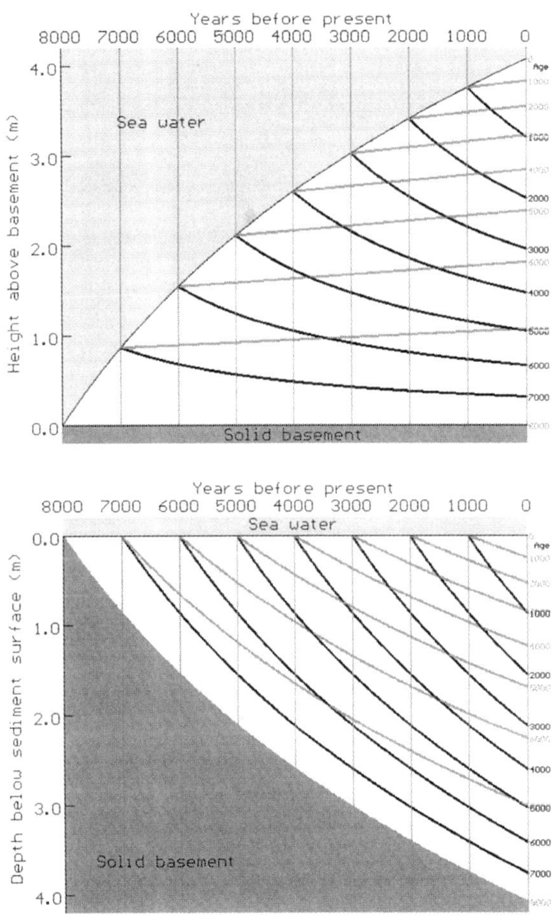

Fig. 8 Coupled model of sedimentation, compaction, and pore-water movement. Reference horizon: solid basement (*above*), sediment surface (*below*); *gray* pore water; *black* sediment particles

8.2 Effects on Pore Water Movement

On interpreting the pore-water movements predicted by the model, one should consider that all diffusion effects within the sedimentary column and between sediment and overlying water have been ignored. The reduction of the water-filled pore space induced by compaction causes an upward movement of pore water in which older pore water rises into younger sedimentary layers and displaces the younger pore fluids. This process leads to different burial velocities of solid and liquid sediment components and to increasing distances of pore fluids to their source layers with burial. Assuming a complete water displacement, finally, at the end of the process there will be complete separation of solids and liquids.

Figure 8 outlines the extent of this effect for the sedimentary sequence of interest. The shade of the numbers marks the age of pore water (gray) and sedimentary layers (black). According to this scenario, one can expect differences of up to 3000 years between the age of pore water and that of sediment particles. Expressed in length units, this is equivalent to distances up to 2 m between pore water and its source layer. One should keep this fact in mind when interpreting pore-water analysis with respect to the accumulation environment of the apparent source layer.

9 Summary and Conclusions

It is our experience that the classical exponential compaction curve is not a good model for the description of porosity decrease in near-surface muds. We have been able to find that, for the Arkona Basin, integration of sedimentological factors into a pressure-based exponential model produces a satisfactory account of measured porosity variations.

This new model was used for the reconstruction of initial porosities from sedimentological proxy data. The good coincidence of modeled and measured data confirmed the high quality of the model and facilitated the generation of a dense mesh of data points to be laid over the Arkona Basin. On this basis the lateral distribution of initial porosities could be approximated by a second-order trend surface. Modeling a scenario of compaction effects on an idealized, homogeneous mud sequence showed significant differences between the rate of burial of solids and pore fluid, caused by the compaction-driven, upward-directed movement of pore water.

Although the age boundaries of pore water are not sharp due to diffusion processes, the estimated difference in age and distance between pore water and source layer should be kept in mind when interpreting results of pore-water analysis and relating them to the environmental conditions of sediment accumulation of the apparent source layer.

Acknowledgements

The Author wishes to express gratitude to Rudolf Endler and Thomas Leipe for providing numerous unpublished data that formed an important part of this study. I am also grateful to Ricardo A. Olea (Kansas Geological Survey) for critically reviewing the manuscript and to Franz Tauber for constructive discussions and for helping with the illustrations.

References

Athy, L. F. (1930): Density, porosity and compaction of sedimentary rocks. *AAPG Bull.* **14:** 1-24, Tulsa/Okl.

Baldwin, B., and Butler, C. O. (1985): Compaction curves. *AAPG Bull.* **69** (4): 622-626, Tulsa/Okl.

Huang, Z., and Gradstein, F. M. (1990): Depth-porosity relationship from deep-sea sediments. *Scientific drilling* **1**: 57–162, Berlin (Springer).

Gustavs, I. (1992): Rechnergestützte Inversion und Modellierung des Kompaktionsprozesses von Sedimenten. Diploma thesis, Ernst Moritz Arndt University, Greifswald.

Endler, R. (1990): Akustische Eigenschaften des seebodennahen Schlicks im Arkonabecken. Dissertation, Institut für Meeresforschung der AdW der DDR.

Roll, A. (1974): Langfristige Reduktion der Mächtigkeit von Sedimentgesteinen und ihre Auswirkung - eine Übersicht. *Geol. Jahrb. (a)* 14: 1–76, Hannover.

Sclater, J. G., and Christie, P. A. F. (1980): Continental stretching: An explanation of the Post-Mid-Cretaceous subsidence of the central North Sea basin. *J. Geophys. Res. (B)* **85** (7): 3711-3739, Richmond.

Seifert, T., and Kayser, B. (1995): A high resolution spherical grid topography of the Baltic Sea. *Meereswissenschaftliche Berichte / Marine Science Reports*, Institut für Ostseeforschung, Warnemünde.

Weller, J. M. (1959): Compaction of sediments. *AAPG-Bull.* **43**: 273-310, Tulsa/Okl.

Wold, C. N. (1992): Paleobathymetry and sediment accumulation in the northern North Atlantic and southern Greenland-Iceland-Norwegian Sea. Diss. Univ. Kiel, 1992.

Classification of Compositional Data Using Mixture Models: a Case Study Using Granulometric Data

C. Barceló, V. Pawlowsky and G. Bohling

1 Introduction

Classification of samples is a fundamental problem in the geosciences. Usually, an initial unsupervised classification is used to define natural groups of samples. Then, discriminant analysis is applied to obtain posterior probabilities that a given sample belongs to each group. Those probabilities may be mapped onto geographic space, resulting in a regionalized classification. Almost always this procedure implicitly assumes that the observations forming the different groups come from multivariate normal distributions. Although the majority of classification procedures based on minimum Mahalanobis distance are robust with respect to deviations from the hypothesis of normality, the above-mentioned associated posterior probabilities, calculated to draw regionalized classification maps, are more sensitive to departures from the assumed distribution model.

For those cases where a hypothesis of normality can be assumed, McLachlan and Basford (1988) suggest mixture models as a clustering technique. With a normal mixture model, a clustering is performed assigning the data to groups on the basis of estimated posterior probabilities, but it requires maximum-likelihood estimates for the parameters of the model. The EM (expectation/maximization) algorithm provides a convenient way for the iterative computations of solutions of the likelihood equations. Initial values for the mixture parameters needed to start the EM algorithm are calculated using memberships derived from a standard clustering method.

In the case of compositional data, the hypothesis of normality has to be directly rejected because the natural sample space for this kind of data is the simplex. Thus, strictly speaking, the data should be transformed prior to application of the previously discussed methodology. Two approaches have been considered: transforming the data using the log-ratio approach (Aitchison 1986) or applying multivariate Box-Cox transformations (Barceló et al. 1996). Assuming transformed data to come from a mixture of multivariate distributions, the clustering procedure proposed by McLachlan and Basford (1988) can be applied. It was expected that the resulting groups, the posterior probabilities of group membership and the associated regionalized classification maps obtained by either of these procedures were, from a statistical point of view, more consistent with the assumed model and thus in principle more reliable than those obtained applying traditional clustering procedures to untransformed compositional data.

2
Case Study

We applied the clustering procedure based on a mixture model using both types of transformations to granulometric data from the Darss Sill area of the Baltic Sea (Fig. 1). As stated in Davis et al. (1995) and in Bohling et al. (1998), the samples of Darss Sill data set represent the weight percentage of each sample within each of eight grain size classes: gravel (>2000 μmm), very coarse sand (2000–1000 is μmm), coarse sand (1000–630 μmm), medium sand (630–400 μmm), medium fine sand (400–200 μmm), fine sand (200–100 μmm), very find sand (100–63 μmm) and silt (<63 μmm). The Darss Sill area represents the primary bottleneck in the exchange of water between the North Sea and the Baltic Sea. In the study area the predominant current flow and the sediment transport direction is from the southwest across the Darss Sill and northeast and east into the Baltic. There is evidence that current patterns in the eastern portion of the study area follow a counterclockwise circulating pattern. The Darss Sill acts as a barrier between the Mecklenburgian Bight in the southwest and the Arkona Basin in the east. It consists of glacial tills and is covered in areas with lag sediments. The Kadet Channel was incised in the glacial tills by postglacial drainage.

Sections 2.1, 2.2 and 2.3 show the results obtained applying different clustering procedures to the Darss Sill data set. Results obtained applying the clustering procedure based on mixture models to data transformed by the additive logratio transformation are shown in Section 2.2, while the case for data transformed by a multivariate Box-Cox transformation is presented in Section 2.3. These results

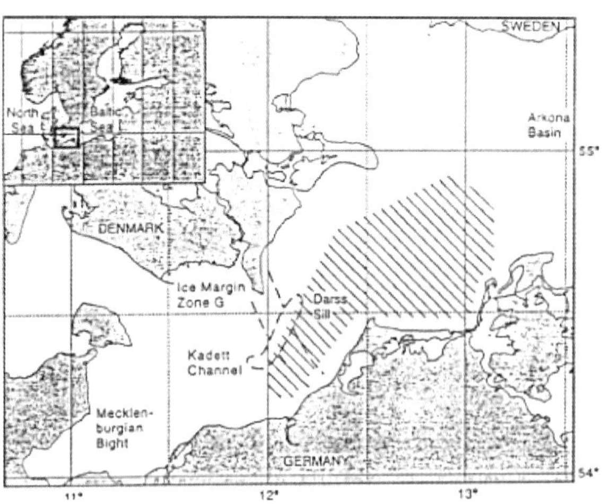

Fig. 1 Darss Sill and surrounding area of the eastern part of the Baltic Sea. (After Davis et al. 1995)

can be compared with those derived from a traditional clustering procedure (Ward's method) applied to untransformed data, which are displayed in Section 2.1.

Tables 1, 3 and 5 and the histograms of Fig. 2, 5 and 8 summarize the typical mean weight percents (centroids) in the size classes of the groups. Figures 3, 6 and 9 show the ternary diagrams of data amalgamated into three grain-size divisions: gravel through medium-fine sand (grv-mfs), fine sand (fs), and very fine sand and silt (vfs-slt). Differents colors are used to represent different groups. Figures 4, 7 and 10 show classification maps obtained interpolating Mahalanobis distances in a regular grid with a 300-m spacing in each direction using a two-step scaled inverse-distance squared interpolation scheme that incorporates local estimates of slope. All these maps were plotted with Surface III® (Sampson, 1988).

2.1 Hierarchical Clustering Applied to Raw Granulometric Data

Davis et al. (1995) and Bohling et al. (1996) apply a hierarchical clustering analysis using Ward's method (Everitt 1993) to group untransformed samples into classes that are as homogeneous as possible in a minimum variance sense. Because the minimum within-groups sums-of-squares criterion employed in Ward's method corresponds to the assumption of a spherical, homoscedastic (equal-covariance) normal model for the groups, this method tends to create hyperspherical clusters (Statistical Sciences 1995).

Inspection of the results shows – as stated by these authors – that the grain-size distributions corresponding to the centroids of seven groups (see Table 1 and Fig. 2) reflect the well to moderately sorted nature of groups 1 through 6 and the poorly sorted nature of group 7.

Table 1 Grain-size distribution of centroids of each group (hierarchical clustering)

Gr	N	Gravel	vcsand	csand	msand	mfsand	fsand	vfsand	Silt
1	161 12.6%	2.3	6.9	14.2	26.4	39.0	8.4	1.6	1.2
2	381 29.7%	0.3	0.7	2.1	7.2	64.7	21.3	3.1	0.6
3	196 15.3%	0.4	0.6	1.1	3.1	36.9	46.8	6.8	4.2
4	222 17.3%	0.1	0.4	0.7	1.9	22.6	66.2	6.1	1.9
5	219 17.1%	0.0	0.0	0.0	0.5	5.6	86.2	6.7	0.9
6	55 4.3%	0.0	0.0	0.0	0.2	1.8	60.0	34.2	3.7
7	47 3.7%	2.1	2.2	3.4	4.0	6.1	20.8	24.4	37.0

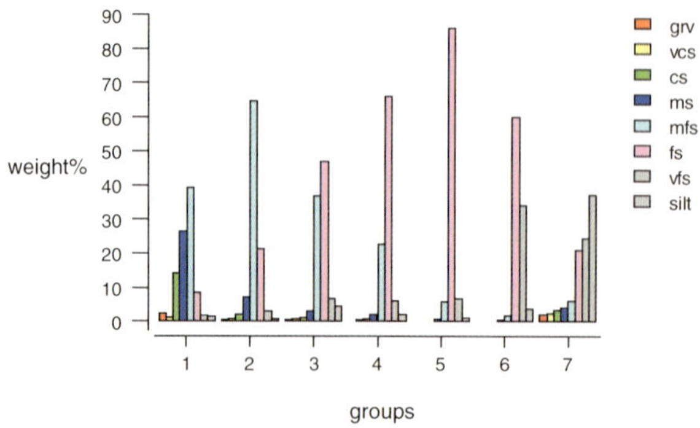

Fig. 2 Grain-size distribution of centroids of each group (hierarchical clustering)

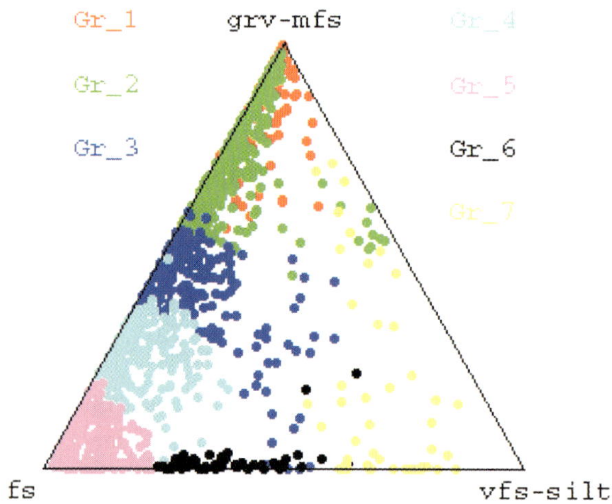

Fig. 3 Ternary diagram for three-part amalgamation (hierarchical clustering)

A genetic interpretation can be assigned to the results of sediment classification. The poorly sorted group 1, with an unusually high proportion of gravel and coarse sand, represents the remnants from submarine erosion of a glacial till. Group 2 is also erosional in origin, but shows a transitional character. Group 3 belongs to a depositional facies but its relatively poor sorting suggests that it is a transitional sediment. The more typical depositional facies include group 4, which has transitional characteristics, group 5, which is typical of depocenters, and group 6, which represents a sediment that has bypassed the depocenter and has been deposited in a distal position. Group 7 represents a channel environment where erosion, transportation, and deposition change very rapidly. The ternary diagram of Fig. 3 also reflects the sequence from group 1 to group 6 and the erratic location in the diagram of samples belonging to group 7. Note the difficulty to distinguish samples from groups 1 and 2 in this ternary diagram.

In spite of the fact that group distributions deviate from the assumed model of multivariate normality with a common covariance matrix, Bohling et al. (1998) contend that the classification map shown in Fig. 4 makes geologic sense:

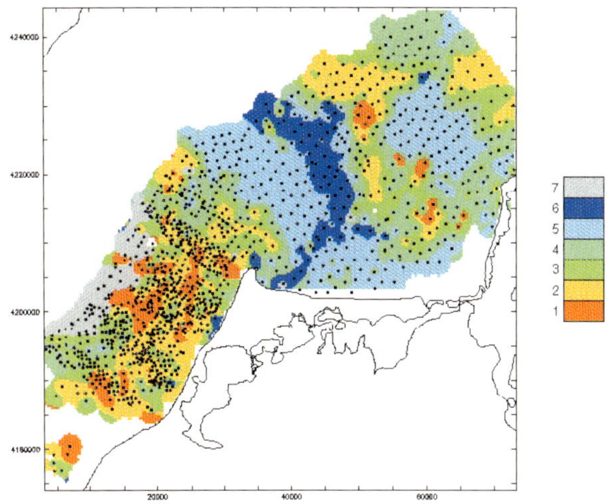

Fig. 4 Regionalized classification based on hierarchical clustering procedure applied to raw granulometric data

the groups form geographic sequences corresponding to the source area to depocenter sequence represented by groups 1 to 6.

The overall error rate (see Table 2) of the linear discriminant rules associated to this classification under the assumption of equal prior probabilities is equal to 7.3% and can be considered reasonably low. Samples of group 5 are the best characterized, whereas there is a significant confusion between samples of groups 1 and 2. Although the confusion between samples belonging to adjacent groups seems to be natural, it is surprising that almost 8% of samples from group 3 were wrongly classified in group 7.

Table 2 Resubstitution results for linear discriminant analysis (hierarchical clustering)

Assigned group	True group						
	1	2	3	4	5	6	7
1	150	14	0	0	0	0	3
2	11	335	0	0	0	0	0
3	0	32	171	13	0	0	0
4	0	0	8	198	1	0	0
5	0	0	0	8	218	0	0
6	0	0	2	2	0	54	0
7	0	0	15	1	0	0	44
Total N	161	381	196	222	219	55	47
N wrong	11	46	25	24	1	1	3
Error rate	6.8%	12.1%	12.8%	10.8%	0.5%	1.8%	6.4%
					Overall error rate:	7.3%	

2.2 Mixture Clustering Model over the *alr*-Transformed Data

Since granulometric data are compositional, attention should be paid more to ratios of components than to their absolute values. For this reason, Aitchison (1986) suggests applying the additive log-ratio transformation (*alr*) to compositional data. This transformation breaks the constant sum constraint which distinguishes this kind of data.

Since one cannot compute the logarithm of zero, zero data values must be adjusted to some small positive value before the *alr* transformation is performed (Aitchison 1986, Chap. 11). This adjustment was done using a value of $\delta=0.05$ as the estimated maximum rounding error (Bohling et al. 1997). Bohling et al. (1998) demonstrate that the computed results do not vary greatly when δ is varied over a reasonable range, despite the large number of zeros in the data set.

Then, assuming that *alr*-transformed data come frome a mixture of seven homoscedastic normal multivariate distributions, the EM-algorithm is applied as a clustering technique. The sample is considered to be a completely unclassified, although the initial estimations of the parameters of the mixture, needed to start the EM algorithm, are calculated from the seven groups which are obtained when Ward's method is applied to raw granulometric data (see Section 2.1). This procedure provides a new classification of data in seven groups (see Table 3 and Fig. 5). These new groups are, from a statistical point of view, very well delimited, because the overall error rate of the associated linear discrimination rule applied to transformed data is equal to 1.2%. However, the ternary diagram of Fig. 6 does not reflect the separation of the groups.

The new groups are quite different from the initial hierarchic cluster groups of Subsection 2.1. The new clustering tends to move samples from initial hierar-

Table 3 Grain-size distribution of centroids of each group (*alr*-clustering)

Gr	N	Gravel	vcsand	csand	msand	mfsand	fsand	vfsand	Silt
1	125 9.8%	5.6	6.4	9.9	16.6	34.6	17.9	4.1	4.2
2	218 17.0%	0.0	4.1	8.1	15.8	43.7	22.7	3.4	2.2
3	316 24.7%	0.0	0.0	2.1	5.7	50.5	35.3	5.3	1.1
4	413 32.2%	0.0	0.0	0.0	2.4	34.7	57.2	4.6	1.1
5	151 11.8%	0.0	0.0	0.0	0.3	1.1	79.4	16.4	2.7
6	13 1.0%	0.0	0.0	0.0	0.0	0.0	26.9	34.8	38.3
7	45 3.5%	0.0	0.0	0.0	0.0	10.4	48.9	18.2	22.5

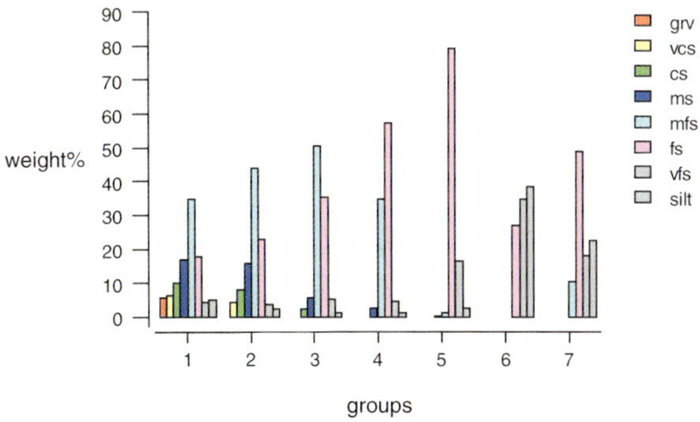

Fig. 5 Grain-size distribution of centroids of each group (*alr*-clustering)

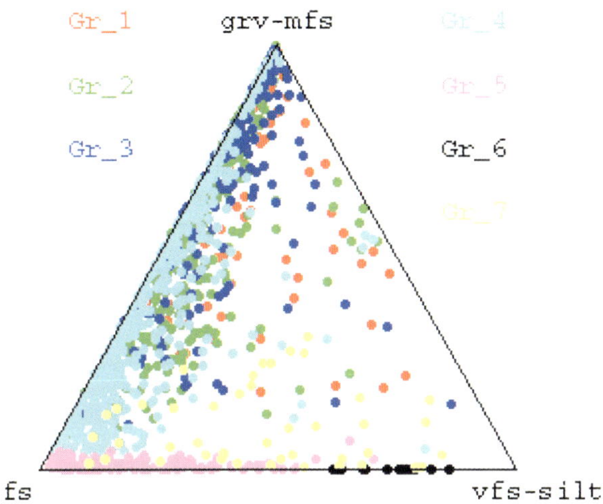

Fig. 6 Ternary diagram for three-part amalgamation (*alr*-clustering)

chic groups to new central groups 3 and 4 (see Table 4). Tests of normality applied to marginal distributions in each of the seven groups show that the hypothesis of multivariate normality cannot be supported a posteriori. In spite of this, the classification map of Fig. 7 has a pattern globally similar to the map in Fig. 4, although less differentiated in the eastern part, while more chaotic in the Kadet channel. A geologic interpretation of obtained results would be necessary to determine the validity of the applied procedure.

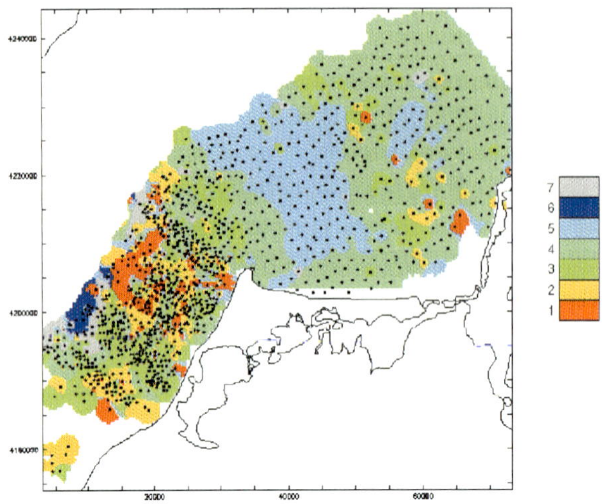

Fig. 7 Regionalized classification based on a mixture model as a clustering procedure applied to data transformed by the *alr*-transformation

Table 4 Hierarchical clustering groups versus *alr*-groups

Hierarchic (initial) groups	Put into new *alr*-groups							
	1	2	3	4	5	6	7	Total
1	61	89	8	3	0	0	0	161
2	27	75	172	107	0	0	0	381
3	18	25	62	79	3	0	11	196
4	9	21	61	121	1	0	9	222
5	0	4	10	100	99	0	6	219
6	0	0	2	1	45	0	7	55
7	12	4	1	2	3	13	12	47
Total	125	218	316	413	151	13	45	1281

2.3
Mixture Clustering Model over the BC-Transformed Data

In some cases, if the log-ratio transformation cannot adjust compositional data to a mixture of multivariate normal distributions, a Box-Cox (BC) transformation can be a good alternative to obtain a better adjustment. In such cases, it is necessary to choose a denominator to form ratios and to estimate the vector λ of the transformation (Barceló et al. 1996). For the Darss Sill data set, the variable "fine sand %" was used as the denominator of the ratios and the vector λ of the BC-transformation was estimated as

$$\lambda=(-0.76,-0.33,0.61,0.08,0.13,0.02,-0.02)^{t} .$$

Table 5 Grain-size distribution of centroids of each group (*BC*-clustering)

Gr	N	Gravel	vcsand	csand	msand	mf-sand	fsand	vfsand	Silt
1	276 21.5%	0.7	4.4	8.8	17.0	43.6	20.8	3.1	1.7
2	84 6.6%	0.0	0.0	2.6	11.7	72.3	7.8	4.9	0.7
3	209 16.3%	0.0	0.0	0.9	4.1	62.3	29.2	2.8	0.8
4	505 39.4%	0.0	0.0	0.4	1.5	21.4	69.8	5.8	1.1
5	103 8.0%	0.0	0.0	0.0	0.3	4.7	59.8	25.4	9.9
6	23 1.8%	0.0	0.0	0.0	0.1	1.7	25.2	34.6	40.4
7	81 6.3%	6.4	5.9	7.8	12.3	28.9	25.0	5.8	7.9

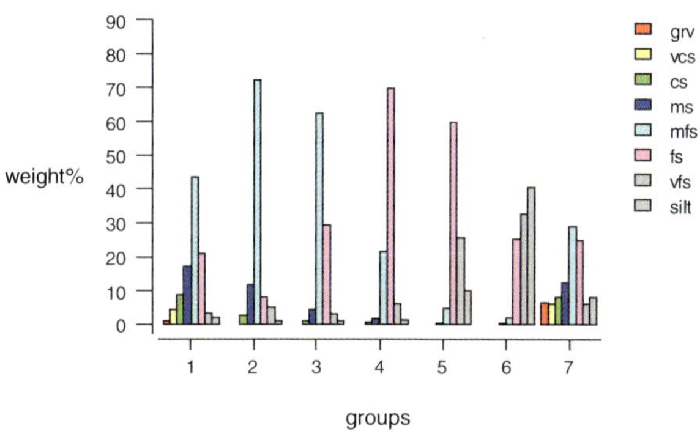

Fig. 8 Grain-size distribution of centroids of each group (*BC*-clustering)

The seven groups obtained from hierarchical clustering analysis were used to estimate λ with the additional hypothesis of homoscedasticity of transformed groups. After the BC-transformation was applied to the granulometric data, the EM-algorithm was executed in the same fashion as in the *alr* case (Section 2.2).

In Table 5 and Fig. 8 it can be seen that the new groups 1 to 6 follow again a genetic sequence, from coarser to finer granulometry, whereas group 7 resembles a residual clustering group. From a statistical point of view, new groups seem to be well defined because a small overall error rate of 2.5% is associated with the corresponding linear discriminant rule when applied to transformed data. However, neither the genetic sequence nor the separation of groups is visible in the ternary diagram of Fig. 9.

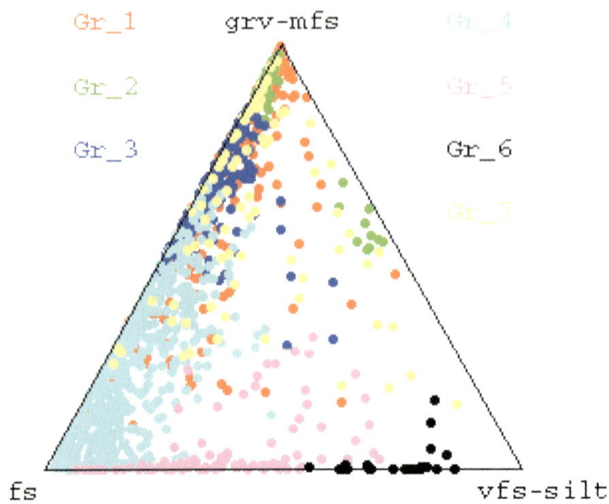

Fig. 9 Ternary diagram for three-part amalgamation (*BC*-clustering)

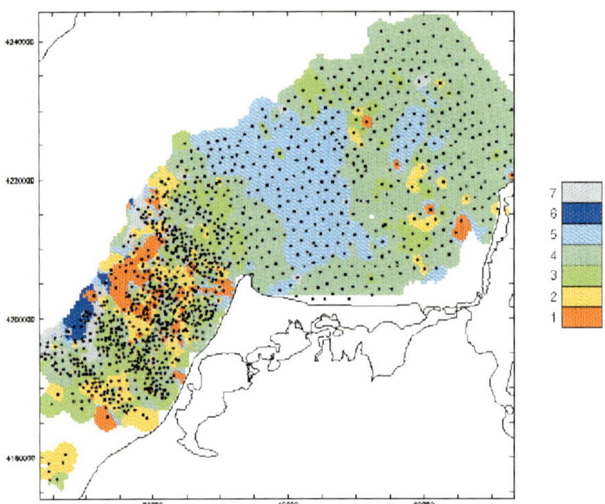

Fig. 10 Regionalized classification based on a mixture model as a clustering procedure applied to data transformed by the *BC*-transformation

Table 6 makes evident that samples from hierarchic group 2 are split and distributed among new groups 1 and 3, samples from hierarchic groups 3 and 5 move to new group 4, while samples of new group 5 come mainly from hierarchic group 6. The new groups 5 and 6 look like groups 6 and 7 of the hierarchic cluster. It seems that only six groups would be necessary to obtain a reasonable clustering of available data. Some of these facts are evident when the maps in Fig. 4 and Fig. 10 are compared.

Table 6 Hierarchical clustering groups versus *BC*-groups

Hierarchical (initial) groups	Put into new *BC*-groups 1	2	3	4	5	6	7	Total
1	127	7	2	0	0	0	25	161
2	88	76	180	18	0	0	19	381
3	28	0	27	111	14	1	15	196
4	26	0	0	178	9	0	9	222
5	4	0	0	196	19	0	0	219
6	0	0	0	2	53	0	0	55
7	3	1	0	0	8	22	13	47
Total	125	218	316	505	103	23	81	1281

2.4. Comparison of the Three Clustering Models

From the graphical and numerical display of the centroids of each group it can be seen that the three approaches admit an ordering of coarser to finer predominant components for the centroids, except for group 7. This gradient can also be observed in the regionalized classification with hierarchical clustering and with the Box-Cox transformation, while with the *alr*-transformation some spatial "discontinuities" can be observed, specially for data assigned to group 1 surrounded by data belonging to either group 5 or group 4. On the other hand, resubstitution results for linear discriminant analysis show that the overall error rate is smallest for the *alr*-approach (1.2%), followed by Box-Cox clustering (2.5%), with hierarchical clustering coming last (7.3%). If we look at the error rates for each group, it can be seen that in the first two cases (*alr* and Box-Cox transformations) the most important concentration of misclassifications is in group 4, while in the hierarchical clustering procedure misclassification appears to be significant by comparison, although acceptable from a statistical point of view, for groups 2, 3 and 4, and less important for group 1.

To check if the two approaches using transformed data are just an improvement of hierarchical clustering, cross tabulation of group memberships are useful. Tables 4 and 6 clearly evidence the breakup of the initial clusters, forming new groups in two clearly different manners except for the initial group 6, which is completely merged into group 5 in both cases.

Finally, some common features can be observed in the regionalized classification: a smooth variation in the eastern part of the study area, and a certainly erratic behaviour in the Darss Sill.

3 Conclusions

Interpretation of results from a cluster-analysis procedure is never the result of merely looking at some numbers. Groups have to be reasonable in our case from a geological point of view, and thus the opinion of a geologist is lacking at this point. Nevertheless, natural groups that are clearly different will be quite stable if subjected to different clustering methods. This is not the case of the Darss Sill data set, where the three different approaches presented clearly lead to different groupings of data. In the ternary diagram of Fig. 3, the data seem to follow a fairly continous distribution. This fact suggests that the data might be better treated with a continuous approach, possibly in an appropiate transformed space. In such a case, available methods have to be carefully checked for underlying hypotheses, particularly concerning assumed models. The necessity of a more profound study of available information is a must in the present case.

References

Aitchison, J. 1986, *The Statistical Analysis of Compositional Data*: Chapmann and Hall, London (GB), 416 p.

Barceló, C., Pawlowsky, V., and Grunsky, E., 1996, Some aspects of transformations of compositional data and the identification of outliers: Mathematical Geology, vol. 28, no. 4, pp. 501–518.

Bohling, G. C., Davis, J. C., Olea, R. A., and Harff, J., 1998, Singularity and nonnormality in the classification of compositional data: submitted to *Mathematical Geology*, vol. 30, no. 1, pp.5–20.

Davis, J. C., Harff, J., Lemke, W., Olea, R. M., Tauber, F., Bohling, G.C., and Zhou Di, 1995, Analysis of sedimentary facies by regionalized classification: *Kansas Geological Survey Open File Report* 95–4, 28 p.

Everitt, B.S., 1993, *Cluster Analysis*: Edward Arnold, New York (USA), 3rd ed., 170 p.

Harff, J., and Davis, J.C. 1990, Regionalization in geology by multivariate classification: *Mathematical Geology*, vol. 22, no. 5, pp. 573–588.

McLachlan, G. J., and Basford, K.E. 1988, *Mixture models. Inference and applications to clustering*: Marcel Dekker, New York (USA), 253 p.

Sampson, R. J. 1988, *Surface III® User's Manual*, Kansas Geological Survey.

Statistical Sciences 1995, *S-PLUS® Guide to Statistical and Mathematical Analysis*, Version 3.3, Seattle: StatSci, a division of MathSoft, Inc.

Statistical Analysis of Dispersion and Geochemical Patterns of Sedimentary System in Northern Shelf of the South China Sea

Di Zhou

1 Introduction

The northern shelf of the South China Sea is the site of dispersion for a large quantity of discharges from the Southeast China continent. Today, the shelf borders the most prosperous economic zone in China; it is also the site of a flourishing development of the Chinese marine industry, including offshore petroleum exploration and exploitation, fishery, aquaculture, and shipping. This chapter presents a statistical analysis of dispersion and geochemical patterns of the seafloor sediments of the shelf based on data accumulated during the period 1964–1975. Thus the patterns represent the background configuration of the area with respect to the present patterns; the latter may be contaminated by the fast economic development in the area since the mid 1980s.

2 Study Area

The South China Sea is a tropic marginal sea of the West Pacific. It covers over 17° latitude, with a total area of 3.5×10^6 km^2. The surface water of the sea has a temperature of 19–29 °C in winter and 27–31 °C in summer, and a salinity of 31–34‰ in the open sea and of 28–15% near river mouths. The water circulation of the South China Sea is controlled mainly by SE-Asian monsoon, which is northeasterly and strong in winter, but southwesterly and mild in summer.

The northern shelf of the South China Sea includes the sea area bordered by the shelf break at about 150–300 m water depths (Fig.1). It is the largest shelf of the South China Sea, about 150–310 km wide and 700 km long, gently dipping southeastward with an average slope of about 0°03′–0°04′. The shelf widens southwestward until it is blocked by Hainan Island.

A large portion of the central shelf is occupied by the underwater Zhujiang River (Pearl River) delta system, wich mooves southeastward up to 200 m isobath. The system consists of four ancient deltas superimposed upon each other, corresponding to four steps of terraces on the shelf, and indicating four stages of sea level variations during Late Pleistocene to Quaternary (Feng et al. 1988).

A large quantity of land discharges are transported onto the shelf from the Southeast China through numerous rivers. Among these the Zhujiang River

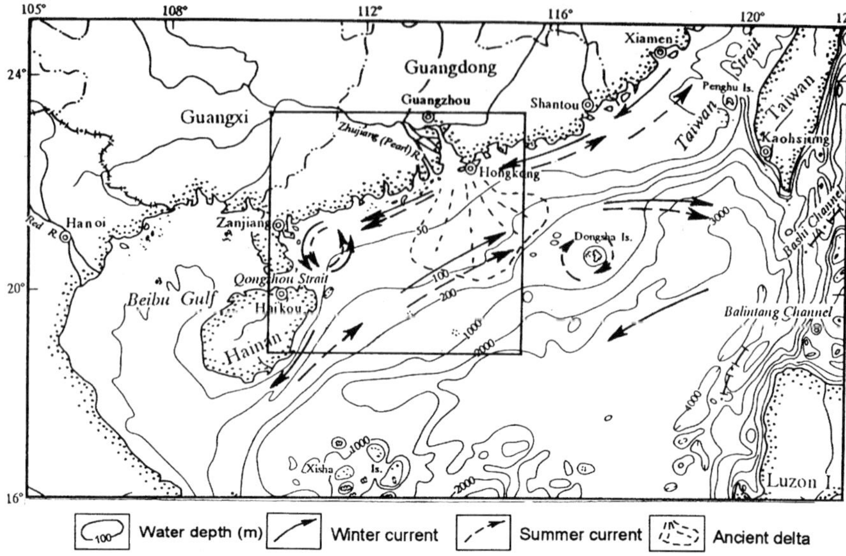

Fig. 1 Map of the northern South China Sea, showing major geographic features and currents

alone supplies 8.3 million tons annually, the third largest supplier in China. Twenty percent of this deposited within the Zhujiang Delta, and 80% is carried onto the shelf (Huang et al. 1982).

The current system of the shelf consists of the Zhujiang River Diluted Current (ZRDC), the East and the West Guangdong Coastal Current (EGCC and WGCC), and the South China Sea Warm Current (SCSWC) (Fig. 1; Huang et al. 1994). The ZRWC may cover a large area north of 21°N during the summer monsoon, but flows only near to and west of the river mouth during the winter monsoon. The EGCC follows the direction of monsoons, but the WGCC flows southwestward all year round. The SCSWC appears adjacent to and south of the EGCC and WGCC, and flows always northeastward, even against the monsoon in winter.

3 Data Base

The South China Sea Institute of Oceanology (Chinese Academy of Sciences) conducted composite scientific surveys over the northern South China Sea between 1964 and 1971. This chapter uses part of the grain-size data and geochemical data collected during these surveys, covering the central and western shelf. The grain-size data came from 723 sampling stations, with an average density of one station per 10 km^2. These include percentages of 12 size fractions from −1 to 10 ϕ, among which the fractions of −1 to 4 ϕ are obtained by sieving, and the finer fractions by pipette. A portion of these sediment specimens was used for chemical analysis. The geochemical data of 248 sampling stations include concentrations of nine compounds (N, P, Mg, Fe, Mn, Ti, organic carbon, $CaCO_3$, and Al_2O_3,) obtained by chemical analysis. The average density of geochemical data is about one station per 36 km^2.

4 Grain-Size Classification of Sediments

The statistical procedure for sediment classification includes two basic steps: (1) recognition of statistical correlations between attributes (grain-size fractions or chemical concentrations) and statistical associations between objects (sediment specimen) by means of RQ-mode principal components analysis; (2) classification of the objects by means of cluster analysis based on selected principal components.

The method of RQ-mode principal component analysis is used to examine the correlations between variables (grain-size fractions), the clustering of objects (sediment specimens), and the correspondence between object clusters and variables (Zhou et al. 1983). Robust procedures were used to reduce the bias caused by outliers.

As the grain-size data are compositional in nature, they are 'closed' in the sense that entries in each row of the data sum to unity. This will induce negative correlation and spurious correlation between variables and thus mask the true picture of covariance structure (Aitchison 1982). Another problem is that compositional data are confined to the range of [0,1] and thus do not have a normal distribution, which is the prerequisite of many classical statistical procedures.

Aitchison (1982, 1986) proposes logratio transformation to overcome the problems of compositional data analysis. In this study, we transformed the raw grain-size data to their centered logratios before performing RQ-mode principal component analysis. For the entry x_{ij} the centered logratio transform has the following form

$$z_{ij} = \log[x_{ij} / g(x_i)], \quad \text{for } i=1,2,\ldots,n;\ j=1,2,\ldots,D,$$

where $g(x_i)$ is the geometric mean for the ith row.

As the data have zero entries but one can not take logarithms of zeros, we follow Aitchison (1986) by replacing the zeros with a small positive number $\delta(C+1)(D-C)/D^2$, where δ is the maximum rounding-off error, C the number of zeros in the row, and D the number of entries in the row. Meanwhile, each positive entries in the row is reduced by $\delta C(C+1)/D^2$ to keep the row sum to 1. For our data, δ=0.005, D=12, and C varies between 0 and 5. This procedure implies that we regard the zeros in the data as traces under a given detection limit.

For comparison, we also conducted robust principal component analysis without logratio transformation. The program ROPCA (Zhou 1989) was used to complete both tasks. Figure 2 shows the loadings of the first four principal components from the two runs. For the logratio ROPCA the first four principal components represent 94% of total variance. Clear clustering of sediments is seen in

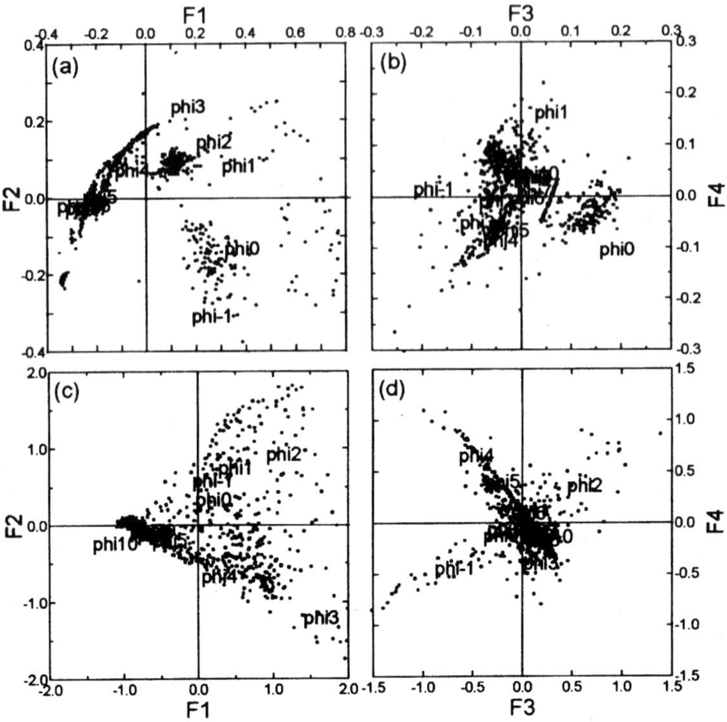

Fig. 2 Scatter plot of the R- and Q-loadings of the first four principal components given by ROPCA based on logratio-transformed grain-size data (a and b), and ROPCA based on raw grain-size data (c and d)

their Q-loading diagrams. In contrast, the first four principal components given by ROPCA on raw data yield horn- and dragonfly-shaped Q-loading diagrams, in which no clusters are distinguishable.

The cluster analysis using the first four principal components of log ratio ROPCA yields a hierarchic classification of three main types and seven subtypes (Table 1, Fig.3). This classification is compared with that given by the cluster analysis using raw data. Since 'no universally acceptable criterion of optimality of classification is available' (Sneath and Sokal 1973, p.276), we define an index of clustering by distance as (Zhou et al. 1991)

$$ICD = (\Sigma\, d_i / d_i^*) / k,$$

which is the average ratio of the maximum distance between the objects within the ith cluster (d_i) over the minimum distance between the ith cluster and any of the (k–1) other clusters (d_i^*). The value of ICD ranges in (0,1), and a classification with lower ICD is better than another with higher ICD. For our example,

Table 1 Average grain-size parameters (in ϕ) of the sediment types for the northern South China Sea

Type	Description	Mean size	Standard deviation	Skewness	Sample size
I1	Silt and clay with very fine sand	7.95	2.12	−0.35	66
I2	Silt and clay with fine sand	7.49	1.96	−0.30	134
I3	Silt and clay with medium and fine sand	7.17	2.24	−0.36	96
II1	Poorly sorted clayey fine sand	5.36	2.80	0.40	174
II2	Poorly sorted medium-fine sand with fine gravel and clay	4.20	2.86	−0.71	141
III1	Gravel-bearing coarse-medium sand, very well sorted	1.91	1.85	1.36	66
III2	Gravel-bearing sand and clay, very poorly sorted	3.14	3.79	0.65	46

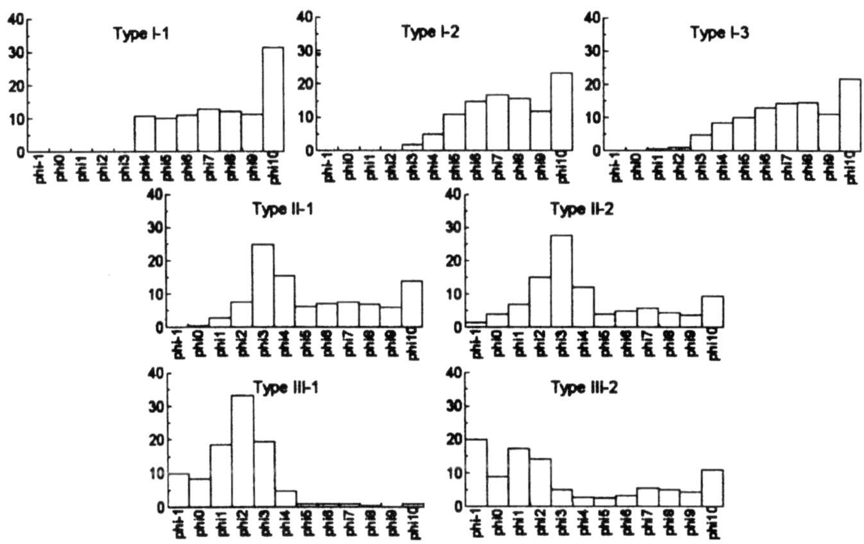

Fig. 3 Histograms of grain-size distributions of sediment types

ICD=0.5 for the 7 clusters given by the logratio procedure; this means that on average the within-cluster distance is half of the inter-cluster distance. For the cluster analysis of raw data, however, the best classification yields 22 clusters with ICD=0.71, or the classification of 9 clusters with ICD=0.74. The logratio procedure improved the classification remarkably.

5
Geochemical Classification of Sediments

The statistical analysis of geochemical data follows the procedure stated in previous section only without logratio transformation, as these data do not contain complete compositions.

As the first step, ROPCA were applied to the data. A five percent multivariate trimming has improved the analysis, as evidenced by the trimming of three samples with unusually high phosphorus, three with unusually high Mn, and several with very high Fe. The correlation coefficients of P, Mn, and Fe with other constituents, except of $CaCO_3$, were improved compared with results given by ordinary least squares PCA (OLS); the percentage of variance for the first principal component was increased from 50.3 to 60.9%; and the 3rd and 4th principal components in OLS, which were generated mainly by outliers in P and Mn, were removed (Zhou et. al. 1990).

The first four principal components given by ROPCA of geochemical data represent 88% of the total variance (Fig.4). The first principal component (F1), representing 60% of the total variance, is composed of Al_2O_3, organic carbon, N, Ti, Fe, Mn, and P. These chemical constituents are carried onto the shelf mainly by terrigenous clay minerals and organic matter. $CaCO_3$ has low loading in F1. As the contents of all chemical constituents in any specimen must sum to unity, and in our data SiO_2 is the major constituent not determined, the negative end of F1 may be considered as an indication of high SiO_2. F2 represents 15.6% of the total variance and is composed of mainly $CaCO_3$ and secondarily Mg. Its variation is independent of F1 and indicative of the intensity of biochemical sedimentation. F3 reflects the diversity between elements carried by heavy minerals (Mn, Fe, Ti) and those carried by organic matter (N, Corg, Al , P). F4 indicates the difference between Ca and Mg in biochemical sedimentation.

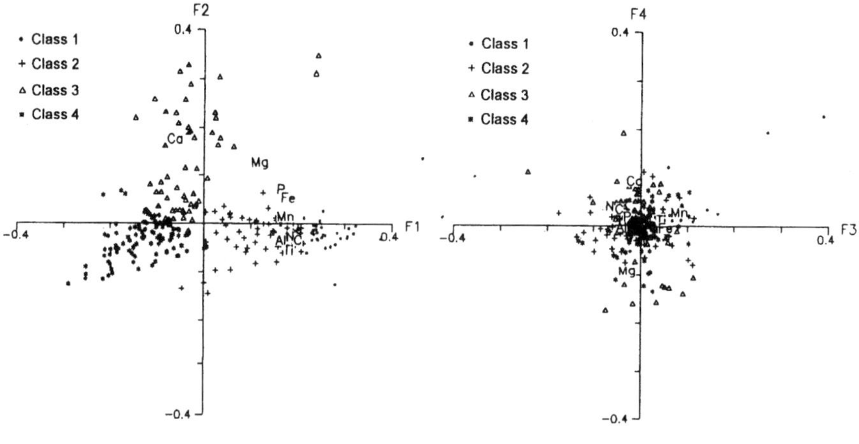

Fig. 4 Scatter plot of the R- and Q-loadings of the first four principal components given by ROPCA of geochemical data

Table 2. Average concentrations (5% trimming estimation, in percentage) of geochemical constituents in sediments from the northern shelf of the South China Sea

Class	Size	N	P	Mg	Fe	Mn	Ti	C_{org} [a]	$CaCO_3$	Al_2O_3
Class 1	31	0.097	0.066	0.769	3.88	0.111	0.596	1.445	6.49	14.19
Class 2	64	0.083	0.050	0.724	2.89	0.063	0.450	1.147	9.18	11.77
Class 3	58	0.040	0.049	0.837	2.53	0.042	0.201	0.522	31.11	6.36
Class 4	95	0.030	0.029	0.415	1.45	0.031	0.201	0.417	13.55	5.35

[a] Organic carbon.

In Fig.4, sediment specimens show clustering only at F1 and F2. Thus Q-loadings of these two principal components are used in cluster analysis. This resulted in four geochemical classes as shown in Table 2 and Fig.4. Classes 1 and 2 are enriched in all chemical constituents except $CaCO_3$ and represent terrigenous sedimentation. They are located along the positive portion of F1, transitional to each other. Class 1 has a higher proportion of clayey constituents relative to class 2. Class 3 is characterized by very high $CaCO_3$, concentrated near the positive end of F2, and represents biochemical sedimentation (grains of biotests and/or authogenic calcites). Class 4 is distributed along the negative portion of F1. It has the second-highest concentration of $CaCO_3$, and the lowest concentration of all other chemical constituents, suggesting that these sediments are enriched in quartz and then in calcareous grains.

6

Sedimentary Regionalization

The geographic distributions of grain-size types show clear pattern and may be regionalized visually (Fig. 5). Four types of regions are identified, corresponding to four types of sedimentary-hydrodynamic environments. The geochemical regionalization (Fig.6) bears some analogy to the grain-size regionalization on the inner shelf, but different on middle and outer shelves. Combining of grain-size and geochemical characteristics, the study area may be divided into four types of sedimentary regions:

1. A terrigenous sedimentary region lies basically on the inner shelf within the 50 m isobath (region A in Fig.5, also region I in Fig.6). Its sediments are almost exclusively of Type I fine-grained silt and clay. Geochemically, these are sediments of Classes 1 and 2, characterized by high terrigenous chemicals such as Al_2O_3, organic carbon, N, P, Fe, Mn, and Ti. Class 1 sediments have higher terrigenous chemicals and appear closer to the Zhujiang River Mouth and the west Guangdong coast than Class 2. This indicates the westward dispersion of terrigenous supply.
   Satellite photos and other evidence show that these sediments are supplied mainly by the Zhujiang River and then carried mostly southwestward onto the shelf by Guangdong Coastal Current and Zhujiang River Diluted Current.

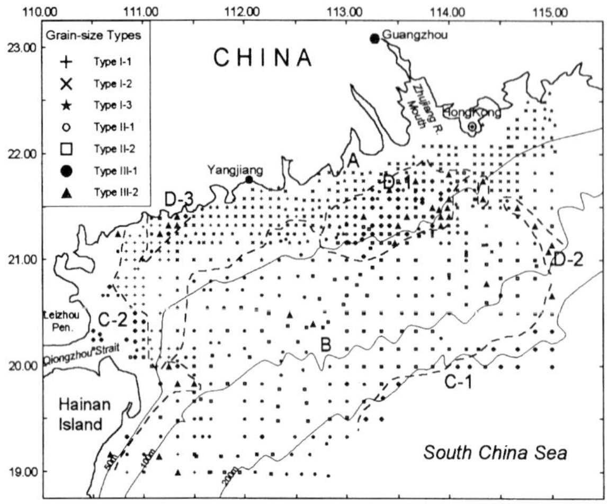

Fig. 5 Geographic distributions of grain-size types of sediments and grain-size regionalization

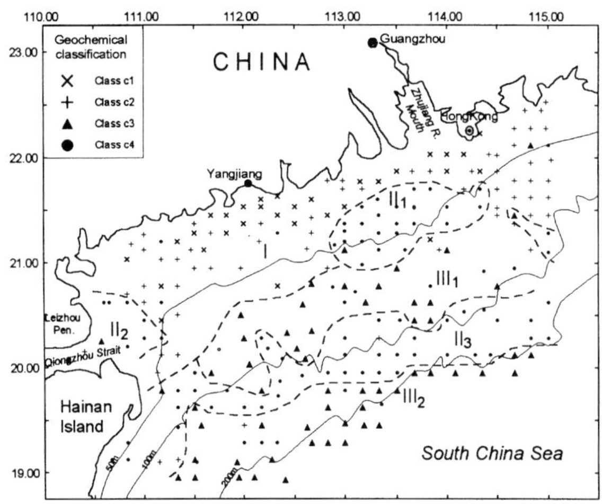

Fig. 6 Geographic distribution of geochemical classes of sediments and geochemical regionalization

The western most portion of this region is unique in having only Subtype I-1 sediments, the finest sediments on the shelf with no grain coarser than fine sand. The deposition of such sediments is controlled by the specific hydrodynamic condition: the southwestward-moving WGCC is bent and slowed down near the Leizhou Peninsula, and a counterclockwise eddy is formed by the meeting of this bent current and the eastward current from the Qiongzhou Strait (Huang et al. 1994).

2. A modified relict littoral sedimentary region lies in the present middle and outer shelf between the 50 m and 200 m isobaths (region B in Fig.5, also regions II_3 , III_1 and part of III_2 in Fig. 6). In this region, sediments are dominated by the Type II clayey fine sand with very poor sorting. The sandy fractions of these sediments deposited in a littoral condition during the Pleis-

tocene Würm Glaciation, as evidenced by the intertidal shells with a radiocarbon age of 15–20 Ka B.P. (Yan and Jin 1985). The finer parts of the sediments are mainly modern terrigenous origin, mixed in from region A.
In this region the spatial variation in geochemical composition does not coincide with the variation in grain-size features. The geochemical regionalization includes regions III_1, III_2, and sandwiched region II_3 (Fig.6). Boundaries of the regions are subparallel to the coastline. Region III is characterized by predominant class 3 sediments, and differs from region II by higher $CaCO_3$ and lower SiO_2. The geochemical composition of these relict sediments may be controlled by the sedimentary environment of their formation. For example, regions III_1 and III_2 might represent the neritic zone during the climax of the Wurm Glaciation and some later glaciation, respectively. The lowest coastline of the Wurm Glaciation is suggested to be the underwater terrace at 130–155 m present water depth (Yang and Jin 1985).

3. An ancient delta is exposed at the seabed about 50 km southwest of the Zhujiang River Mouth (region D-1 in Fig. 5, also region II_1 in Fig. 6). It is a small area of 160 km long and 50 km wide southwest of the Zhujiang River mouth. There exist all types of grain-size, with type III coarse-sediments dominating. Geochemically these sediments are SiO_2-enriched class 4. We interpret these as ancient deltaic sediments, similar to the Lower Holocene encountered by drilling in the Zhujiang River delta. These relict sediments have been protected from modern terrigenous sedimentation by an upwelling current in the area (Huang and Huang 1985). A similar environment is found in a very small area near the coast SW of Yangjiang (region D-2), and is also observed along the 50-km-long, SE-striking underwater extension of the Zhujiang River (Region D-3), indicating an uncovered ancient river channel.
4. A modern high-energy sedimentation environment is characterized by subtype III-1 well-sorted medium-coarse sand. There are two districts of such an environment: The shelf break sedimentation zone (region C-1 in Fig. 5, also part of region III_2 in Fig. 6), where sediments are calcareous class 3. The finger delta of the Qiongzhou Strait and the underwater extension of the sand beach east of the Hainan Island (Region C-2 in Fig. 5, also region II_2 and part of II_3 in Fig. 6), where sediments are relatively siliceous class 4.

7 Dominant Directions of Sediment Dispersion

To determine the predominant direction of sediment dispersion, the method of the one-tail Z test proposed by McLaren and Bowles (1985) was adopted. For n(n-1) pairs of n specimens in a possible direction of dispersion, the parameters of grain-size distribution are compared. Denoting the mean size decrease as F, increase as C; standard deviation increase as P, decrease as B, and skewness increase as +, decrease as –, then for each pair of sediment samples there are eight possible events FB–, FB+, FP–, FP+, CB–, CB+, CP–, and CP+. Based on stochastic modeling, only FB– and CB+ are indicative for transportation direction. FB–

indicates a common event in which sediment becomes finer, better sorted and negatively skewed during the transportation. Event CB+ represents an event where sediment becomes coarser but also better-sorted and negatively skewed; which appears only in a particular dynamic condition where sediment is transported from a high-energy regime (flow velocity 0.7 m/s) to a regime of low energy. For a binomial distribution, define the statistic

$$Z=(M-Np)/\sqrt{Np(1-p)},$$

where N is the number of tests; p the probability of given event; q=1–p; and M the number of observations for the event. In our problem, let n be the number of sediment pairs, then N=n(n–1)/2, p=0.125, q=0.875, and the critical value for Z at 0.05 significance levels $Z_{0.05} = 1.645$. If for a given direction the observed Z is greater than the critical value, the zero hypothesis is rejected, and we say that the direction is predominant.

This method is based on two assumptions: the sediment specimens are from the same source; and the sample size is large (N≥30, which equals n≥8). Thus in our application we used only type I terrigenous sediments. As possible directions of sediment dispersion, 18 lines with at least 8 sediment specimens were selected. The Z statistic was computed for the two directions of each line, and predominant directions were obtained when Z exceeds 1.645.

Resulted predominant directions of sediment dispersion (Fig. 7) agree with the predominant directions of currents. From the east border to offshore Yangjiang, sediments are transported mostly southwestward, as controlled by the constant southwestward current WGCC, and the mostly southwestward currents ZRDC and EGCC. The transport direction becomes variable west of offshore Yangjiang, corresponding to the eddy current in the area.

In all lines the predominant grain-size variation is of FB-type, indicating a low-energy sedimentary environment. This agrees with the observation that the current velocity in the area is 0.15–0.5 m/s (Zhao 1984).

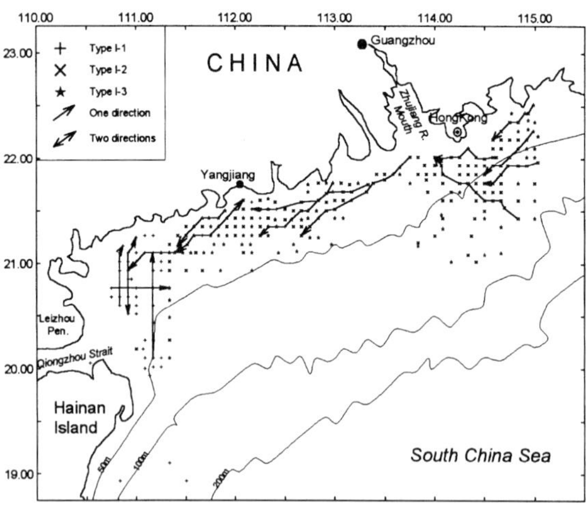

Fig. 7 Dominant direction of sediment transportation suggested by the Z-test of the variations of grain-size distributions along the direction of transport

8
Conclusions

Statistic analysis of grain-size and geochemical data of seabed sediments from the northern South China has provided the classification of sediments, regionalization of sedimentary environment, and pattern of dispersion of the sedimentary system.

Grain-size distribution and chemical composition of sediments are controlled jointly by hydrodynamic condition and material supply. In particular, Guangdong Coastal currents and Zhujiang River Diluted Current confine most terrigenous fine sediments to the inner shelf. Local upwelling may prevent the area from terrigenous sedimentation. Sediments on middle and outer shelf are mostly relict fine sand, mixed with terrigenous silt and clay and biogenic/authogenic calcareous clasts. Well sorted medium-coarse sand appear in high-energy shelf break, finger delta, and beach.

References

Aitchison, J., 1982. The statistical analysis of compositional data (with discussion). J. Royal Statist. Soc., B., 44, 139–177.

Aitchison, J., 1986. The Statistical Analysis of Compositional Data. Champman and Hall, London, 416p.

Wenke, F., Wanjun, X., and Dayuan, Y., 1988. The Geological Environment of Late Quaternary in the Northern South China Sea. Guangdong Science and Technology Publishing House, Guangzhou, 261 p. In Chinese with English abstract.

Jianchong, H. and Qizhou, H., 1985. Classification of seasonal variations of water masses. In: Reports pf the Comprehensiv Investigation of the South China Sea, v..2, China Science Publishing House, 183–204, in Chinese.

Huang, Q.Zh., Wang,W.Zh., Li,Y.S., and Li,C.W., 1994. Current characteristics of South China Sea. In Zhou,Di et al. (eds.), Oceanology of China Seas, Volume 1, Kluwer Acad. Pub., Dordrecht, 39–47.

Zhenguo, H., Pingri, L., Zhongying, Z., Konghong, L. and Pengnian, Q., 1982. Formation, Development, and Evolution of the Zhujiang River Delta. Guangdong Bran of Science Popularization Publishing House, Guangzhou, 274 pp. In Chinese.

McLaren,P. and Bowles,D., 1985. The effects of sediment transport on grain-size distributions. J. Sed. Petrol. 55(4), 457–470.

Sneath,P.H.A. and Sokal,R.R., 1973. Numerical Taxonomy. Freeman and Company, San Francisco, 573pp.

Yang,Sh.M. and Jin,Bo, 1985. Geomorphologic and sedimentational characteristics of the underwater delta in Zhujiang River mouth. In: Symposium on Coastal Dynamics, Geomorphology, and Sedimentation. China Science Publ. House, 35–40, in Chinese.

Huanting, L., 1984. Basic process of the evolution of Zhujiang River Mouth. Tropic Oceanology 5(3), 79–81, in Chinese with English abstract.

Zhou Di, Chang,T., and Davis,J.C., 1983. Dual extraction of R-mode and Q-mode factor solutions. Math. Geol. 15, 581–606.

Zhou, D., 1989. ROPCA: a FORTRAN program for robust principal components analysis. Comput. Geosc.i., 15, 59–78.

Zhou, D., Shaomou, C., and Hanzong, C., 1990. The geochemical zonation of sea-bottom sediments of the shelf off Pearl River Mouth. Tropic Oceanology 10(3), 21–28, in Chinese with English abstract.

Zhou, D., Hanzong, C. and Wenzhi, W., 1991. Robust principal component analysis of the sedimentary geochemical data from the northern South China Sea. Tropical Oceanology 9(3), 12–17, in Chinese with English abstract.

Computerized Modeling of Rock-Varnish Dating

N. Gorelikova, N. Patyk-Kara, I. Chizova, J.Plakht, A. Sivtsov

1 Introduction

One type of epigenetic alteration of recent sedimentary formations is desert weathering. Rock varnish is a widespread feature of a stony desert landscape throughout the world. It is a product of complicated physical and chemical processes under arid and semiarid environments. Despite the interest in rock varnish of both archeologists and geologists, the mechanism of varnish formation is controversial. Most workers believe that the formation of rock varnish is responsible for changing pH and Eh conditions on its surface layer, which results in solution and subsequent deposition of matter (Tricart et al. 1960). These researchers consider this process as insolative weathering of deserts which consists of circulation in mineralised solutions to a rock surface, evaporation and mineral deposition. Some investigators advance arguments which indicate a biogenic nature of manganese layers (Krumbein et al. 1981). In addition, it is also believed opinion that patina has a primary eolian origin (Allen 1978; Dorn et al. 1981).

Recently, considerable attention has been focused on new methods for quantitative age determinations of patinated surfaces. At present, there are a dozen new methods for quantitative age dating of rock-varnish. Rock-varnish coatings from arid environments have currently gained importance in the dating of geomorphic events and revealing climatic changes (Dorn et al. 1991).

From a modern viewpoint, rock varnish is formed under conditions of high Eh and pH, which are common to deserts at low altitudes with annual precipitation of no more than 150 mm. Microorganisms, lichens and bacteria, which consume energy in the course of manganese oxidation, play an important part in manganese absorption by clay particles. It was found that coating composition, primarily of manganese and ferrum content, calibrated against other methods, makes the age dating of native phenomena (terraces, fans, pediments, etc.) and artefacts in the range from hundreds to 1 million years possible (Hunt 1961; Hooke et al. 1969; Krumbein et al. 1981).

2 The Goal

Many papers are devoted to investigating the potential of using chemistry rock varnish for the elucidation of environmental formation, climatic changes and dating of geological events in the Quaternary period. For these purposes, researchers have used different parameters, such as pair cation ratio, complicated cation ratio, cosmogenic stable nuclides, organic matter, etc. These attempts and different approaches were summarized by Dorn and Fillips (1991). The problem of dating Quaternary events is very complicated because there are no reliable methods of determination of geological age younger than 2000 years. Furthermore, the evolution of geomorphic events which change in space and time often eludes interpretation. When dating, one cannot always use stratigraphic markers, such as the K-Ar age of lava flows, ^{14}C data of charcoal intercalated with deposits, uranium-series, paleomagnetic data, fissiontrack, stable isotopes, thermoluminescence data, and others. Suitable material for stratigraphic dating is lacking for the most geomorphic processes. Furthermore, stratigraphic age control cannot be performed for erosional landforms. In this study we also attempted to use different approaches and various special ratios, such as Fe:Mg, Fe:Ti, Fe:Mn, Fe+Mg/Mn, etc. However, these operations did not give positive results, i.e. they did not show a clear-cut relation between the chemistry of rock varnish and geological age. Thus only models designed on the basis of linear programming using all possible combinations of ratios and products provided relative and absolute dating of geomorphic surfaces.

Rock-varnish coating on cobbles from geomorphic surfaces and exposed deposits in arid environments is an effective medium for dating over the range of several thousands to some million years. This chapter shows an attempt to use rock varnish from Makhtesh Ramon for dating various geomorphic levels and to establish quantitative models of relationships between rock-varnish coating chemistry and its age-dependent characteristics. The focus of our study is on creating foolproof methods of rock-varnish dating assessment and revealing age relations between geomorphic levels. As compared with rock varnish from other regions, for example Death Valley (Dorn et al. 1987), the patina of Makhtesh Ramon is considerably thinner, which is the result of the lack of maturity of these formations. It has posed procedural difficulties in the analysis of rock-varnish coating. Therefore, a special technique of microprobe analysis for unpolished surfaces was created by G.Nechelustov. This procedure analyzed the chemical composition of extremely thin coatings of rock varnish from Makhtesh Ramon. Element concentrations in rock varnish coating on cobbles from different geomorphic levels of Makhtesh Ramon in the Central Negev (Israel), formed from 2 million years to 3000–4000 years B.P., were determined.

Since the mobility of some elements (Fe, Mn, Ti, Si) making up the bulk of rock varnish is closely connected with environmental change, it is supposed that coating composition can be an indicator of climatic conditions. Among other things, the ratio between Mn and Fe is higher in semiarid environments and increases under arid conditions (Dorn et al. 1988). It is also verified that the chemistry of rock varnish, especially its Mn and Fe content as well as the ratios Fe:Mn, K+Ca:Ti, Fe:Ti, etc., provides a way to date natural phenomena (Dorn et al. 1987; Harrington et al. 1987).

3 Study Area

Makhtesh Ramon is located within a stony desert plateau Negev and represents a spacious deep denudational depression (40 x 10 m) arisen as a result of selective denudation of rocks in the core of the monocline Ramon (Fig. 1). The upper part of the depression slopes is composed of the limestone and dolomite Hazera Formation of Upper Cretaceous age. The lower part of slopes and also the main part of the bottom is represented by loose sands of the Hatira Formation of the Low Cretaceous. In the places where these rocks were destroyed by erosion, volcanic basalt rocks of the Low Cretaceous are discovered. Basal conglomerates of the Cretaceous Formation Arod are the lowest member of the Cretaceous section (Masor 1978; Ben-David et al. 1988). The deepest parts of the depression are Jurassic sands broken by massifs of Cretaceous gabbro and syenite.

The geomorphic history of Makhtesh Ramon has been reflected in its erosional-accumulative and denudational bottom levels. It is supposed that de-

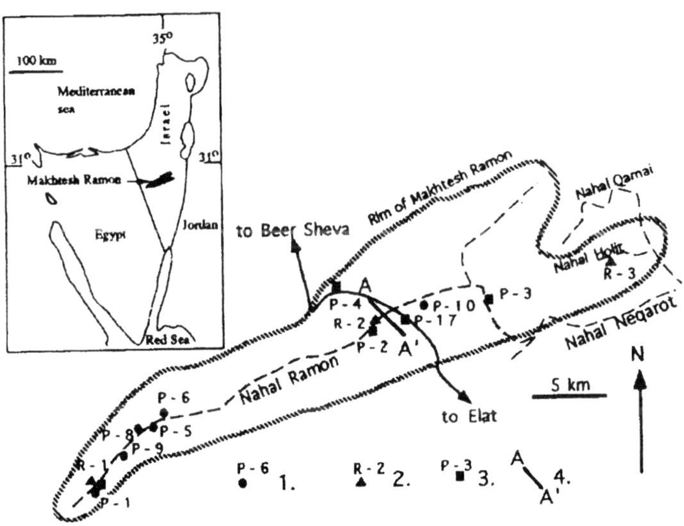

Fig. 1 Map showing the study area and sampling locations:
1 sites of rock varnish data; 2 control rock varnish samples; 3 sites of thermoluminescencal data (TL); 4 geomorphologic cross section (J.Plakht)

Table 1 Characteristics of samples from Makhtesh Ramon and their geomorphic levels

Sample	Number of determinations	Level type	Absolute height[a]	Relative Age in thousands of years[b] (TL analysis)
RVE-1	5	Ancient bottom	400[a]	>2000[b]
RVE-2	10	Ancient bottom	400[a]	>2000[b]
RVE-1	5	Ancient bottom	400[a]	>2000[b]
RVT-9	5	Terrace	30	375 ± 95
RVT-20	8	Terrace	30	443 ± 110
RVP-1	4	Pediment	25	>300[b]
RVP-1	3	Pediment	25	>300[b]
6/MR	5	Pediment	25	>300[b]
RVT-18	5	Terrace	25	240 ± 48
RVT-18KP	4	Terrace	25	240 ± 48
RVT-16	5	Terrace	15–16	111 ± 22
RVT-19	3	Terrace	15–16	143 ± 29
RVT-17	5	Terrace	15–16	48 ± 12
RVP-3	4	Pediment	10	48 ± 12
RVT-15	4	Terrace	5.5	35 ± 9

[a]Absolute height in m above sea level; other values are relative heights above the surrounding area in m.
[b]Estimated age based on stratigraphy, time marks in sediments below, etc.

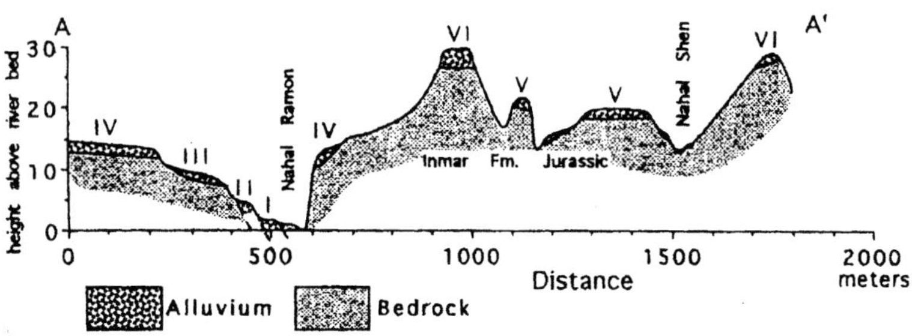

Fig. 2 Geomorphic cross section, the central part of Maktesh Ramon (J. Plakht)

pression formation in its modern form occurred at the boundary of the Miocene and the Pliocene.

It is expected that the major geomorphic level markers for interpretation of Makhtesh Ramon relief evolution are the following: (1) The ancient bottom of Makhtesh; (2) The stairs from six terraces above floodplain; (3) The stairs from five level pediments correlating with terraces (Fig. 2).

Some of these levels were dated by thermoluminescence analysis by O.Kulikov (Vlasov et al.1988). These age data were used for the absolute age scale (Table 1).

The arid and hyperarid climatic environment of Makhtesh Ramon provided intensive formation of coating rock varnish covering more than 70% pebbles on surfaces of terraces, fans and pediments. This is a black thin patina (1–3 μ) on an external surface of pebbles.

For the purpose of using rock varnish as an age marker, the chemistry of the patina was sampled from different geomorphic levels – 5–, 10–, 15–, 25–, 30–m terraces of wadi Ramon, 10–, and 25 m pediments, as well as from the depression bottom at 400 m elevation (see Table 1).

4 Chemical Composition

Rock-varnish chemistry was studied under electron microprobe analysis (Camebax). The special technique developed by G.Nechelustov for unpolished surfaces was used because of the extreme thinness of the patina. The intensity of X-ray radiation for compact samples with rough surface differs from that for polished sections. A rough surface in samples strongly distorts the absorption correction value and is reflected on the coefficient of inverse scattering. Therefore, no standard corrections procedure applies to such samples. In order to correct geometric effects, use was made of a ratio intensity line to an intensity phone as an analytical parameter.

The chemical analysis of Fe-Mn matter was conducted under an electron microprobe procedure using energy-dispersive analysis. Analytical material was presented by 67 analyses, characterizing 15 samples from distinct geomorphic levels (Table 2). The distribution of chemical elements forming the basis of coating – K, Na, Mg, Ca, Al, Si, Ti, Cr, Mn, Fe – was determined. Most differences in the content of elements are common to such elements as Fe, Mn, Ti, Si, Al, Mg. However, there is no clear relationship between initial signs and their age. In our opinion, it is the result of some immaturity of these formations compared to those from other regions. This conclusion is supported by correlative analysis data. Nevertheless, the elemental differences in coating varnish, formed under various conditions, permitted mathematical analysis of the data obtained and showed the relation between the chemistry of the rock varnish and its age.

Table 2 Electron microprobe analyses of rock varnish samples from Makhtesh Ramon (Camebax)

		Element concentration, weight %									
Sample no.	Absolute height[a]	Na	Mg	Al	Si	K	Ca	Ti	Cr	Mn	Fe
1(RVE–1)	400[a]	0.85	2.40	11.66	22.08	1.65	1.3	0.58	0.02	7.06	8.29
"–"	400[a]	0.88	2.34	12.07	19.47	1.44	1.73	0.53	0.02	8.61	9.99
"–"	400[a]	0.24	2.7	13.54	20.57	1.86	1.3	0.5	0.02	6.63	8.58
"–"	400[a]	0.88	2.18	10.21	19.44	2.5	1.58	0.73	0.03	10.37	10.11

Table 2 Electron microprobe analyses of rock varnish samples from Makhtesh Ramon (Camebax)

Sample no.	Abs. height[a]	Na	Mg	Al	Si	K	Ca	Ti	Cr	Mn	Fe
"–"	400[a]	0.4	1.54	7.06	29.04	1.15	1.15	0.63	0.02	5.97	6.85
2(RVE-2)	400[a]	0.8	2.36	13.23	19.5	1.95	0.88	0.7	0.02	8.42	8.98
"–"	400[a]	0.16	2.34	13.6	22.43	1.82	0.68	0.66	0.07	4.39	8.68
"–"	400[a]	0.67	3.23	13.49	23.51	2.1	0.94	0.64	0.07	1.98	7.26
"–"	400[a]	0.43	3.25	13.38	23.65	2.15	0.98	0.5	0.04	3.04	6.57
"–"	400[a]	0.68	2.79	13.43	20.28	1.82	0.94	0.49	0.02	5.03	10.42
"–"	400[a]	0.84	3.37	15.57	22.65	1.53	0.77	0.52	0.02	2.22	6.16
"–"	400[a]	0.36	2.45	14.78	20.39	2.03	0.65	0.81	0.02	4.38	9.47
3(RVE-1)	400[a]	0.44	2.58	12.56	23.34	2.66	1.28	0.49	0.02	4.1	7.25
"–"	400[a]	0.9	2.2	12.21	19.67	1.73	2.01	0.63	0.02	5.86	11.65
"–"	400[a]	1.12	2.89	14	22.07	2.71	0.99	0.43	0.07	0.38	9.84
"–"	400[a]	0.14	2.28	10.63	23.49	1.59	1.4	0.69	0.03	4.22	10.6
"–"	400[a]	0.46	2.79	12.97	20.15	1.74	1.25	0.65	0.02	7.77	8.53
4(RVP-1)	25	0.22	1.77	10.47	21.28	2	1	0.47	0.02	8.17	11.48
"–"	25	0.55	2.2	13.35	19.05	1.84	1.32	0.35	0.02	8.05	10.11
"–"	25	0.51	1.62	9	24.66	1.63	1.03	0.45	0.02	5.34	11.22
"–"	25	0.94	2.29	12.5	17.55	1.74	2.48	0.58	0.02	7.54	10.63
5(RVP-1)	25	0.51	2.14	13.63	19.02	1.71	1.44	0.46	0.02	8.58	9.88
"–"	25	0.55	2.09	12.66	18.69	1.66	1.57	0.4	0.02	8.84	11.5
"–"	25	0.53	2.38	13.55	20.81	1.74	1.13	0.33	0.02	8.19	7.93
6(6/MP)	25	0.75	2.38	8.8	26.6	1.53	1.07	0.37	0.02	2.72	10.06
"–"	25	0.51	1.63	14.23	20.03	1.96	1.17	0.83	0.02	6.00	9.71
"–"	25	0.53	3.42	10.62	24.81	1.87	1.21	0.57	0.02	2.93	8.27
"–"	25	0.36	2.91	9.92	24.11	1.61	1.26	0.67	0.02	5.62	8.61
"–"	25	0.47	2.88	10.08	24.76	1.59	1.33	0.36	0.02	5.33	7.98
7(RVT-19)	15	0.11	2.23	13.37	19.63	2.02	1.48	0.47	0.04	4.44	12.45
"–"	15	0.72	2.3	15.23	18.21	1.63	1.43	1.96	0.02	7.23	7.78
"–"	15	0.21	1.96	13.35	24.78	1.72	0.89	0.66	0.02	2.7	7.25
8(RVT-17)	15	0.79	3.02	13.52	24.72	1.6	1.48	0.35	0.01	0.32	7.33
"–"	15	0.53	2.14	9.29	16.04	1.39	0.88	0.67	0.02	0.14	27.76
"–"	15	0.09	2.21	15.11	25.98	1.81	0.64	1.2	0.04	0.19	4.68
"–"	15	0.48	2.42	13.64	25.55	1.95	1.33	0.5	0.08	0.43	6.43
9(RVP-3)	10	0.7	2.43	12.43	20.67	1.32	1.46	0.6	0.02	4.98	11.67
"–"	10	0.71	1.47	11.08	16.38	0.82	0.96	0.41	0.02	4.84	22.63
"–"	10	0.75	2.34	10.45	21.98	1.29	1.67	0.62	0.02	6.62	11.16
10(RVT-16)	15	0.78	3.69	14.01	22.97	1.86	1.58	0.72	0.04	1.19	6.98
"–"	15	0.05	3.94	13.12	24.41	2.4	1.18	0.59	0.02	0.23	7.4
"–"	15	0.94	4.75	14.39	22.88	2.04	1.54	0.66	0.05	0.24	6.02
"–"	15	0.72	3.87	12.52	22.97	2.3	1.21	0.75	0.01	1.61	8.43
"–"	15	0.38	3.8	13.8	23.3	2.13	1.4	0.45	0.02	1.13	7.33
11(RVT-9)	30	0.82	2.64	11.8	25.12	3.77	1.12	0.43	0.02	1.67	6.68

Table 2 Electron microprobe analyses of rock varnish samples from Makhtesh Ramon (Camebax)

Sample no	Abs. height[a]	Na	Mg	Al	Si	K	Ca	Ti	Cr	Mn	Fe
"–"	30	0.81	3.01	12.06	19.99	1.47	1.48	0.37	0.01	5.22	11.99
"–"	30	0.67	3.17	13.25	25.89	1.57	0.94	0.4	0.01	0.01	6.71
"–"	30	0.21	3.67	12.24	22.19	1.6	1.38	0.61	0.01	3.91	9.3
"–"	30	0.9	2.92	12.72	23.82	1.67	1.39	0.44	0.01	1.95	8.24
12(RVT-15)	5	0.81	3.03	11.5	24.23	1.93	1.12	0.43	0.01	3.94	7.43
"–"	5	0.48	3.43	14.48	24.29	1.78	0.96	0.45	0.02	1.36	5.81
"–"	5	0.67	3.82	15.01	23.23	1.66	0.92	0.36	0.01	2.14	5.66
"–"	5	0.46	2.7	11.23	23.84	1.94	1.27	0.61	0.01	2.87	9.7
13(RVT-20)	30	0.77	2.02	9.38	27.66	1.07	1.06	0.51	0.01	0.17	10.42
"–"	30	0.88	3.36	14.7	23.75	1.77	1.19	0.71	0.03	0.51	6.27
"–"	30	0.95	2.87	14.33	23.64	1.68	1.5	0.47	0.01	0.64	7.42
"–"	30	0.41	2.71	13.23	22.43	1.85	1.42	0.38	0.1	4.42	7.83
"–"	30	0.32	3.31	13.35	22.21	1.53	2.24	0.46	0.01	3.31	8.0
"–"	30	0.54	4.11	12.8	25.21	1.43	0.61	0.52	0.01	0.08	7.54
"–"	30	0.29	2.51	11.09	26.1	1.89	0.7	0.59	0.01	2.69	7.67
"–"	30	0.03	2.03	9.7	29.13	1.21	1.35	0.57	0.03	1.6	6.69
RVT-18	25	0.91	2.91	14.93	20.66	1.21	1.62	0.45	0.08	4.19	8.06
"–"	25	0.29	2.54	11.71	25.56	1.19	1.26	0.45	0.01	3.52	7.09
"–"	25	0.84	3.05	15.65	22.35	1.39	1.14	0.4	0.01	2.5	6.5
"–"	25	0.2	2.95	13.62	25.49	1.51	0.93	0.33	0.05	0.11	7.47
"–"	25	0.75	3.23	14.86	22.9	1.79	1.14	0.43	0.02	1.76	6.86
RVT-18KP	25	1.39	3.69	12.62	24.71	2.34	1.42	0.4	0.01	0.19	6.7
"–"	25	0.68	2.74	13.43	20.14	1.7	1.5	0.45	0.06	6.76	8.66
"–"	25	0.37	2.47	12.55	21.57	1.18	1.47	0.69	0.01	4.39	1.73
"–"	25	1.33	2.78	13.87	18.57	1.19	1.5	0.31	0.1	7.76	9.41
				Control samples							
RVT-21		0.32	1.37	10.85	17.99	1.87	1.82	0.68	0.01	9.16	16.06
		0.22	1.39	12.40	18.33	1.59	1.18	0.43	0.01	12.2	11.07
		0.25	1.42	11.70	18.16	1.77	1.38	0.35	0.01	14.30	10.21
		0.16	1.23	10.73	19.71	1.74	1.74	0.59	0.01	11.50	11.60
RVT-21a		0.18	1.76	12.28	20.88	1.54	1.27	0.44	0.01	7.71	11.76
		0.13	1.61	12.51	20.55	1.48	1.20	0.37	0.01	8.72	10.69
		0.15	1.70	11.15	19.46	1.51	2.68	0.50	0.01	13.47	7.69
RVT-22		0.04	1.36	9.21	19.66	1.15	3.16	0.42	0.01	11.26	12.98
		0.08	1.76	12.56	20.54	1.25	1.83	0.84	0.02	7.83	10.65
RVT-23a		0.63	2.50	13.36	22.08	2.10	1.49	0.52	0.02	6.89	7.78
		0.72	2.40	11.37	21.52	2.12	1.58	0.41	0.01	8.59	8.72
		0.73	2.08	12.01	21.0	1.89	1.78	0.50	0.01	8.50	9.83
RVT-23b		0.19	1.79	12.43	20.47	1.5	1.24	0.38	0.01	8.17	11.65
		0.12	1.68	13.00	19.62	1.54	1.22	0.55	0.05	8.49	11.94
		0.20	1.73	12.59	19.30	1.34	1.20	0.43	0.03	9.83	11.82

Table 2 Electron microprobe analyses of rock varnish samples from Makhtesh Ramon (Camebax)

Sample no.	Abs. height[a]	Na	Mg	Al	Si	K	Ca	Ti	Cr	Mn	Fe
RVT–24		0.16	1.71	12.33	20.17	1.37	1.75	0.59	0.01	10.26	9.64
		0.12	1.64	12.85	23.14	1.35	1.38	0.40	0.04	7.52	8.72
		0.08	1.29	12.03	19.44	1.81	1.55	0.35	0.01	10.83	11.42
		0.10	1.64	12.08	20.60	1.49	1.66	0.43	0.01	8.86	10.80
RVT–25		0.17	1.55	11.81	20.34	1.32	1.52	0.47	0.01	10.20	10.59
		0.13	1.40	10.59	19.48	1.48	1.65	0.41	0.01	13.00	11.11
		0.14	1.33	11.25	19.52	1.12	1.31	0.67	0.02	11.01	12.44
		0.11	1.46	11.37	17.82	1.13	1.43	0.54	0.01	13.97	11.99

[a]Height values in m above sea level; other values are relative heights above the surrounding area in m.

5 Mineral Composition

The mineralogy of desert varnish was first examined by Potter and Rossman (1979). They identified birnessite and hematite as Mn- and Fe-oxide phases in desert varnish by IR spectroscopic and chemical techniques. These oxides are in intimate physical association with the mixed-layer illite-montmorillonite clay minerals of varnish. Since mineral phases of rock varnish have not been identified in detail, the coating of three different samples from various geomorphic levels has been examined with the help of electron microscopy (TEM). Electron microdiffraction pictures have been obtained. Element composition has been controlled with energy-dispersive analysis (KEVEX) (A.Sivtsov). Detailed examination of patina established the following mineral phases in rock varnish: kaolinite, montmorillonite-nontronite (neotokite), litiophorite, goethite, hematite, rutile, Mg-Al asbolan, chlorite, Ti-augite, barite and apatite (Fig. 3). Minerals such as Ti-augite, barite, apatite and others appear to be transported by eolian processes. It was established that the dominant phase in all samples, characterizing different geomorphic levels (400, 25-30 and 5-6 m), is a clay mineral from montmorillonite-nontronite group (Fig. 3a). In this case, two varieties with different Fe content are revealed. The poor structural ordering of this phase suggests that it can also be related to the hizingerite (neotokite) group. Kaolinite is not noted in all samples and always forms fine dispersed aggregates. Litiophorite is represented by the aggregates of the finest plates or fine dispersed particles (Fig. 3b). Hematite is also very fine, and is rarely monocrystal tabular. Goethite is in intimate association with mixed-layer clay minerals (Fig. 3c). Amorphous or tabular particles are typical for chlorite. This mineral is easily identified on the basal reflection range $d_{(001)}$ with the periodicity ~ 14 Å. Mg–Al asbolan is related to the structural asbolan type from Kara-Chagyr (Chukhrov et al. 1980) but it does not have the same composition (Fig. 3d). Al and Mg appear to form octahedron layers with Mn^{4+} occupying the tetrahedron position. The parameters of sublattices are a= 2.84 Å and 3.0 Å. Usually, asbolan occurs in an extreme by ultrafine mixture with kaolinite.

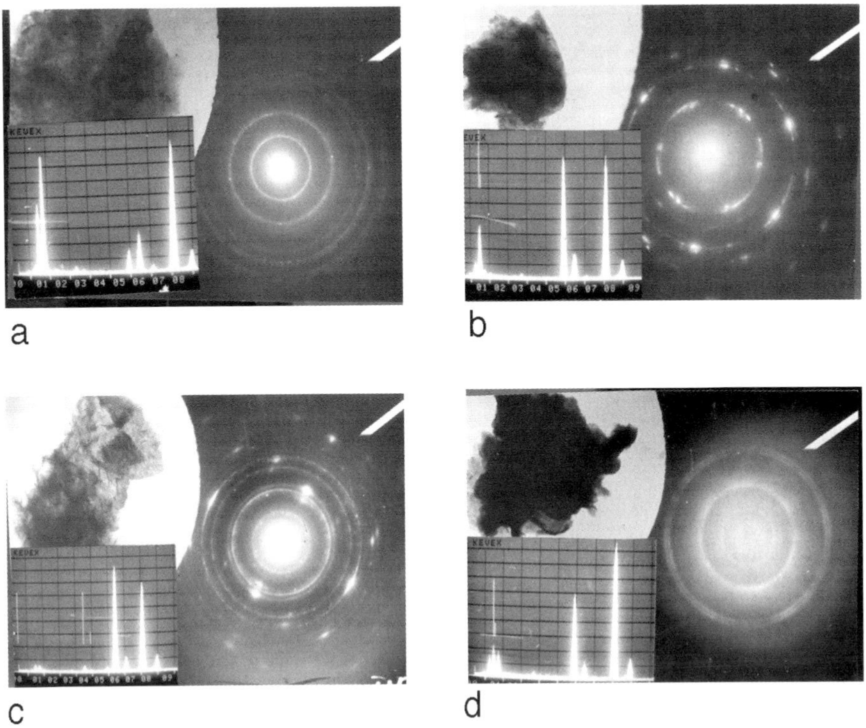

Fig. 3a-d TEM microimages, microdiffraction pictures and dispersive spectra of some minerals from rock varnish coating: a Mn–Fe montmorillonite; b litiophorite; c Al goethite; d Mg–Al asbolan.

6 Mathematical Operations

On the basis of logic informative analysis, it was found that samples having different ages are characterized by dissimilar composition. Logic informative analysis makes it possible to find indicator intervals of elemental content range for every sign characterizing the samples with a distinct age. Using these intervals, mathematical rules have been established, which identify the degree of closeness of every sample to a standard group in by its chemistry.

Analysis of the modification range of indicator intervals for samples with dissimilar ages shows that there is no clear-cut linear tendency to value age from primary signs. This is also confirmed by data of correlation analysis: coefficient values in a coefficient matrix of multiple correlation are no more than 0.5. This testifies to more complicated relationships between rock-varnish chemistry and geological age. For the description of this relationship we used an equation of nonlinear relation of the following form:

$$Y = F(x_1, x_2, \ldots, x_n),$$

which characterizes the relation between absolute age (Y) and concentration values of elements $x_1, x_2, \ldots, x_N$. Since it is impossible to restore a true form of a function, it is possible to design its approximation in a class of the given form. In this connection, the task of search modeling of this relation has been set, taking into account not only elemental concentrations but their ratio and products. For calculating numerical ages, the absolute age scale constructed on the basis of thermoluminescence data was applied. The design of the equation of nonlinear regression is very complicated, and handling it traditionally by the method of the least-squares assures minimal error of determination on the average for a sampling, in this case high deviations are possible for some points. Therefore, the task of an designing equation has been attempted which gives minimal error for every point of determination. The description of the desired model is the following:

$$Y = a_{00} + \sum_{i=1}^{N} a_{0i} x_i + \sum_{i,j=1}^{N} \left(b_{ij} x_i x_j + c_{ij} x_i / x_j + d_{ij} x_j / x_i \right)$$

where $x=(x_1, x_2, \ldots, x_N)$ is a description of a sample in a set of initial signs; Y is the calculated age (million years).

In designing this equation the main task was to find coefficients satisfying set conditions. The study of the desired coefficients *a, b, c, d* is conducted on the condition of distinction minimization $/Y_k - V_K/$ (distinction between calculated values Y_K and initial values V_k, characterizing the age of a sample k for every sample).

Such a method of attack is reasonable if there is plausible dating of samples. Otherwise, the elaborated mathematical rules using logic informative analysis provide identification of the age group to which an analyzed sample is closest. This can be considered as a task of linear programming:

$$\min_{H>0, /Y_k - V_k /< H, k=1,2,\ldots,n} H ,$$

where H = an estimation accuracy of a rated age, n = a number of samples in a sampling.

For the solution of this equation, a simplex method is used. As a result, coefficient values *a, b, c, d* were estimated, which describe a model of a chosen form, and the value *H* characterises the accuracy of this description. For the model built on the provided data H =0.064, that is an error in estimation of absolute age does not exceed 64000 years. The values of elemental concentrations are considered as an argument x_j, (j =1,2,...,10). To facilitate forming the resultant equation, the following designations are used:

x_1=Na - elemental concentration of Na, weight %
x_2=Mg - elemental concentration of Mg, weight %, etc.

The calculated model for dating on rock-varnish chemistry in terms of resultant coefficients is as follows:

$$
\begin{aligned}
Y=&330.1418Na-8.63694Mg+6.604966Al+1.998514Si+39.39983K+\\
&+108.7927Ti+223.1224Cr++5.85118Mn+1.642149Fe+8.874461NaMg+\\
&+88.80886Na/Mg+0.452191Mg/Na-2.092061NaAl-9.667416NaSi-\\
&-3378.058Na/Si-0.012379Si/Na-6.70379NaK-0.809987K/Na+\\
&+1.719279NaCa+10.34105Na/Ca-0.774867Ca/Na-15.83823NaTi-\\
&-1.599621Na/Ti-0.433843Ti/Na+119.9918NaCr+0.020519Na/Cr+\\
&+5.903579Cr/Na-1.789714NaMn+0.289843Na/Mn-0.016879Mn/Na+\\
&+1.262028NaFe+124.2728Na/Fe+0.220534Fe/Na+0.462026Mg/Al+\\
&+2.743229Al/Mg+416.4457Mg/Si-3.283021Si/Mg+1.199073MgK+\\
&+9.779447Mg/K-36.083020K/Mg+5.570085MgCa-0.244911Mg/Ca+\\
&+48.72757Ca/Mg-7.275502MgTi+1.444417Mg/Ti-70.54765Ti/Mg-\\
&-20.59409MgCr-35.18291Cr/Mg+0.544135MgMn+0.171234Mg/Mn+\\
&+2.117619Mn/Mg-10.85525Mg/Fe+3.731436Fe/Mg-119.314500Al/Si+\\
&+91.18935K/Al-0.728109AlCa-0.584765Al/Ca-0.10423Al/Ti-\\
&-0.002248Al/Cr+6707.978Cr/Al-26.432950Mn/Al-3.347590Fe/Al-\\
&-0.569367SiK+0.619844Si/Ca-0.984085SiTi-6.243132SiCr+\\
&+0.001532Si/Cr-0.083056SiMn-1.3789Mn/Si-0.008665SiFe-\\
&+5.199395KCa-4.408683K/Ca-2.2418KTi-97.972950KCr-\\
&-0.008719K/Cr-159.8334Cr/K-0.393301K/Mn-0.862342Mn/K-\\
&-0.458225KFe-2.144309Fe/K-3.705322CaTi-86.225500CaCr-\\
&-0.018887Ca/Cr-0.088510CaMn-0.960711CaFe-126.143700TiCr+\\
&-0.001154Ti/Cr-0.846977TiMn-0.205294Mn/Ti-1.718866TiFe-\\
&-12.71956CrMn-11.0507Cr/Mn-0.000828Mn/Cr-15.80915CrFe-\\
&-0.097761MnFe-11.19832Mn/Fe-69.23621.
\end{aligned}
$$

Calculating the coefficients with a precision of six signs after the point insures estimation of absolute age by this model with an accuracy of several thousand years. The new elementary functions needed to describe this relationship testify to the complicated relation between absolute age and ten variables – the elemental concentrations. An attempt to construct a correct model for rock-varnish dating without considering the pairs ratio of elements revealed great deviations from the initial dating data. Therefore all combinations of primary signs should be considered. As a result of the mathematical analysis, a data base for a system which allows dating of geomorphic levels on the chemistry of varnish coating was built. For a verification of this model, control samples without allocation from different geomorphic levels were analyzed and the geological age was calculated on the basis of developed the model. The results obtained correspond to TL determinations and geological data (Fig. 4). The value calculated by a given programming operation was considered as the generalized geochemical indicator of rock-varnish dating (see Table 3). This indicator reflects the relation be-

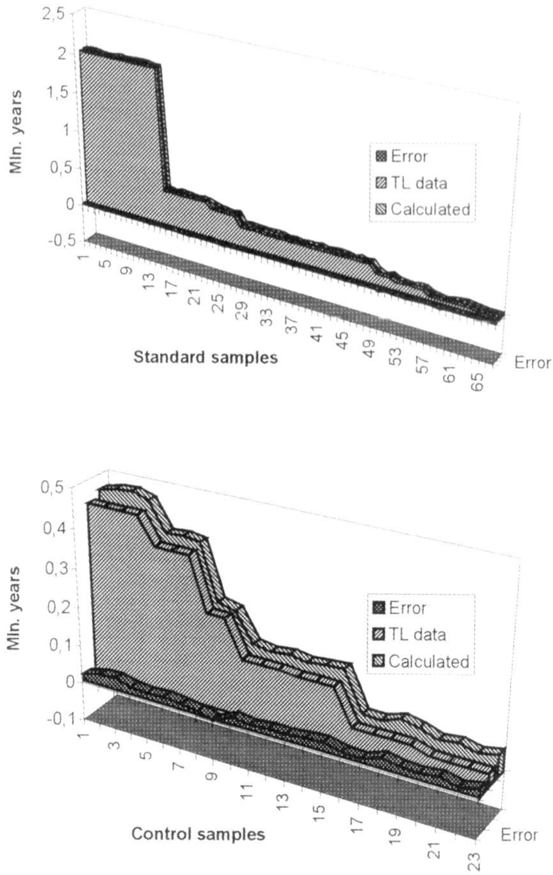

Fig. 4a,b A comparison of age dating for different levels calculated on the generalized geochemical index GI rv and on TL data: *a* a standard group of samples; *b* control samples

tween the coating chemistry and the numerical age of geomorphic levels, and is of great genetic importance.

7 Results and Discussion

The present study shows that rock varnish sampled from various geomorphic surfaces differs in the following characteristics:

1. Concentration limits of elements, which do not overlap each other. This tendency is most clearly pronounced for Mn, Fe and also for Si and Al, less clearly for Al, Ti, Mg, Ca.
2. Values of pairs and multiple correlation of elements. It has been established that values of multiple correlation, both positive and negative, are higher as a whole in rock-varnish samples from young levels, except for the pairs Al-Mg with positive correlation and Mn-Si with negative correlation, which have a

tendency to be ordered at the high levels. A change from positive correlations to negative ones is observed with geological age, for example for Fe-Mn and Fe-Ti pairs.

3. Values of the generalised geochemical indicator GI rv, which change non-continuously from one level to another.

These geochemical differences permit the use of various geochemical coefficients for rock-varnish dating. Considering the multifactor character of geochemical processes and the complex composition of the formations analyzed, this task could be solved only in consideration of all primary signs and their ratios, and using the geochemical indicator mentioned above, which gives the maximal approximation of calculated ages to data obtained by other methods. Fig. 4 shows that the divergence value in age dates is no more than 5–10% .

As a result of this investigation, the data base for the system was created, allowing absolute and relative dating of the geomorphic surfaces of Makhtesh Ramon and adjacent regions of the plateau Negev, by using the chemistry of rock-varnish coating. This approach put forward the complex problem of dating different geomorphic surfaces and revealing climatic changes in different environments.

Acknowledgements

Supported by the Russian Fundamental Research Foundation (RFFI), Grant No 96-05-66197, and in part by the University after Ben-Gurion in Beer-Sheva (Israel). We especially wish to thank our colleagues Dr.G.Nechelustov, for performing electron microprobe analyses of rock varnish and development the special technique for unpolished samples, and Dr.O.Kulikov, for the thermoluminescencal data which we used for the absolute dating scale of different geomorphic levels.

References

Allen, C.C., 1978, Desert varnish of the Sonoran desert: Optical and electron probe microanalysis: Journal of Geology, v.86, p.p. 743-752.

Ben-David R. and Mazor E., 1988, Stages of the evolution of the Makhtesh Ramon and its drainage system: Israel. Journal Earth Sciences, v. 37, No 2-3, p.p. 125-135.

Chukhrov F.V., Gorshkov A.I., Rudnitskaya E.S., Beresovskaya V.V. and Sivtsov A.V., 1980, Manganese Minerals in clays: A Review. Clays and Clay Minerals, v. 28, No 5, p.p. 346-354.

Dorn R.I. and Oberlander T.M., 1981, Rock varnish origin, characteristics, and usage: Zetschrift fur Geomorphologie, v. 25, p.p. 420- 436.

Dorn R.I., DeNiro M.J. and Ajie H.O., 1987, Isotopic evidence for climatic influence on alluvial-fan development in Death Valley, California: Geology, v. 15, No 2, p.p. 108-110.

Dorn R.I., Tanner D.L., Turrin B.D. and Dohrenwend J.,C., 1987, Cation-ratio dating of Quaternary materials in the east central Mojave Desert, California: Physical Geography, v. 8, p.p. 72-81.

Dorn R.I. and Phillips F.M., 1991, Surface exposure dating: Review and critical evaluation: Physical Geography, 12, 4, pp. 303-333.

Harrington C.D. and Whitney J.W., 1987, Scanning electron microscope method for rock - varnish dating: Geology, v. 15, p.p. 967-970.

Hooke R.LeB., Yang H. and Weiblen P.W., 1969, Desert varnish: an electron probe study: Journal of Geology, v. 77, No 3, p.p. 275-288.

Hunt Ch.B., 1961, Stratigraphy of desert varnish: U.S. Geological survey professional paper 424-B, Geological survey Research, short papers, p.p.194-195.

Krumbein W.E. and Jens K., 1981, Biogenetic rock varnishes of the Negev Desert (Israel): an ecological study of iron and magnese transformation by cyano-bacteria and fungi: Oecologia, v. 50, p.p. 25-28.

Masor E., 1978, Field geology - Makhtesh Ramon: Rehovot: Weizmann Institute Science.

Potter R. M., Rossman G.R., 1979, The mangenese- and iron-oxide mineralogy of desert varnish: Chemical Geology, v. 25, No 1/2, p.p. 79-94.

Vlasov V. K., Kulikov O. A., 1988, Radiothermoluminescencal method of dating of loose depositions: Moscow, Publishing House of Moscow State University.

Risk, Uncertainty and Priorities – Quantitative Models

I. Lerche, S. Cao, S. Mallory, K. Petersen, and A. Lowrie

1 Introduction

Quantitative basin analysis has three major components: (i) determination of the spatial and temporal dynamic evolution of the sediments in relation to evolving basinal structure, and with respect to the spatial variations of lithologic and diagenetic fabrics; (ii) determination of the thermal history of sedimentary units in relation to their burial histories; (iii) determination of the hydrocarbon generation, migration and accumulation histories of all hydrocarbons generated in the basin, so that the relative timing of hydrocarbon evolution can be assessed in relation to the development of traps and their seals, and in relation to the continued integrity of sealed traps after hydrocarbon emplacement. In this way a better appreciation can be obtained of potential hydrocarbon sites in a basin. On the path to achieving such an evaluation of hydrocarbon site determination, four major concerns have to be addressed in relation to the output from any basin analysis model. The concerns are: (i) resolution; (ii) sensitivity; (iii) uniqueness; (iv) precision. These four factors are controlled to a greater or lesser extent by three overriding inputs: (i) intrinsic assumptions; (ii) parameter values; (iii) quality, quantity and sample spacing of data.

A detailed analysis of the evaluation of basins taking into account the three input factors and the four output concerns has been given elsewhere (Lerche 1989, 1990) for dynamical, thermal and hydrocarbon assessments. Every case history investigated should be evaluated for its rugged stability (or lack thereof) in relation to the points made above.

Thus the development of quantitative basin analysis can be separated into three stages: the first stage of *model development* concentrates on the construction and use of mathematical and numerical models of basinal evolution; the second stage is *uncertainty and sensitivity* in which the main aim is to examine the stability of model results in relation to the uncertainties and rages of variations in assumptions, parameters and input data; the third stage, and the one we are dominantly concerned with here, is related to the uncertainties and sensitivities of the models used, as in stage two above.

The purpose, then, is to provide quantitative probability measures to outcomes of model calculations in order to have objective, reproducible estimates of the likelihood of uncertainty. From such measures strategic exploration decisions can be made, tied also to the economic probability of successful exploration for hydrocarbon accumulations (Lerche 1992).

2
Cumulative Probability and Error Assessment

In principle, Monte Carlo procedures should be applied to the study of risk assessment and most probable outcomes of basins analysis models, i.e. the probability distribution of each model output result should be evaluated by considering all possible values of each assumption, each parameter, and each datum point, within determined ranges, and then an intrinsic probability distribution of finding each such input control assigned within its given range (e.g. uniform, Gaussian, triangular, log-normal, etc.), and then, using a multi-Monte Carlo approach, run through a basin modeling calculation to produce a set of outputs. The probability distributions of the respective outputs can then be recorded and used in risk assessment (Fig. 2). While the Monte Carlo procedure is completely general and can be applied to any basin analysis model, once the range and underlying probability distributions for each parameter and for the measurement errors in data have been provided, the problem is that such a numerical investigation can be extremely computer intensive, perhaps one of the reasons that such methods are not routinely applied to the results derived from basin analysis models (Cao and Lerche 1989).

It would, therefore, be helpful if alternative procedures could be used, which would provide the same probability outcome, but which would not demand such massive computer time investment as do the Monte Carlo methods. This paper will demonstrate how to use cumulative probability techniques to achieve such a goal; the technical mathematical details are given elsewhere (Lerche 1997).

Examples of the procedure are probably more useful illustrations than generic mathematical formulae. Accordingly, we now illustrate use of the cumulative procedure with two case histories.

3
An Illustration of Cumulative Probability: the Kugrua No. 1 Well, Alaska

The Kugrua No. 1 well is located in the western section of the National Petroleum Reserve, Alaska, just to the south of Peard Bay. The well was drilled to a total depth of 4127 m (12580 ft.). The present-day oil window in the well, based on vitrinite reflectance, encompasses all formations from the Sadlerochit Group through the Torok Formation as depicted in Fig. 1.

The major problem is that only six vitrinite reflectance measurements are available for the well, so that it is difficult to provide any reasonable measure of paleoheat flux based on thermal inversion methods with such sparse data. Fortunately three other wells, Tunalik, Peard and S. Meade share the general regional area, and all three also have vitrinite reflectance values which can be used with

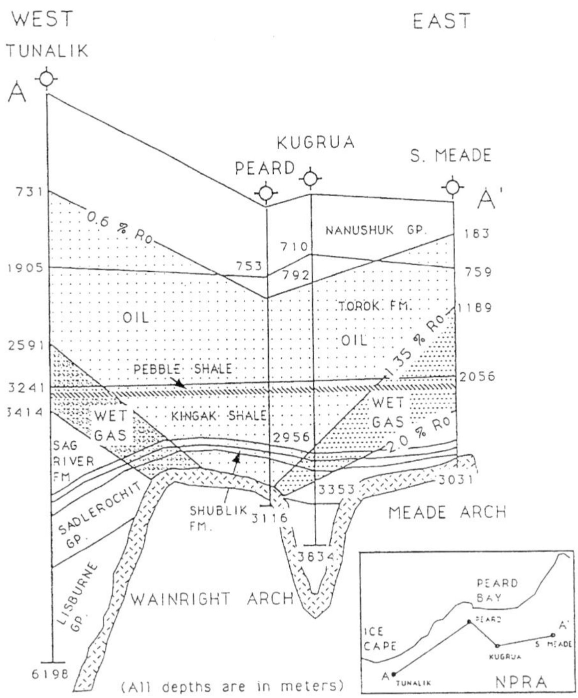

Fig. 1 Present-day oil window in the northwest of the National Petroleum Reserve, Alaska based on the vitrinite reflectance data from four wells (including Kugrua No. 1). (Magoon and Bird 1986)

thermal inversion procedures to obtain paleoheat flux estimates for each well. While each well has only of order 6–10 measurement values, one would naively anticipate that a more-or-less average consistent paleoheat flux behavior would result from each well. Fig. 2 shows the predicted "best fit" vitrinite reflectance for each well, while Fig. 3 shows that the individual paleoheat fluxes are fairly different, suggesting that the reflectance data from any single well are not sufficient to have a high degree of confidence in the quality of the "best" paleoheat fluxes. In turn, this uncertainty leads to an uncertainty in the thermal maturation onset for any one, and all, of the wells in the area. In order to illustrate the degree of risk and uncertainty, three heat flow patterns with time were used to encompass in a broad sense the possible paleoheat flux values based on regional geological and geothermal information. These three patterns are drawn in Fig. 6 and labeled Decrease, Constant and Increase, respectively.

All other dynamical and thermal parameters were then held constant and only the paleoheat flux allowed to vary with time. The Decrease pattern was taken as representing the most likely minimum heat flux at each instant of time, the Constant pattern as representing the likely value, and the Increase pattern as representing the most likely maximum value. It would appear that this range of values generously encompasses the best estimate made for the Kugrua No. 1 well based on sparse data.

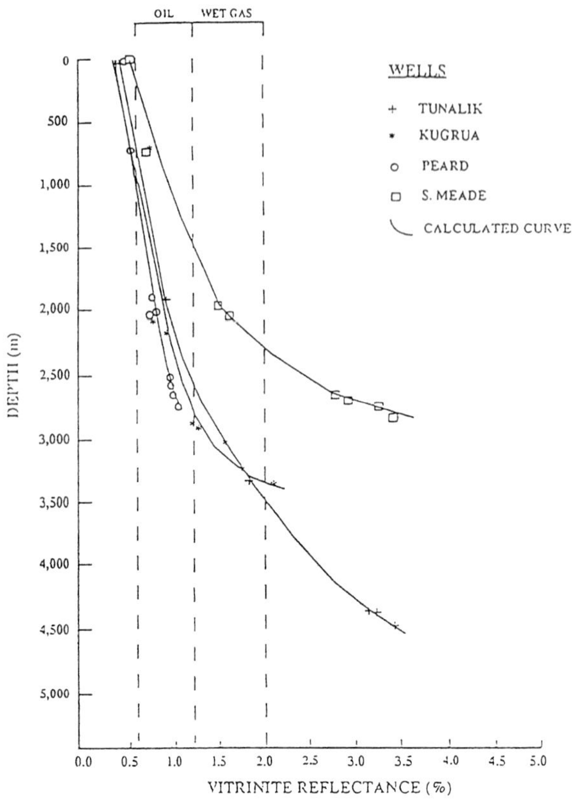

Fig. 2 Predicted "best fit" vitrinite reflectance for the four wells in the northwest NPRA

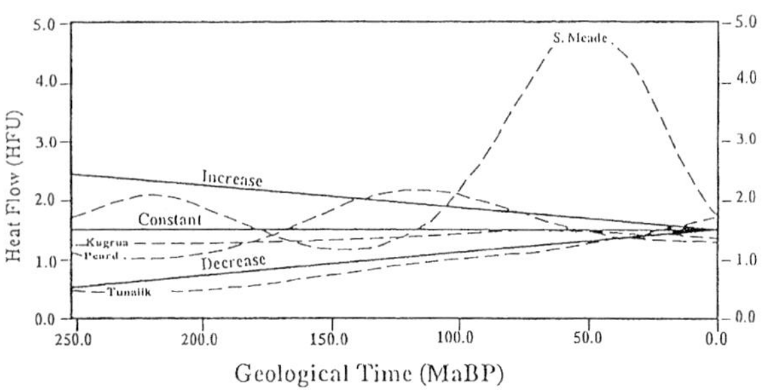

Fig. 3 Modeled "best" paleoheat flux for the four wells in the northwest NPRA.

We then followed the cumulative probability (Lerche 1991) procedure laid out elsewhere. Presented in Fig. 4 are the observed vitrinite reflectance measurements with depth with superposed cumulative probability curves based on the three paleoheat flux estimates described above. Inspection of Fig. 4 suggests that

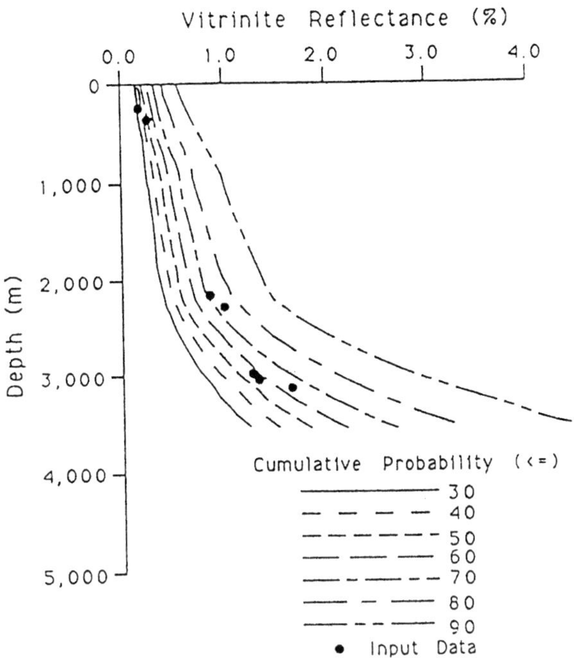

Fig. 4 Vitrinite reflectance versus depth with different cumulative probability values for Kugrua No. 1 well

the two vitrinite reflectance measurements at shallow depth (less than 500 m) may need to be re-evaluated to higher values because the cumulative probability that they are correct is less than about 50% for each.

Likewise the deepest three vitrinite measurements also fall at about the 60% cumulative probability, suggesting that they, too, may be slightly low. The cumulative probability expected value is at the 68% point (see Sect. 2) so that the combined information would indicate that the paleoheat flux in the Kugrua No. 1 well needs to be higher than the value most consistent with the sparse downhole vitrinite measurements. There is about a 40% overall chance that the total thermal maturity is underestimated.

To provide a different presentation of information, cumulative probability values with depth are given in Fig. 5 for different fixed values of vitrinite reflectance. Note from Fig. 5 that there is an 80% chance (not 100%) that the surface reflectance value is less than 0.4%, suggesting that there is a 20% chance of a surface reflectance value greater than 0.4% most likely due to late Cretaceous and Tertiary erosion. Also note that at 3000 m depth there is a 70% chance that the reflectance should be less than about 1.4%, and a 40% chance of a value *greater* than 1.2%. Observations (Fig. 4) indicate reflectance values of between about 1.1 to 1.7% at 3000 m, suggesting that the high values (1.7%) are probably too high and, equally, that the low values (1.1%) are likely to low – or that the intrinsic uncertainty is about ± 0.3% in any one measurement – or both of course.

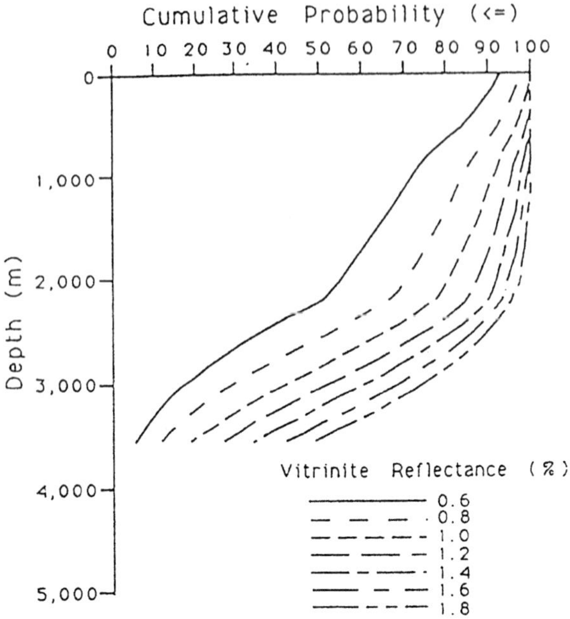

Fig. 5 Cumulative probability of vitrinite reflectance with depth for Kugrua No. 1 well

In order to complete the investigation, we also looked at the probability of obtaining particular values of vitrinite reflectance in relation to the burial history of the sediments. Thus: Fig. 6a,b,c, shows the cumulative probability behaviors with time for obtaining vitrinite reflectance values of less than 0.6% (Fig. 6a), 1.2% (Fig. 6b), and 2.0% (Fig. 6c); while Fig. 7a,b,c, show the conjugate variation of vitrinite reflectance behaviors with depth and time at fixed estimates for the cumulative probability of 30, 70 and 90%. From Fig. 7 c it is very likely (90%) that at no time or any depth was there any value of reflectance in excess of 2.5%; indeed, it was most likely (70% chance) that reflectance was always less than 2.0%. Thus: while the vitrinite reflectance data from the Kugrua No. 1 well are themselves too sparse to provide a good and precise measure of paleoheat flux, data from neighboring wells can be used to assess probable ranges of uncertainty of the paleoheat flux for the Kugrua No. 1 well. These ranges of paleoheat flux, in turn, can be used to assess likely regimes of timing of thermal maturity either at a given level of probability or at a given value of vitrinite reflectance.

The results from this example indicate that only with further well information can an appreciation be obtained of the likelihood of the correctness (or lack of correctness) of any particular choice for thermal maturity.

Perhaps the salient point from the examples is an understanding for how probability procedures can be used to assess risk for hydrocarbon proneness in the face of uncertainty of parameter values, or in the face of uncertainty in geological factors caused by a lack of good-quality, high-quantity, well-sampled data.

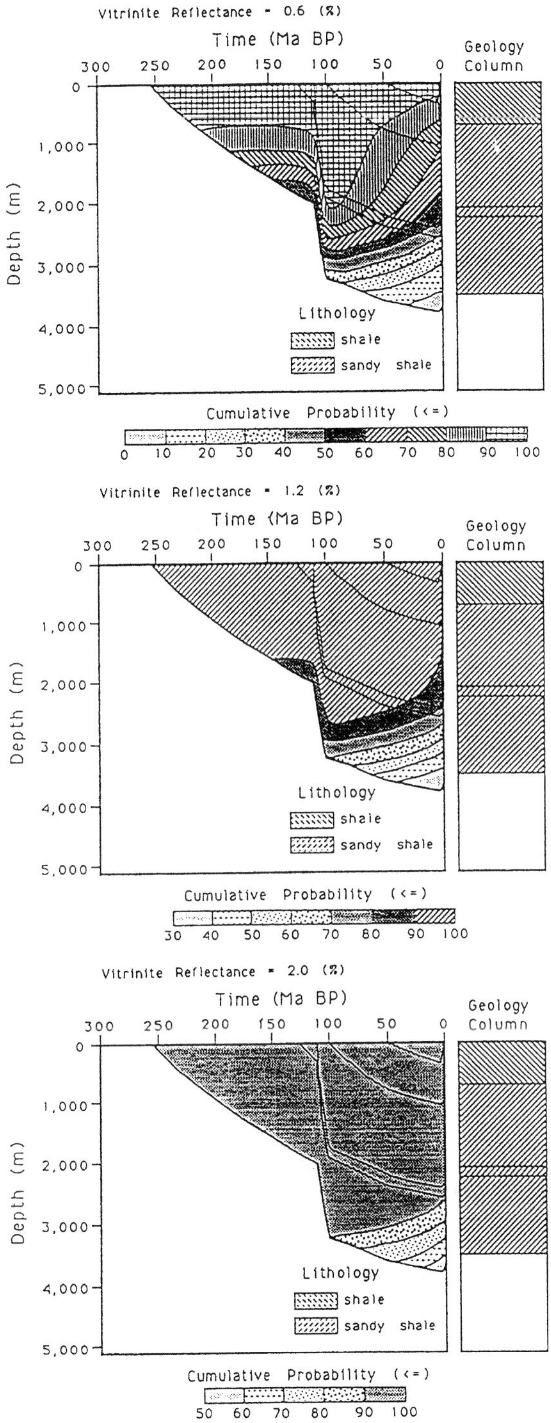

Fig. 6a–c Burial history with cumulative probability for different vitrinite reflectance values . **a** $R_o = 0.6\%$; **b** $R_o = 1.2\%$; **c** $R_o = 2.0\%$

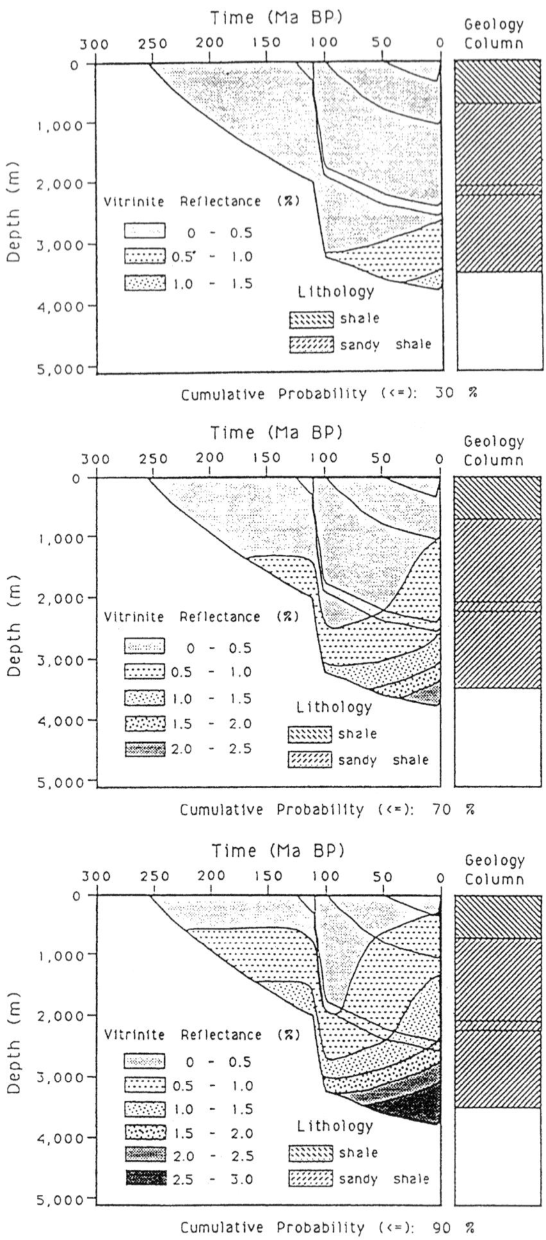

Fig. 7a–c Burial history with vitrinite reflectance for different cumulative probability values. **a** 30%; **b** 70%; **c** 90%

4 An Illustration of Relative Importance: Lateral Variations of Sub-Salt Overpressure buildup

The presence of laterally extensive sheets of salt in the Gulf of Mexico has been well documented from both seismic and direct drilling through salt sheets (see e.g. many of the references in Lerche and Petersen 1995).

In addition, the lateral variations with depth of different stratigraphic formation thicknesses in relation to current salt sheet positions, and the juxtapositioning of the same formations both above and below salt at different lateral locations, have been used to infer rapid lateral motion of salt sheets, with speeds of order 1–20 cm a^{-1} (West 1989; Lowrie 1994).

Because salt has an extremely low permeability, a salt sheet provides an effective seal after emplacement at a given location, trapping sub-salt fluids (including hydrocarbons) and retaining porosity in sub-salt formations. Because of the weight of the salt and of formations above the salt, the fluid pressure in sub-salt formations is enhanced by the salt emplacement, as has been detailed elsewhere from simple model considerations as well as by observational information, where fluid pressure increases at sub-salt breakout are about 100–300 atm/ km^{-1} thickness of salt (O´Brien and Lerche 1994).

And yet the evolution of sub-salt overpressure development with both time and spatial location in relation to the motion of a salt sheet remains an outstanding concern. The point here is that there is a degree of uncertainty on the assessment of the lateral speed of salt, and there is also a degree of uncertainty on the lateral variations of thickness of a salt sheet; accordingly, both the rate at which sub-salt overpressure builds up with time at any given location, as well as the spatial pattern of variation of sub-salt overpressure at any instance of time, have corresponding degrees of uncertainty.

The ability to bracket the spatial and temporal structure of sub-salt overpressure development is of use in a variety of contexts. First, both the earliest and latest times for the sealing of sub-salt formations by emplaced salt, with respect to lateral position in the basin, could be bracketed, thereby providing both maximum and minimum estimates for the retention of hydrocarbons in formations underlying salt.

Second, the deeper the emplacement of salt at a given location, the more the formations are compacted prior to the appearance of salt. In this case the porosity available for hydrocarbon storage in sub-salt sediments is reduced relative to early salt emplacement. Further, the more compacted the sediments are, the greater is the effective stress (and so the lower the excess fluid pressure) and the greater is the sedimentary consolidation strength, so that salt speed and direction of motion can be influenced more significantly by the corresponding higher sedimentary resistance.

Third, early development of excess pressure implies a higher porosity retention and, accordingly, a higher geothermal gradient below salt because of the lower formation thermal conductivity. The maturation of hydrocarbons entrapped below salt can be hastened by this increase in temperature.

Fourth, the dissolution of salt and the formation of collapse features, as well as accommodation space creation by both salt motion and dissolution, are all influenced by lateral salt speed, salt thickness, and porosity retention in sub-salt formations, leading to shifts in depositional and fluid-flow patterns.

Fifth, the lateral speed of salt can itself be influenced by the water content of the sub-salt formations on which the salt mass glissades. A high porosity implies less friction at the salt/sediment contact due to the lowering of contact viscosity (as well as higher salt dissolution) than does a drier, more compacted, sub-salt formation, thereby yielding a lowering of dynamic friction for salt motion through sediments.

For all of these reasons it is important to be able to determine quantitatively the degree of uncertainty of overpressure buildup in respect of magnitude, time and sub-salt location. Here we consider the two salt masses exhibited in the right hand side of Fig. 8, as interpreted from a large-scale, regional seismic study in the Gulf of Mexico.

Each of the salt masses is allowed to move laterally with time so that different lateral locations (labeled A-H in Fig. 8 and progressively measured from the current, most-basinward (southernmost), position towards the (northern) shoreline are subject to the effects of the salt at different times and depths. For instance, at location A, the salt "nose" passed this location very recently so that effects of the salt are minor, the salt is also thinner at location A than at location C, for instance, so that pressure buildup is less. At location D, on the other hand, the salt was emplaced early, gradually increased its thickness at location D as the salt mass moved basinward and then, quite recently, the salt vacated location D. thus the pressure buildup at location D must have started quite early, must have been maintained for some considerable time at a value higher than at location A, before being reduced recently.

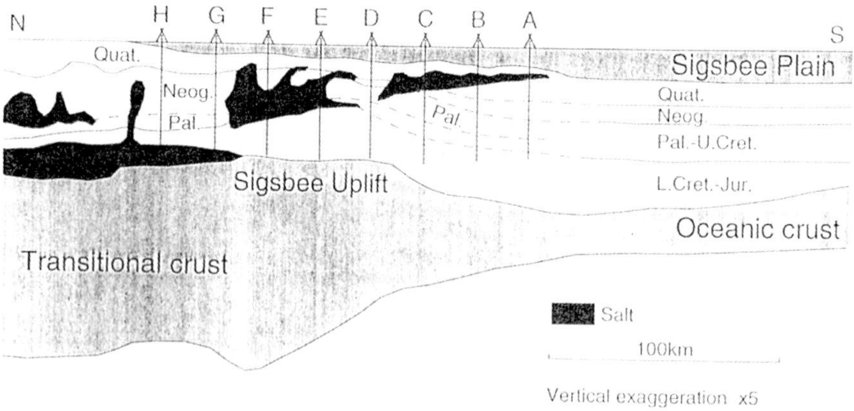

Fig. 8 Sketch of the 2-D section across the Gulf of Mexico with well locations **A-H** indicated

The pattern of sub-salt pressure development and associated porosity retention must therefore be substantially different at location D compared to location A. It is this difference which is of interest here. However, the actual speed of the salt sheet is uncertain, as is the thickness, so that some degree of uncertainty must also be present in the pressure buildup, in both magnitude and time at each location.

Thus, both an estimate of average pressure buildup and of the range of uncertainty of the estimate are concern. These twin problems are addressed here.

4.1 Computer Model Manipulations

4.1.1 *General Considerations*

The dynamical control on the modeled development is that sedimentary thicknesses calculated forwards in time to the present day should agree with current observations. This constraint is used to adjust parameters describing equations of state behavior for each lithology until a minimum disagreement is reached between calculations and observations (Lerche 1991).

For instance, for a fixed present day salt thickness, a range of salt speeds generates a set of overpressure build-up calculations at a given location. At each time-step the overpressure with depth is recorded for each computer run, so that the cumulative probability of obtaining less than a particular value of overpressure at each depth can be calculated. Thus, Fig. 9 shows cumulative iso-probability contours of excess fluid pressure with depth for location A at the present day. As an example, from Fig. 9 we see that there is a 90% chance of an excess pressure less than about 50 kg/cm at a depth of 2000 m, while there is only a 10% chance of obtaining an excess pressure less than about 30 kg/cm^{-2} at the same depth. The median value (50% chance) for excess pressure at 2000 m depth is about 40 kg/cm^{-2}.

The same sort of diagram can be generated at each time-step at each location across the section of figure 8, allowing for uncertainty in both the salt speed and present-day salt thickness. Three measures provide indications of the influence of moving salt sheets on lateral variations of excess pressure in formations below the salt. First is a measure of the mean excess fluid pressure with depth at each location and for each time step; we use the excess pressure (p_{50}) at 50% cumulative probability as this indicator. Second is the volatility, v, in the mean pressure estimate p_{50}, which provides a measure of the uncertainty in the 50% pressure value; a definition of volatility is

$$v = (p_{90} - p_{10})/p_{50}, \qquad (1)$$

where $p_{90}(p_{10})$ is the excess pressure value at 90% (10%) cumulative probability. A low volatility, $v \ll 1$, implies but little uncertainty in p_{50} so that there is not a dramatic influence on excess pressure evolution due to the ranges of variation in

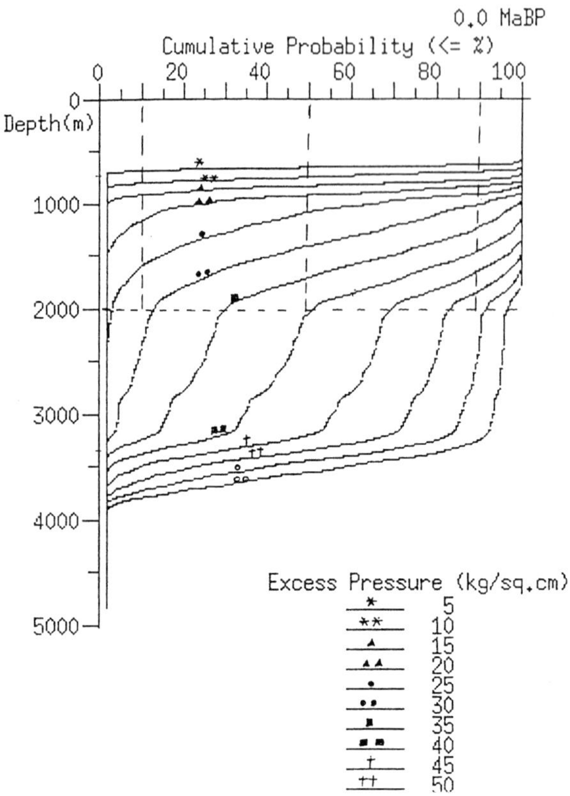

Fig. 9 Contours of cumulative iso-probability at present day at location A for excess fluid pressure variations with depth

salt speed or salt thickness around their central values; on the other hand a high volatility, v>>1, implies a significant influence on pressure development by either (or both) of variations in salt speed and salt thickness around central values. Thus, for the numerical illustration above we have a volatility of 50% at 2000 m, representing a considerable uncertainty in the excess pressure.

Third is the relative importance, RI, which measures the relative contributions to the total volatility of excess pressure produced by the range of variations of salt speed and salt thickness, respectively, around central values. RI provides a measure of which factor (salt speed/salt thickness) is causing the largest uncertainties in excess pressure at each location. RI is, therefore, a useful device in helping to focus attention on the priority with which each of the two factors needs to have its range of uncertainty narrowed if the total volatility of excess pressure is to be reduced. Note that the RI does not provide a measure of necessity for improving the total volatility, but does provide a measure of where to focus attention should it be considered that the total volatility is too large to allow an accurate measure of likely excess pressure. Effectively, one takes the distribution of excess pressure results for each of the two parameters (salt speed/salt thickness) with the other parameter held fixed, and computes the equivalent volatility. Then the fixed value is systematically increased, and the volatility again calculated for a suite of runs. The average of the volatility values then provides a measure of the contribution of uncertainty of one parameter irrespective of the value of the other parameter. The procedure is then reversed so that the volatility, v_j, of each parameter on its own is calculated. The relative importance of the jth parameter is then defined by

$$RI(i) = v_j / \sum_{J=1}^{N} v_i , \tag{2}$$

where N is the number of parameters being varied (2 in the present case).

We use the measures p_{50}, v, and RI in the next sub-section to illustrate excess pressure development across the section of Fig. 8.

4.1.2
Specific Model Results

The present-day section of Fig. 8 can be interpreted (Lowrie 1994) as providing the spatial domain of an approximately wedge-shaped salt mass currently intersecting locations A, B and C, and which recently intersected location D; together with a roughly block-shaped slab of salt currently intersecting locations E, F and G, and which recently intersected location H. As each separated salt mass moved laterally, the sedimentary infill behind the salt masses, at locations D and H respectively, is apparent from Fig. 8. The exact thickness of the salt at any location is not too well determined from Fig. 8 due to both uncertainty in the seismic velocity used to convert a two-way travel-time section into a physical depth section, plus the difficulty of imaging the base of salt on the seismic section.

For instance, Lowrie (1994) has estimated that the salt thickness at location A could be as large as 1500 m, is likely to be around 900 m, and could be as thin as 600 m. Equally, the lateral speed of motion is not precisely inferred, ranging from a low estimate of about 1 cm a^{-1}, with a likely value of about 10 cm^{a-1}, but with a high value of possibly 20 cm a^{-1}.

At each of the locations B-H, the corresponding salt thicknesses today are taken to be equally as uncertain fractionally as at location A, so that each of the salt masses can range from extremes of slow moving, relatively thin, bodies to rapidly moving, thick bodies across the section. Originally we did consider omitting calculations for the salt mass intersecting wells E-F at the present day, because we thought it unlikely that the mass has moved southward with the shape it currently enjoys; it would seem more likely that the two "tongues" originating in the south of the salt mass are what is currently moving. However, we have included the calculation as though the salt mass kept its present-day shape because we were unable to prove unequivocally that the tongues were indeed the only moving components of the salt. In any event the comparison between the two salt masses provides an interesting end-member comparison of the range of possible results, despite our misgivings on the northern salt mass conserving its shape with time.

At each geological time-step the salt bodies are taken to move across the section preserving their shapes with time to within the constraints of thickness uncertainty. Thus the salt masses cross-cut sedimentary beds, and sedimentary deposition continues across the section both prior to, during, and post-emplacement of salt at each location.

4.1.3
Mean Excess Pressure Results

As each salt sheet moves across the section there is a change in fluid excess pressure at each location. Shown on Fig. 10 are the values of mean excess fluid pres-

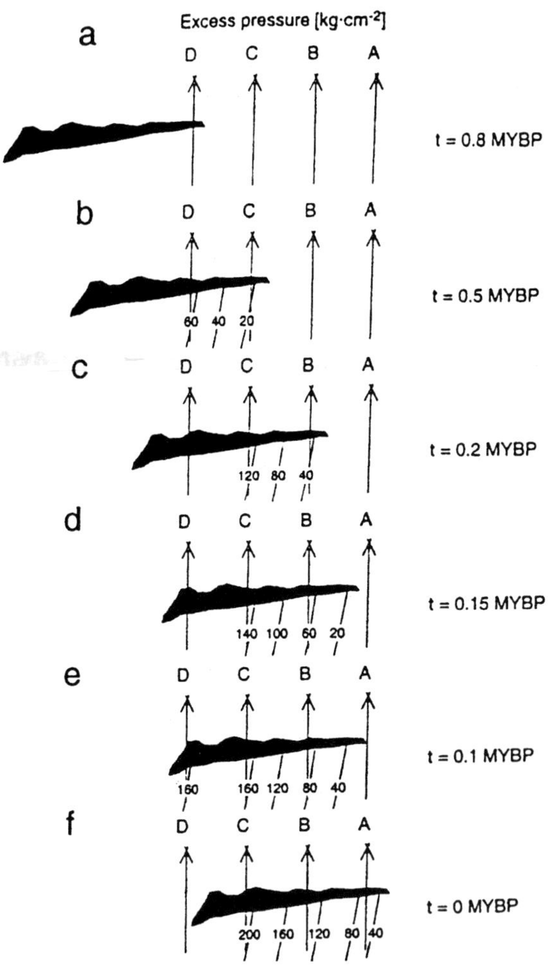

Fig. 10 Mean excess pressure variations with time directly under the southerly salt mass at locations A-D

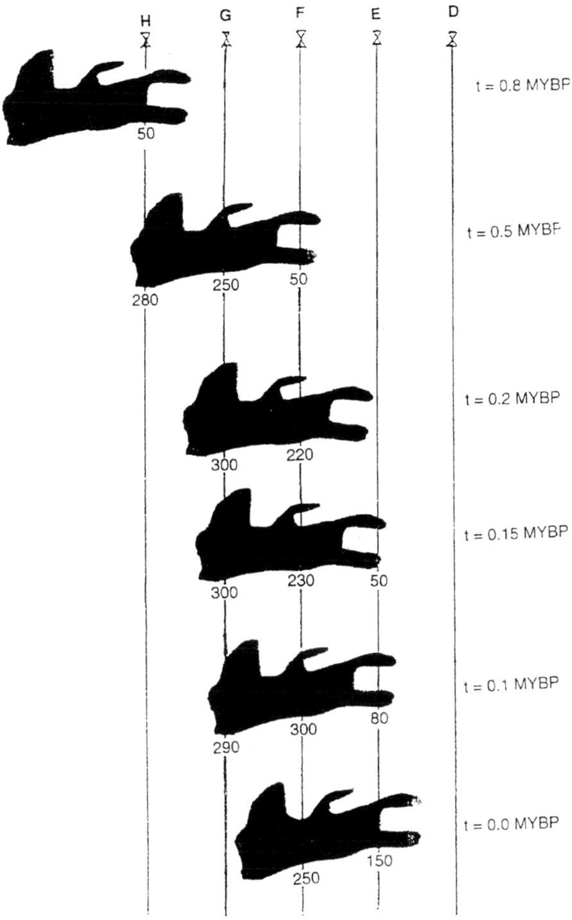

Fig. 11 Mean excess pressure variations with time directly under the northerly salt mass at locations E-H

sure directly under the salt at locations A, B, C and D as the more southernly of the two salt masses moves across the domain; while Fig. 11 depicts the corresponding mean excess fluid pressure directly under the more coastward salt mass as it transits locations E, F, G and H with time. The mean excess pressure results are obtained from the average of all runs done with variable salt thickness and salt insertion speed. To be noted from Figs. 10, 11 is the rapid rise in sub-salt mean overpressure as the salt masses progress laterally basinward to occupy specific sites. Equally as important is the rapid diminution of overpressure at locations D and H from 0.8 Ma**B.P.** through to present day as the salt vacates those locations. The increase in excess fluid pressure directly under the salt masses at the different locations would seem to be approximately proportional to the local salt thickness at each location, with a mean excess pressure of about 200 ± 100 atm corresponding to a salt thickness of about 1 km.

4.1.4
Fluctuations in Mean Excess Pressure

Because both the lateral salt insertion speed and the salt sheet thickness are quite variable, both the timing and emplaced thickness of salt at each location carry some degree of uncertainty. As a consequence, the anticipated overpressure at breakout through a salt mass will also have some degree of uncertainty. We have quantified this degree of uncertainty in terms of the volatility, v. In essence the code is run with a range of salt insertion speeds and salt thickness, to record the excess pressure in each case below salt, and then to order (in terms of increasing values of excess pressure) the histogram of results so that, by normalizing to unit area, one can then draw the cumulative probability of obtaining excess pressure values less than specific amounts. As the salt sheets move across the section, at a given location site the volatility for excess pressure will itself vary in concert with the mean excess pressure.

In addition, from the compendium of results for different salt speeds and different salt thickness, it is also possible to determine whether uncertainty of speed or thickness is providing the largest contribution to the uncertainty in excess fluid pressure. Thus, for given ranges of uncertainty of salt speed, and given ranges of uncertainty of salt thickness, one can determine the Relative Importance (RI) of each variation to the uncertainty in excess pressure.

Shown in Table 1a and b is the volatility (in %) of mean excess pressure with depth for both the salt masses at the present day at each location under two conditions: Table 1a presents volatility for the salt at its median thickness, but with a range of salt speeds from 1–20 cm a^{-1}; while table 1b presents volatility for a

Table 1 Present-day volatilities (%)

a) Median thickness/variable velocities

Depth (m)	Well A	Well B	Well C	Well E	Well F
1000	30	10	7		10
1500	10	5	8		5
2000	9	10	4	9	3
2500	9	10	7	9	2
3000	9	4	5	6	2
3500	9		3	2	2

b) Median velocity/variable thickness

Depth (m)	Well A	Well B	Well C	Well E	Well F
500					22
1000	57	17	7		17
1500	57	40	5		2
2000	66	4	6	29	4
2500	61	4	9	5	9
3000	56	2	3	6	3
3500	33	2	4	16	3

Table 2 Present-day relative importance (5) for velocity/thickness

Depth (m)	Well A	Well B	Well C	Well E	Well F
1000	66/34	94/6	39/61		63/37
1500	85/15	89/11	58/42		29/71
2000	88/12	29/71	39/62	76/24	57/43
2500	87/13	29/71	60/40	36/64	75/25
3000	86/14	33/67	56/44	50/50	60/40
3500	79/21			89/11	60/40

median salt speed of 10 cm a^{-1} but with a range of salt thickness varying to ± 30% around the median values at each location site. Note from both components of Table 1 that, with the exception of location A, the volatility is usually much less than about 10% at nearly all depths and locations under the salt masses, so that the values of mean excess pressure, recorded in Fig. 10, 11, are likely determined at sub-salt drill break-out to better than 20%. If drilling mud weight can be geared to handle a 20% uncertainty on the breakout pressures, then there is no need to improve matters further.

However, if it is required that a better job be done on estimating the uncertainty on mean break-out excess fluid pressure, then it is important to know whether it is the uncertainty in salt speed or the uncertainty in salt thickness which is causing the greatest contribution to volatility so that effort can be expended to narrow the range of uncertainty of the most important factor first. The relative importance, RI, provides the relevant measure. Note again that the RI does *not* determine that an improvement is necessary, but does determine which factor to concentrate on if it is decided that improvement is required.

Plotted on Table 2 are the RI values (in %) with depth for uncertainty in salt thickness and salt speed across the section at the present day. In this case, because only salt speed and salt thickness are being varied, the RI for salt speed is just 100% - RI (thickness)%, so that it is easy to see which of the two factors dominates overpressure uncertainty.

Note from Table 2 that there is an approximately equal balance of importance overall of variations in salt velocity and salt thickness in contributing to the sub-salt excess pressure uncertainty. There are, to be sure, considerable fluctuations in RI with depth at different locations, but there is no systematic trend apparent indicating dominance of either contribution. The fluctuations in salt velocity are from 1–20 cm a^{-1} around a median of 10 cm a^{-1}, while the salt thickness fluctuations are around ± 30%.

It would seem that if one were required to improve the volatility on sub-salt excess pressure at break-out then, for the ranges of uncertainty used, a roughly equal effort should be expended on attempting to narrow the range of uncertainty salt thickness to better than ± 30%, and salt speed to better than ± 100% around the estimated median of 10 cm a^{-1}.

4.2
Inclusion of Sub-Salt Lithology and Salt Burial Depth

For particular and specified parameter values for moving salt sheets, calculations have been given (O´Brien and Lerche 1994) which illuminate the excess fluid pressure (i.e. fluid pressure in excess of hydrostatic pressure) values to be expected. Four factors confound the use of the calculations in assessing drilling risk. First, the lithology under a salt sheet is often imperfectly known, so that the permeability of formations underlying a salt sheet is unclear. Because of the many orders of magnitude variation in permeability between sands and shales (which are the dominant lithologies in the Gulf of Mexico), the overpressure beneath a salt mass could then vary considerably depending on lithology variations. Second, it is often difficult to be precise about the salt thickness ahead of drilling, due to the problem of determining the base of salt from seismic information. The thicker the moving salt, the greater the load produced by the salt and so the greater the underlying fluid pressure needs to be in order to support the load. Thus the uncertainty in salt thickness also influences the degree of overpressure in a sub-salt formation. Third, the timing of salt emplacement at a given location in a basin is also not too well determined; a high speed salt mass presumably overshadows a given location much earlier geologically than a slower speed salt, thereby trapping fluids (including hydrocarbons) at that location for a longer period of time. Thus a more rapid, and earlier, emplacement of salt is expected to be responsible for a higher degree of overpressure. Fourth, the sub-sea bed depth of the salt top also influences the overpressure of sub-salt formations. For a given salt sheet thickness, a shallow salt top means that the salt sheet has thinner sediments of higher porosity above it than a more deeply buried salt sheet. Thus the weight of the sediments overlying the salt is less in the case of a shallow salt top. Because the weight of such sediments is transferred through the salt to be balanced by overpressure in the sub-salt formations, a shallower buried salt sheet will contribute less sub-salt overpressure, all other factors being equal.

The influence of uncertainties in the above factors on the sub-salt overpressure is by no means equal, due to the non-linearity of response of compaction and fluid-flow in sediments to geological conditions, and their variations: (a) to parameters in equations of state controlling the intrinsic evolution of sediments under applied loads; and (b) to the basic non-linear dynamical response of the geologic system. Thus it is important to determine the relative importance (RI) of uncertainties in each factor (salt thickness, sub-salt lithology, salt speed or emplacement time, salt top depth) in contributing to estimates of sub-salt overpressure.

West (1989) and Lowrie (1994) have argued that the salt sheet displayed on the cross-section has moved to its currently observed location at a rate of between 1and 20 cm a^{-1}. At location A, the emplacement of the salt, which is interpreted to be currently situated in the midst of Quaternary sediments, is treated as a wedge of inserted salt, currently of thickness 1700–2300 m. The top-of-salt depth below the sea bed is not clearly delineated, but original seismic information leads to the suggestion of a salt top at 300–500 m depth. The lithology under the salt at location A is not well known[1)]; accordingly we have run computations with the underlying formations treated as 100% sand, 50% sand and 50% shale, and 100% shale, respectively.

4.2.1
Burial History and Excess Pressure

Prior to salt insertion, each formation is sequentially deposited and continuous evolution of porosity, permeability, fluid-flow and overpressure development computed for prescribed lithologies for each for the formations.

At salt insertion time (determined by taking the "nose" of the salt sheet leftwards at a constant speed until the "nose" intersects location A) the salt is considered to develop as an ever increasing thickness unit inserted at a precise depth. [For instance if a lateral salt speed of 10 cm/a is taken, with the present day salt top at 500 m below the present-day seabed, then for a salt "nose" 60 km to the right of location A, the salt would have been emplaced at location A at 0.6 MaBP. At that time, if the sediment deposition rate above the salt is 100 m/Ma, then 60 m of the present overburden above salt would need to be removed. Thus the emplacement depth of the salt would be 440 m (500–60 m)]. In fact, the calculations are a little more complex, because porosity evolution with time and depth of each formation, including the influence of salt, needs to be included in the calculation of salt insertion depth.

Two cases were examined: case 1 in which the present-day salt top is at 300 m below the seabed; case 2 in which the present-day salt top is at 500 m below the seabed. For each case, the salt thickness was varied, the lateral insertion speed of the salt was varied, and the lithology underlying the salt was varied from 100% sand through a 50% sand/50% shale mix, to 100% shale. Table 3 provides the ranges of values used for each of the two cases. Values for each case.

Fig. 12a,b shows, as an example, the predicted excess fluid pressure (in units of $kgcm^{-2}$ = 1.03 atm) with depth for a salt insertion speed of 10 cm a^{-1}, with the current salt top at 300 m sub-sea bed, for a 50% sand/50% shale lithology under the salt and with the salt thickness set at 2000 m (Fig. 12a) and 2200 m (Fig. 12b) respectively. Fig. 12c shows the difference (higher sub-salt excess pressure) from figure 12a when the lithology is changed to 100% sand and all other parameter values are held fixed[2)]. For each case similar figures can be drawn, each providing an excess fluid pressure value at breakout just below the salt sheet at the present-day.

1 One could, of course, tie seismically to a nearby well if one exists, but one does not know, as opposed to guess or model, the actual lithologic sequence under the allochthonous salt sheet

2 Because the salt sheets depicted are of order 100 km in lateral extent, there is no flow of fluids around the "ends" of the sheets in the short time available. Thus the sand formations are overpressured.

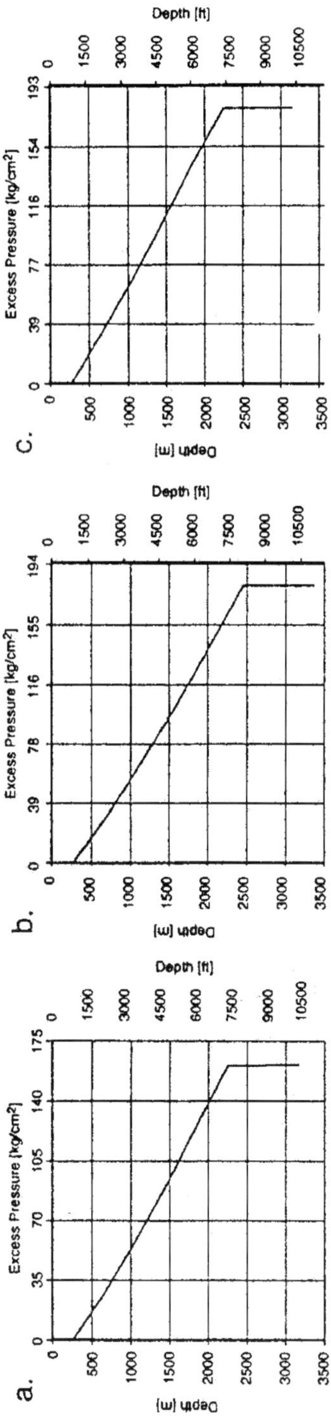

Fig. 12a–c Excess pressure (in kg/cm^{-2} = 1.03 atm) variations with present-day depth at location A for an intrusive salt sill inserted at a speed of 10 cm a^{-1} and present-day salt top at 300 m sub-sea bed. **a** Salt thickness of 2000 m. **b** Salt thickness of 2200 m. c As for **a** but when the underlying formation lithology is changed to 100% sand from the 50% sand/50% shale lithology used in **a** and **b**

4.3 Probability, Uncertainty and Relative Importance

Because the exact salt thickness, salt speed and underlying lithology are not known, it follows that the estimated excess fluid-pressure at sub-salt break-out is uncertain. What is needed are two factors: (i) the probability of encountering an excess pressure less than (greater than) a prescribed value at break-out; and (ii) the degree of uncertainty of the estimate made, together with information on which parameter variation (salt thickness, salt speed, underlying lithology) is providing the greatest fraction of the uncertainty in excess pressure variability, i.e. the relative importance (RI) of variations in each contributory factor to excess pressure uncertainty. In this way one can assess which factor first needs to have its range of uncertainty narrowed if a better job of estimating excess pressure is needed.

This RI then enables one to concentrate on the significant parameters, without having to expend effort on improving parameter values which have a lesser effect on uncertainties in excess pressure. Consider each component in turn.

Cumulative Probability and Uncertainty. Suppose that, for each case, the individual parameters (salt thickness, salt speed, underlying lithology) are held at their most likely values of Table 3, with the exception of each one of the parameters in turn, which is allowed to vary from its minimum value through to its maximum value. Each such parameter value then produces an excess pressure value at sub-salt breakout. For each parameter variation, the excess pressure values are then ordered from smallest to largest and cumulative probability plots of obtaining an excess pressure less than or equal to a prescribed value are obtained, as shown in Fig. 13a,b and c for each case ad with each of the three parameters varying. For instance, in Fig. 13a, case 1–Excess Pressure (Thickness) means present-day excess pressure at sub-salt break-out due to variations in salt thickness (as prescribed in Table 3) when salt speed and lithology are held at 10 cm/a and 50/50 sand/shale, respectively.

Table 3
Case 1 – present-day depth of burial = 300 m

	Minimum	Most likely	Maximum
Lateral speed (cm a^{-1})	8	10	12
Thickness (m)	1800	2000	2200
Lithology (% sand)	0	50	100
Case 2 – present-day depth of burial = 500 m			
Lateral speed (cm a^{-1})	5	10	15
Thickness (m)	1700	2000	2300
Lithology (% sand)	0	50	100

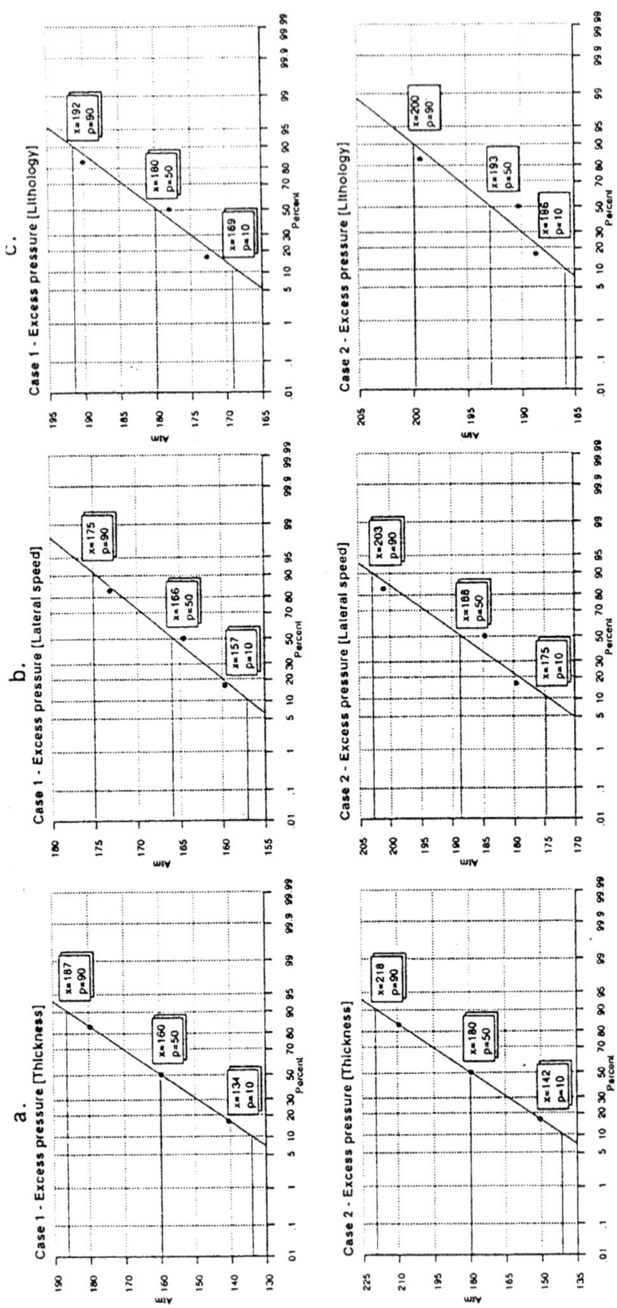

Fig. 13a–c Cumulative probability excess pressure plots as each parameter, **a** thickness, **b** lateral speed, **c** sub-salt lithology) is varied individually for the two cases of salt top at 300 m and 500 m sub-sea bed depth, respectively

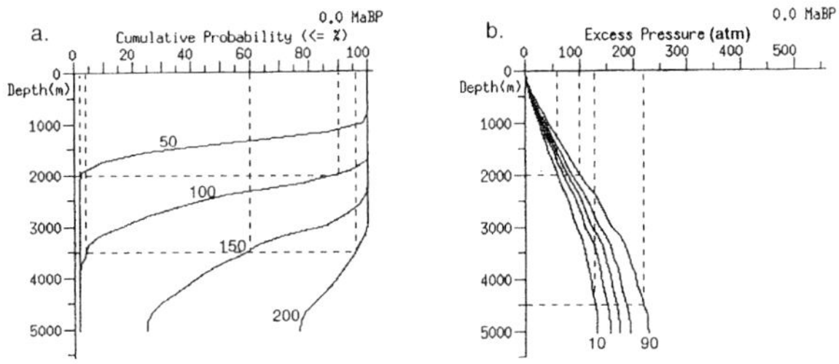

Fig. 14 **a** Cumulative probability versus depth plot of isobaric excess pressure (atm) contours. **b** Excess pressure versus depth plot of constant cumulative probability values

The values x = 134, p = 10; x = 160, p = 50; x = 187, p = 90 on figure 13a, Case 1, mean an excess pressure of 134 atmospheres is found at 10% cumulative probability chance; 160 atm at 50% chance; and 187 atmospheres at a cumulative probability of 90%. For example, it is 90% likely that for the situation being described the break-out excess pressure will be less than 187 atm.

At each sub-surface depth (and not just at salt breakout depth) one can then provide two different perspectives of excess pressure development as shown in Fig. 14a,b. A cumulative probability versus depth behavior, as shown in Fig. 14a for different isobaric excess pressure values, gives the cumulative probability of obtaining an excess pressure less than or equal to the posted values. Thus: at 2000 m depth there is a 90% chance the excess pressure will be less than 100 atm and only about a 5% chance of an excess pressure less than 50 atm; while at 3500 m depth there is only about a 5% chance of an excess pressure less than 100 atm, and about a 95% chance of an excess pressure less than 150 atm.

Equally, one can plot the converse information as shown in figure 14b where, at different depths, one plots the excess pressure for fixed iso-probability values. Thus: there is a 90% chance that at 4500 m the excess pressure will be less than 220 atm but only a 10% chance of less than 130 atm, as shown on Fig. 14b. Alternatively: a 130 atm isobar has less than 10% chance of *not* being found at 4500 atm or greater; and a 90% chance of *not* being found shallower than 2000 m. Thus, both the probable depth range of achieving given excess pressure values, and the probability of achieving a specified range of excess pressure at a given depth are determined.

Relative Importance. As each of the parameters – salt thickness, lateral speed, and sub-salt lithology – are varied independently, each causes a different variation of excess pressure with depth. The concern is to determine which of the factors is providing the largest uncertainty in the excess pressure, so that one has an appreciation of which factor(s) to focus on first in attempts to narrow the range of uncertainty in excess pressure. To answer this question, proceed as follows: Each factor as it varies produces an uncertainty in the excess pressure and, for the cumulative probability, the distribution of excess pressure values recorded is close to a log-normal distribution. For a precisely log-normal distribution, where E_2 is the second moment of the excess pressure distribution, p, and E_1 is the first moment (mean value), the scale value

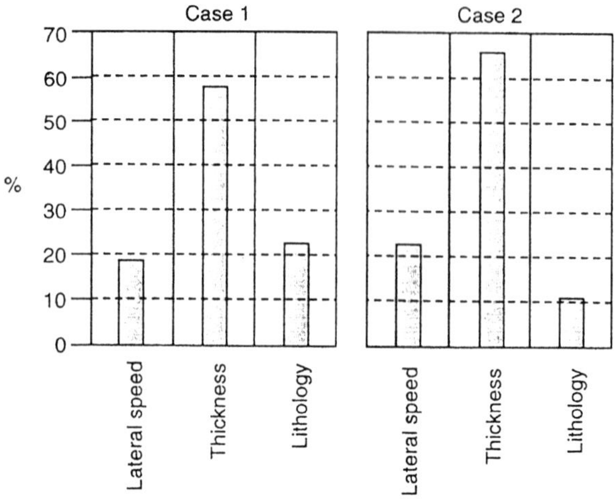

Fig. 15 Relative importance (%) of the uncertainty in salt thickness, salt speed and sub-salt lithology in contributing to excess pressure uncertainty. Case 1 salt top at 300 m sub-sea bed. Case 2 salt top at 500 m sub-sea bed

$$\mu^2 = \ln\{E_2(p)/E_1(p)^2\} \tag{3}$$

describes the spread in uncertainty around the mean $E_1(p)$, because the variance, δ^2, is given by

$$\delta^2 = E_2 - E_1^2 (\exp(\mu^2)-1). \tag{4}$$

The variation in each individual parameter then provides a contribution μ_i to the total μ for excess pressure. The relative importance, RI(%), of each factor in contributing to the uncertainty in excess pressure is defined by

$$RI_i(\%) = 100\mu_i / \sum_{j=1}^{N} \mu_j, \tag{5}$$

where N is the number of parameters being varied (3 in the present situation, for each of cases 1 and 2 respectively).

Plotted on Fig. 15 are the RI values for the two cases where the salt top today is 300 and 500 m, respectively, below the sediment surface.

In both cases note that it is the uncertainty in salt thickness which contributes between 58 – 65% of the uncertainty in excess fluid pressure directly beneath the salt, with the lateral speed of salt and lithology uncertainty below salt contributing between 18 and 23%, and 11 and 23%, respectively.

Overall, then, if one wishes to narrow the range of uncertainty on excess fluid pressure at salt break-out, effort should predominantly be placed on attempting to cut the estimated uncertainty of ± (200–300)m on present-day salt thickness, with a secondary, but lesser, effort on salt speed delineation and lithology improvement. The dominance of the salt thickness uncertainty is about a factor three more important to resolve than the uncertainties in either salt speed or lithology.

5 Summary

Several major points result from the study presented here:

1. Perhaps the dominant point of the numerical experiments is that they indicate the development of significant overpressure under both salt sheets in the last 0.8 MaB.P. through to present day. The exact degree of overpressure and its precise buildup rate are controlled by variations in salt speed and salt thickness, but the buildup of overpressure always occurs for a laterally extensive salt sheet.
2. An overpressure increase (relative to regional) of about 100–300 atmospheres occurs under and, in the vicinity of the salt, corresponding to about 200 atm per km thickness of salt.

In the present situation it is clear that the ±(200–300)m uncertainty on a salt thickness of around 2000 m, is causing over half of the uncertainty in excess pressure values, with salt speed and sub-salt lithology uncertainties providing roughly equal contributions of about 10–25% each to the total uncertainty in excess pressure depending on the individual case. The variations in excess pressure estimates at break-out are about ± 30 atm on a mean value of about 170 atm. Thus, if one can accept a high break-out value of 200 atm while drilling, then there is no need to attempt to improve the uncertainty on salt thickness prior to drilling, which is responsible for about 58–65% of the ± 30 atm. On the other hand, if one cannot accept that level of uncertainty on excess pressure, then a concerted effort should be made first to improve salt thickness estimates rather than concentrating on salt speed or sub-salt lithology

Basin modeling has become a useful tool in petroleum exploration, especially in those areas where not many wells are available. However, when data availability decreases, the uncertainty involved will increase and so will the risk. Therefore the uncertainty and associated risk have to be assessed when applying basin modeling techniques to such areas. Risk analysis in basin modeling should include the resolution, sensitivity, uniqueness and precision of modeling results. The Monte Carlo approach is not appropriate in practice because of the complexity of the system being modeled; too much computer time is taken to perform risk analysis for any basin analysis model due to the multiplicity of processes related to the tectonic and strata development, sediment deposition and diagenesis, petroleum generation, migration and accumulation in a sedimentary basin.

The probability method for risk analysis in basin modeling is numerically quick without loss of accuracy in risk assessment. The problem does not require massive Monte Carlo computer runs and can be performed on a personal computer or workstation in a timely manner, which is of importance in risky and competitive petroleum exploration.

Acknowledgements

The work reported here was supported by the Industrial Associates of the Basin Analysis Group at USC.

References

Bour, O., and Lerche, I. (1994): Numerical modeling of abnormal fluid pressure in the Navarin Basin, Bering Sea. *Marine and Petroleum Geology* **11**, 491–500.

Cao, S., Abbott, A.E., and Lerche, I. (1996): Risk and probability in resource assessment as functions of parameter uncertainty in basin analysis exploration models, in: *Quantification and Prediction of Hydrocarbon Resources* (ed. A. Doré). Norwegian Petroleum Society/Elsevier Publishing Co., 199–218.

Cao, S., and Lerche, I. (1989): Sensitivity analysis of basin modeling with applications, in: Applications of statistics in the Earth sciences (ed. F. Agterberg). Can. Geol. Soc. Paper 89–9, 489–504.

Carter, K., and Lerche, I. (1992): Moving salt sills and hydrocarbon maturity: Pure and Appl. Geophys., v. 138, 478–487.

Kai, Z., and Lerche, I. (1993): Inversion of dynamical indicators in quantitative Basin analysis models II. Synthetic tests and a case history using dynamical indicator tomography. *Mathematical Geology* **25**, 107–123.

Lerche, I., and O´Brien, J.J. (1987): *Dynamical Geology of salt and related structures*, Academic Press, San Diego, 832 p.

Lerche, I. (1989): *Basin analysis: Quantitative methods, Vol. 1*, Academic Press, San Diego, 562 p.

Lerche, I. (1990): *Basin analysis: Quantitative methods, Vol. 2*, Academic Press, San Diego, 570 p.

Lerche, I. (1991): Inversion of dynamical indicators in quantitative basin analysis Models. I. Theoretical considerations: Math. Geol., v. 23, 817–832.

Lerche, I. (1992): *Oil exploration: Basin analysis and economics*, Academic Press, San Diego,178 p.

Lerche, I. (1997): *Geolocical Risk and uncertainty in oil exploration.* Academic Press, San Diego, 658 p.

Lerche, I., and Petersen, K. (1995): *Salt and sediment dynamics*, CRC Press, Boca Raton, 322 p.

Lowrie, A. (1994): *Seismic stratigraphy and hydrocarbon traps: Louisiana onshore and offshore*,. Soc. Explor. Geophys. Course Notes Series, Vol. 5 (Series editor S.N. Domenico), 213 p.

Magoon, L.B., and Bird, K.J. (1986): Organic carbon content, hydrocarbon, content, visual kerogen and vitrinite reflectance data within NPRA (National Petroleum Reserve in Alaska) – contour maps to evaluate petroleum source rock richness, type, and thermal maturity, in: Gryc, G., Geology of the National Petroleum Reserve in Alaska: U.S. Geological Survey Professional Paper, 139 p.

O´Brien, J.J., and Lerche, I. (1994): Understanding subsalt overpressure may reduce drilling risks: Oil and Gas Journal, Issue of January 24, 1994, 28–34.

West, D.B. (1989): Model for salt deformation on deep margin of Central Gulf of Mexico Basin: *AAPG Bull*, v. 73, 1472–1482.